SCALE INSECTS OF CENTRAL EUROPE

SERIES ENTOMOLOGICA

EDITOR

K.A. SPENCER

VOLUME 41

Scale Insects of Central Europe

By

M. KOSZTARAB and F. KOZÁR

1988
SPRINGER-SCIENCE+BUSINESS MEDIA, B.V.

Library of Congress Cataloging in Publication Data

Kosztarab, Michael, 1927-
Scale insects of Central Europe.

(Series entomologica: v. 41)
Includes indexes.
1.Scale-insects—Central Europe. 2. Insects—
Central Europe. I. Kozár, F. II. Title. III. Series.
QL526.4.C36K67 1988 595.7'52 86-27415

ISBN 978-94-010-8295-2 ISBN 978-94-009-4045-1 (eBook)
DOI 10.1007/978-94-009-4045-1

Book information

Joint edition published by: Dr W. Junk Publishers, Dordrecht, The Netherlands, and Akadémiai Kiadó, Budapest, Hungary

CONTENTS

PREFACE

This book is dedicated to the scientists whose professional devotion and accomplishments in research on scale insects of the area made the compilation of this book possible.

This book has two primary objectives: (1) to enable interested workers, who are not specialists of scale insects, to recognize and indentify these to families, genera and species; and (2) to provide information on the life cycle, host-plant range, natural enemies, geographical distribution, economic importance for each known species from Central Europe, and to provide a bibliography on each taxa covered.

Scale insects are important pests of fruit and nut trees, forest vegetation, woody ornamentals, greenhouse and indoor plants. World-wide losses and increased production costs attributed to scale insects are estimated to reach $5 billion annually. It is difficult to recognize them because of their microscopic size and hidden habits. Identification of pest species enables plant growers to check appropriate reference sources for biological information and to select control methods.

For the first time since L. Lindinger's 1912 outdated book on scale insects of this large geographical area, the authors are providing keys for their identification, descriptions, and illustrations to assist individuals with species identification. Also provided are records on the life cycle, host plant selection, and geographical distribution of each of 233 species, 95 genera and 12 families treated. With resistance to insecticides shown by a number of species, special efforts are needed to reduce the use of chemicals and to employ integrated pest management emphasizing biological control methods. Thus, the authors are providing information on the known natural enemies of each species, and on the parasites and predators which could be considered for biological control of pest scales. This work is an essential reference for the economic entomologists, plant production and protection consultants, plant quarantine inspectors, biologists, and especially for zoologists and coccidologists.

The authors have cited over 700 references which are listed under the Cited References. In addition, to facilitate readers with easy retrieval of desired information, there are four different indices provided. These include indices of host plants, natural enemies and other associated organisms, distribution in Central Europe and in other regions, and scale insect names. Several new species were discovered during this study and a number of new faunistic records were found. In addition, one new synonym was established. It is the authors' hope that this compilation will stimulate interest in the scale insects of Europe. Many new species remain to be discovered and described, especially in the incompletely studied families of Eriococcidae, Margarodidae, and Pseudococcidae.

In the General Part (I), a general introduction to coccidology, the methods utilized for collecting, preserving and mounting scale insects on microscope slides for study and rearing are given. Among the 154 illustrations, there are 72 full plate-size drawings, 38 smaller ones and 9 photographs placed in the text; also, 35 color photographs were combined into 16 color plates, and one full-page map.

Both authors have completed doctoral degree research on scale insects and are continuing to study them. They have 55 years combined experience with this group. Research samples for this work were collected in most of the countries of Europe with the exception of countries in the border areas of the continent, such as Albania, Belgium, Finland and Portugal. To assist individuals in these areas with their work, the authors paid special attention to compiling information from over 700 reference sources to supplement their own research data.

ACKNOWLEDGEMENTS

We are grateful to our colleagues who read sections of the first draft of the manuscript and made valuable comments for improvements. These persons include: D. R. Miller (General Part and Eriococcidae), S. Nakahara (Coccidae and Diaspididae), L. M. Russell (Asterolecaniidae), each with the Systematic Entomology Laboratory, Agricultural Research Service of the U. S. Department of Agriculture, Beltsville, Maryland; and D. J. Williams (Pseudococcidae), with the Commonwealth Institute of Entomology in London.

The entire manuscript was reviewed by T. Jermy, former Director of the Research Institute for Plant Protection, Budapest, and J. Koteja, Professor of Applied Zoology, Cracow. Their valuable suggestions have been incorporated in the manuscript. Some valuable "Reviewer's remarks" provided by J. Koteja have been included in the book at the general discussions of some families. The authors take sole responsibility for the final text of this manual.

Faunal lists were provided for the Federal Republic of Germany by H. Schmutterer, and for the Ukraine (USSR) by E. M. Tereznikova. Supplementary faunal records from Switzerland were sent to us by F. Bachmann and K. Boratynski, and from Poland by J. Koteja. Assistance with some species determinations and confirmations was given by E. Danzig, J. Koteja, and D. Matile-Ferrero. Some species were collected and provided for our project from Hungary by B. Nagy and G. Vinis; from Austria, Luxembourg and Poland, by K. Boratynski. D. R. Miller, S. Nakahara and M. B. Stoetzel allowed us full use of the scale insect card files of the U. S. Department of Agriculture. In addition, colleagues, too many to be listed here, provided reprints and books for the completion of the literature review. Our literature search was aided by the bibliographies compiled by Morrison and Renk (1957), Morrison and Morrison (1965); Russell, Kosztarab and Kosztarab (1974).

M. P. Kosztarab assisted with the literature search and checked our cited reference list. B. Martin typed parts of the manuscript, as well as M. Rhoades, who also provided substantial editorial assistance, as did M. C. Holliman, who reviewed the final manuscript of the general part. The host plant index was reviewed by P. Solymosi.

Supplementary illustrations were provided by a number of colleagues. The figure numbers of color prints provided by each are given in parenthesis after their names. These include: G. M. Buxton (9, 10), I. Foldi (2, 3, 4, 7b, 8b, 11, 18a, 26, 28b), R. J. Gill (8a, c, d, 24, 28a, 32), A. B. Hamon (96), H. H. Lyon (15a, b, 19, 21, 23, 29, 33, 34b), D. R. Miller (34a), B. Nagy (14, 31a), J. A. Weidhaas (34d), and R. F. Wilkey (4). A number of the black and white habitat illustrations are from the authors' earlier work

1

2

3

4

Plate I

Figs 1—4: 1. Mealybug females visited by ant for honeydew (Photo R. F. Wilkey). — 2. *Orthezia urticae* (Linnaeus), female with ovisac, dorsal view (Photo I. Foldi). — 3. *Porphyrophora polonica* (Linnaeus), female (Photo I. Foldi). — 4. *Porphyrophora polonica* (Linnaeus), adult male (Photo I. Foldi)

5

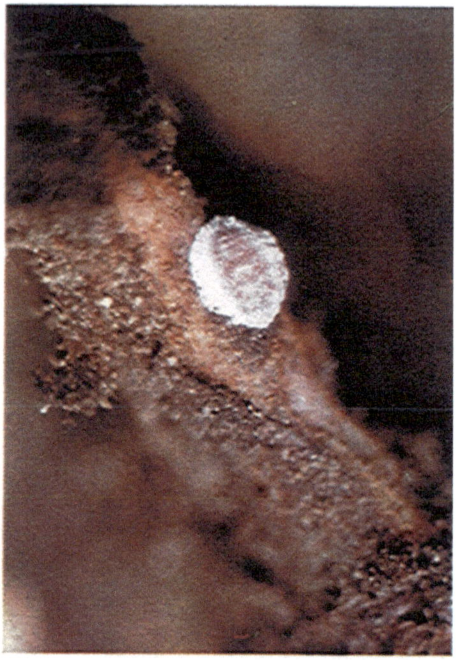

6

7/a

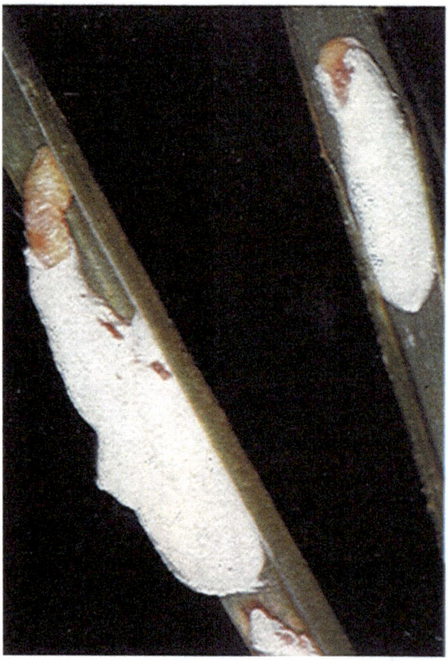

7/b

Plate II

Figs 5–7: 5. *Allococcus vovae* (Nassonov), ovisacs and females on twigs of *Juniperus communis* (Photo M. Kosztarab and F. Kozár). — 6. *Coccura comari* (Künow), female on root of *Sanguisorba* sp. (Photo M. Kosztarab and F. Kozár). — 7. *Eriopeltis festucae* (Fonscolombe), ovisacs *a)* with and *b)* without protruding females on leaves of *Agropyron repens* (Photo L. Móczár and M. Kosztarab *a)*, I. Foldi *b)*)

8/a

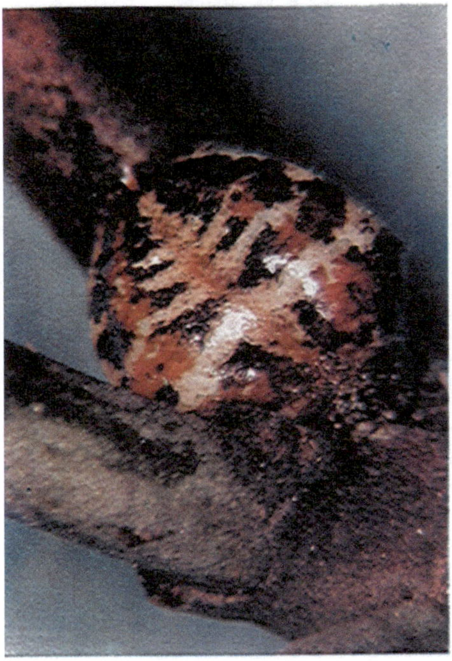

8/b

8/c

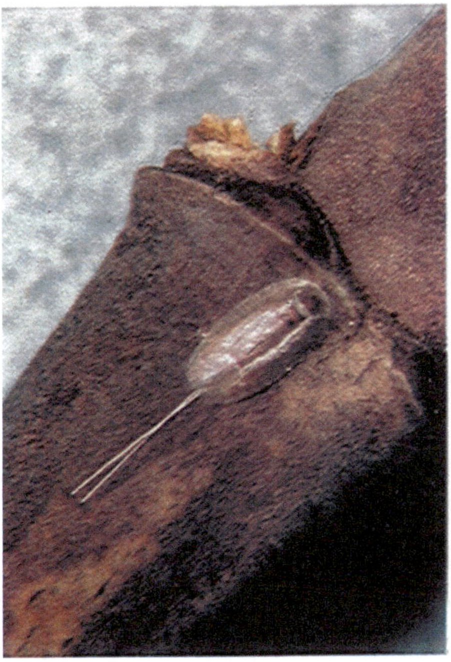

8/d

Plate III

Fig. 8. *Eulecanium tiliae* (Linnaeus), on twigs *a)* young adult female, *b)* old adult female, *c)* male test with wax threads of male, *d)* adult male (Photo R. J. Gill *a)*, *c)*, *d)*, I. Foldi *b)*)

9

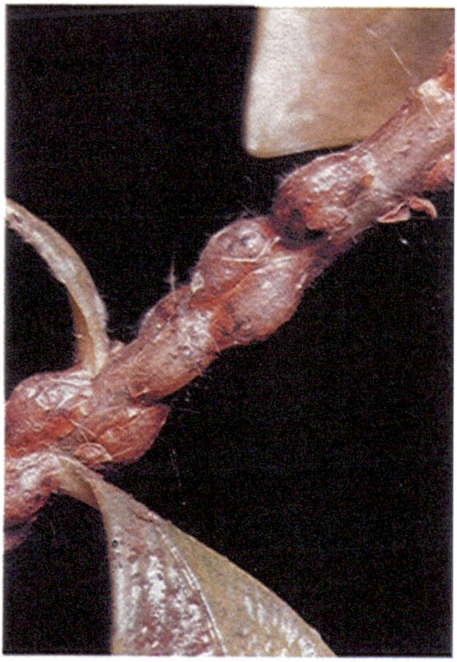

10

11

Plate IV

Figs 9–11: 9. *Parthenolecanium fletcheri* (Cockerell), females on leaves of *Thuja occidentalis* (Photo G. M. Buxton). — 10. *Parthenolecanium persicae* (Fabricius), females on twig of *Daphne* sp. (Photo G. M. Buxton). — 11. *Parthenolecanium rufulum* (Cockerell), adult female on twig of *Quercus* sp. (Photo I. Foldi)

12/a

12/b

12/c

Plate V

Fig. 12. *Physokermes inopinatus* Danzig & Kozár, on twigs and needles of *Picea abies*, *a)* females (Photo B. Koroknai and F. Kozár), *b)* first instars (crawlers) *c)* damage: shoots and needles reduced in size during the year of heavy infestation (Photo B. Koroknai and F. Kozár)

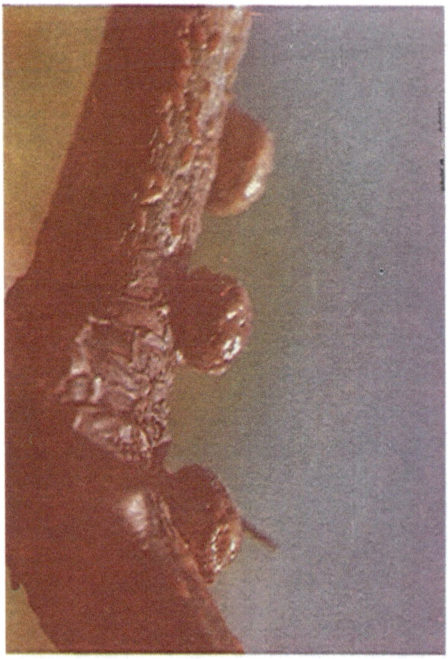

13

14

15/a

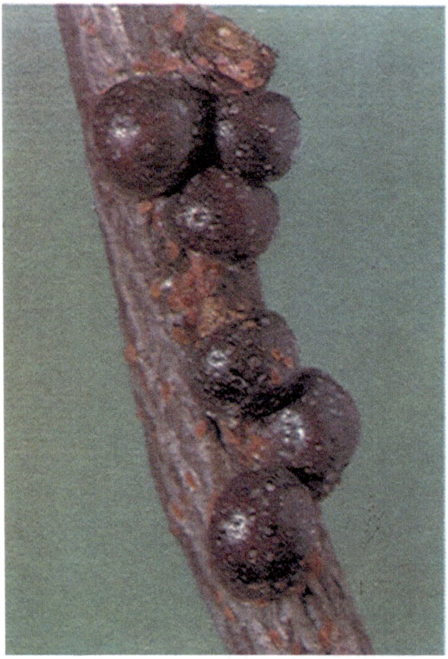

15/b

Plate VI

Figs 13–15: 13. *Rhodococcus perornatus* (Cockerell & Parrott), females and first instars on twig of *Rosa* sp. (Photo B. Koroknai and F. Kozár). — 14. *Scythia craniumequinum* (Kiritchenko), ovisac on leaf of *Stipa* sp. with crawlers (Photo B. Nagy). — 15. *Sphaerolecanium prunastri* (Fonscolombe), on branches of *Prunus domestica*, *a)* postreproductive females with emerging first instars, *b)* male tests (white) and nymphs (Photo H. H. Lyon)

16

17

18/a

18/b

Plate VII

Figs 16—18: 16. *Kermes gibbosus* Signoret, parasitized female, on twig of *Quercus* sp. (Photo B. Koroknai and F. Kozár). — 17. *Kermes quercus* (Linnaeus), females in bark crevice of *Quercus* sp. (Photo B. Koroknai and F. Kozár). — 18. *Kermes roboris* (Fourcroy), young *a)* and old *b)* females on twig of *Quercus* sp. (Photo I. Foldi *a)*, B. Koroknai and F. Kozár *b)*)

19

20

21

22

Plate VIII

Figs 19—22: 19. *Cryptococcus fagisuga* Lindinger, females in ovisacs on trunk of *Fagus* sp. (Photo H. H. Lyon). — 20. *Acanthococcus aceris* Signoret, ovisacs containing females and eggs on twig of *Acer campestre* (Photo M. Kosztarab and F. Kozár). — 21. *Gossyparia spuria* (Modeer), females on branch of *Ulmus* sp. (Photo H. H. Lyon). — 22. *Rhizococcus agropyri* (Borchsenius), females with ovisacs on leaves of *Agropyron repens* (Photo M. Kosztarab and F. Kozár)

23

24

Plate IX

Figs 23—24: 23. *Asterodiaspis variolosa* (Ratzeburg), adult females on bark of *Quercus* sp. (Photo H. H. Lyon). — 24. *Planchonia arabidis* Signoret, young females on twig of *Pittosporum* sp. (Photo R. J. Gill)

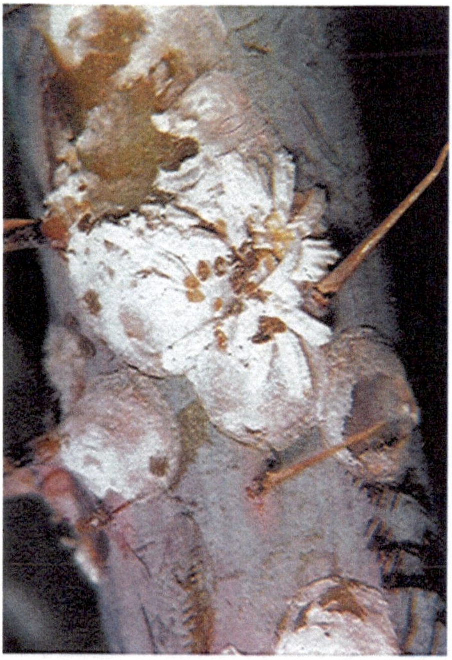

25/a

25/b

26

27

Plate X

Figs 25–27: 25. *Aulacaspis rosae* (Bouché), on stem of *Rubus* sp., *a)* a group of male tests overlapping adult female tests, *b)* female test enlarged (Photo R. J. Gill). — 26. *Carulaspis juniperi* (Bouché), female and male tests on berry of *Juniperus* sp. (Photo I. Foldi). — 27. *Chionaspis salicis* (Linnaeus), female tests chewed by predators on branch of *Salix* sp. (Photo M. Rhoades and M. Kosztarab)

28/a

28/b

Plate XI

Fig. 28. *Epidiaspis leperii* (Signoret), *a)* female (circular) and male tests (elongate) on bark of *Juglans regia* (Photo R. J. Gill); *b)* red females with tests removed on *Prunus* sp. (Photo I. Foldi)

29

30

31/a

31/b

Plate XII

Figs 29–31: 29. *Lepidosaphes ulmi* (Linnaeus), adult female tests (dark and large) and second instar tests (light and small) on branch of *Syringa* sp. (Photo H. H. Lyon) — 30. *Leucaspis pini* (Hartig), adult female tests on needles of *Pinus* sp. (Photo M. Rhoades and M. Kosztarab). — 31. *Mohelnaspis massiliensis* (Goux), *a)* female tests lined up inside grass blade, *b)* same enlarged (Photo B. Nagy *a)*, A. Meszlény and F. Kozár *b)*)

32/a

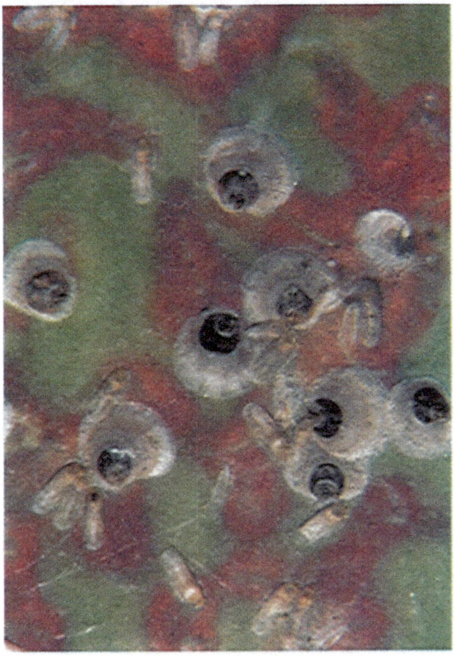

32/b

33/a

33/b

Plate XIII

Figs 32—33: 32. *Parlatoria oleae* Colvée, *a)* female tests circular, male tests elongate, *b)* same enlarged (Photos R. J. Gill). — 33. *Pseudaulacaspis pentagona* (Targioni-Tozzetti), *a)* female tests on branch of *Morus* sp., *b)* female and male tests (elongate) on branch of *Prunus* sp. (Photo M. Rhoades and M. Kosztarab *a)*, H. H. Lyon *b)*)

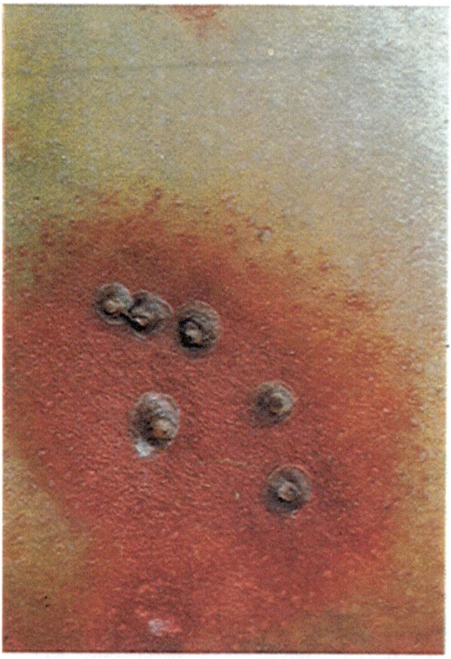

34/a

34/b

34/c

Plates XIV–XV

Fig. 34. *Quadraspidiotus perniciosus* (Comstock), *a)* female tests on fruit, with red discoloration produced, *b)* first instar tests on a twig, *c)* sticky trap card for capturing adult males, *d)* dead adult females after application of dormant oil spray, *e)* apple tree destroyed by this scale insect (Photo: D. R. Miller *a)*, H. H. Lyon *b)*, F. Kozár *c)*, *e)*, J. A. Weidhaas *d)*)

34/d

34/e

35/a

35/b

Plate XVI

Fig. 35. *Unaspis euonymi* (Comstock), female and male tests on *Euonymus*, *a)* on twig and leaves, *b)* enlargement with 2 adult female tests (large brown) and several male tests (elongate white), also a few first and second instars (Photo H. H. Lyon)

(Kosztarab and Kozár, 1978) and were prepared by M. Csiby; while a number of enlargements are from earlier works by M. Kosztarab with J. O. Howell, P. L. Lambdin and M. L. Williams. Illustrations by the two authors and those borrowed from others, including the ones prepared by Mrs. F. Kozár, have been drawn in the final format for this book by M. Dobiecki.

I. GENERAL PART

INTRODUCTION

State of knowledge. This book is the first to treat all of the families, genera, and species in Central Europe since Lindinger's (1912) summary. Our goal is to provide tools for recognizing and identifying species of scale insects encountered in Central Europe (CE); to present for each species information on distribution, biology, and natural enemies; to provide literature sources for additional information on the families, genera, and species included; to foster interest in this economically important group of insects; and to encourage scientists to study the less well known species so as to increase our understanding of their taxonomy. The data given on life cycles, parasites, and predators should assist in the selection and timing of control programs against pest species.

The task of completing this book was more difficult than originally anticipated. The difficulties stem from the vast size of CE, its diversity and the number of the species to be treated, and the cursory and fragmentary knowledge of many species present in CE. Unfortunately, there is no network of amateurs working on scale insects as there are for Coleoptera and Lepidoptera. In about half the countries of CE there are no active professional coccidologists. Therefore it is common for many species to be known only from one locality and one host, suggesting that much is to be learned about the scale insect fauna of CE.

Because of space constraints, it is impossible to include all the literature citations on each of the 233 species treated. We have given what we consider to be the most important literature sources available, and these are often marked with an asterisk. In selecting and citing the more than 700 literature sources, we gave priority to: (1) publications with descriptions and illustrations rather than those merely listing species, unless the paper included relevant new faunal records; (2) more recent rather than older works; (3) papers published in lesser known journals and in infrequently used languages, thus being inaccessible to the average user.

To fill gaps in the distribution of species in CE, the authors made a concerted effort to collect in each of the Central European countries. Even with these collections and with the special efforts of collaborators residing in the countries of CE, there still remain conspicuous gaps in distribution patterns of species, and many species remain to be discovered.

This publication would not have been possible without the research of colleagues; we have freely extracted information from their publications. Some of the important studies on Coccinea systematics in CE were completed during the past half century by Baccetti, Bachmann, Balachowsky, Bazarov, Boratynski, Borchsenius, Canard, Cox, Danzig, Dziedzicka, Foldi, Geier, Gerneck, Goidanich, Goux, Jakubski, Kawecki, Kiritchenko, Komosinska, Konstantinova, Kosztarab, Koteja, Kozár, Kozarzhewskaya, Köhler, Kuznetsov, Lellakova-Duskova, Lindinger, Lupo, Matile-Ferrero,

Ördögh, Panis, Podsiadlo, Rasina, Řehaček, Reyne, Rieux, Rogojanu, Sachtleben, Săvescu, Schmidt, Schmutterer, Siewniak, Smolyannikow, Šulc, Szelényi, Ter-Grigorian, Tereznikova, Thiem, Tranfaglia, Tremblay, Tsalev, Vinis, Williams, Wünn, Zahradnik, Zak-Ogaza and others.

A number of unpublished faunistic records are included here. New synonymy is established for *Rhodococcus perornatus*.

New species discovered during this study were described in separate publications and then included in this manual.

In a few cases type material of some CE species is lost or unattainable, making it impossible to authenticate their identities. In such cases the descriptions are based on the original or subsequent published descriptions and illustrations. Attention is drawn to these species in their respective sections, under "Notes", in the hope that future workers will attempt to recollect topotype material, designate neotypes, and prepare more complete descriptions and illustrations.

Concept of Central Europe. Central Europe zoogeographically includes a wider area (fig. 36) than many would recognize using political boundaries as criteria. The southern border of the region passes from east to west through the middle of Bulgaria, into the central part of Yugoslavia, south of Belgrade, across to Northern Italy, through the high mountainous part of Northern Italy, into the Le Pontin Alps of Southern Switzerland, and into the highlands of Eastern France. The western border extends through northeastern France, along the border between Germany, Luxembourg and the lowland countries of Belgium and Holland. The northern limit extends eastward just south of the Baltic Sea coast of Germany and Poland excluding the coastal belt and reaching its most northern extension in northeastern Poland. The eastern border extends along the southeastern corner of White Russia and western Ukraine and south along the Romanian and Bulgarian Black Sea coast.

In our concept, Central Europe does not include the Mediterranean and Atlantic coastal areas, Scandinavia and Finland, or the Baltic seacoast.

The borders of this biogeographical subdivision are basically those of Emeljanov (1974), who called the region the "Subatlantic Sector," of the Palearctic Realm. According to Emeljanov (1974), the region is characterized by a transitional subatlantic climate and a sub-boreal biogeography. It is comprised in the west of the European nemoral zone, in the east by the semi-arid woodland or forest-steppe zone, with a semi-arid component, the Hungarian or Pannon Steppe.

Scale insects that occur only on house plants or in greenhouses in the area, and as such do not belong to the CE fauna, are listed under the pertinent family but are not treated in detail. Their inclusion would be more appropriate in a second manual that would also treat the Mediterranean scale insects, even when not all the greenhouse scales in CE are of Mediterranean origin.

Techniques in general. In this work the English language terms *scales, scale insects* and *coccoids* are used interchangeably and refer to all members of the suborder Coccinea, while the term *coccid* refers only to members of the family Coccidae, or soft scales.

Authors of scientific names for parasites, predators, ants and other associated organisms are given in the index; they are not repeated in the text.

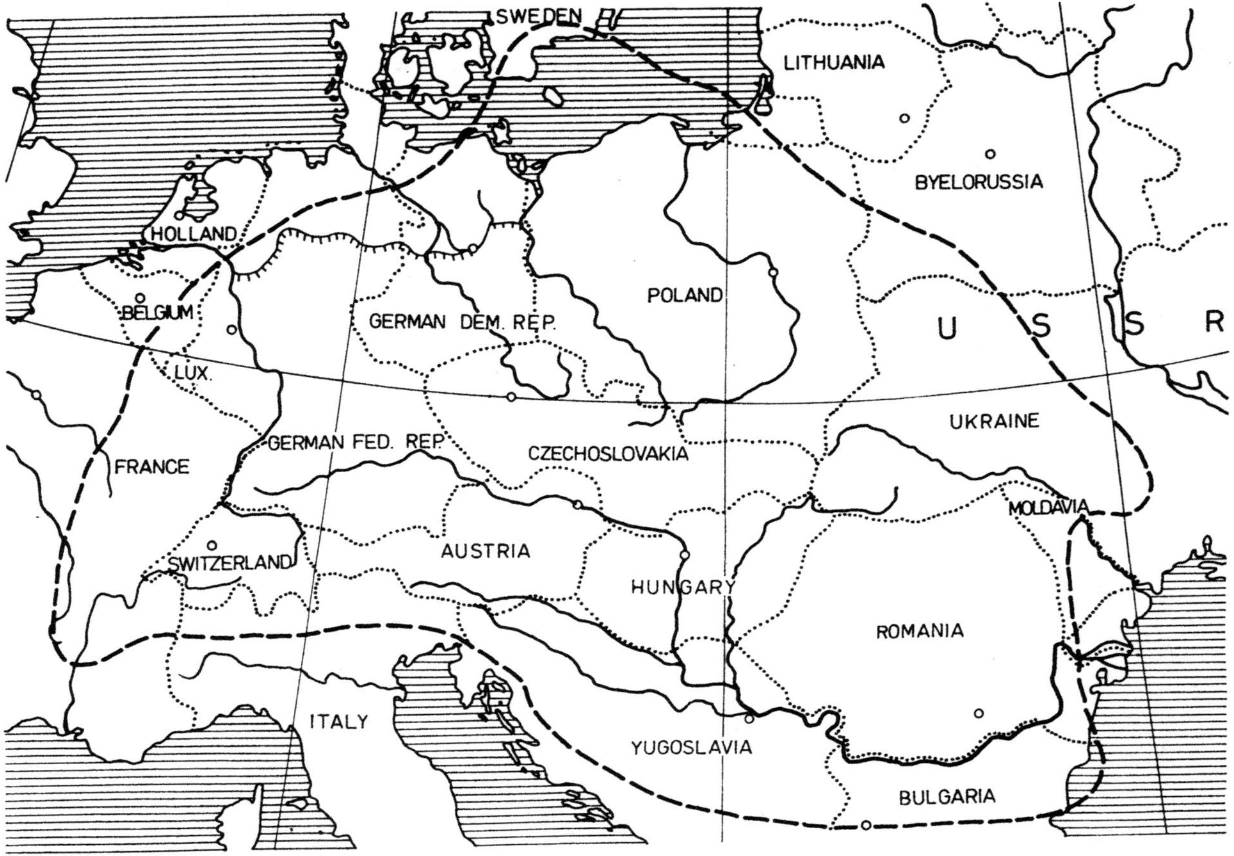

Fig. 36. Map of Central European Zoogeographical Area (after Emeljanov, 1974)

Measurements of adult females are of slide-mounted specimens, and color descriptions are based on live specimens, unless stated otherwise. The term *prosoma* is often used alternately with *cephalo-thorax*, especially for Diaspididae in our text.

For European-English common names, Gozmány's (1979) dictionary was used as our guide. Common names are proposed for those scale insects that do not have such a name in the European-English language literature. A few alternative common names are suggested when an earlier name is not sufficiently descriptive or perpetuates misleading information. In the case of some cosmopolitan species, the common name used in North America is given in addition to the European name.

Families are given in phylogenetic order; genera and species within each family are listed alphabetically since their phylogenetic placement often is unsettled. The authors followed, with minor modifications, the higher classification given by Borchsenius (1950) and Ferris (1955).

Descriptions. The general descriptions of families apply to the entire family, and are not limited to CE genera and species. We have made a concerted effort to include two or more characters in each key couplet to assist with identification of difficult taxa.

For most of the coccoid species in CE only the adult females have been described. For this reason, with few exceptions, our keys and descriptions include only the slide-mounted adult females. When available, short descriptions are given of postreproductive females and of ovisacs or tests of adults. Primary literature sources of descriptions of immature stages and adult males are given after the adult female descriptions.

In addition, we have attempted to list for each species, genus, and family all the frequently used scientific synonym names, and the most important references under "Syn. & Lit.". In only a few cases, where there is an overwhelming number of synonyms and literature (e. g. *Parthenolecanium corni*), did we deviate from this principle. The page number given in our literature listing is the page on which the species is mentioned or on which its description starts. The first page of an article is listed when the article deals with only one species. In either case our "References Cited" list provides the full bibliographic citation. An asterisk (*) has been used for literature sources that provide additional information.

The Synonym and Literature (Syn. & Lit.) section found at the beginning of each species description includes both the synonyms and the valid name with pertinent literature citations. The first name listed is always the one under which the species was originally described and is followed by subsequent literature citations under currently used name combinations in chronological sequence. A comma following a name indicates a valid name or a name combination, rather than a synonym.

Morphological descriptions are in telegraphic style; numbers are given as numerals, and measurements are rounded off to the nearest decimal or even number.

Names of countries in CE are abbreviated by using the first two letters for each, except for USSR. Biological information (life cycle, natural enemies) is included to assist growers and specialists with the preparation of control recommendations. Known host-plant genera are listed for each scale species, but the host species is given only for monophagous species. In a few cases of polyphagous species, the listing of all host genera would have been impractical (e. g., San José scale, with 239 host genera);

therefore, only the commonly collected host genera are given, but the cited references provide more complete host lists.

The list of natural enemies is normally restricted to those known from Europe; rarely are important natural enemies given from other parts of the world. Caution is in order when utilizing the list of natural enemies, because it is based almost entirely on literature records. It is very likely that a number of literature records were based on misidentifications in the past. Also, it is almost certain that some of the natural enemies have been reared from host plant samples that were infested with more than one species of scale insects.

Under "Note" at the end of the descriptions, we point out taxonomic, nomenclatural and other problems with the species, so that future workers may clarify the problems.

Illustrations. In most cases an illustration is given of one species in each genus treated; in the case of pest species, color photographs are also provided to assist in field recognition. Full-page line illustrations include a central drawing of the insect, with the left half representing the dorsal surface and the right half the ventral surface. Important details are enlarged and placed around the perimeter of the figure. The drawings are not to the same scale in all species, nor are the dermal structures and enlargements in direct proportion to each other. Setal patterns are given in a generalized form in the figures.

References. The "Cited References" section includes many articles that were written in languages unfamiliar to the average reader; therefore, these titles are translated into English. Russian titles were taken from the *Entomological Review* (Entomol. Obozr.), when available, and were cited from both Russian and English language sources. Because some articles came to our attention late during the preparation of this manual, the first article of an author for a certain year may be marked either with the year only or with year and the letter "a" (e. g., Šulc 1908, 1908a; 1944a, 1944b).

Indices. Four indices are provided: (1) distribution of scale insects by countries in CE with species references by numbers; (2) for host plants, using both scientific and common names, when available; (3) for parasites, predators, ants and other associated organisms; (4) for scale insect taxa. (Valid names are in Roman type; synonyms, homonyms and misidentifications are in *italics*. Page numbers of descriptions are in **bold face.**)

The authors kept the locality names as were given in the cited references. This is why records from the United Kingdom may be listed under British Isles, England, Great Britain, Scotland, UK and Wales. Spelling throughout this publication is in accordance with American usage.

LIST OF ABBREVIATIONS

Au = Austria
Bu = Bulgaria
C_{1-18} = Cerarius 1–18
°C = Centigrade
ca = circa, about
CE = Central Europe
Cz = Czechoslovakia
diam. = diameter
e. g. = for example
etc. = and so forth
fig. = figure
Fr = France

Ge = Germany
Hu = Hungary
Isl = island(s)
It = Italy
L_1, L_2, L_3 = 1st, 2nd and 3rd lobe (Diaspididae)
L_{1a}, L_{1b} = 1st and 2nd lobule of 1st lobe (same for L_2 and L_3)
Lit. = literature
m = meter
mm = millimeter
Mtn = mountain
N = North
N_1, N_2, N_3 = 1st, 2nd, 3rd nymphal stage
NE = Northeast
No = number
Nos = numbers
NW = Northwest
p = page
Po = Poland
pp = pages
Ro = Romania
S = South
s. = sensu
SE = Southeast
s. l. = sensu lato, or in the wide sense
sp. = species
spp. = species (plural)
ssp. = subspecies
s. str. = sensu stricto, or in the strict sense
Sw = Switzerland
SW = Southwest
Syn. = synonym
UK = United Kingdom
US = United States
USA = United States of America
USSR = Union of Soviet Socialist Republics
Yu = Yugoslavia
μ = micron
& = and

GENERAL MORPHOLOGY OF COCCINEA

This section includes only general features that are present in most families of Coccinea and that separate them from other Homoptera.

Adult females. These are variable in shape but normally are ovoid, elongate or circular. The main body regions often are indistinct and fused. The dorsum is usually convex, with the venter flat, concave, or rarely slightly convex. Those species that live on the stems of grasses under the leaf sheaths are flat bodied, e. g., species in the genera *Aclerda* and *Antonina*, also *Chaetococcus phragmitis* (fig. 56). Species of *Kermes* are gall-like and spherical (figs 16, 17, 18), while members of *Physokermes* on spruce are bud-shaped, kidney-shaped, or heart-shaped (fig. 12a). In both genera the cuticle is hardened and sclerotized to various degrees. The body size of scales varies from 0.5 to 35 mm, but most representatives from CE are between 1 and 5 mm in length. The antennae are absent or 1- to 16-segmented. In the families Asterolecaniidae, Cryptococcidae and Diaspididae, the antennae are reduced to unsegmented stubs, while in some Margarodidae the segments rarely number up to 16. The eyes are simple, and normally reduced to two small pigmented areas. In Ortheziidae and some Pseudococcidae, the eyes are normally on sclerotized bulging eyestalks. The eyes are often absent in Diaspididae. The mouthparts include the clypeolabral shield and the unsegmented to 4-segmented conical labium with four piercing-sucking stylets. The labium normally is located between the front coxae. The labial stylets, when not in use, are enclosed in a sac-like crumena and form a stylet loop. The mouthparts rarely are reduced or entirely absent in some adult female Margarodidae. Labial characters were found to be useful in higher classification (Koteja, 1974a). The legs in scale insects vary from well-developed and 5-segmented to entirely lacking (e. g. Diaspididae). In some species the leg segments are fused, e. g. the tibia with tarsus, and the trochanter with the femur in all Ortheziidae. The front legs are enlarged and adapted for digging in some Margarodidae. The tarsi are usually unsegmented with only one claw. At base of the claw are two setose or spatulate claw digitules, of adhesive function. The claw often has a small denticle on its inner or plantar surface. The trochanter is normally triangular with 2 or more sensory pores on each surface; a number of small translucent pores often occur on the hind coxa, on the femur of some Eriococcidae, and on all or any of the coxa, femur or tibia of some Pseudococcidae. Morphological characters of leg segments, and also the arrangement, size and number of setae and spines, are often useful in separating species (Dziedzicka, 1977). The four large thoracic spiracles are present in all families, but the small abdominal spiracles are only found in Margarodidae and Ortheziidae. Normally, a number of discoidal pores are associated with each thoracic spiracle, and such pores often form rows or bands from the spiracle to the margin of the body, where some thick spiracular setae are situated in some families.

The body margin and derm surface normally bear setae of various sizes and shapes. Wax-producing discoidal derm pores are of different types, from simple monolocular to multilocular; these types and their arrangements often are characteristic of certain higher taxa. Most dermal ducts are tubular and of various shapes and sizes, and are useful characters in the classification of coccoids.

The abdominal segmentation is not always distinct. There are only 8 segments in the Diaspididae and 10 in the Pseudococcidae (McKenzie, 1967). In the Diaspididae the terminal segments from the 4th or 5th to the 8th are fused into a pygidium, which contains various species-specific lobes and plates on its posterior margin. The anal opening normally is located on the last abdominal segment, but in the Kermesidae it appears to be shifted on the venter. In most of the families the anal opening is surrounded by a sclerotized anal ring which bears setae and wax pores. In the family Margarodidae the anal ring is small and lacks setae or pores. The anal opening normally is covered by two anal plates in the Coccidae and by one plate in the Aclerdidae. In these families, an anal cleft is present. The vulva is located on the ventral side of the abdomen between the borders of the 8th and 9th segment in Pseudococcidae. Its presence identifies mature females. In the Diaspididae it is located between the 7th and 8th segments and usually is surrounded anteriorly and laterally by 4 or 5 groups of perivulvar pores. The abdomen usually terminates in two protruding anal lobes (e. g., Pseudococcidae, Eriococcidae,Cerococcidae) which bear setae. The longest seta is distinguished as the anal-lobe seta.

The female's body and her eggs are usually protected by a felt-like ovisac (e. g. Eriococcidae), by wax powder and loose wax threads (e. g. Pseudococcidae), or by a secreted test or shield (e. g. Asterolecaniidae, Diaspididae). In some scales (e. g. Coccidae, Kermesidae) the eggs and the hatching young are protected by the hardened exoskeleton of the dead female. The Ortheziidae females carry their eggs attached to the end of their abdomen, in a tube-like ovisac made of wax plates.

Immature stages. Immatures have not been studied intensively until recently. Because the first instars or crawlers are the more commonly-encountered immature stage, a short general description is presented here. This stage normally is recognized by its elongate elliptical or ovoid shape; small size, 0.3 to 1 mm length in most; well-developed 5-segmented legs, with 1 claw; and usually long 5- or 6-segmented setaceous antennae, which may possess several fleshy setae. Their derm normally bears 2 single-faceted eyes, 4 thoracic spiracles, a large clypeolabral shield, and a labium with a crumena and 4 stylets; various disc pores, tubular ducts, and setae; an anal ring which has a band or row of translucent pores, and 6 setae. The anal lobes usually are well-developed, protruding, and have long, anal-lobe setae.

Adult males. The three main body regions are usually distinct (fig. 37). The head often is sclerotized and is well separated from the thorax with a "neck" region (e. g. Coccidae, Pseudococcidae), or is fused to the thorax (e. g. Diaspididae). The antennae are thread-like, in most species in CE, and are 9–10-segmented, with many setae. The eyes are of several different types depending on the family. A pair of multifaceted

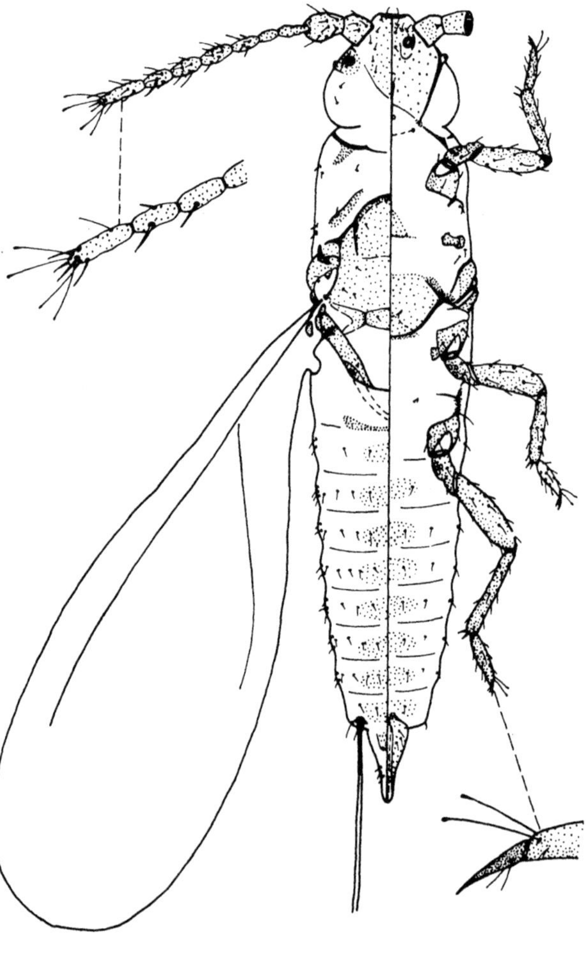

Fig. 37. Male of *Chaetococcus phragmitis* (Marchal) (after Afifi and Kosztarab, 1967)

compound eyes is common in some Margarodidae; 5 to 7 pairs of simple eyes, also a pair of ocelli are found in *Puto* (Pseudococcidae); 2 to 5 pairs of simple eyes with 1 pair of lateral ocelli are present in many Coccidae; and only 2 pairs of simple eyes (1 pair located dorsally and 1 pair ventrally on the head) are common in most Coccidae, Eriococcidae and Diaspididae, except that "ocelli" are also present in nearly all Aspidiotini. The tentorium may be present or very reduced. The mouth is non-functional, its opening is usually surrounded by a small ventral plate. The thorax is variable, usually partly sclerotized with well-defined sclerites and ridges separating it into pro-, meso- and metathorax. The mesothorax normally has the main sclerites (prescutum, scutum, and scutellum) well differentiated. When present, only 1 pair of wings develops; these usually are long and slender (macropterous), or are short (brachypterous). The wings are thin, membranous, normally transparent, usually with only 2 obvious veins. The second wing pair is reduced to hamulohalteres, each bearing apically one or more hooked setae. Hamulohalteres are absent in Aclerdidae and certain groups within other families. The legs are long and slender, with many setae and 1- or 2-segmented tarsus and a single claw. The trochanter and femur are rarely fused, e. g. Lecanodiaspididae. Four thoracic spiracles are present. The abdomen is 9-segmented and terminates in a well-developed bivalved penial sheath. Disc pores are present or absent. Their presence is characteristic of the Orthezioidea, and primitive members of Coccoidea.

Some species, especially in Diaspididae, are dimorphic. In such cases, the winged forms are usually abundant in the first generation, while wingless forms are predominant in the second generation of the same species (Knipscher et al., 1976). The mealybug, *Polystomophora ostiaplurima,* was found trimorphic by Vinis and Kozár (1981). In this species there are winged and wingless forms and a neotenic form that resembles the second instar stage, except for the longer legs and antennae and the presence of a short penial sheath. The species has two partly overlapping generations in Hungary; neotenic males were observed in the first generation only. This condition, the presence of macropterous, brachypterous and apterous forms in the same species, is probably not uncommon in other taxa.

LIFE CYCLE, BIOLOGY AND DISPERSAL

The number of generations per year varies by species.Most are limited to one generation in CE, while in the tropics the same species may have several generations. There are two generations each year in *Quadraspidiotus perniciosus* in CE, and in *Xylococcus filiferus* the life cycle requires several years.

Different species overwinter in different life stages: *Lepidosaphes ulmi* overwinters in the egg stage, *Q. perniciosus* in N_1 stage, *Parthenolecanium corni* and *Q. ostreaeformis* in N_2 stage, *Epidiaspis leperii* as adult females, while *Leucaspis loewi* in both the adult and N_2 stage.

Reproduction in most species is biparental, in others parthenogenetic. The same species may reproduce biparentally and or parthenogenetically. Most species lay eggs, but some give birth to N_1, for example, *Q.perniciosus*. Eggs are laid under the test or shield in armored scales, in which case the body of the female shrinks to provide room for the deposited eggs. Kermesidae and Coccini lay eggs under their venter and their hardened dorsum provides protection for the eggs and newly hatched nymphs. Many others produce ovisacs. These are made of loose wax threads, as in most Pseudococcidae and in Pulvinariini, and of wax plates in Ortheziidae. In *Pulvinaria betulae* the eggs laid are found both under the female and in the cottony ovisac attached to the end of abdomen. The number of eggs laid per coccoid female varies from 5 to over 8,000. The lowest number was observed in *Syngenaspis parlatoriae*, the highest in *P. corni* and *P. betulae*.

The sperm of Coccinea are filamentous without distinct regions. They have no flagellum, acrosome, centrioles, or mitochondria, and show no signs of a 9+2 microtubule arrangement. Scale sperm are packed into bundles containing 8–64 individual sperm. Nur (1962) and Robinson (1978) provide more details on the subject.

Egg production probably varies according to host plant and climatic conditions. The length of the egg stage, or embryonic development, is quite variable. In some species, e. g. *Sphaerolecanium prunastri*, this developmental period lasts only a few hours, while in the species that overwinter in the egg stage (e. g. *L. ulmi*) it can last 8–10 months.

Females normally develop through two, rarely three nymphal stages. Three nymphal stages are known in *S. prunastri* (Ben-Dov, 1968). Males usually develop through five life stages: N_1, N_2, prepupa, pupa, and adult; but sometimes only through three. There are no functional mouthparts in adult males, and they live from a few hours to a few days. Their activities are restricted to finding females and mating. It has been demonstrated in a few species that virgin females release sex pheromones that attract the searching males.

Coccinea disperse passively with the aid of wind, water, soil, humans, and domestic and wild animals. Winged adult males and first instars (crawlers) of most species disperse actively. The adult females and various nymphal stages of species that retain their vagility throughout their entire life (e. g. Ortheziidae) may also move limited distances. Many important pest species have been accidentally introduced into various parts of the world with plant material, e. g. *Pseudaulacaspis pentagona* and *Q. perniciosus*.

Crawlers may accidentally be transported on the body surfaces of larger animals. Ants often disperse honeydew-producing species to new host plants.

GENETICS AND ENDOSYMBIONTS

Coccinea have nine different sexual chromosome systems. These are XX female-XO male, 2N female-2N male, 2N female-N male (which includes hermaphroditism), and the lecanoid, comstockiella, and diaspidoid chromosome system. Hermaphroditism is known in five species of margarodids in the genus *Icerya* including *I. purchasi*. The lecanoid system differs radically from others, because it involves a haploid set and a deeply pigmented or heavily condensed (heterochromatic) set of chromosomes in the males. The comstockiella system is somewhat similar to the lecanoid system, except during spermatogenesis. The diaspidoid system is unique and involves a diploid female and haploid male. Mycetomes and mycetocytes are frequently polyploid.

The variety of parthenogenetic types is great, indicating independent evolution in numerous lineages. The high incidence of parthenogenesis is probably caused by the short-lived, fragile nature of males. Seven kinds of parthenogenesis occur: haploid arrhenotoky, diploid arrhenotoky, facultative deuterotoky, obligate deuterotoky, facultative thelytoky, obligate automictic thelytoky, obligate apomictic thelytoky. For a more detailed summary on coccoid genetics, see Brown (1977), Miller and Kosztarab (1979), Nur (1980), and other sources listed by them.

Coccoid endosymbionts exhibit extraordinary diversity. These organisms may be bacteroids or yeastlike. Some species have one symbiont type, others have two. They are transferred to the next generation transovarially. More information on coccoid endosymbionts is found in Buchner (1965), and Tremblay (1977).

PHYLOGENY

Scale insects are widely accepted as a sister group of aphids. Although fossilized members of the latter are known from Permian and the scale insects, as a distinct group, must have existed also at that time, the first evident coccoid has been found in amber of Upper Cretaceous age (Beardsley, 1969a). The level of morphological organization of this form is more advanced than that of some recent groups, suggesting that the primary radiation occurred in the Lower Cretaceous age or earlier. More than 200 specimens from Eocene and Oligocene ambers are known that represent all the main phylogenetic branches of coccoids; however, about half of them belong to the pine-infesting Matsucoccidae. A few fossils have have been found also in younger deposits. For a more comprehensive discussion see Koteja (1984, 1985).

Phylogenetic assessment of the family categories have been studied briefly by Borchsenius (1958), Hoy (1962), Boratynski and Davies (1971), Lambdin and Kosztarab (1973), Koteja (1974b), Danzig (1980a), Koteja (1984) and Miller (1984), and are based on characteristics of adult females, males, adult female mouthparts, and an integration of these and other character systems. The Orthezioidea (Archaeococcoidea), (ensign scales and giant scales), are considered most primitive, the Coccoidea (Neococcoidea) (including the mealybugs, eriococcids, soft scales, pit scales, and associated families) are more advanced, and the armored scales are the most advanced. As coccoids have become more specialized, their mobility has decreased. Waxy secretions are of two functional groups, one that covers the body and one that covers the eggs, but the secretion often covers both the body and eggs (eriococcids, pit scales, etc.). There is a sequential eye reduction in coccoid adult males. The Orthezioidea generally have multi-faceted compound eyes, like aphids. In annectent forms a row of as many as 8 unicorneal eyes occurs on each side of the head. Further reduction occurs in most Coccidae and Diaspididae. Other evolutionary sequences include: a gradual tendency for reduction in body size, development of a unique "pupal" stage in the males, decrease in the number of female and rarely of male instars, reduction in the number of labial (Koteja, 1974a) and antennal segments (Koteja, 1980), etc.

ECOLOGY OF SCALE INSECTS IN CENTRAL EUROPE

Only a few ecological studies have been completed on scale insects in CE. The comprehensive studies by Schmutterer (1952b, 1965) on northern Bavarian scale insects provide most of the published ecological information for the area. Kozár and Kosztarab (1982) analyzed host relationships among 64 Coccinea and the 14 tree genera which are the major components of Central European forests. Others who have published ecological studies on scale insects from the region include: Danzig (1972c) and Tereznikova (1959a) from USSR, Koteja and Zak-Ogaza (1966, 1969) from Poland. The large land mass that comprises the Central European zoogeographical area is diverse topographically and climatically. It includes elevations from sea level to the peaks of the Alps. The southern border is Mediterranean climatically, while the northern border is very cold.

Danzig (1962a) analyzed the scale insect species composition and habitats in the Leningrad area, just above the northern limits of the Central European fauna. She noted that species that normally live in the crowns of broad-leaved shade trees in CE, will, in the Leningrad area at the northern limit of their range, select and develop in dry, warm and sunny locations. She found *Lepidosaphes ulmi* near the ground on heath shrubs, and *Parthenolecanium corni* ♀ on root collars of raspberry.

Many gaps exist in our knowledge of the ecology of scale insects in CE. It is the authors' hope that the short compilation of the available records given for each species will provide the impetus for further work.

Abiotic conditions. Both temperature and humidity limit the range and abundance of scale insects. Many species are adversely affected by high humidity which permits the development of parasitic fungi. On the other hand, extensive summer drought was shown by Szelényi (1936) to be the main cause of mortality among the unprotected first instars (crawlers) of *Quadraspidiotus pyri*. Scale insect populations disperse mainly through the motile first instar or crawler stage. Scale crawlers generally prefer to settle on the undersides of branches and leaves and on the side of the trunk that provides maximal protection from direct sunlight, wind, and rain. In many instances crawler emergence is synchronized with the rainy season. Eggs hatch after the rains subside, so that crawlers are prevented from washing away.

The population build-up of *Fiorinia externa* was directly related to the abundance of foliar nitrogen in hemlocks (McClure, 1980).

Biotic conditions. Scale insects maintain various types of relationships with both plants and animals. Host-scale interactions have been summarized with special emphasis on North American conditions by Miller and Kosztarab (1979). Coccoids occur in most terrestrial habitats of CE having plant life. They occur on nearly all parts of the host, but most Coccinea live on the above-ground parts of plants, and many species are specific to certain parts of plants. Some species infest roots only, e. g. *Chortinaspis subterraneus, Rhizopulvinaria artemisiae, Rhizoecus* spp.; while species of *Eriopeltis* are always on leaves of grasses. Other species, e. g., *Quadraspidiotus perniciosus,* infest all the above-ground parts of trees and shrubs.

Micro-habitats. Many scale insects show preference for tight protected areas, such as cracks and crevices of bark, leaf axils, root crowns, or nodes of grass stems. Mealybugs often seek out tunnels bored by insect larvae in the roots or stems. Those feeding under leaf sheaths of grass stems are normally flattened dorsoventrally, as in *Chaetococcus phragmitis*. Those feeding on plant roots often have robust, enlarged front legs with heavily sclerotized claws for digging, such as *Neomargarodes festucae*.

Host-specificity. Many species that move in the adult stage often produce ovisacs on plants other than their host, this phenomenon causes problems in true host determination. Coccinea display a wide range of host-specificity. Most of the important pest species are polyphagous: e. g., *Q. perniciosus* is known to infest more than 700 plant species. Others are oligophagous and feed on plants belonging to one family: e. g., *Antonina* spp. and *Chaetococcus* spp. feed on Poaceae only. Or they feed on plants belonging to one genus; e. g., *Kermes* spp. on oaks of the genus *Quercus*. There are very few monophagous species; e. g., *Xylococcus filiferus*, which is known only from linden.

A number of plants in Central Europe have their own characteristic scale insect fauna. The genera *Quercus* (with 17 species of scales), *Pinus* (10) and *Picea* (8) have the highest proportion of mono- and oligophagous species, while most coccoids infesting deciduous trees (except oaks) are polyphagous (Kozár and Kosztarab, 1982). From among the herbaceous plant genera, *Carex* and *Thymus*, apparently have the largest number of specialized species (Kosztarab and Kozár, 1978).

Plant-scale insect associations. Species associated with evergreens or gymnosperms (Cupressaceae, Pinaceae, Taxaceae) include: *Carulaspis juniperi, Chionaspis austriaca, Eulecanium sericeum, Lepidosaphes juniperi, L. newsteadi*, 4 *Leucaspis* spp., *Matsucoccus matsumurae, Nemolecanium graniformis, Nuculaspis abietis, Parthenolecanium fletcheri, P. pomeranicum, Phenacoccus piceae*, 3 *Physokermes* spp., *Puto antennatus* and *Syngenaspis parlatoriae*.

The grass family, (Poaceae or Gramineae), serves as host for the largest number of scale insect species in Central Europe. These are listed in the host index rather than here, but two grass genera, *Agropyron* and *Festuca*, are noteworthy with a diverse array of coccoid species. Almost all of the grass-infesting species are members of families Pseudococcidae, Eriococcidae, and Coccidae.

On Rosaceae the most common coccoid species is *Quadraspidiotus perniciosus*; the two associated with roses are *Aulacaspis rosae* and *Rhodococcus perornatus*. We often find *Quadraspidiotus ostrearformis* on apples (*Malus*); *Epidiaspis leperii, Quadraspidiotus marani*, and *Q. pyri* on pears (*Pyrus*), and *Sphaerolecanium prunastri* on plums and peaches (*Prunus* spp.).

Species associated with oaks (Fagaceae) are: *Acanthococcus roboris, Asterodiaspis* spp., *Asterodiaspis roboris, Chionaspis lepineyi, C. salicis, Diaspidiotus hungaricus, D. wuenni, Eulecanium ciliatum*, 5 *Kermes* spp., *Parthenolecanium rufulum, Quadraspidiotus lenticularis, Q. zonatus*, and *Targionia vitis*.

The bark of ash trees, *Fraxinus* spp., in Karst forests is frequently infested with *Pseudochermes fraxini*; of elms, *Ulmus* spp., with *Gossyparia spuria* and rarely with *Ritsemia pupifera*.

Common species in moist mountain forests on Ericaceae (*Arbutus, Calluna, Erica, Rhododendron* and *Vaccinium*) are: *Chionaspis salicis, Diaspidiotus bavaricus, Eulecanium franconicum, Phyllostroma myrtilli*, and *Spinococcus calluneti*.

We find the following scale insects in hygrophilous habitats, associated with Salicaceae (*Salix* and/or *Populus*): *Acanthococcus aceris, Chionaspis salicis*, 5

Eulecanium spp., *Lepidosaphes ulmi, Pulvinaria betulae, P. populi,* and *Quadraspidiotus gigas.*

Some common species on Betulaceae (*Alnus* and/or *Betula*) are: *Chionaspis salicis, Lepidosaphes ulmi, Phenacoccus aceris, Pulvinaria betulae, Quadraspidiotus ostreaeformis, Q. perniciosus* and *Steingelia gorodetskia.*

At high elevations, primarily on grasses, we find *Rhizococcus pseudinsignis*; also often above the tree line, in the moss ground cover, are: *Arctorthezia cataphracta,* on plant roots, *Puto janetscheki,* and on wet meadows in mosses and in wet forest litter, *Newsteadia floccosa.*

Hygrophilous species that are common at lake or pond shores on stems of the reed, *Phragmites communis,* are: *Chaetococcus phragmitis, Eriopeltis lichtensteini,* and *Trionymus phragmitis*; on *Carex* spp.: *Boreococcus ingricus, Chnaurococcus danzigae, Luzulaspis frontalis, L. montana, Parafairmairia gracilis* (latter also on other hosts), and *Vittacoccus longicornis.*

The xerophilous species in warm, rocky habitats that are often associated with *Thymus* spp. are *Cerococcus cycliger* and *Quadraspidiotus labiatarum,* while *Juniperus communis* on warm sand dunes is often infested with *Allococcus vovae* and *Carulaspis juniperi.*

Species common on grass roots in warm habitats that are often tended by ants include: *Euripersia europea* and *E. tomlini, Lecanopsis festucae, L. formicarum, Rhizoecus albidus,* and *Rhodania porifera.* On underground stems of *Calluna* we find *Spinococcus calluneti.*

Mesophilous grass-infesting species, characteristic to meadows, include *Acanthococcus munroi,* several *Heterococcus* spp., *Phenacoccus* spp., *Rhizococcus insignis,* and *Trionymus* spp. The grasses in warm steppe habitats of sand dunes and sea beaches are frequented by *Chaetococcus sulci, Neomargarodes festucae* and *Scythia* spp., while *Euphorbia* serves as host for *Spinococcus marrubii.*

Pest species of broad distribution, that occur in abundance in almost all habitats in CE, are: *Chionaspis salicis, Lepidosaphes ulmi, Parthenolecanium corni,* and *Quadraspidiotus perniciosus.* Each of these species is polyphagous.

The common pest scales in artificial habitats on ornamental or shade trees are: *Lepidosaphes ulmi,* 2 *Leucaspis* spp., *Parthenolecanium corni, Phenacoccus aceris, Quadraspidiotus perniciosus,* and *Sphaerolecanium prunastri.* Physiologically weakened and thus predisposed trees in stress-habitats often attract the above species.

Plant resistance. Host plants often show varying degrees of susceptibility to a particular scale insect. Many plants have been found with environmentally induced or with physiological resistance to a particular scale insect species. Certain cultivars of economic plants appear to have genetic resistance to scale insect infestations (Kosztarab, 1981).

Host-induced biotypes. Scales of the same species that infest different host plants sometimes develop different morphologies (shape, color, size) (Danzig, 1961). Some species in CE also have different stem and leaf forms that differ in their biology and morphology and often were described as separate species (Lupo, 1943; Danzig, 1970). Knipsher et al. (1976) demonstrated experimentally and conclusively this phenomenon in North America.

Parasites, predators and ants. Coccoids have a diverse complex of natural enemies including hundreds of species of parasites and predators. The largest number of parasites are chalcidoid wasps, most of which belong to the families Encyrtidae, Aphelinidae and Eulophidae. The best known predators of coccoids are Coleoptera: Coccinellidae, Anthribidae, Nitidulidae; Lepidoptera: Sesiidae; Diptera: Chamaemyiidae; and Acarina: Hemisarcoptidae. Some lesser known predators belong to Heteroptera and Neuroptera.

Trjapitzin (1978) summarized information on chalcids parasitizing scale insects in the European parts of USSR. The natural enemy complex in northern Bavaria was described in detail by Schmutterer (1952b). The latter author also summarized the mutualistic relationship between coccoids and ants for the same area, while Way (1963) reviewed this relationship for all Homoptera, including many interesting examples on coccoids, especially from the tropics.

Mealybugs (fig. 1), soft scales, and Kermesidae benefit from ant associations both directly and indirectly. When the ants remove the honeydew produced by these insects, they eliminate the substrate necessary for the growth of sooty-mold. This fungus could contaminate and destroy the scale insects. The ants protect also coccoids from their natural enemies (Bartlett, 1961). Further protection is provided by ants which keep the scales in underground galleries during harsh winter conditions; construct tents over mealybug and soft scale colonies to protect them from summer heat and sunlight; and transport the scales to new plants when host conditions become unsuitable. Ants are also known to actively regulate the population size of certain scales, an activity that could be beneficial to the species in case of scarcity of host plants. Ants in turn benefit from scale insect associations by being provided with honeydew for a constant food supply. They also obtain protein when they feed on dead or live coccoids, especially while regulating scale population size.

GEOGRAPHICAL DISTRIBUTION

Although 233 species are known from Central Europe (fig. 36), the fauna still needs more detailed studies. It is estimated that as many as 100 new species are waiting for discovery, especially species with cryptic life habits in the families of Pseudococcidae, Eriococcidae and Coccidae — those that live primarily on roots, in leaf sheaths or in bark crevices. A number of species are known from few localities and from only one or two countries in Central Europe. This phenomenon apparently reflects not the rarity of species, but rather our incomplete knowledge of the fauna. There are many species that have large and questionable gaps in their distribution; for example, *Balanococcus boratynskii* is known only from England and USSR (European Parts and the Far East). Because of this lack of information, it is premature to report on the zoogeographical distribution patterns, although preliminary reports on this subject have been prepared by Schmidt (1956), Zahradnik (1957), Tereznikova (1959a, 1959b), Schmutterer (1980), and Kozár and Kosztarab (1982). Species associated with certain zoogeographic areas are given below.

Relatively few species are considered truly **montane,** such as *Eulecanium sericeum* and *Nemolecanium graniformis*. **Boreo–montane** or **alpine** species are: *Arctorthezia* *cataphracta, Cryptococcus fagisuga, Diaspidiotus bavaricus, Eulecanium franconicum, Heliococcus nivearum, Newsteadia floccosa, Phyllostroma myrtilli, Physokermes hemicryphus, P. piceae, Puto antennatus, P. janetscheki,* and *Spinococcus calluneti.* Species characteristic to **deciduous forests** of temperate climate are: *Asterodiaspis variolosa, Eulecanium tiliae, Kermes quercus,* and *Pseudochermes fraxini.* **Hygrophilous** species in steppe habitats are: *Chaetococcus phragmitis, Luzulaspis scotica,* and *Parafairmairia gracilis.* **Xerophilous** species in steppe habitats are: *Acanthococcus desertus, Cerococcus cycliger, Heliococcus sulci, Luzulaspis kosztarabi, Rhizopulvinaria spinifera,* and *Scythia craniumequinum.* These species are also called **thermophilous** and normally inhabit sunny and rocky slopes.

A large portion of the coccoids in Central Europe are **mesophilous** species. These are found throughout the faunal area, inhabiting mostly grasslands (Kozár, 1981a) with approximate average conditions of moisture and temperature. Because of their large number, their listing here is needless.

Only relatively few species in Central Europe are known to exhibit **Holarctic** distribution, e.g. *Arctorthezia cataphracta, Eriopeltis festucae, Gossyparia spuria, Heterococcus nudus, Physokermes hemicryphus, Pulvinaria betulae, Quadraspidiotus gigas, Q. ostreaeformis.*

Recent studies by Danzig (1980a) revealed that there are a number of species with a **Trans-Palearctic** distribution pattern, such as: *Acanthococcus greeni, Allococcus vovae, Atrococcus achilleae, A. cracens, A. paludinus, Chionaspis salicis, Dysmicoccus walkeri, Eulecanium ciliatum, E. douglasi, E. franconicum, Greenisca gouxi, Heliococcus bohemicus, Kaweckia glyceriae, Matsucoccus matsumurae, Newsteadia floccosa, Orthezia urticae, Parafairmairia bipartita, Phenacoccus piceae, P. interruptus, Pseudochermes fraxini, Puto pilosellae, Rhizococcus insignis, Rhizoecus albidus, Steingelia gorodetskia, Trionymus perrissi,* and *Vittacoccus longicornis.*

Many of the species known from the faunal area have an **Euro-Siberian** distribution pattern. The majority of species with boreal, montane or steppe habitat distribution belong here. Tereznikova (1975) considered as **Black Sea–Caucasus** elements: *Mirococcopsis stipae, Rhizopulvinaria spinifera, Spinococcus marrubii, Scythia craniumequinum*; as **Kazakhstani** (or Southwest Siberian) elements: *Eulecanium caraganae, Kaweckia rubra, Rhodococcus spiraeae* and *Trionymus multivorus.* Species apparently of **Mediterranean** origin are: *Antonina purpurea, Chaetococcus phragmitis, Chionaspis lepineyi, Diaspidiotus distinctus, Eriococcus buxi, Lecanodiaspis sardoa, Lepidosaphes granati, Lichtensia viburni, Parlatoria oleae, Puto superbus, Quadraspidiotus labiatarum, Q. lenticularis, Rhizaspidiotus canariensis,* and *Targionia vitis.*

A number of pest species are **cosmopolitan** in distribution, such as: *Aulacaspis rosae, Lepidosaphes ulmi, Parthenolecanium corni, P. persicae, Quadraspidiotus ostreaeformis,* and *Unaspis euonymi*; and some were accidentally **introduced** to Central Europe with plant material, e.g. *Pseudaulacaspis pentagona,* and *Quadraspidiotus perniciosus.* There is a large number (73 spp.) of introduced species on ornamental plants in the **greenhouses and homes** that originated from the subtropical and tropical parts of the world. These are listed at the end of each general family description.

ECONOMIC IMPORTANCE

Scale insects are notorious plant pests, particularly on perennials. They are especially serious pests on fruit and nut trees, woody ornamentals, forest trees and shrubs, and indoor plantings. They cause damage to plants by feeding on plant sap. Vigor is reduced, leaves may drop prematurely, and unchecked heavy infestation may kill trees (fig. 34e). Discoloration often is produced on infested leaves (Euonymus scale) and fruits (San José Scale, fig. 34a), or twigs and branches are deformed (Pit Scales, Italian Pear Scale); rarely plant galls are also produced (many Australian coccoids). The sugar-rich honeydew produced by mealybugs and soft scales covers the foliage and serves as a breeding medium for sooty mold (fig. 116). When the latter covers the leaves with a black coating of mycelia, it interferes with photosynthesis; as a result less sugar is produced in the fruits, and ornamental plants lose their aesthetic value. Indirect damage may be caused by scale insects injecting toxins into the host, and some mealybugs serve as vectors of viruses. The losses and extra production costs attributed to coccoids in the USA was estimated to exceed $500 million yearly (Kosztarab, 1977). The most frequent pest species in CE are *Quadraspidiotus perniciosus, Epidiaspis leperii, Lepidosaphes ulmi, Parthenolecanium corni, Chionaspis salicis, Cryptococcus fagisuga, Pseudochermes fraxini, Phenacoccus aceris,* and *Asterodiaspis variolosa.* Under optimal conditions in local areas, many other species can build to damaging population levels.

Coccoids and their products have been widely utilized by men since ancient times (Miller & Kosztarab, 1979). The mealybug, *Trabutina mannipara,* is believed to be responsible for the production of the honeydew exudate "manna" consumed by the Israelites in the Sinai desert. Dyes were provided by a variety of unrelated scale insect groups, such as *Kermes,* lac insects, Polish cochineal (*Porphyrophora polonica*), and cochineal scale. The latter is still in cultivation as a commercial dye source in the Canary Islands, Peru, and Mexico. The excretions of lac insects, besides their use as dye, are also widely used in the production of waxes and resins; the latter is most significant as a precursor of natural shellac. Wax produced by *Ericerus pela* is used to make candles in China. *Cerococcus quercus* was collected by Indians of North America and chewed like gum. The "fat" extracted from adult females of the giant scale, *Llaveia axin,* has been used by the natives of Mexico and Central America as a lacquerlike coating for waterproofing wood and gourds. It also has been used as a base for medicines and cosmetic paints. The cysts, or "ground pearls", of certain giant scales were strung for necklaces in the Mediterranean Region. Scale insects have also been utilized for biological control of noxious weeds. *Dactylopius opuntiae* was responsible for the near elimination of *Opuntia* sp. on the Santa Cruz Island, near mainland California.

The honeydew produced by many scale insects is an important source of food for ants and other foraging insects, and also serves as the basis for honey production by bees. Because scale insects contain an array of chromosome systems, they are ideal subjects for genetic studies.

There are a number of methods for establishing economic thresholds. In only a few species have ranking systems been developed for establishing economic thresholds. In the case of *Parthenolecanium corni*, according to Saakyan-Baranova et al. (1971), control is justified when small colonies of 2 to 10 are noted; while Bailey (1964) suggests control in California when an average of one nymph is found in the autumn per 2.5 cm section of twig. In the latter case mass infestation is expected the following year. Economic thresholds for scale insects are often not a consideration because of their quarantine importance.

Pheromone-baited traps can be used to detect the presence and abundance of certain species (Miller and Kosztarab, 1979) in the area. Rice and Jones (1977) found the pheromone traps to accurately monitor flight activities of male San José scale and to determine discrete generations of this pest under field conditions. Similar traps used by Rice and Moreno (1970) assisted in establishing the effect of temperature and relative humidity on the flight activities of *Aonidiella aurantii*. Huba (1962) used temperature sums and temperature extremes to predict distribution and pest status of San José scale in Europe.

Several methods can be used to calculate timing of control measures. Kozár (1975) evaluated about ten of these. The most frequently used methods include trapping (suction, sex pheromone, and white painted card traps, fig. 34c), utilization of temperature sums, plant phenology, and dissection of gravid females. Each of these serves as a guide for forecasting crawler emergence and its duration and gives important information for making control decisions.

There are a number of methods for forecasting the size of expected populations. Most often the rate of winter mortality is used as a primary basis. In many species the rate of parasitization provides important guidelines. Talickij et al. (1966) believed that no significant population increase is to be expected after a 30% parasitization rate in *P. corni*. Kozár (1978) summarized such forecasting factors for the San José scale.

Quarantines are used by most countries to prevent the introduction of pest species to new areas. Many plant quarantine regulations deal specifically with scale insects. However, the lack of stringent plant quarantine measures in earlier years has resulted in the world-wide distribution of important fruit and ornamental pests such as *Quadraspidiotus perniciosus* and *Pseudaulacaspis pentagona*.

Cultural practices are important in the prevention of infestation and in the control of scale insects. These may include: planting of resistant varieties; use of certified, pest-free nursery stock for propagation; or providing optimal conditions for plant growth by such techniques as proper pruning. For example, apple trees with a high trunk and a dense crown may provide ideal conditions for a San José scale population increase by protecting the scale from sun, wind, and sprays. Consideration should be given to the native and ornamental vegetation surrounding crop systems. These plants frequently serve as reservoirs for reinfestation of orchards.

Mechanical control of scale insects may include scraping the colonies and layered infestations from the trunk and large branches with a wire brush. The scrapings should be removed and destroyed if eggs or crawlers are present. Indoor plants can be cleaned with a strong spray from a hose or with a fine brush under a flowing faucet.

Biological control has proved successful in CE against some introduced scale insects. In Hungary, *P. pentagona* was controlled with *Prospaltella berlesei*, while *P. perniciosi* was utilized in several countries in CE against *Q. perniciosus*. Many coccoid species have effective native parasites and predators. It is important to understand the effect of these natural enemies for successful long-term pest management systems. Through the application of *integrated pest management* (IPM) programs the number of necessary chemical sprays can be reduced and be integrated with the beneficial effects of the natural enemies of scale insects. New possibilities opened in IPM with the introduction and use of *insect growth regulators* that have *juvenile hormone* activity and are often harmless to beneficial insects and the environment.

Chemical control should be applied only when it is justified by the presence or predicted occurrence of economic levels of scale infestations. For coccoids feeding on roots, systemics may be applied in granular form or as a drench. Those scale insects that live in leaf sheaths of Gramineae are controlled with systemic sprays. On woody plants, dormant oil sprays (fig. 34d) may be applied against overwintering life stages during early spring, while contact insecticides, or systemics, are used during the crawler stage against those that overwinter as eggs or adults. Eggs and adults are also controlled with summer oil sprays. Most scale insect species can be successfully controlled at the time of first instar emergence (crawler stage). Infested plant propagation material and fruits are usually fumigated before marketing or use. The adult males of *Q. perniciosus* can be controlled by insecticides with the help of sex pheromone baited traps, which also provide information on population size, and for timing of control applications.

The more recent literature on scale insect control has been reviewed by Miller and Kosztarab (1979).

COLLECTING, PRESERVING, MOUNTING AND REARING OF COCCINEA

Because the present classification of scale insects is almost entirely based on adult females, most of the **collecting** effort is directed to obtain these (Borchsenius, 1950a). In the case of Coccidae and Kermesidae, young adults must be collected while their exoskeleton is still soft and is not distended due to the development of eggs. For the majority of species in CE, May and June in Hungary and August in Poland, are the best months for collecting. Some species develop into young adults at the end of the summer and overwinter in the adult or egg stage. Often the presence of honeydew-seeking ants on the bark or of sooty mold on leaves draws attention to the presence of coccoid colonies. On trees, coccoids often infest areas around wounds, and are common on the underside of branches or on protected and shaded parts of the trunk, in bark crevices, twig forkings, or other sheltered areas of plants. Their color often is cryptic, allowing them to blend into the environment. During collecting expeditions it is useful to pay special attention to often-neglected areas, such as plant roots, root crowns, leaf sheaths, both sides of leaves of grasses, fruits, galls, and areas under bark flakes (on *Pinus, Platanus*). Egglaying females of mealybugs and eriococcids sometimes may be found in the leaf litter, on dried plants, under stones, or attached to mosses or lichens. Mealybugs can be collected by beating branches over an inverted umbrella, or over canvas. Mealybugs, ensign scales, giant scales and some other coccoids can be obtained from leaf litter, bark, soil or moss samples with a Berlese funnel. Some coccidologists have succesfully collected many new species (Asterolecaniidae, Diaspididae, etc.) from preserved plant specimens in herbaria. Adult males can be collected with color cards or light sources and with sex pheromone-baited or suction traps.

All scale insects can be **preserved** in vials of 70–75% ethyl alcohol. To fix tissues for quality slide mounts, the live scales can be placed in heated alcohol for a few minutes. Preservation by drying infested bark pieces in paper envelopes is practiced with representatives of Asterolecaniidae and Diaspididae. Postreproductive Kermesidae females, egg sacs of Eriococcidae, and adult female Ortheziidae with egg sacs are often dry-preserved in the collections.

Live adult males and females are fixed in a fresh solution of 4 parts chloroform, 3 parts of 100% ethyl alcohol and 1 part glacial acetic acid for 48 hours and stored in 75% ethyl alcohol. Such fixed material can also be used for cytogenetic studies.

There are a number of good methods for **mounting** coccoids on slides (Wilkey, 1962; Kozarzhevskaya 1968; Howell and Kosztarab, 1972; and others). We are describing here a slight modification of the method usable for different family representatives (Wilkey, 1962).

1. Dry or alcohol-preserved specimens should be placed directly into a 10% solution of KOH. Live specimens should be fixed in preheated 70–75% alcohol for a few minutes first and left in cool alcohol for about 2 hours before transferring them into 10% KOH.
2. KOH should be heated gently (never boiling) until the specimens soften and begin to clear. Delicate specimens can be kept overnight in 5% KOH and treated without heating.
3. A small incision should be made on the lateral side or between the antennae; the body contents are teased out by gently pressing the specimens with a small spatula.
4. When the specimens are transparent, they are transferred to distilled water or to 95% ethyl alcohol for 10–15 minutes.
5. The specimens are placed in Essig's Aphid Fluid* with staining solution** and gently heated for 5–60 minutes (normally Coccidae 5–20, Diaspididae 10–30, Pseudococcidae for 10–60 minutes).
6. They are returned to 95% ethyl alcohol for 10 minutes or until excess stain is removed.
7. The final step is to transfer specimens to clove oil for 5–15 minutes and mount them in Canada balsam on microscope slides.
8. The slides are placed in a drying oven or warming tray at 40°C for about 2 weeks before labeling and study.

* Essig's Aphid Fluid is made of 20 parts of lactic acid (85%), 4 parts glacial acetic acid, 2 parts phenol (saturated in distilled water), and 1 part distilled water.

** Staining solution is made of 15 ml. Essig's Aphid Solution with 2% aquaeous solutions of: 20 drops acid fuchsin, 20 drops lignin pink, and 20 drops of erythrosin.

Rearing of coccoids is relatively easy in the case of houseplant-infesting species. Colonies living on plants can be started by introduction of egg masses or first instars (crawlers) of the desired species to the selected plant. It is important to duplicate the environmental conditions (temperature, humidity, shading, etc.) of the original colony on the new plant, also to avoid mixed infestations of more than one coccoid species.

Adult males seldom are collected in the field. They are best collected by rearing them from plant parts that contain male pupae. They may be placed in a tightly closed container, such as glass cylinder, petri dish, cellophane bag, etc. Similar procedures can be used to sample emerging females, first instars and parasites.

II. SYSTEMATIC PART

CLASSIFICATION

The classification of scale insects is unsettled and currently is evolving. It has been based almost entirely on the morphology of the degenerate adult females and as such is somewhat inadequate. At present there are serious discrepancies even at the familial and generic levels. More intensive studies on the poorly-known immature stages and on the adult males will assist in improving our present classification system.

There are about 6,000 described species of scale insects, belonging to about 800 genera world-wide, with probably twice as many new species waiting for discovery, especially in the tropical areas of the world. The number of families recognized by different coccidologists varies between 15 and 22. We recognize and discuss 12 families from Central Europe.

Reviewer's remark. Only two consistent major-group systems of scale insects have been proposed that stimulated studies on evolution and phylogeny of these insects. Balachowsky (1942) divided them into three branches (phyla) which represented also three levels of morphological organization: the primitive margaroids with abdominal spiracles and compound eyes in males, the diverse and abundant lecanoids (= coccoids) lacking abdominal spiracles and compound eyes but with head separated from thorax, and the diaspidoids with head and thorax fused in males, with particular organization of abdomen in nymphs and females and with peculiar test. On the basis of this concept a number of valuable studies on males, initiated by Theron (1958), have been undertaken.

Borchsenius (1950) and Bodenheimer (1953) divided the scale insects into two major groups: the archaeococcids and neococcids. Also in this system, further developed by other students and adopted in the

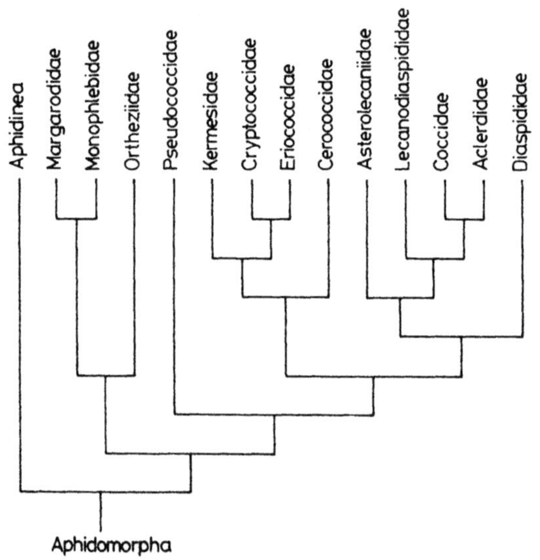

book, the former were considered as ancestral, the latter as derived, and both as monophyletic, until Boratynski and Davies (1971), also Brown (1977) postulated a polyphyletic origin of the neococcids, with *Phenacoleachia* and *Steingelia* giving rise to various groups of families. The concept of polyphyletic origin of the neococcids has not been accepted by coccidologists (Koteja, 1974b; Nur, 1980; Danzig, 1980a). Recently Miller (1984) basically modified the system by transferring the Phenacoleachiidae, Ortheziidae and some margarodids from the archaeococcids to the neococcids (see Reviewer's remark at Margarodidae).

Among the families traditionally assigned to the neococcids the Pseudococcidae are the most primitive, these are followed by the Eriococcidae from which other groups derived. The proposed relationships of European Scale Insects (Based on Hennig's sister-group concept, as prepared by J. Koteja) is presented on page 32.

KEY TO SUBORDERS OF HOMOPTERA IN CENTRAL EUROPE

(Based on adults of both sexes unless stated otherwise)

1. Tarsi 3-segmented; antennae short and bristle-like; legs well-developed, often adapted for jumping; beak arises from back of head .... (cicadas and hoppers) *Auchenorrhyncha*

   Tarsi 1 or 2-segmented; antennae normally long and filiform, rarely reduced or absent; legs moderately developed or often reduced, not adapted for jumping (except in Psyllinea); beak when present, arises between front coxae 2

2. Tarsi 2-segmented, with 2 claws; adult females normally winged and with well-developed legs except in some aphids; male (when winged) always with 2 pairs of wings (psyllids, white-flies, aphids) .... *Sternorrhyncha*

   Tarsi 1-segmented (often 2-segmented in males and some female margarodids), with only 1 claw; adult females always wingless; rarely legless (Diaspididae, Asterolecaniidae); males normally winged and with only 1 pair functional (scale insects) .... *Coccinea*

KEY TO SUPERFAMILIES AND FAMILIES OF COCCINEA IN CENTRAL EUROPE

(Based on adult females)

1. Abdominal spiracles present (fig. 40); usually without tubular ducts; labium without apical setae (Superfamily Orthezioidea) 2

   Abdominal spiracles absent (fig. 55); normally with tubular ducts; labium with apical setae (Superfamily Coccoidea) 3

2. Anal ring well-developed, with numerous pores and 6 setae; eyes usually stalked; antennae 3–8-segmented (fig. 40); in nature body covered with waxy plates, which also form a tubular ovisac (fig. 41) ..... *Ortheziidae*, p. 35

   Anal ring absent or reduced, without apparent pores and setae; eyes rarely stalked; antennae 6–13-segmented (fig. 46); body not covered with waxy plates .... *Margarodidae*, p. 46

3. Abdomen terminating in fused segments forming a pygidium; anal opening simple; legs absent or reduced to remnants; antennae unsegmented (fig. 138); on host body covered with a secreted thin shield-like scale formed around first and second molted skin (fig. 138a) .... *Diaspididae*, p. 321

sheath which curves back and down. Adult males were described by Morrison (1925), Jancke (1955), and Koteja (1986).

Immatures. Poorly known. Nymphs similar to adult females, but without ovisac band and usually with fewer antennal segments.

Distribution. Present on all continents. Most species found in the Nearctic and Neotropical Regions. World-wide 90 species, belonging to 6 genera. In CE 4 genera with 4 species.

Biology. Schmutterer (1952b) found on wet meadows, in mosses, and in soil on slopes and shaded areas *Ortheziola vejdovskyi* and *Newsteadia floccosa,* while latter with *Orthezia urticae* also was present in forest biotopes; *N. floccosa* in leaf litter and on ground cover of wet deciduous mixed forests and in swamps, while *O. urticae* in herbaceous and dwarf shrub layers. They infest and feed on all parts of plants; many host-specific, others polyphagous. Some of the plants listed as host are often incidental objects. They develop through 4 instars in females, 5 in males. Most species appear to have more than 1 yearly generation. No parasites observed in this family. Only few species of economic importance, e. g. Greenhouse Orthezia or Kew Bug, *Orthezia insignis* Browne, which in CE is common indoors only.

Reviewer's remark. The Ortheziidae share some of the most primitive features with specialized ones. Miller (1984) removed this family from the archaeococcids and placed it within the neococcids because of presence of a setigerous anal ring and other characteristics.

Key to genera and species of Ortheziidae
(Each genus with one species)

1. Tibia and tarsus not fused; antennae usually 8-segmented (sometimes 7); body without ovisac ca. 3–5 mm long. 2
   Tibia and tarsus fused; antennae mostly with fewer than 7 segments; body without ovisac ca 1.5–2.5 mm long 3
2. No disc pores in atrium of thoracic spiracles; claws usually with 2 (seldom 3) denticles; eye stalks elongate, conical, usually with tubercle; unpaired wax plates on mid-dorsum absent . . . . *Orthezia urticae,* p. 41
   With disc pores in atrium of thoracic spiracles; claws without denticles; eye stalks curved, thumblike, without tubercle; 3 unpaired wax plates on mid-dorsum semicircular . . . . *Arctorthezia cataphracta,* p. 37
3. Antennae 6- or 7-segmented, with the 2 basal segments enlarged; anterior 5 pairs of abdominal spiracles developed; entire dorsum covered with wax plates. . . . . *Newsteadia floccosa,* p. 39
   Antennae 3-segmented, but appear 4-segmented; abdominal spiracles not clearly defined; center of dorsum bare, without wax plates . . . . *Ortheziola vejdovskyi,* p. 44

Genus **Arctorthezia** Cockerell
Fig. 38

Syn. & Lit.: *Arctorthezia* Cockerell, 1902a:259, 1902b:114; Morrison 1925:143, 1952:53; Gomez-Menor, 1937:386;Hadzibejli, 1963:611; Danzig, 1964:806, 1980a:105; Kosztarab & Kozár, 1978:12.

Type-species: *Orthezia occidentalis* Douglas, 1891.

Adult females elongate circular, without ovisac 3–4 mm long; ovisac short; body covered with 4 longitudinal rows of wax plates; plates in median rows semicircular, not elongate; latter enclose small triangular plates; eyes large, stalked, thumblike; tarsal claws without tooth; thoracic spiracles with circle or band of multilocular disc pores inside atrium; 3 more or less distinct triangular median clusters of spines on anterior of thorax; abdomen with 7 or 8 pairs of spiracles. Adult males unknown, immatures poorly known.

Distribution. Holarctic, with 4 species; 1 in CE at high elevations.

Biology. Found on a variety of plants, that include mosses, lichens, grasses, shrubs, and trees; often under stones or on roots.

1. **Arctorthezia cataphracta** (Olafsen) – Alpine Ensign Scale
Figs 38, 39

Syn. & Lit.: *Pediculus cataphractus* Olafsen, 1772:610. *Dorthezia chiton* Zetterstedt, 1828:314, *Orthezia signoreti* F. B. White, 1877:160; List, 1886:201 (A monograph); Douglas, 1881b:172, 203. *O. uva* Blanchard, 1896:681. *A. cataphracta*, Morrison, 1925:143, 1952:54; Goux, 1941:31; Balachowsky, 1953a:93; Schmutterer, 1955:160; Rogojanu, 1958:308; Hadzibejli, 1963:612; Koteja & Zak-Ogaza, 1966:316; Zahradnik, 1974:141; Tereznikova, 1975:114; Olmi & Sampo, 1975:146; Kosztarab & Kozár, 1978:12; Danzig 1980a:105.

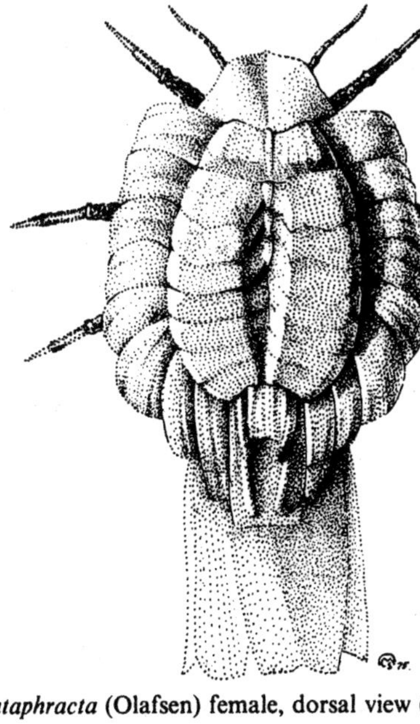

Fig. 38. *Arctorthezia cataphracta* (Olafsen) female, dorsal view (after Kosztarab and Kozár, 1978)

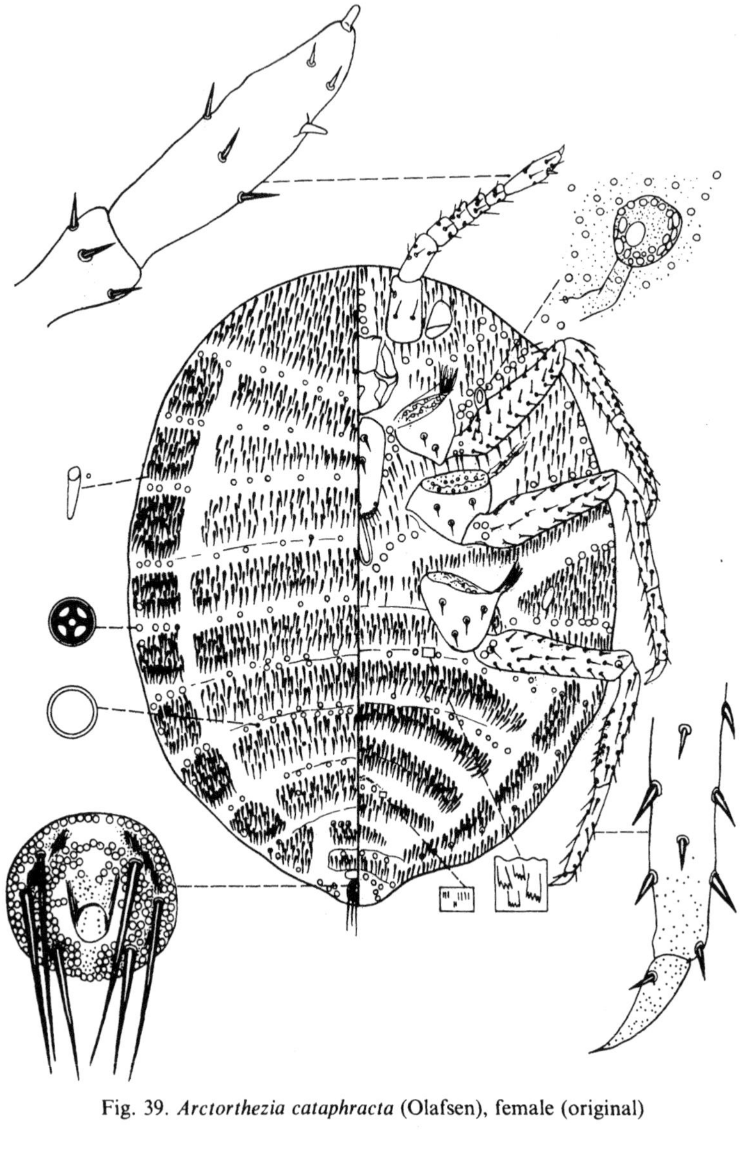

Fig. 39. *Arctorthezia cataphracta* (Olafsen), female (original)

Adult female with 9 paired dorsal wax plates in each marginal row, 8 in each median row and 3 small triangular or shield-shaped in center of thorax; ovisac parallel-sided, short, about half length of body. Adult female 2.5–3.5 mm long, 2.5–3 mm wide; antennae 7-, rarely 6- or 8-segmented; eyes curved, thumblike; legs normal, claws without denticles; thoracic spiracles with wide band of pores inside atrium; 8 pairs of abdominal spiracles; derm with quadrilocular disc pores, and short thick spines; latter form bands and clusters; ovisac band narrow; ventral posterior spine clusters enclose 3 spiracles on each side; anal ring oval, pointed posteriorly.

Distribution. Holarctic, circumpolar; boreo-alpine in CE, found at higher elevations only. It was reported by Olmi and Sampo (1975) from 1600–2600 m elevation in the Italian Alps. In CE: Au, Cz, Fr, Ge, It, Po, Ro, Sw, USSR, and other countries, especially in N Europe.

Biology. Prefer moist habitats, not uncommon on roots, under stones or logs and in mosses or lichens. Found on a variety of plant species, including fungi, grasses, and other herbaceous plants, also shrubs. Host genera: *Calluna, Caltha, Carex, Chrysanthemum, Dryas, Geranium, Gramineae, Hymogyne, Iris, Rhacomitrium, Rhododendron, Saxifraga, Sphagnum, Soldanella, Vaccinium* (Balachowsky, 1953a). Overwinter in different stages, lay up to 120 eggs per female (Tereznikova, 1975).

Genus **Newsteadia** Green
Fig. 40

Syn. & Lit.: *Newsteadia* Green, 1902:285; Morrison 1925:146, 1952:57; Gomez-Menor, 1937:388; Borchsenius, 1950:28; Tereznikova, 1975:116; Kosztarab & Kozár, 1978:12; Danzig, 1980a:105.

Type-species: *Coccus floccosus* De Geer, 1778.

Entire dorsum covered with wax tufts or plates; fully developed eggsac as long as body. Adult females oval, 1–2.5 mm long; antennae 3- to 7-segmented; basal 2 segments much enlarged; eye stalks elongate, curved; tibia and tarsus fused; thoracic spiracles without multilocular disc pores inside atrium; only 5 anterior pairs of abdominal spiracles developed; most of marginal abdominal clusters of body spine clusters fused into a wide band; anal ring with short stout setae. Adult males and immatures poorly known. First instar with 4-segmented antennae only.

Distribution. Cosmopolitan, from temperate zones to tropics, with 14 species; 1 in CE.

Biology. Host plant preferences vary, some may feed on fungus hyphae; commonly collected in forest leaf litter, under logs, from mosses, herbaceous plants, and trees.

2. **Newsteadia floccosa** (De Geer) – Boreal Ensign Scale
Fig. 40

Syn. & Lit.: *Coccus floccosus* De Geer, 1778:604. *Orthezia normani* Douglas, 1881a:300. *Newsteadia collarti* Ghesquiere, 1934:27. *N. floccosa,* Morrison, 1925:151, 1952:59; Goux, 1941:31; Jancke, 1955:270; Boratynski, 1955:66; Rogojanu, 1958:310; Koteja & Zak-Ogaza, 1966:312; Zahradnik, 1972:392; Tereznikova, 1975:116; Kosztarab & Kozár, 1978:12; Danzig, 1980a:106.

Dorsum of female covered with large semicircular wax plates. Body oval, about 1.6 mm long; antennae usually 6-segmented; eye stalks slightly bent, fingerlike; tarsal claws without teeth; marginal abdominal spine clusters 6 to 9 fused: 5 pairs of abdominal spiracles; spine clusters of abdominal margin fused into a band; anal ring with 6 stout setae. Adult male described by Jancke (1955). A short nymphal description was given by Gomez-Menor (1937).

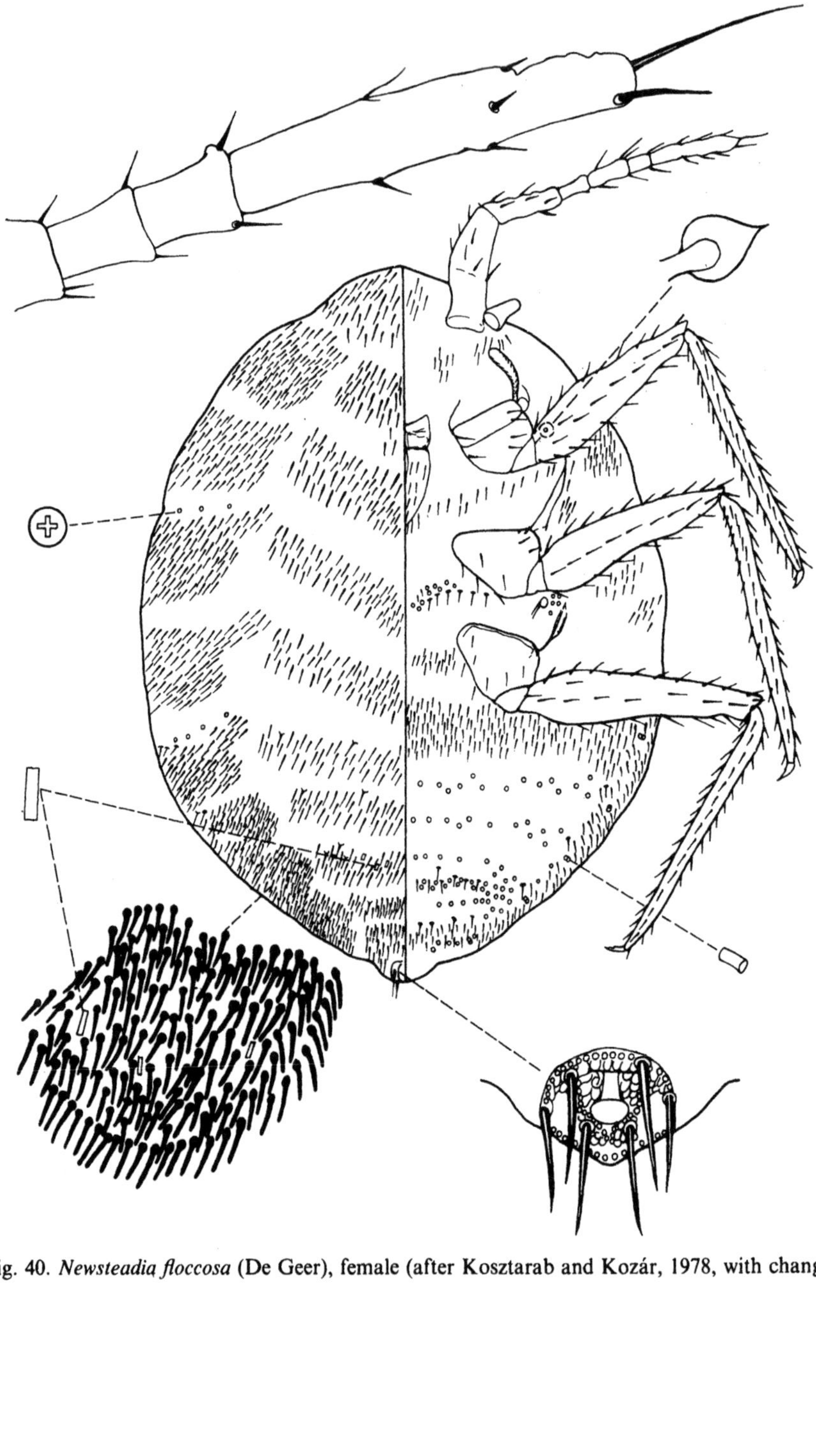

Fig. 40. *Newsteadia floccosa* (De Geer), female (after Kosztarab and Kozár, 1978, with changes)

Distribution. Trans-Palearctic; known in CE from; Au, Bu, Cz, Fr, Ge, Hu, It, Po, Ro, USSR also from Belgium, England, Spain and Sweden.

Biology. A boreal species; apparently it is common in moist and cool habitats, also in leaf litter of mixed, deciduous and coniferous forests. Polyphagous. It was collected in CE on various herbaceous and woody vegetation, often in soil from roots, from mosses and fungi. Host genera: *Acer, Aspidistra, Carex, Carpinus, Crataegus, Fagus, Fragaria, Glecoma,* Gramineae (undet.) *Helianthemum, Iris, Luzula,* Musci (undet.), *Pinus, Populus, Quercus, Rhododendron, Sphagnum, Thymus,* and others (Morrison, 1925, 1952; Schmutterer, 1952b; Zahradnik, 1972). A biparental species with 1 yearly generation; all stages overwinter, except first instars; lay 40–80 eggs per female; adult males in July and August (Schmutterer, 1952b).

Genus **Orthezia** Bosc d'Antic
Fig. 2

Syn. & Lit.: *Orthezia* Bosc d'Antic, 1784:173. *Dorthezia* Bosc d'Antic, 1784:173; *Cionops* Leach., 1815:126; *Cyphoma* Gistel, 1848:151; (Fernald, 1903:33). *Polyocellaria* Imhof, 1900:527. *Douglasia* Mac Gillivray, 1921:254 (preoccupied). *Douglariella* MacGillivray, 1921:254. *Orthezia,* Morrison, 1925:102, 1952:3; Gomez-Menor, 1937:383; Borchsenius, 1950:26; Tereznikova, 1975:111; Kosztarab & Kozár, 1978:13; Danzig, 1980a:103.

Type-species: *Orthezia characias* Bosc d'Antic, 1784 (= *Aphis urticae* Linnaeus, 1758).

Female entirely covered by wax plates; ovisac long, obvious, often longer than body. Adult female oval, 2.5–4 mm long without ovisac; antennae 7- or 8-segmented, its 3 basal segments not conspicuously enlarged; eye stalks variable; tibia and tarsus well divided; tarsal claws usually with denticles; thoracic spiracles without atrial disc pores; disc pores on derm of quadrilocular type only; spines grouped in definite bands and clusters; ovisac band usually broad; anal ring flat, oval, with 6 long setae. Adult male of *O. insignis* described by Jancke (1955).

Distribution. Cosmopolitan, but most known from the Nearctic and Neotropical Regions. World-wide 61 species; 1 in CE.

Biology. Apparently most species of *Orthezia* prefer warm and moist habitats but some are known from dry zones of subtropics and of temperate zone. No host preferences found. Some feed on roots, others on all parts of plants.

3. **Orthezia urticae** (Linnaeus) – Nettle Ensign Scale, Ensign Coccid
Figs 2, 41, 42

Syn. & Lit: *Coccus urticae* Linnaeus, 1758:453. *O. characias* Bosc d'Antic, 1784:173. *Coccus dubius* Fabricius, 1794:228. *Dorthezia dispar* Kaltenbach, 1874:484; Fernald, 1903:36. *O. martelli* Leonardi, 1908:150. *O. arenariae* Vayssierre, 1924:28. *O. urticae,* Morrison, 1925:140, 1952:52; Rogojanu, 1958:305; Tereznikova, 1975:112; Kosztarab & Kozár, 1978:14; Danzig, 1980a: 104.

Adult female covered with 6 rows of wax plates but no plates at midline; fully developed ovisac longer than body, strongly ribbed dorsally. Adult female oval, 3–5 mm

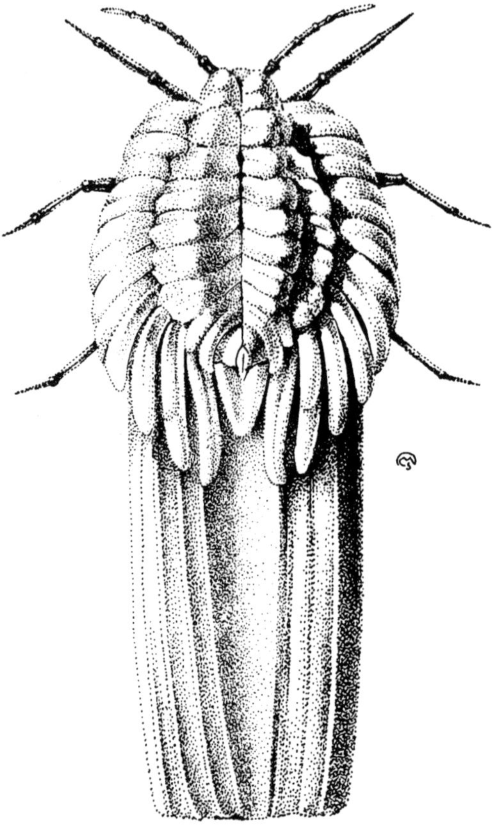

Fig. 41. *Orthezia urticae* (Linnaeus), female, dorsal view (after Kosztarab and Kozár, 1978)

long, 2.5 mm wide; antennae 8-segmented; eye stalks elongate, conical, usually with 1, rarely with 2 lateral tubercles; tarsal claws usually with 2, rarely 3 small denticles on inner face; conspicuous collar of spines encircling thoracic spiracles; 8 pairs of abdominal spiracles. Adult males and immatures poorly known.

Distribution. Trans-Palearctic. In CE: Au, Bu, Cz, Fr, Ge, Hu, It, Po, Ro, USSR and Yu.

Biology. According to Schmutterer (1952b) it is common in moist deciduous and mixed forest biotopes, also in marshes in the herbaceous and dwarf shrub layers. Often in anthropogenic habitats. Polyphagous. Usually feed on underside of leaves of herbaceous plants; adults common on stems. Its host genera: *Achillea, Aegopodium, Artemisia, Aster, Ballota, Bergenia, Caltha, Centaurea, Coronilla, Cuscuta, Erica, Euphorbia, Galium, Geranium, Glaux, Glecoma,* Gramineae (undet.), *Hedera, Heracleum, Hieraceum, Humulus, Lathyrus, Leontodon, Matricaria, Melampyrum, Melittis, Onobrychis, Parietaria, Phlomis, Potentilla, Rosa, Rubus, Spiraea, Stellaria, Symphytum, Taraxacum, Teucrium, Trifolium, Tunica, Urtica, Vinca* (Morrison, 1925; Matesova, 1968; Danzig, 1980a). It was also found on cultivated plants, such as *Pyrus*

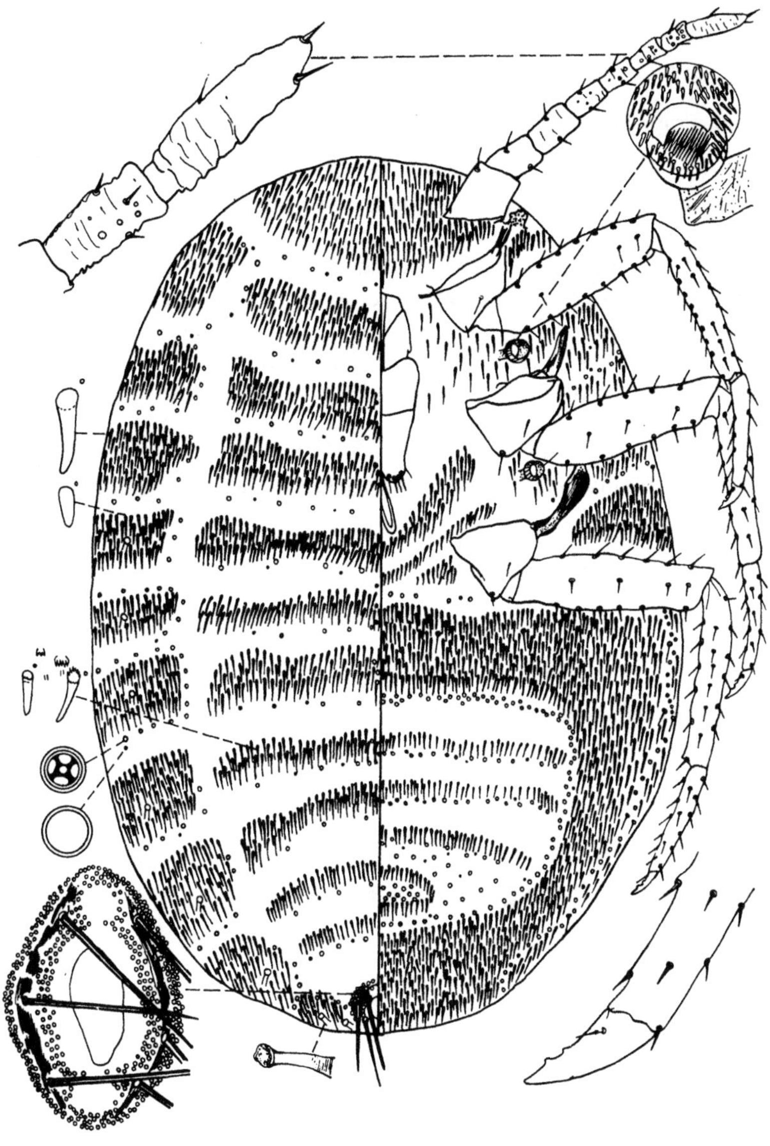

Fig. 42. *Orthezia urticae* (Linnaeus), female (original)

and *Vitis* (Morrison, 1952). In addition, it has been recorded in Hungary from: *Anthemis, Cardaminopsis, Dorycnium, Filipendula, Odontites, Sanguisorba* and *Satureja* (Kosztarab, 1959). Except first instars, all other stages overwinter. Females lay 150 to 500 eggs. It has one yearly generation. Probably can reproduce with or without fertilization (Schmutterer, 1952b; Köhler, 1983).

Genus **Ortheziola** Šulc
Fig. 43

Syn. & Lit.: *Ortheziola* Šulc, 1895:1,5; Morrison, 1925:152, 1952:75, 1954:120; Ghesquiere, 1946:236.

Type-species: *Ortheziola vejdovskyi* Šulc, 1895 (= O. *signoreti* Haller).

Most of adult female body covered with wax plates, forming 4 longitudinal rows. Adult female stout, 1.5—2.5 mm long; antennae actually 3-segmented, appear 4-segmented because eye stalks fused with pseudo-basal antennal segment; apical spine of antenna slender; eye stalks elongate thumblike; thoracic spiracle openings with a spine cluster; some abdominal spiracles not clearly defined, but fewer than 8 pairs; multilocular disc pores with 8 or more loculi on venter in addition to normal quadrilocular ones; derm spines swollen at base, most capitate at tip; anal ring with 6 subequal or equal setae.

Distribution. Irregular; only 2 species, 1 in S America, and 1 in Europe.

Biology. Host plants include mosses and roots of herbaceous and woody plants.

4. **Ortheziola vejdovskyi** Šulc – Moss Ensign Scale
Figs 43, 44

Syn. & Lit.: *Ortheziola vejdovskyi* Šulc, 1895:2, 1898:13. *Orthezia signoreti* Haller (preoccupied name). *Ortheziola vejdovskyi,* Morrison, 1925:153, 1952:75; Reyne, 1953:235; Schmutterer, 1955a:160; Rogojanu, 1958:312; Varshney, 1964:280; Sampo & Olmi, 1976:217; Kosztarab & Kozár, 1978:15.

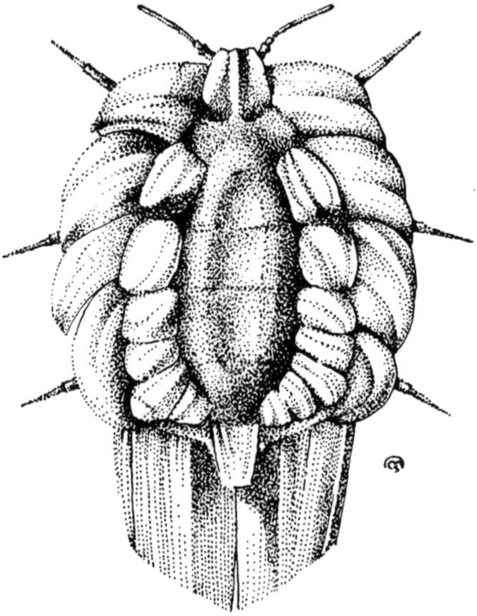

Fig. 43. *Ortheziola vejdovskyi* Šulc, female, dorsal view (after Kosztarab and Kozár, 1978)

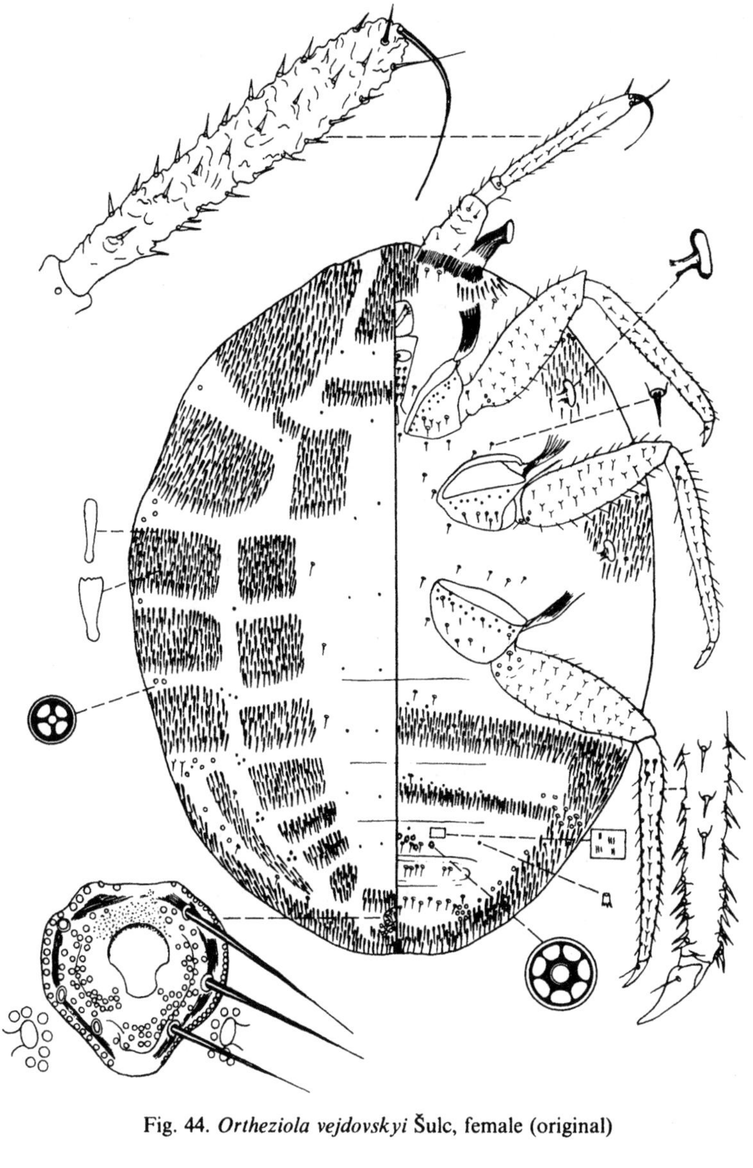

Fig. 44. *Ortheziola vejdovskyi* Šulc, female (original)

Dorsal wax plates of adult female forming 4 longitudinal rows; center of dorsum yellowish with segmentation visible; ovisac short with V-shaped tip. Adult female about 2 mm long; antenna 3-segmented, tip with long slender spine; tibia and tarsus fused, with spines forming obvious rows; claws without denticles; abdominal spiracles hard to distinguish; derm with quadrilocular pores, but some 8-locular disc pores on venter; mid-dorsum without spines, leaving this area exposed; 2 small clusters of spines on each side of anal ring; anal ring with pores and 6 subequal setae. First instar was described by Tereznikova (1977).

Distribution. Palearctic; in CE: Au, Cz, Fr, Ge, Hu, It, Po, Ro, Sw, USSR, Yu; also known from Belgium, Luxembourg (Boratynski's unpublished data), Sweden and the Island of Azores, Hebrides and Madeira.

Biology. A rare boreal species. Often found by plant quarantine inspectors in mosses and other packing materials used for shipping live plants. Adult females noted from July through September in Poland by Zak-Ogaza and Koteja (1964). Number of generations uncertain. All stages overwinter; no males observed; each female lays 20 to 40 eggs (Schmutterer, 1952b). Schmutterer (1952a) reported it from wet meadows, in mosses, in soil of shaded areas in hills of Bavaria feeding on grass roots, especially on *Festuca ovina*. In Poland and Sweden it was collected in xerothermic habitats. Other host genera: *Arrhenaterum, Begonia, Calamagrostis, Calluna, Centaurea, Cyperus, Dactylis, Erica, Rosa,* etc.

FAMILY **MARGARODIDAE** – GIANT SCALES

Figs 3, 4, 45, 46

Syn. & Lit.: Margarodinae Cockerell, 1899b:390. Margarodidae, Morrison, 1928:1; Borchsenius, 1950:28; Zahradnik, 1959b:527, 1972:392; Danzig, 1964:621, 1980a:85; Jakubski, 1965:1; Tereznikova, 1975:118; DeKlerk et al., 1982:61.–Jakubski (1965) divided Margarodini into two families; we recognize here only Margarodidae. Porphyrophoridae Signoret, 1875:346; Williams, 1969:335; Koteja, 1974b:288; Kawecki, 1985:11.

During oviposition, they often produce white waxy filaments to cover their body and eggs. Adult females of variable shape, up to 35 mm long, most between 10–25 mm long; body usually yellowish-brown, seldom red as in Polish Cochineal Scale, *Porphyrophora polonica*; with antennae 6–13-segmented; beak usually present, and normally 1–3-segmented, with setae variable in size and number; legs normally 5-segmented, seldom reduced or absent; tarsus may be divided into 2 segments, sometimes with numerous claw digitules; forelegs sometimes fossorial, adapted to digging in soil; with 2–8 pairs of abdominal spiracles; anal ring absent, except for its internal structure; anal ring pores present only internally and its setae entirely absent; in some species anal tube distinctly developed. Tubular ducts absent; disc pores vary from simple to multilocular type; cicatrices often present on derm as large porelike structures and are often surrounded by chitinized rim.

Theron (1958) described males of 3 species. Morrison (1928) described adult males and immatures of several species. In addition adult males were also described by Jancke (1955). Males of *Callipappus* in Australia have long wax tails, and are called "bird of paradise flies."

Distribution. In all zoogeographical regions, but most abundant in tropical and subtropical areas. World-wide about 250 species in about 50 genera. In CE 6 genera with 6 species.

Biology. Females develop through 3 or 4 instars, males through 5. Some form cysts ("ground pearls") during development, which are used as beads in some parts of the world. Cysts may endure several years of quiescence under drought conditions. The biologies are very diverse. Most species develop in one year, but *Xylococcus filiferus* develops in several years. They feed on a variety of herbaceous and woody plants. Some are monophagous, many oligophagous, most polyphagous.

Note. Goidanich (1963) described the ontogenetic variability in the females. *Icerya purchasi,* the cottony cushion scale, was found in greenhouses only.

Reviewer's remark. The family name "Margarodidae" is used in several definitely different meanings which causes numerous misunderstandings. Many coccidologists (also the authors of the book) accept the broad concept of Morrison (1928) according to which all archaeococcids, except Ortheziidae and Phenacoleachiidae, form only one family. Some students (e.g. Danzig, 1980a) divide this group into two families: the Margarodidae (that includes forms with apodous preadult stage and non-feeding adults) and Monophlebidae. Koteja (1974b and records cited herein) recognizes within the margarodids s. l. the family Margarodidae s. str. (with *Margarodes, Porphyrophora, Neomargarodes,* etc.) and about 7 other families, as Monophlebidae, Matsucoccidae, etc.. Jakubski (1965) divided the "ground pearls" (Margarodiae s. str. = Margarodini s. Morrison) into two families with several subfamilies and tribes, without taking into consideration the other groups of margarodids s. l. Miller (1984) affiliated the Ortheziidae and Phenacoleachiidae, and also *Pityococcus* originally assigned in the Margarodidae, with the neococcids, but retained Morrison's concept of Margarodidae, as a result this family become synonymous with the Archaeococcoidea (= Orthezioidea, but without *Orthezia*).

It seems likely that the recent margarodids s. l. are relics of an ancient, abundant and diverse group that is not well known. There is only one comprehensive monograph by Morrison (1928). The usage of the name Margarodidae in a broad sense is justified, particularly in Central Europe, only for practical reasons (small number of species), but it obscures the poor biological and phylogenetical knowledge, that we now have.

Key to genera and species of Margarodidae
(Each genus with one species)

1. Claws, when present, with 2 digitules; abdominal spiracles not in 6 pairs; on other hosts than birch 2
   Claws with 6–12 knobbed digitules; abdominal spiracles in 6 pairs; on *Betula* spp. (fig. 46) .... *Steingelia gorodetskia,* p. 53
2. Legs and antennae many segmented; on other hosts than *Tilia* spp. 3
   Legs and antennae unsegmented or lacking; on *Tilia* spp. .... *Xylococcus filiferus,* p. 56
3. Abdominal spiracles in 7 pairs; forelegs not enlarged; not on roots 4
   Abdominal spiracles in 2 or 8 pairs; forelegs enlarged, fossorial; on roots of herbaceous plants 5
4. Claws with 3 denticles; antennae 11-segmented (rarely 10); dermal multilocular pores of two types, some with large, others with small loculi; ventral cicatrices present; polyphagous, including pines .... *Palaeococcus fuscipennis,* p. 50
   Claws without denticles; antennae 9-segmented; dermal multilocular pores of one-type, with bilocular center; ventral cicatrices lacking; only on *Pinus* spp. .... *Matsucoccus matsumurae,* p. 48

5(3). Abdominal spiracles 8 pairs; base of claws on forelegs wide, forming a heel; antennae 6-segmented; on roots of grasses .... *Neomargarodes festucae,* p. 50
   Abdominal spiracles poorly developed, at most 2 pairs only; base of claws on forelegs not especially widened, claws uniformly tapering toward apex; antennae 7- or 8-segmented; polyphagous, on roots of herbaceous plants, including grasses (fig. 45) .... *Porphyrophora polonica,* p. 52

Genus **Matsucoccus** Cockerell

Syn. & Lit.: *Matsucoccus* Cockerell, 1909:56; Morrison, 1928:48, 1939: 1;McKenzie, 1943a:42, 1943b:53; Borchsenius, 1950:33; Boratynski, 1952a:285, 1952b:507, 1960:240; Danzig, 1964:622, 1980a:95; Beardsley, 1968:1449; Tereznikova, 1975:124; Kosztarab & Kozár, 1978:16; Koteja, 1984:467.

Type species: *Xylococcus matsumurae* Kuwana, 1905.

Insects are usually hidden in cracks and crevices, in galls, under bark flakes of the host, *Pinus* spp. Some are found at the base of pine needles. Adult females produce an ovisac or a mass of white woolly threads during oviposition. Their presence is sometimes indicated by resin exudation on the infested twigs. Adult female elongate, broader near posterior end, about 3–6 mm long; antennae normally 9-segmented, seldomb 6–7- or 2–3-segmented; tarsus 2-segmented, its basal segment short, claws short and thick; abdominal spiracles in 7 pairs; disc pores both simple and multilocular type; short body setae scattered, long setae in segmental rows on abdomen; anal tube reduced. Adult males described by Morrison (1928), Beardsley (1968), Siewniak (1976), and Koteja (1984); immature stages by Morrison (1928) and McKenzie (1943b).

Distribution. Holarctic; with about 17 species, 2 of these in CE. A few species with economic importance. *M. feytaudi* Ducasse, 1941, because of its distribution restricted to the Mediterranean area, was left out from this study (Ducasse, 1941).

Biology. Genus restricted to *Pinus* spp. Life cycle and number of yearly generations varies by species. Biology of some species described by McKenzie (1943a), Boratynski (1952a), Siewniak (1976), and others.

5. **Matsucoccus matsumurae** (Kuwana) – Pine Bark Scale

Syn. & Lit.: *Xylococcus matsumurae* Kuwana, 1905:91. *Kuwania pini* Green, 1925a:34. *M. insignis* Borchsenius, 1955a:222. *M. matsumurae,* Morrison,* 1928:48; Taketani, 1972:1; Danzig, 1972b:261. Tereznikova,* 1975:124; Young et al., 1976:199; Kosztarab & Kozár, 1978:17; Vinis, 1981:206.

M. pini was separated from *M. matsumurae* by Boratynski (1952a: 285, 1952b: 570, 1960: 240) who believed that the European populations belong to *M. pini.* His finding was accepted by Cadahia (1971: 81), Zahradnik (1972: 393), Siewniak (1976: 337) and others. Rieux (1975: 157, 1976: 231) found intraspecific variations in the European populations of *Matsucoccus pini.* Therefore, its taxonomic status could be clarified only by studying specimens from its entire geographical range in the Palearctic Region. Such a study would also clarify the status of *M. mugo* Siewniak (1970: 168), known from Ge, not treated separately in our study.

Ovisac fluffy, silky. Adult female elongate, broader near posterior end, with characteristic opaque shine, 2.5–4.5 mm long, 1–2 mm wide (females of overwintering generation larger than individuals of summer generation); antennae close together at base, 9-segmented, with 2 basal segments enlarged, 2nd segment with 3 small sensory pores; mouthparts absent; tarsi distinctly 2-segmented, basal segment short, apical segment with a row of strong setae on inner margin, claws stout, claw digitules strongly knobbed and longer than claws; thoracic spiracles larger than abdominal ones, without disc pores; disc pores with bilocular center, scattered at bases of coxae, along thoracic margin and in transverse segmental rows on abdomen; sievelike pores form transverse rows or bands on most abdominal tergites; disc pores with bilocular and quadrilocular center form a large group on apex of abdominal venter; ventral cicatrices lacking.

Distribution. Trans-Palearctic; in CE: Au, Bu, Cz, Fr, Ge, Hu, Po, Ro, and USSR; also known from England, Spain and Japan; it is probably generally distributed in the entire area.

Biology. It infests pines: *Pinus halepensis, P. mugo,* (=*pumilio, uncinata*), *P. nigra, P. pinaster, P. pinea,* and *P. sylvestris.* Its biology varies; reproduction is parthenogenetic in England (Boratynski, 1952a); biparental in CE; it may alternate these (e.g. in Spain) between the two yearly generations; overwinter as first instars. Females found in crevices on the sunny side of tree trunks in April–May and July–August, or other times under the bark flakes; each female lays 60–250 eggs. There are 4 instars in females, 5 in males (Siewniak, 1976).

The species may be easily recognized by the presence of rotund, brown, shiny cast skins of the preadult females which remain under the bark flakes through the years. Occasionally becomes a pest (Siewniak, 1976). Boratynski (1952a) noted the possibility that the growing insects may exert pressure to cause rapid premature peeling of bark flakes.

Natural enemies. Only predators observed: Acarina, *Cheyletia flabellifera* and *Hemisarcoptes malus* (Zahradnik, 1972); larvae of Neuroptera, Raphidiidae, Hemerobiidae; also *Dromius* sp., a Carabidae (Siewniak, 1976); birds: *Aegithalos caudatus, Parus ater,* and *P. cristatus* (Cadahia, 1971).

Genus **Neomargarodes** Green

Syn. & Lit.: *Neomargarodes* Green, 1914b:263; Morrison, 1928:78; Borchsenius, 1950:36; Danzig, 1964:623; Jakubski, 1965:57; Tereznikova, 1975:138; Kosztarab & Kozár, 1978:17.

Type-species: *Neomargarodes erythrocephala* Green, 1914b: 263. Unfortunately, only the adult male was available for the type-species description.

Older nymphs found in a hyaline, spherical cyst in soil; adult females produce a white woolly ovisac (Danzig, 1964). Adult female broad, oval, almost circular, derm membranous; antennae usually 6-segmented, short and thick, wide at base; mouthparts absent; all legs thick, anterior much enlarged, claws with wide base; abdominal spiracles 8 pairs; disc pores multilocular type; slender hairlike setae covering derm. Adult males and nymphs described by Green (1914b), Morrison (1928) and Jakubski (1965).

Distribution. In S Palearctic, extending into Ethiopian and Oriental Regions, with a total of 12 species, 1 in CE.

Biology. Found on roots of grasses.

Note. Genus is related to *Eumargarodes* Jakubski and *Margarodes* Guilding (Jakubski, 1965).

6. **Neomargarodes festucae** Archangelskaja – Grass Pearl Scale

Syn. & Lit.: *Neomargarodes festucae* Archangelskaja, 1935:21; Jakubski, 1965:64; Tereznikova, 1975:140; Kosztarab & Kozár, 1978:17.

Adult female emerges from a brown cyst and produces white cottony ovisac. Adult female oval, yellowish-white, 3½–6 mm long, 2½–4 mm wide; antennae 6-segmented (usually 7-segmented according to Jakubski) short and thick, with wide base; eyes and mouthparts lacking; thoracic and abdominal spiracles with multilocular and distinctive unilocular pores; multilocular disc pores form double transverse rows on venter of abdominal segments 1–6, rows interrupted on segments 1–3 (Kosztarab and Kozár, 1978); anal opening semicircular, anal tube present.

Distribution. In CE: Cz, Fr, Hu, Po, and USSR.

Biology. Develops on roots of grasses, *Agropyron cristatum, Dactylis* sp., *Festuca ovina, F. pallens, F. sulcata, F. vaginata, Koeleria gracilis* and *Stipa capillata* (Jakubski, 1965; Tereznikova, 1975). A xerophilous steppe inhabiting species. Reproduction biparental; one yearly generation. Adults appear in August depending on temperature and moisture. Female lays 50–200 eggs; overwinter as nymphs (Tereznikova, 1975).

Genus **Palaeococcus** Cockerell

Syn. & Lit.: *Palaeococcus* Cockerell, 1894:36. *Leachia* Signoret, 1875:359 (a preoccupied name). *Palaeococcus*, Cockerell, 1902:233; Fernald, 1903:21; Vayssière, 1926:266; Morrison, 1928:128.

Type-species: *Monophlebus fuscipennis* Burmeister, 1835.

Adult females elongate oval; antennae 11-segmented (rarely 10); claws with 3 small denticles and a pair of digitules; abdominal spiracles 7 pairs; dermal multilocular pores of two types; ventral cicatrices and vulva present.

Distribution. The 11 species included in the Fernald (1903) catalog represent all major zoogeographical areas of the world, except the Oriental. The generic assignment of most species is questionable; 1 species in CE.

Biology. Very little is known on the biology of the species assigned to this genus, except for the type species; females move freely through their lives on woody and herbaceous plants.

Note. Fernald (1903) included 11 species in the genus. Vayssière (1926) considered this genus inseparable from *Monophlebus.* Morrison (1928) suggested its close relationship to *Monophleboides,* and the possibility that the latter genus and *Palaeococcus* are indentical. Because the genus needs a revision, with many questionable species included, our description is based on the type species. See for more details the species description.

7. **Palaeococcus fuscipennis** (Burmeister) – Ancient Giant Scale

Syn. & Lit.: *Monophlebus fuscipennis* Burmeister, 1835:80. *M. hellenicus* Gennadius 1883:32. *Monophlebus fuscipennis*, Vayssière,* 1926:267; Lindinger, 1912:257; Szulczewski, 1921:81; Scheidter, 1930:91; Gomez-Menor, 1937:402: *P. fuscipennis*, Morrison, 1928:129; Zahradnik, 1972:396.

Adult female with distinct segmentation on both body surfaces, covered with cottony or mealy wax secretion; body elongate-oval, reddish, about 7 mm long and 3 mm wide, 2½–3 mm thick; antennae black or brownish black, 11-segmented, segments 10 and 11 often combined; a sensory pore in membrane between 2nd and 3rd segment; eyes not prominent; labium faintly sclerotized and located between front legs; legs black or brownish black, trochanter with 4 sensory pores; each claw with 3 small denticles and a pair of digitules; 7 pairs of abdominal spiracles; both ventral and dorsal derm covered with multilocular disc pores, some with large, others with small loculi; ventral cicatrices present; vulva prominent, located between 7th and 8th abdominal segments. Adult males figured and described by Vayssière (1926), while Gomez Menor (1937) described first instars and redescribed adult females.

Distribution. A rarely collected insect; in CE: Fr, Ge, Po, and Sw. It was also reported from Spain and the Princes Islands in the Sea of Marmara (Turkey).

Biology. It is found on trunk and branches of trees where it moves freely; host genera: *Abies, Acer, Pinus,* and *Quercus.* Its economic importance is unknown.

Natural enemies. Parasite: *Cryptochaetum buccatum* in Germany (Fulmek, 1943). Predator: *Novius cruentatus* reported by Lindinger (1923).

Genus **Porphyrophora** Brandt

Syn. & Lit.: *Porphyrophora* Brandt & Ratzeburg, 1883:355; Borchsenius, 1950:37; Jakubski, 1965:5; Hadzibejli, 1966:693; Tereznikova, 1975:142; Kosztarab & Kozár, 1978:19; Danzig, 1980a:99.

Type-species: *Coccus polonicus* Linnaeus, 1758.

During oviposition female produces a white cottony ovisac. Cysts found in the soil. Adult female oval or almost circular, rarely pear-shaped; size varies by species; antennae 7- to 16-segmented, most with 7 or 8 segments, apical segment with many sensory setae in addition to regular ones; mouthparts absent; legs with fine thin setae, fore legs enlarged and usually twice as far from middle legs as latter from hind legs; claws remarkably enlarged and pointed; thoracic spiracles include several multlilocular disc pores, have a circular opening with thickened border; abdominal spiracles usually lacking or only 2–3 anterior pairs developed. Jakubski (1965) described adult males, nymphs and cyst stage.

Distribution. The genus with 23 species is of Holarctic and Ethiopian distribution, but the largest number of species (10) are in the Mediterranean Region; 1 species (*P. bolivari*) is of montane distribution; 1 species in CE.

Biology. They live on roots of grasses and other herbaceous plants. Occasionally damage wheat, strawberries and alfalfa.

Note. The genus is related to *Neomargarodes.* Morrison (1928) combined it with *Margarodes,* while in the more recent literature it was treated as a separate genus. Jakubski (1965) provided a detailed redescription of the genus and the species included in it.

8. **Porphyrophora polonica** (Linnaeus) – Polish Cochineal Scale, Ground Pearl
Figs 3, 4, 45

Syn. & Lit.: *Coccus polonicus* Linnaeus, 1758:465. *Coccus radicum* Beckman, 1790:1. *P. frischii* Brandt & Ratzeburg, 1838:225 (Since the species is very variable, it was redescribed under a variety of other synonym names). *P. polonica,* Jakubski, 1934:1, 1965:18; Tereznikova, 1975:143; Kosztarab and Kozár, 1978:19.

Adult female oval, deep red to cherry-violet, from 1.5 to 6.5 mm long; antennae 7- or 8-segmented; mouthparts lacking; forelegs enlarged, coxa very broad, femur and trochanter fused, tarsus fused with claw; thoracic spiracles with 3–6 multilocular disc

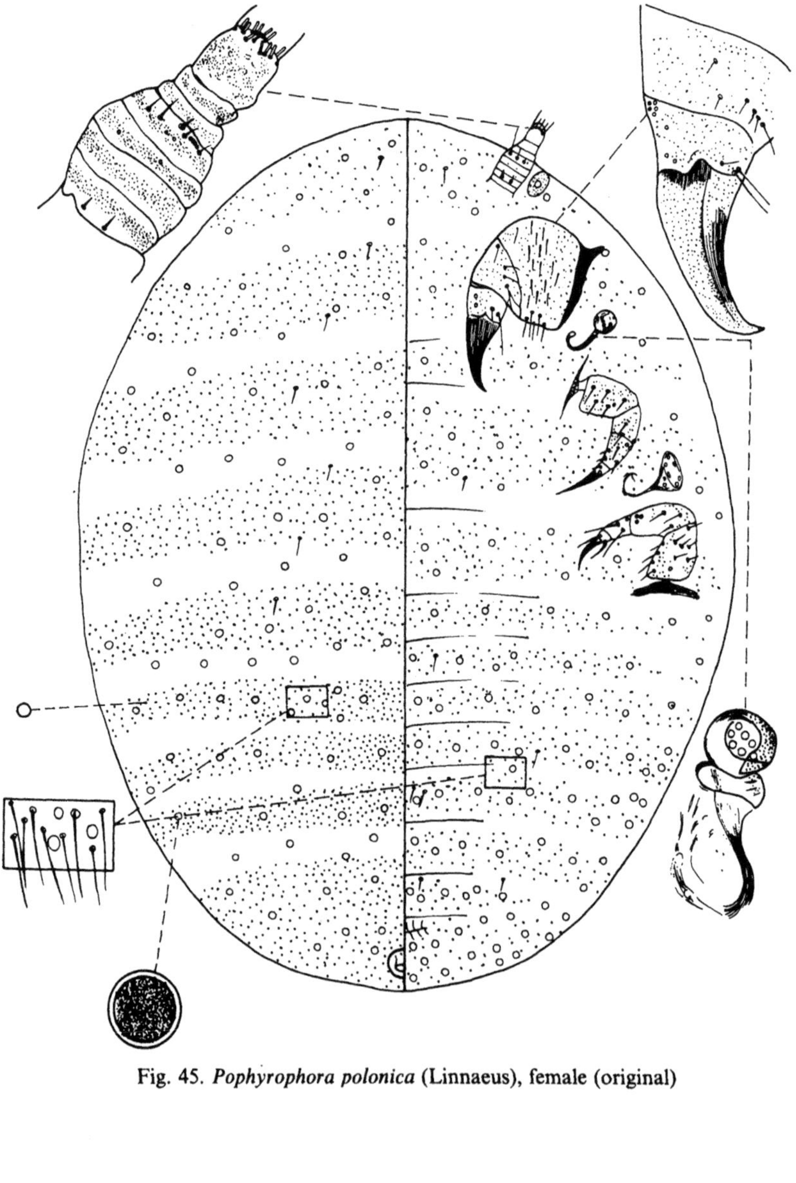

Fig. 45. *Pophyrophora polonica* (Linnaeus), female (original)

pores; first 2 abdominal spiracles often faintly developed; both derm surfaces covered with wide transverse bands of thin long setae and multilocular disc pores; anal tube and anal opening developed. Other instars described by Jakubski (1965) and Tereznikova (1975).

Distribution. In CE: Cz, Fr, Ge, Hu, Po, Sw and USSR. It is also known from Europe. Apparently it was more common during the past centuries, when it was collected in large quantities as a basis for carmin dye (Kawecki and Wernerowna, 1975).

Biology. The insect is polyphagous and is found in CE most often on the following host plant genera: *Alkanna, Cerastium, Festuca, Fragaria, Herniaria, Melampyrum, Potentilla, Spergularia,* and others. Apparently *Scleranthus perrenis* is a favorite host, but its hosts include over 60 plant species belonging to 12 families (Jakubski, 1965). It feeds on roots, root collars and on underground stems and prefers loose sandy soils; with heavy infestation may kill the plants. It has one yearly generation; first instars overwinter in egg sacs; second instars form cysts; adults of both sexes appear during second half of June and first part of July; females crawl to top of plants for mating; each lays underground 600—700 eggs (Jakubski, 1965).

Genus **Steingelia** Nassonov

Syn. & Lit.: *Steingelia* Nassonov, 1908:345; Green, 1914:197; Morrison, 1928:55; Borchsenius, 1950:34; Theron, 1958:9; Danzig, 1964:623; 1980a:979 Koteja, 1974b:286; Kosztarab and Kozár, 1978:19; Koteja and Zak-Ogaza, 1981a:171.

Type-species: *Steingelia gorodetskia* Nassonov, 1908.

Ovisac woolly, 4–6 mm long 3 mm wide. Adult female elongate, almost parallel-sided, tapering toward prosoma; 3.5–5 mm long, 1.5 mm wide; antennae set close together at their base, normally 8-segmented (rarely 7 or 9), basal segment enlarged; mouthparts absent; legs thick, 5-segmented; no pores in spiracular atrium; only multilocular pores on derm; both anal tube and anal opening reduced.

Distribution. Palearctic; with 1 species in CE.

Biology. Oligophagous, on *Betula* spp. See for more details at *S. gorodetskia.*

Note. Morrison (1928) and Tereznikova (1975) provided more detailed redescription of the type species. Latter author placed it in the Steingeliinae with *Kuwania* Cockerell.

9. **Steingelia gorodetskia** Nassonov – Birch Bark Scale
Fig. 46

Syn. & Lit.: *Steingelia gorodetskia* Nassonov, 1908:345, *Kuwania britannica* Green 1914:197. *S. orientalis* Borchsenius 1949:334. *S. gorodetskia,* Morrison 1928:55; Ossiannilsson 1959:193; Koteja & Zak-Ogaza 1981a:171; Zahradnik 1972:394; Tereznikova 1975:148; Kosztarab & Kozár 1978:21; Danzig 1980a:97.

Old dried females, blackish or grayish-brown; normally body elongate with marked segmentation, 3.5–5.0 mm long. Antennae arise at extreme anterior apex of head; legs with long spines, trochanter short, with 3 sensory pores on each side; claws stout, wide at base, curved, without denticle, with ca 10 knobbed digitules, these longer than claw;

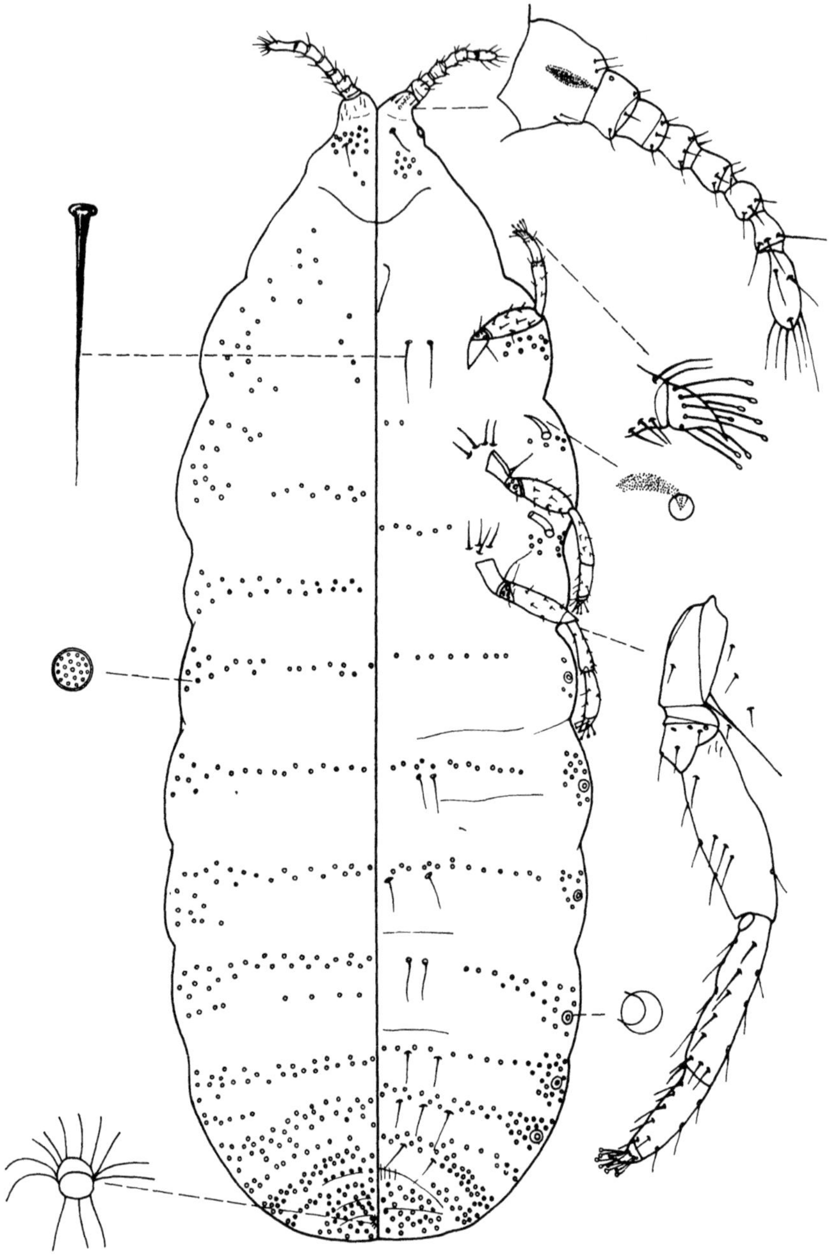

Fig. 46. *Steingelia gorodetskia* Nassonov, female (modified after Kosztarab and Kozár, 1978)

6 pairs of abdominal spiracles; body setae of 2 types: large ones, ca 200 μ long, forming transverse rows on abdominal venter and located in groups at base of coxae; smaller setae, ca 20 μ long, scattered on both body surfaces; larger setae form rows on tibia, 4 located on venter of each abdominal segment 3 and 4; multilocular disc pores form transverse bands or rows on segments, more dense toward abdominal apex; anal opening small, circular, located close to abdominal apex; anal tube short. Adult male described by Morrison (1928) and Theron (1958). A short description of first instars was given by Nassonov (1908) and Green (1914). A detailed description of female and male nymphs at Koteja and Zak-Ogaza (1981a).

Distribution. A boreal species; in CE: Ge, Po, and USSR. Also known from England, Sweden. Far East USSR and Korea.

Biology. A rarely collected species, probably due to its cryptic life habits. Bisexual. Earlier records state that during June the ovipositing females crawl to the base of birch trees (*Betula* spp.) and others, such as *Acer, Malus, Phellodendron, Quercus, Sorbus,* and into grass, leaf litter and moss, where they lay their eggs in white oval ovisacs that cover the body of the female. Apparently eggs overwinter. Nassonov (1908) found eggs at end of August and in May.

Reviewer's remark. Third instar male nymphs appear in forest litter in early spring, form puparia and moult into pupae, these into adults (April, May). At that time adult females appear. After mating, females lay eggs in loose ovisacs, usually among fallen leaves, dry litter, etc. After 2–3 weeks crawlers appear and move to the roots of *Betula* (CE) where they feed and transform into spherical second and third instars. These stages overwinter. In the spring the fourth female instar (adult) and third male instar (corresponding with prepupa in other coccoids) moult and crawl above the ground.

Genus **Xylococcus** Löw

Syn & Lit.: *Xylococcus* Löw, 1882:271; Morrison, 1928:41; Borchsenius, 1950:33; Zahradnik, 1959b:527; Koteja, 1974a:288; Tereznikova, 1975:126; Kosztarab & Kozár, 1978:21; Danzig, 1980a:89.

Type-species: *Xylococcus filiferus* Löw, 1882.

Insects hidden in bark crevices or in cells, inside plant tissue.

Body of adult females ovate, tapering and sclerotized posteriorly; antennae and legs short, unsegmented; 8 pairs of abdominal spiracles larger than the thoracic ones; multilocular and simple disc pores on derm; multilocular pores with characteristic quadrilocular or trilocular center; anal tube present in a sclerotized area.

Adult males poorly known; have compound eyes, long bushy antennae, and 2-segmented tarsi in *X. japonicus* Oguma (Morrison, 1928; Beardsley, 1968).

Distribution. Palearctic; includes 3 species, 1 in CE.

Biology. Apparently oligophagous host preference for deciduous trees.

Note. Zahradnik (1959) transferred this genus into a new family, Xylococcidae. Koteja (1974b:288) also found it distinct and confirmed the new family assignment.

10. **Xylococcus filiferus** Löw – Linden Pearl Scale

Figs 47, 48

Syn. & Lit.: *Xylococcus filiferus* Löw, 1882:271; Morrison, 1928:41; Kawecki, 1935: 74, 1948:1; Šulc, 1936:60; Danzig, 1964:623; Zahradnik, 1972:392; Tereznikova, 1975:126; Kosztarab & Kozár, 1978:21.

Insects found in cells, inside plant tissue at the base of branch forkings of *Tilia* spp., where they produce swellings. Only fine white wax threads are visible to assist in locating the live insects. Body of adult females oval, without segmentation, red, about 2 mm long; antenna reduced to unsegmented cylindrical tubercle with setae on apex and 4 sensory pores at its base; legs absent, except for small dermal protuberances in their place; 8 pairs of abdominal spiracles, these larger than thoracic ones; atrium of spiracles with multilocular pores that have bilocular structure in center; derm with simple and multilocular pores, latter with 1 to 4 locular structure in center, usually with 16 loculi in a marginal band; body setae short and rare; anal tube obvious and located in a sclerotized area. Adult males not known. Immatures described by Morrison (1928) and Tereznikova (1975).

Distribution. In CE only: Au, Cz, Ge, Po, and Sw, and USSR.

Biology. Lives on *Tilia cordata* and *T. platyphyllos*. It develops through several years (Šulc, 1936; Kawecki, 1948; Danzig, 1964).

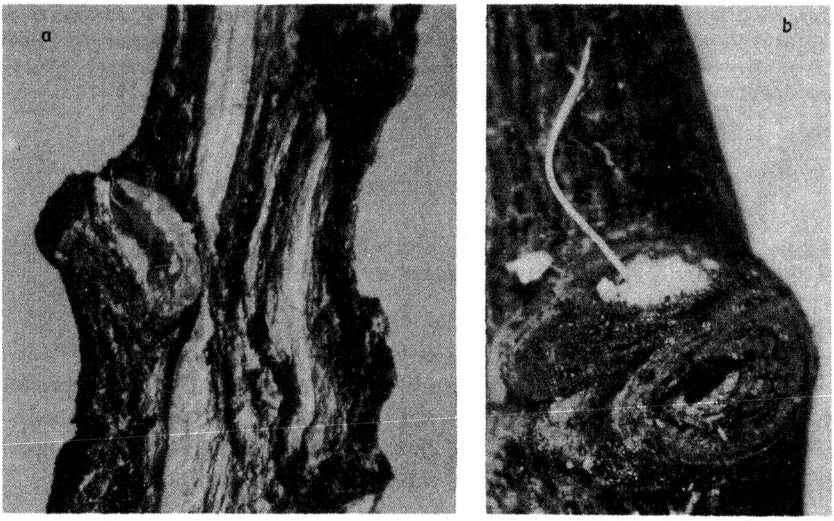

Fig. 47. *Xylococcus filiferus* Löw, (a) female in *Tilia cordata* twig gall, (b) with long white wax tube (Photo M. Kosztarab)

SUPERFAMILY **COCCOIDEA** – ADVANCED COCCOIDS

Syn. & Lit.: Neococcoidea Borchsenius, 1950:46; Bodenheimer, 1952:317. Neococcomorpha, Borchsenius, 1965:362. Coccoidea, Koteja, 1974b:289; Danzig, 1980:58.

It includes about 13 families belonging to 7 morphologically distinct groups or branches. It is a very diverse group, and comprises families where the adult females lack abdominal spiracles; tubular ducts usually present; labium with distinct apical setae; have specialized ducts and pores for secretion of protective covers or tests for eggs and their body. Adult males lack faceted compound eyes.

Distribution. World-wide. Ten families represented in CE.

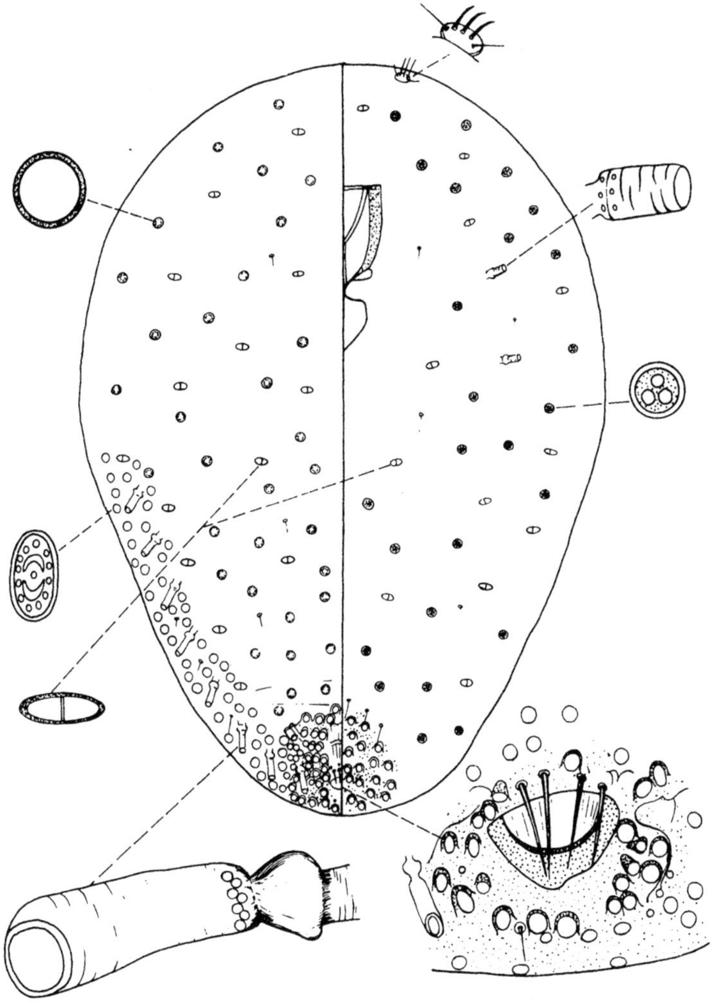

Fig. 48. *Xylococcus filiferus* Löw, second instar (original)

FAMILY **PSEUDOCOCCIDAE** – MEALYBUGS

Fig. 1

Syn. & Lit.: Dactylopiinae (in part), Newstead, 1903:77; Fernald, 1903:39. Pseudococcini, Balachowsky, 1948:243. Pseudococcinae, Borchsenius, 1949:84. Pseudococcidae, Ferris 1950:1, 1953:279; Schmutterer, 1952b:382; Ezzat & McConnell, 1956:3; Williams, 1962:3; McKenzie, 1967:1; Afifi, 1968:1; Ter-Grigorian, 1973:5; Koteja, 1974b:290; Tereznikova, 1975:159; Kosztarab & Kozár, 1978:21; Danzig, 1980:107.

Unfortunately, this family is one that was most neglected in CE. We are attempting here to put together for the first time available information for the often poorly understood species. This is the reason for the disproportion between individual descriptions and for the apparent lack of biological information on many species. We are hoping that the gaps found in our knowledge on Central European mealybugs will soon be bridged after further research by our colleagues in the area. Among recent studies in Central Europe that stand out are Borchsenius (1949) who described many species from the USSR, Schmutterer (1952b) who redescribed 7 species from S. Germany, and Tereznikova who compiled information on the Ukrainian species. The works of Williams (1962) although dealing with the British mealybugs, or of Ter-Grigorian (1973) on the Armenian fauna, often provide needed information for a number of Central European species.

The Pseudococcidae are extremely diverse, and as such make the work of systematists truly challenging, forcing us to often use such qualifying terms as, "almost, normally, most, often, usually, etc." Their common name is derived from the mealy wax secretion which usually covers their bodies. It is produced by different types of disc pores but mainly the trilocular pores. During oviposition, the insects normally produce an ovisac of cottony wax threads.

Adult females. Usually ovoid, with head, thorax and abdomen closely fused; cuticle in most species membranous; abdominal segmentation normally indicated (abdominal segmentation numbered after Ferris, and as such differs from that of Borchsenius, Danzig, etc). Body color varies by species, often yellowish, gray or pink. Antennae 2–9-segmented; but reduced to 1–3 segments in legless mealybugs, such as in *Antonina, Chaetococcus*; only medium sized, 5–6-segmented, and geniculate in *Rhizoecus*; eyes absent or domelike, often surrounded by small pores. Labium between anterior coxae, cone-shaped, 3-segmented, with 13–17 pairs of setae. Legs normally 5-segmented, but entirely lacking in *Antonina, Chaetococcus*; 2 translucent sensory pores common on each side of trochanter, while other pores often appear in clusters on hind coxa, femur or tibia, and are used in classification; claw often with a tooth or denticle on plantar surface, and with a pair of claw digitules of various shapes and lengths; width of legs often of specific value. Only 2 thoracic pairs of spiracles; anteriors in membrane between pro- and mesothorax, posteriors between meso- and metathorax; often with associated disc pores especially around the margin of atrium; in some grass infesting species, e. g. *Antonina,* the spiracles are remarkably enlarged and sclerotized. Dorsum usually with two pairs of slitlike openings or ostioles, these with lips that often bear trilocular pores and/or small setae; anterior pair of ostioles at border of head and prothorax, posterior pair on 7th abdominal segment. Cerarii, the most distinctive characters of mealybugs, are normally present along dorsal body margin and are composed of segmental clusters of trilocular pores and 2 or more stout setae, often with a few slender auxiliary setae, all often on sclerotized surface. Anal ring on 10th abdominal segment normally sclerotized, lying on dorsal side of body, with 6 or more setae and a row of irregular pores, ring seldom reduced. Anal lobes forming part of protruding 9th abdominal segment, often sclerotized, dorsal side often with a cerarius that is usually more prominent and more heavily sclerotized than others along body margin; ventral side with a long and stout apical "anal lobe seta" and 2 or more slender auxiliary setae. Presence of circulus distinctive to family; when present, usually a sclerotized ring or hourglass-like structure of variable size, located between 4th and 5th or on 4th sternite of abdomen. Vulva between 8th and 9th sternite. Derm pores of different types: simple, bitubular, tritubular, trilocular (swirled type) quinquelocular, multilocular (with 6 or more loculi) and sievelike. Some discoidal pores near eyes. Most ducts tubular, more heavily sclerotized at opening, with filamentous prolongation and as such may be called invaginated (bottle-shaped), others as oral-collar, oral-rim (mushroom-shaped), ducts. Last 2 often with a distinct sclerotized flange. Invaginated ducts rounded, or dome-shaped internally. Crateriform duct characteristic of *Heliococcus*. Tubular ducts produce wax threads for ovisacs. Derm pores and ducts are important in separation of genera and species. Body setae usually shorter in dorsum, these normally forming transverse rows on abdominal venter.

Eggs. Oval, laid in a filamentous ovisac, color similar to egg-laying female.

Immature stages. Only in a few species described. For more extensive descriptions see Yang and Kosztarab (1967), Panis (1969a) and Miller and Appleby (1971).

First instars elongate-oval, 0.5–0.7 mm long, 0.2–0.3 mm wide, with well-developed, usually 6-segmented antennae and long legs; derm with simple, trilocular, seldom quinquelocular pores; body setae of various sizes. Sexes not distinguishable in this stage. *Second instar female* similar to first instar in most groups (except in *Antonina* and *Chaetococcus,* where legs lacking and antennae already reduced to 2–3-segmented stubs for all stages, except first instars), a few tubular ducts developed on both surfaces, and usually posterior dorsal ostioles present. *Third instar female* similar to second instar female but with more pores and ducts; anterior dorsal ostioles developed and with one additional antennal segment, but with 1 fewer than adult female. *Second instar male* similar to second instar female, but usually with more tubular ducts and an additional antennal segment. *Third instar male* or prepupa with legs and antennae reduced, with very few short setae only, wing pads short and stublike. *Fourth instar male* or pupa with well-segmented, elongate appendages; wings elongate pad-shaped; tenth segment elongate and sclerotized; cephalic and thoracic sclerotizations partly formed.

Fifth instar male, or adult. Elongate spindle-shaped, normally with well-developed mesothoracic wings; metathoracic wings developed into hamulo-halterae; usually with 2 pairs of simple eyes, and a pair of lateral ocelli; antennae and legs long and with many setae; penial sheath short, sclerotized ventrally only, aedeagus evident; with 1 or 2 pairs of caudal filaments. At least four different types of adult males known: e. g. with wings normal, reduced or absent, and a larviform type. For more comprehensive description of mealybug males see Beardsley (1960), Afifi and Kosztarab (1967) and Afifi (1968).

Distribution. This is the second largest family, world-wide with over 1100 species belonging to ca 210 genera. From Central Europe we are discussing 32 genera and 86 species that are permanent members of the CE fauna. The following 14 introduced species are known only from greenhouses and homes, and are not treated here: *Geococcus coffeae* (Green), *Nipaecoccus nipae* (Maskell), *N. vastator* (Maskell), *Planococcus citri* (Risso), *Pseudococcus amaryllidis* (Bouché), *P. calceolariae* (Maskell), *P. longispinus* (Targ.-Toz.), *P. mamillariae* (Bouché), *P. maritimus* (Ehrhorn), *Rhizoecus cacticans* (Hambleton), *R. falcifer* Künckel d'Herculais, *R. dianthi* Green, *Spilococcus cactearum* McKenzie and *Trionymus diminutus* (Leonardi). For further information on some of the afore-mentioned species consult the work of Zahradnik (1968) and the other sources listed by him.

Biology. Pseudococcidae feed on a variety of woody and herbaceous plants including deciduous trees and conifers, but most live on herbaceous plants especially grasses. Some are polyphagous, especially those with cosmopolitan distribution; others are oligophagous or monophagous, such as *Chaetococcus phragmitis* known only from reed, or *Allococcus vovae* which is specific to juniper. Often they are specific to a feeding site, e. g. leaves, stem, roots, etc. A number of species are found only under the leaf sheaths of Gramineae, while species of *Rhizoecus* are confined to roots.

Life cycles have been studied only in a few economically important species. A generalized life cycle of an oviparous species will start with the eggs; often ca 500 are laid in a filamentous ovisac; the sac is attached to the female, also to substrate. In ovoviviparous species, as in *Puto pilosellae*, the first instars hatch from eggs inside the female and are "born alive." Eggs usually hatch in 3–16 days; the first instars often crawl long distances before settling, often to leaves as in *Phenacoccus aceris* where they feed and molt; second instars of this species return to crevices of the host tree trunk for overwintering in small waxy cocoons; the nymphs and the fully developed females feed during the spring on trunks and branches; females lay up to 1,800 eggs at end of May or first part of June; eggs hatch in June and the nymphs move to leaves and the life cycle is repeated. Native species in CE usually have 1–3 yearly generations, those living indoors often have 6–8 generations per year. There are some species where full development requires more than 1 year. Some species reproduce parthenogenetically. Males are not known as yet in many species; but they are easily overlooked because they are small and live for only a few days. Mealybug colonies are normally attended by ants for their honeydew. Natural enemies commonly include chalcids, ladybird beetles, birds, fungi, etc.

Reviewer's remark. Although several groups of genera have been already recognized within the Pseudococcidae (Trabutinini, Sphaerococcini, Planococcini, Rhizoecini, Ritsemiini etc.) a comprehensive classification of the whole group has not yet been proposed, except that by Afifi (1968) based on males of a few species. Generally, the Pseudococcidae seem to form three natural groups, the Phenacoccinae, Pseudococcinae and Rhizoecinae.

The Phenacoccinae comprise both very primitive and specialized forms. The primitive ones bear spine-like dorsal setae, quinquelocular pores, numerous cerarii, long appendages, a denticle on claw (but digitules are poorly developed) labium with numerous setae, various body colors, also males have several pairs of simple eyes, and the most primitive members (*Puto;* see also "Reviewer's remark" at this genus) have an XX–XO chromosome system. The advanced forms lost some or most of these features, others acquired new devices (e. g. crateriform ducts, particular arrangement and structure of multilocular pores, combined distribution of pores and ducts etc.). *Puto, Phenacoccus, Spinococcus, Peliococcus, Heliococcus, Heterococcus, Rhodania, Euripersia* and others belong to this group.

The Pseudococcinae (Planococcini also included here) are more homogeneous; they have quinquelocular pores, additional cerarii, ungual denticle (radially arranged simple eyes in male are always absent), dorsal setae hairlike, labium bears 14 pairs of setae, the body color is uniformly dark (brown to violet black). Some genera developed particular features (e. g. oral rim ducts). *Pseudococcus, Planococcus, Dysmicoccus, Allococcus, Trionymus, Saccharicoccus, Chnaurococcus* etc. belong to this subfamily.

The Rhizoecinae, a relatively small group and restricted mainly to the Nearctic, are well defined by the structure of appendages and anal lobes, peculiar bi- and tritubular pores, hypogeic life behavior etc. Their relationships with other Pseudococcidae are currently unknown.

Chaetococcus and *Antonina* (and possibly some other genera from outside of Europe), the few pseudococcids devoid of legs, are extremely specialized parasites of Poaceae, but essentially they may be quite close to the Phenacoccinae.

Key to genera of Pseudococcidae (Morphological characters given often apply to species present only in Central Europe)

1. Legs fully developed, 5-segmented, although occasionally small; antennae fully developed, at least 5-segmented; feeding on different plants including Gramineae 3
 Legs entirely absent or as unsegmented stubs; antennae reduced to unsegmented or at most 2–4-segmented stubs; feeding on Gramineae only. 2
2. Although sievelike pores scattered on abdominal sternites, never form a large cluster posterior of each hind spiracle; antennae as 2–4-segmented stubs (fig. 50) .... *Antonina*, p. 66
 Sievelike pores form a large cluster posterior of each hind spiracle; antennae reduced to unsegmented stubs (fig. 55) .... *Chaetococcus*, p. 79
3 (1). Dorsum with longitudinal rows of groups of conical setae, each group on a convex sclerotized plate (similar in *Puto*, but with 4 ostioles in latter, these absent here) (fig. 59) .... *Coccidohystrix*, p. 89
 Dorsum without longitudinal rows of groups of conical setae situated on sclerotized plates (except in *Puto*) 4
4. All dorsal ostioles absent 5
 Dorsal ostioles, at least 1 pair, present 6
5. Oral collar tubular ducts present on both body surfaces; only 1 type of disc pore (quinquelocular) on both body surfaces; all cerarii absent (fig. 78) .... *Rhodania*, p. 146
 Without any tubular ducts; disc pores with 3, 4, 5 and 6 loculi on both body surfaces; anal lobe cerarii poorly developed (fig. 79) ... *Ritsemia*, p. 147
6 (4). All (17–26 pairs) marginal cerarii on sclerotized plates, each with numerous conical setae; often with similar cerarii in median area of dorsum (figs 75, 76) .... *Puto*, p. 135
 Normally only 1–3 pairs of terminal cerarii (except in *Heliococcus*) on sclerotized plate and each cerarius with only 2–3 conical setae; cerarii never extend into median area of dorsum 7
7. Dorsum with at least a few enlarged crateriform tubular ducts with orifices borne at apex of sclerotized, projecting tubercles (fig. 63) .... *Heliococcus*, p. 99
 Without such ducts on dorsum 8

8. Dorsal multilocular pores normally arranged in groups of 3–6 pores, each group with a tubular duct at its center; if multilocular pores occur singly or in pairs, these form distinct transverse rows on both body surfaces 9
Dorsal multilocular pores, if present, not arranged in such clusters 10

9. Circulus absent; quinquelocular pores on both body surfaces (fig. 69) . . . . *Peliococcopsis*, p. 115
Circulus present, large, oval; quinquelocular pores on venter only (fig. 70) . . . . *Peliococcus*, p. 117

10 (8). Antennae geniculate, short, 5- or 6-segmented; at least inner row of anal ring pores of elongate pores giving a cellular-like appearance; normally with bitubular and/or tritubular pores (except in *poltavae*); cerarii absent in all Central European species (fig. 77) . . . . *Rhizoecus*, p. 140
Antennae never geniculate, normally 7- to 9-, rarely 6-segmented; anal ring pores circular, seldom oval; never with bitubular and/or tritubular pores; usually some abdominal cerarii present 11

11. Each cerarius (17–18 pairs) on elevated sclerotized tubercle; auxiliary cerarii on midline of body; dorsal setae conical, similar to cerarian setae (fig. 81). . . . . *Spinococcus*, p. 153
Cerarii not on elevated sclerotized tubercles; no auxiliary cerarii on midline of body; dorsal setae seldom conical and similar to cerarian setae 12

12. Trilocular pores entirely absent 13
Trilocular pores present on both body surfaces, rarely on venter only (in *Antoninella*) 15

13. Cerarii and quinquelocular pores absent; only posterior ostioles developed (fig. 66) . . . . *Metadenopus*, p. 108
Some cerarii and quinquelocular pores always present; all 4 ostioles developed 14

14. Quinquelocular pores few, on dorsum only; multilocular pores numerous and distributed widely on both body surfaces (fig. 54) . . . *Boreococcus*, p. 76
Quinquelocular pores numerous, on both body surfaces; multilocular pores normally on abdominal venter only, never in middle of both sides of prosoma (fig. 64) . . . . *Heterococcus*, p. 104

15(12). Oral rim tubular ducts present, usually on both body surfaces 16
Oral rim tubular ducts absent, except in *Allococcus* on dorsum 18

16. Only anal lobe cerarii present in single species (*rostellum*) of Central European distribution (fig. 58) . . . . *Chorizococcus*, p. 86
With 3 to 7 pairs of cerarii present in species of Central European distribution 17

17. Circulus absent (except in *bejbienkoi* where 3 basal segments of antennae twice as long as segments 4, 5 and 6 combined);body with black or blue-black body contents when in potassium hydroxide (fig. 53) . *Atrococcus*, p. 69
Circulus present in single CE species; body with yellow or pinkish contents when in potassium hydroxide . . . . *Spilococcus*, p. 151

18(15). Antennae normally 9-segmented (rarely 8 in *Phenacoccus hordei*); hind coxae without translucent pores (except in *Polystomophora*); claws always with a denticle 19

Antennae normally 6- to 8-segmented; hind coxae usually with translucent pores; claws with or without denticle 20

19. With 9–18 pairs of cerarii (except. *P. bicerarius, incertus* and *phenacoccoides*); quinquelocular pores absent or present on venter only; both pairs of dorsal ostioles developed (fig. 71) .... *Phenacoccus,* p. 122

Only 1 pair of cerarii (C_{18}) developed; quinquelocular pores present on both body surfaces; only posterior pair of ostioles developed for single species in Central Europe (fig. 74) .... *Polystomophora,* p. 133

20(18). Multilocular pores form wide bands on both dorsal and ventral body margin; ducts short, oral collar flange occupies about half length of duct .... *Balanococcus,* p. 75

Multilocular pores do not form wide bands on both dorsal and ventral body margin; oral collar flange of ducts shorter than half length of duct (except in *Trionymus radicum*) 21

21. With oral rim tubular ducts on dorsum; with 18 pairs (rarely 14) of cerarii .22

Without oral rim tubular ducts on body; with fewer than 5 pairs of cerarii (except in *Dysmicoccus* up to 17 pairs)............................23

22. Hind coxae with translucent pores; invaginated (bottle-shaped) ducts absent; quinquelocular pores absent; 1 large slightly quadrangular circulus present (fig. 49) .... *Allococcus,* p. 64

Hind coxae without translucent pores; invaginated (bottle-shaped) ducts present; with quinquelocular pores on venter; 3 oval circuli present (fig. 60) .... *Coccura,* p. 90

23(21). With 0 to 4 pairs of cerarii; but none on head 24

With 5 or more pairs of cerarii; if less than 5 pairs, then at least 1 pair on head 32

24. Without any distinct cerarii, often each C_{18} replaced by 2–4 long hairs . 25

With at least anal lobe cerarii distinctly developed 27

25. Claws without denticle; all anal ring pores large, oval; each anal lobe with 1–2 apical setae and 3 or more shorter setae (fig. 67) . *Mirococcopsis,* p. 110

Claws with denticle; anal ring pores small, circular, sometimes mixed with oval pores; with 1 long apical seta on each anal lobe................... 26

26. Hind legs without translucent pores; circulus large, oval; with some quinquelocular pores on venter (fig. 68)........... *Mirococcus,* p. 114

Hind legs with translucent pores; circulus absent; quinquelocular pores absent (fig. 65) .... *Longicoccus,* p. 107

27(24). Quinquelocular pores always absent; with 1–5 pairs of cerarii (fig. 82)..... .... part of *Trionymus,* p. 157

Quinquelocular pores present (absent only in *Saccharicoccus*); part of them with 1–4 pairs of cerarii 28

28. With only anal lobe cerarii...29

With 2–4 pairs of cerarii...31

29. Only posterior pair of dorsal ostioles developed; spiracles with associated multilocular pores; quinquelocular pores present on venter (fig. 51)..... .... *Antoninella,* p. 67

Both pairs of dorsal ostioles developed; spiracles without associated multilocular pores; quinquelocular pores absent 30

30. Circulus, if present, small, oval; without minute simple pores around hind legs (fig. 57) .... *Chnaurococcus*, p. 83

Circulus large, hourglass-shaped, longer than wide, rarely smaller or absent; with minute simple pores around hind legs (fig. 80). .... *Saccharicoccus*, p. 149

31(28). Quinquelocular pores on both body surfaces; only posterior pair of ostioles developed .... *Brevennia*, p. 78

Quinquelocular pores on venter only; all 4 dorsal ostioles developed (fig. 62) .... *Euripersia*, p. 94

32(23). With 15–17 pairs of cerarii; C_{18} borne on a sclerotized plate, larger than anal ring (fig. 61) .... *Dysmicoccus*, p. 92

With 6–10 pairs of cerarii; usually C_{18} not borne on an unusually large sclerotized anal plate .... part of *Trionymus*, p. 157

Genus **Allococcus** Ezzat & McConnell

Fig. 5

Syn. & Lit.: *Allococcus* Ezzat & McConnell, 1956:13; Danzig, 1964:628; Ter-Grigorian, 1973:80; Kosztarab & Kozár, 1978:29.

Type species: *Pseudococcus inamabilis* Hambleton, 1935.

Adult female oval, antennae long, 8-segmented; labium long, conical; all 4 ostioles well developed, legs slender, hind coxa and tibia usually with translucent pores; normally with 18 pairs of cerarii, each with 2 cerarian setae, without auxiliary setae, except sometimes on anal lobe pair; circulus present or absent; anal ring with 6 setae; anal lobes protruding with long apical setae; trilocular pores over both surfaces; multilocular pores forming transverse rows or bands on abdominal venter; dorsum with large oral collar and oral rim tubular ducts; latter may be present on venter also; small oral collar tubular ducts confined to venter. Body setae short and stout on dorsum, long and flagellate on venter.

Distribution. World-wide ca 10 species; 1 in CE.

Biology. Species known are from woody plants. Biology unknown.

Note. The genus is related to *Planococcus* Ferris, but it can be separated in possessing oral rim tubular ducts.

11. **Allococcus vovae** (Nassonov) – Nassonov's Mealybug

Figs 5, 49

Syn. & Lit.: *Pseudococcus vovae* Nassonov, 1908b:484; Borchsenius, 1949:134; Schmutterer, 1955a:161. *Planococcus vovae*, Tereznikova, 1975:229; Danzig, 1980a:168. *A. vovae*, Danzig, 1964:628; Koteja & Zak-Ogaza, 166:318; 169:359; Zahradnik, 1956:50, 1972:398; Ter-Grigorian,* 1973:80; Kosztarab & Kozár, 1978:29. The *P. gossypifer* (Rondani) record of Lindinger (1921:434) from *Juniperus communis* in Central Hungary (Soroksár), could have applied to this species.

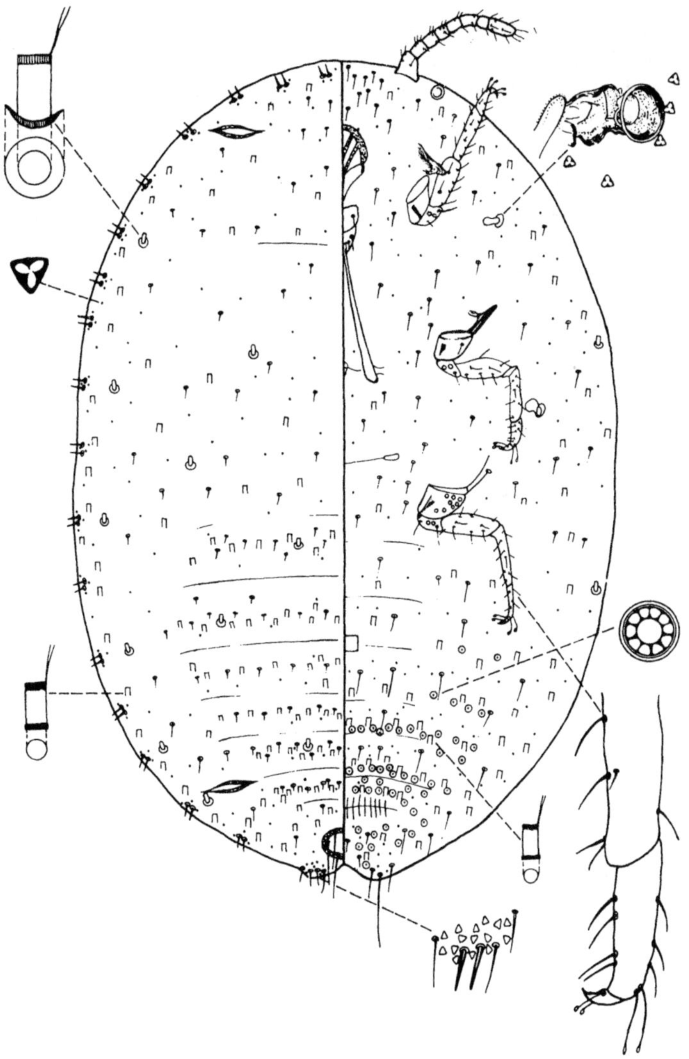

Fig. 49. *Allococcus vovae* (Nassonov), female (original)

Body of adult female always visible at end of ovisac, a characteristic of this species only. Adult female oval, brownish, ca 3.0 mm long and 1.7 mm wide; antennae 8-segmented, apical segment longest; stylet loop extends to line between posterior spiracles; ostioles large; legs slender, hind coxae with translucent pores; dorsal margin with 18 pairs of cerarii, seldom with as few as 14 pairs, each with 2 setae, without auxiliary setae; 1 circulus, large, elongate, quadrangular; anal ring with 6 setae, each 157 μ long; anal lobe protruding apical seta 218 μ long; multilocular pores form transverse rows or bands on abdominal sternites; large oral collar and oral rim tubular ducts on dorsum; venter with small oral collar tubular ducts; setae short and stout on dorsum, on venter long, with bent tip and forming transverse rows. First instar described briefly by Nassonov (1908a).

Distribution. Trans-Palearctic. In CE: Au, Cz, Fr, Ge, Hu, Po, USSR (European part to Far East) and Yu. Also in Corsica (Zahradnik, 1972; Kosztarab and Kozár, 1978).

Biology. On *Juniperus* spp. Often form large colonies on branches and are tended by ants. Mature females seek out bark crevices, as noted in June in Hungary by Kosztarab and Kozár (1978).

Natural enemies. Parasites: *Allotropa mecrida,* (Podsiadlo, 1981b); *Leptomastidea bifasciata* (Zahradnik, 1972); and *Lygoceus* sp. (Matesova, 1968a). Predator: *Pullus* sp. (Matesova, 1968a).

Genus **Antonina** Signoret

Fig. 50

Syn. & Lit.: *Antonia* Signoret, 1872:36 (preoccupied name). *Antonina* Signoret, 1875:322; Goux, 1935:93; Borchsenius, 1949:311; Ferris, 1953:289; McKenzie, 1967:73; Yang & Kosztarab, 1967:1; Afifi & Kosztarab, 1967:1. *Laboulbenia* Lichtenstein, 1877:299.

Type-species: *Antonina purpurea* Signoret, 1872.

Adult females completely enclosed in a white felted wax sac. Adult females oval, with sclerotized derm, especially at posterior end; antennae reduced to 2–3-segmented stubs; ostioles reduced, often obscure, only posterior pair present; legs normally entirely absent, seldom stublike; spiracles large, often sclerotized and with a band of trilocular or multilocular pores in atrium; cerarii and circuli absent; anal ring at end of an anal tube retracted into abdomen, strongly sclerotized, with 6 setae longer than ring; dorsum with trilocular, seldom multilocular pores; venter with both kinds of disc pores; quinquelocular pores absent; often sievelike pores on abdominal sternites; tubular ducts of a modified oral-collar type, with sclerotized internal extension, dome-shaped at end, seldom with a secondary ring around orifice, on both derms; body setae short and scarce, normally some stout, long setae along abdominal posterior margin. Adult males described by Afifi and Kosztarab (1967), nymphs by Yang and Kosztarab (1967).

Distribution. World-wide, with ca 20 species; 1 species in CE.

Biology. On Gramineae only, occurring at base of grass culms, or under leaf sheaths. One generation a year; adult females usually overwintering, laying an average of 100 eggs per female. Females developing through 4 instars, males through 5 (Yang and Kosztarab, 1967).

Note. *A. crawi* Cockerell was introduced to USSR and is found only in Southern Crimea, Odessa and the Black Sea coast of Caucasus. We deleted this species. A description and illustration was provided by Tereznikova (1975). The first author intends to publish a revision of the legless mealybugs of the world (genus *Antonina* and its relatives) in the future.

12. **Antonina purpurea** Signoret – Red Legless Mealybug

Fig. 50

Syn. & Lit.: *Antonia purpurea* Signoret, 1872:36. *Laboulbenia brachypodii* Lichtenstein, 1877:299. *A. purpurea,* Lindinger, 1912:169; Green, 1934:509; Borchsenius, 1949:312; Parker, 1960:174; Afifi & Kosztarab, 1967:24; Yang & Kosztarab, 1967:33; Kosztarab & Kozár, 1978:31.

Adult female flat, elongate-ovoid ca. 1.8–3.4 mm long, 0.9–2.7 wide, red; antennae 3-segmented, rarely 4, 55–78 μ long, with 4–5 apical setae; labium broad triangular, ca 92 μ long, 103 μ wide, stylet loop extends to area of hind spiracles; each leg reduced to clawlike structure with 3–4 small setae; spiracles large, with a double row of multilocular pores inside atrium; anal ring at end of anal tube with two fingerlike apodemes at base; anal ring setae thick, extending to apex of abdomen; anal lobes slightly protruding, with 66–117 μ long apical setae; trilocular pores scattered on both derms; multilocular pores on venter, clustered around spiracles; sievelike pores scattered on abdominal sternites; oral collar tubular ducts and short setae on both derms; long spinelike setae along margin of abdominal apex. Adult male described by Afifi and Kosztarab (1967), first instar by Yang and Kosztarab (1967).

Distribution. A rare Mediterranean species. Known from Fr, It, Yu, and Spain.

Biology. On grasses: *Agropyron, Brachypodium,* and *Milium.* Found at the root crown of grasses. Reproduction biparental. Parker (1960) found adult females overwintering; laid about 100 eggs; crawlers hatched in May.

Natural enemies. Parasites: *Anagyrus diversicornis, Metanotalia maderensis, M. hispanica, Timberlakia europea* (Parker, 1960; Trjapitzin, 1978). Predators: *Scymnus nigrinus; Chamaemyia herbarum* and *C. juncorum* (Parker, 1960).

Genus **Antoninella** Kiritchenko

Syn. & Lit.: *Antoninella* Kiritchenko, 1938:233; Borchsenius, 1949:173; Danzig, 1964:627; Kosztarab & Kozár, 1978:31. *Antoniella,* Neave, 1950:18 (a misspelling). *Antoninella* Šulc, 1944a:148 (a homonym).

Type-species: *Antoninella inaudita* Kiritchenko, 1938.

Females covered with powdery wax. Adult female oval; antennae 6-segmented; posterior dorsal ostioles present; legs well developed; spiracles with associated multilocular pores; circulus absent, only last pair of cerarii, these with tubular ducts and pores; anal ring with 6 setae and a wide band of pores; apical setae thick; trilocular and multilocular pores on both surfaces; quinquelocular pores scattered on venter; tubular ducts usually forming transverse rows on venter; setae slender, on both surfaces.

Distribution and biology. See included species.

Note. It is related to *Rhizoecus.*

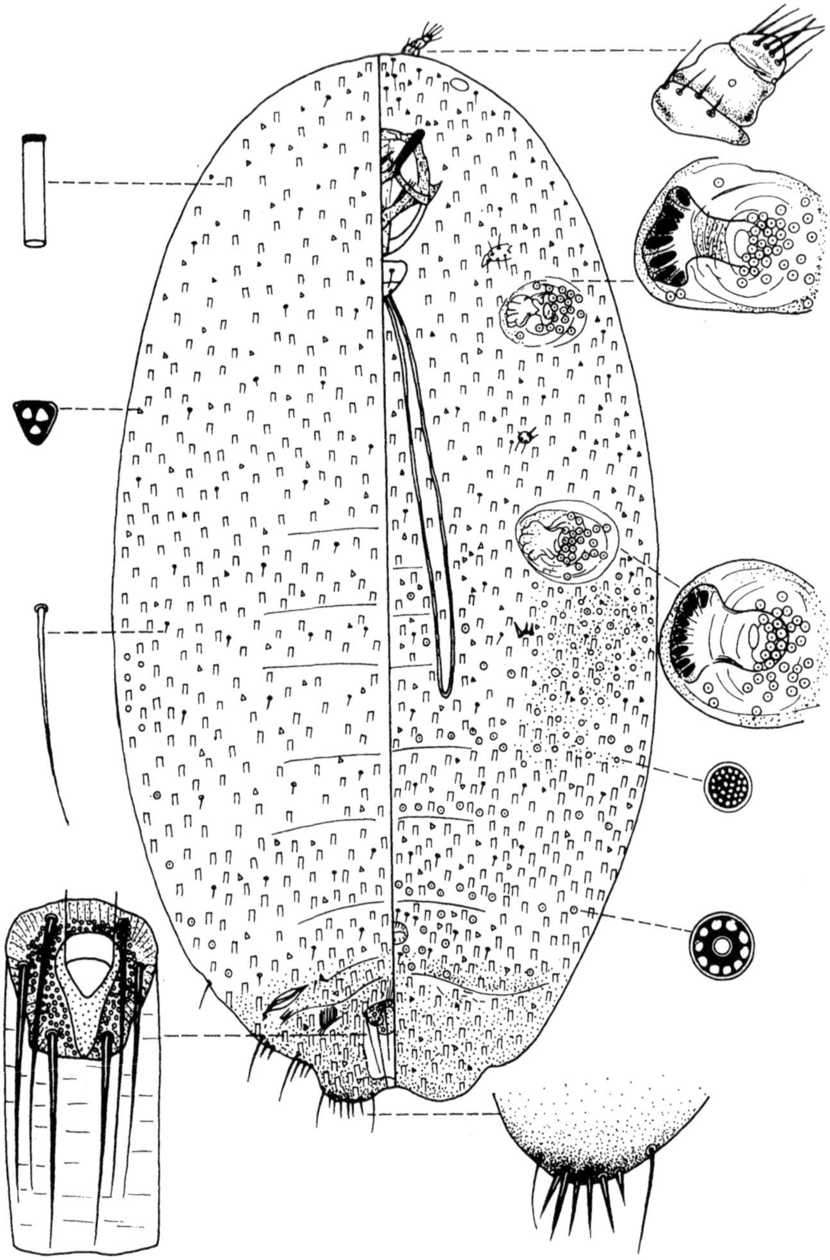

Fig. 50. *Antonina purpurea* Signoret, female (original)

13. **Antoninella inaudita** Kiritchenko – Fescue Mealybug
Fig. 51

Syn. & Lit.: *Antoninella inaudita* Kiritchenko, 1937:233; Goux, 1945:38; Borchsenius, 1949:173; Schmutterer, 1956a:66; Danzig, 1964:627; Tereznikova, 1975:162; Kosztarab & Kozár, 1978:32.

Female covered with white waxy powder. Adult female oval, greyish, becomes heavily sclerotized during egg-laying, ca 3.0 mm long, 2.0 mm wide; abdominal segments distinct; antennae 6-segmented; no anterior ostioles; stylet loop 2–2.5 times longer than labium; legs relatively short and thick, no claw denticles; spiracles surrounded by multilocular pores; only a pair of cerarius-like structures developed; each on a sclerotized plate with 2 thick setae, 7–11 tubular ducts and a few trilocular or quinquelocular pores; no circulus; anal ring wide, with 2 or 3 rows of pores and 6 setae; anal lobes poorly developed each with a thick apical seta; trilocular and multilocular pores on both surfaces; quinquelocular pores and oral collar tubular ducts on venter; setae on both surfaces.

Distribution. A rare species; in CE: Fr, Ge and USSR (Crimea and S Ukraine).

Biology. On roots of *Festuca ovina.* A thermophil species, inhabiting sunny, rocky slopes. Schmutterer (1956a) found adult females in mid-May, first instars on June 10 in Germany. Probably ovoviviparous.

Genus **Atrococcus** Goux

Syn. & Lit.: *Atrococcus* Goux, 1941a:69; Williams, 1962:8; McKenzie, 1967:88; Ter-Grigorian, 1973:55; Koteja, 1974a:100; Tereznikova, 1975:163; Kosztarab & Kozár, 1978:32: Danzig, 1980a:160.

Type-species: *Atrococcus melanovirens* Goux, 1941.

Adult females usually covered with white powdery secretion; during oviposition they form white cottony ovisacs. Adult female oval; antennae 8-segmented; dorsal ostioles present; legs well developed, hind coxae usually with translucent pores, claws without denticles; spiracles usually with associated trilocular pores; cerarii on abdomen only 1 to 7 pairs; circuli absent except in *A. bejbienkoi*; anal ring with 6 setae; anal lobes only slightly protruding and with apical setae; quinquelocular pores absent; trilocular and multilocular pores present on lateral area of venter; oral collar ducts and oral rim ducts numerous on both surfaces; hairlike setae on both surfaces, longer on venter. First instar described by Goux (1941a).

Distribution. Trans-Palearctic; with ca 20 species, some probably mistakenly assigned to the genus; 4 in CE.

Biology. They feed primarily on herbaceous plants, but a few are known from woody plants; probably adult females overwinter; 1 yearly generation.

Note. Williams (1962) found *Atrococcus* very close to *Chorizococcus* and *Spilococcus* in the number of cerarii and other characters except for the black color after death in *Atrococcus.*

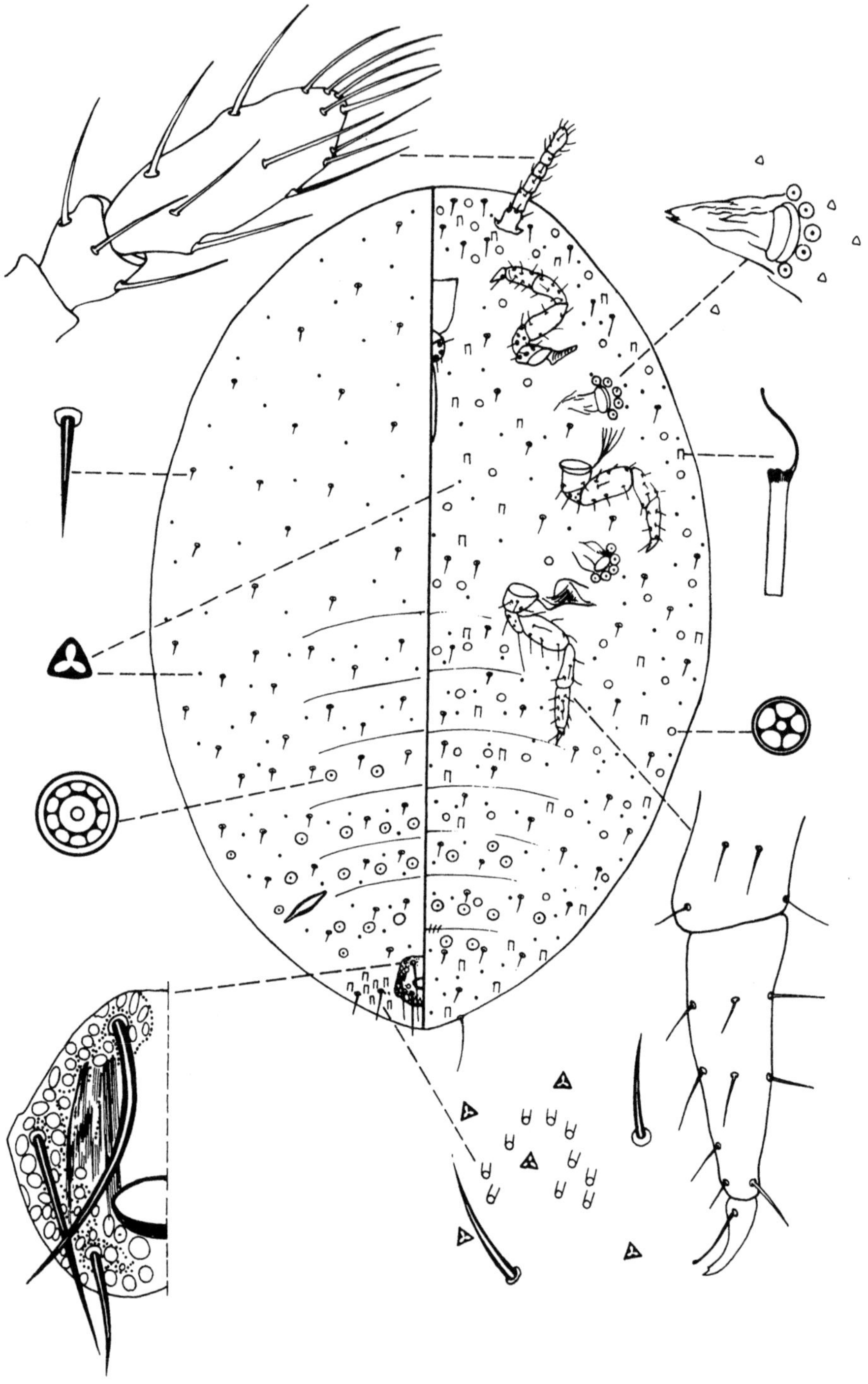

Fig. 51. *Antoninella inaudita* Kiritchenko, female (original)

Key to species of *Atrococcus*

1. Circulus present; 3 basal segments of antennae twice as long as segments 4, 5 and 6 combined .... *bejbienkoi*, p. 71
   Circulus absent; 3 basal segments of antennae shorter than segments 4, 5 and 6 combined..2
2. Oral collar tubular ducts present on dorsum of prosoma; cerarii 3 pairs ... . *achilleae*, p. 71
   Oral collar tubular ducts absent from dorsum of prosoma; cerarii 3–7 pairs 3
3. Multilocular pores form broken transverse rows on dorsum; cerarii 3 or 4 pairs; legs 530–610 μ long .... *cracens*, p. 73
   Multilocular pores few, occur singly and do not form rows on dorsum; cerarii 6 or 7 pairs; legs 740–910 μ long .... *paludinus*, p. 74

14. **Atrococcus achilleae** (Kiritchenko) – Yarrow Mealybug

Syn. & Lit.: *Pseudococcus achilleae* Kiritchenko, 1936:146; Borchsenius, 1949:127. *P. altaicus* Matesova, 1968b:154. *A. achilleae*, Danzig, 1972a:329; *1980a:162; Ter-Grigorian, 1973:60; Tereznikova, 1975:164; Kosztarab & Kozár, 1978:34.

Body of female covered with white wax powder. Adult female oval, dark red, ca 2.5 mm long, 1.5–2.0 mm wide; antennae 8-segmented, according to Ter-Grigorian (1973) ca 377 μ long; stylet loop long, normally extends to line of posterior coxae; ostioles poorly developed; hind coxae with small group of translucent pores, claw digitules slightly longer than claws; only 3 pairs of cerarii on abdominal segments; ventral circuli absent; anal ring setae ca 133 μ long, about twice as long as ring diameter; apical setae ca 215 μ long; few trilocular pores on both surfaces; multilocular pores form groups on submargin of both surfaces, and transverse bands on abdominal sternites; oral rim ducts and oral collar ducts present; setae scattered and/or form transverse rows on entire body.

Distribution. Trans-Palearctic. In CE: Bu, Hu, It, USSR (Transcarpathian Region) and Yu, also in Mongolia and Korea. Rare.

Biology. Usually found on roots of *Achillea, Euphorbia,* and *Pyrethrum.* Kiritchenko (1936) collected all stages in mid-July, 1929. Egg-laying females were frequently found during August–October. Biology poorly known.

Note. Danzig (1972a) noted that the number of multilocular pores and of tubular ducts varies greatly and this makes it difficult to differentiate this species from *A. paludinus*.

15. **Atrococcus bejbienkoi** Kozár & Danzig – Bei-Bienko's Mealybug
Figs 52, 53

Syn. & Lit.: *A. bejbienkoi* Kozár & Danzig, 1976: 65; Kosztarab & Kozár, 1978: 32; Danzig, 1980a: 168.

Females covered with white wax powder and form cottony ovisacs. Adult females oval, lead colored, 2.5–4.0 mm long; antennae 8-segmented, 3 basal segments twice as long as segments 4, 5 and 6 combined; eyes dome-shaped; stylet loop short, extends to line of anterior spiracles only; ostioles present; leg setae thick, numerous; 3–4 pairs of cerarii developed, each with 2 spines, a group of trilocular pores (6 to 11 pores, except anal lobe with 30), and 4 hairlike setae; 1 circulus, small, oval; anal ring with 6 setae; anal lobes slightly protruding; trilocular pores distributed on both surfaces; multilocular pores few on abdominal dorsum, forming small groups on prosomal venter and bands on terminal 4–5 sternites of abdomen; oral collar ducts of two sizes mixed with trilocular pores on abdominal dorsum and on venter; oral rim ducts form transverse rows on dorsum, and distributed on submargin of venter; setae scattered on both derms, in medial area of venter twice as long as on dorsum.

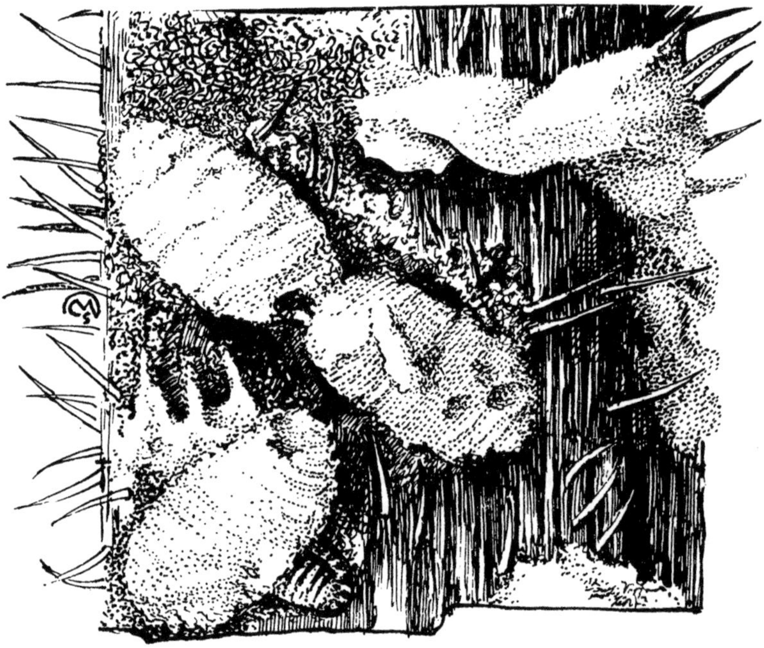

Fig. 52. *Atrococcus bejbienkoi* Kozár & Danzig, ovisacs on root of *Rubus* sp. (after Kosztarab and Kozár, 1978)

Distribution. Rare, known only from Hu and USSR (Siberia).

Biology. Hosts: *Rubus idaeus* in Hungary, also *Arctostaphylos, Pedicularis* and *Vaccinium* in the Far East of USSR. Adult females were collected during September in Hungary, where it was considered a pest of raspberry.

Note. This species is related to *A. indigens* (Borchsenius), but differs from it by the absence of multilocular pores on the dorsum of prosoma.

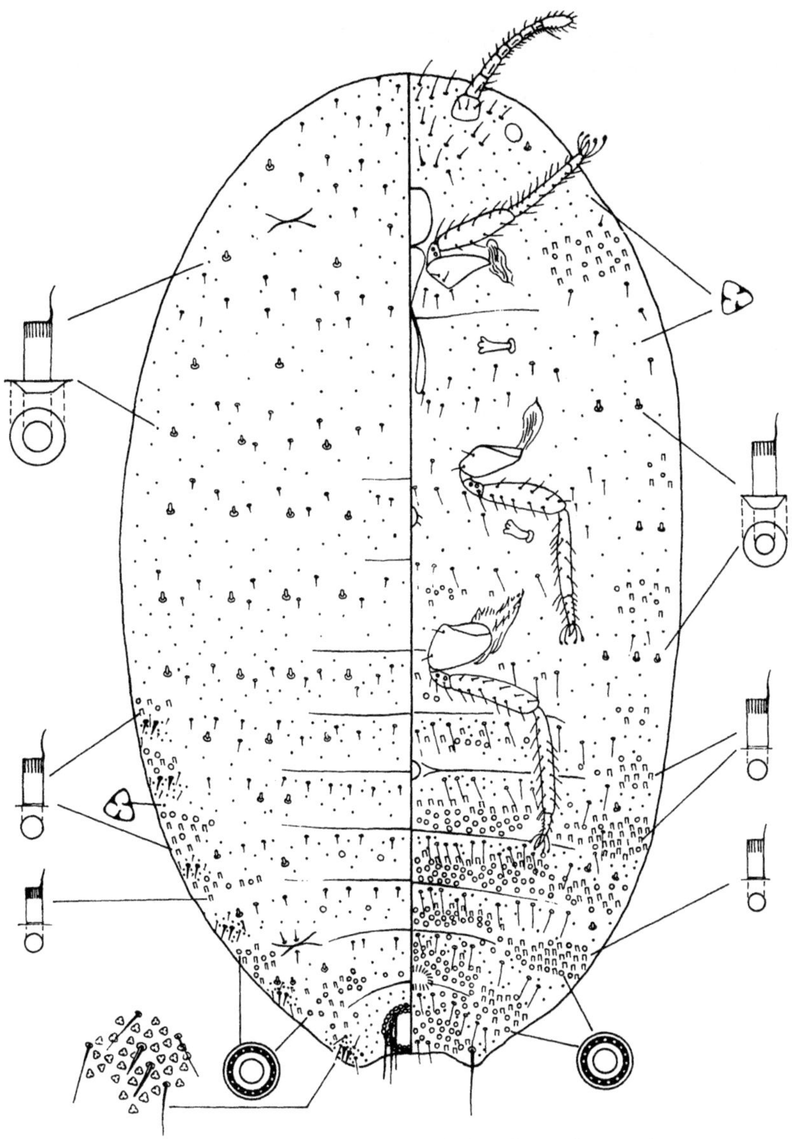

Fig. 53. *Atrococcus bejbienkoi* Kozár & Danzig, female (modified after Kozár and Danzig, 1976)

16. **Atrococcus cracens** Williams – Slender Mealybug

Syn. & Lit.: *A. cracens* Williams,* 1962:9; Koteja & Zak-Ogaza, 1966:318, 1969:359; Koteja, 1971b:321, 1972:566, 1974a:100; Kozár & Danzig, 1976:67; Dziedzicka, 1977:68; Kosztarab & Kozár, 1978:34; Danzig, 1978 :7, 1980a:167. *Pseudococcus borchsenii* Rasina, 1966:24.

Adult female elongate oval, 2.2–2.5 mm long, 1.1 mm wide; antennae 8-segmented 355–365 μ long; labium cone-shaped, ca 95 μ long; stylet loop reaches line between mid-coxae; ostioles well developed, each lip with about 8 trilocular pores and 1 or 2 setae; legs slender, 530, 570, 610 μ long respectively (Dziedzicka, 1977); hind coxae with translucent pores; cerarii on last 3 or 4 segments of abdomen; anal ring setae about twice as long as diameter of ring; anal lobes only slightly protruding with slender apical setae; trilocular pores distributed on both derms; multilocular pores in small submarginal groups on abdominal dorsum, form transverse rows on dorsum of prosoma, and wide bands on abdominal sternites; oral collar ducts of two sizes distributed on abdominal dorsum and venter; oral rim ducts arranged in transverse rows on dorsum, and in pairs on ventral margin of most abdominal segments; setae on dorsum short and slender, longer on venter.

Distribution. A rare Trans-Palearctic species. In CE: Hu, Po and USSR. Also known from England and the Far East USSR.

Biology. On roots and leaves of herbaceous plants. Polyphagous. Host genera: *Achillea, Angelica, Calamagrostis, Centaurea, Chrysanthemum, Crepis, Deschampsia, Glechoma, Hieracium, Leontodon, Melampyrum, Plantago, Taraxacum, Trifolium, Veronica, Vicia* and various undetermined Gramineae. Apparently adult females are present from June through September. Biology not known.

Note. Williams (1962) and Danzig (1972a) noted a close relationship with *A. paludinus*. See key for separation.

17. **Atrococcus paludinus** (Green) – Marsh Mealybug

Syn. & Lit.: *Pseudococcus paludinus* Green, 1921:190, 1922:22. *P. impeditus* Borchsenius, 1949:124. *P. alfredi* Rasina, 1966:24. *A. paludinus*, Goux, 1941a:80; Williams,* 1962:11; Reyne, 1965;96; Koteja, 1971b:321; 1974a:68; Dziedzicka, 1977:68; Ossiannilsson, 1977:13; Kosztarab & Kozár, 1978:33; Danzig, 1978:7, 1980a:164.

Female covered with thin coat of white wax powder; occupies shelters in angles of vein on underside of leaves (Green, 1922). Adult female elongate oval, brownish pink, ca 2.4 mm long, 1.2 mm wide; antennae 8-segmented, 395–475 μ long; labium conical, 130–140 μ long; each lip of ostioles with few trilocular pores and 1 or 2 setae; legs slender, 740, 790, 910 μ long respectively (Dziedzicka, 1977), hind coxae with few translucent pores; 6 or 7 pairs of abdominal cerarii, anal lobe pair with 2 slender conical setae, few trilocular pores and auxiliary setae; circuli absent; anal ring setae about twice length of diameter of ring; trilocular pores distributed on both surfaces; a few multilocular pores in median areas of posterior abdominal segments and in small marginal groups on dorsum, form wide transverse bands on venter of abdominal segments and small submarginal groups on prosomal venter; oral collar ducts of two sizes on both surfaces of abdomen; oral rim ducts form single transverse rows on most segments of dorsum, few scattered on venter, most in submarginal areas; dorsal setae short and slender, ventral setae longer than dorsal ones.

Distribution. A Trans-Palearctic species. In CE: Hu, Po, Ro and USSR. Also known from England, Holland, Korea, Norway and Sweden.

Biology. A polyphagous species; most common in CE on *Rubus*, but also on a wide variety of herbaceous and woody plants, including *Artemisia, Convolvulus, Spiraea* and *Urtica*. Besides infesting leaves, it also feeds at base of stems. Adult females and males were collected in August, but females were also collected during June in Hungary.

Genus **Balanococcus** Williams

Syn. & Lit.: *Balanococcus* Williams, 1962:13; McKenzie, 1964:217; Danzig, 1964:627; 1980a:186.

Type-species: *Ripersia scirpi* Green, 1921.

Adult female elongate oval, almost parallel-sided; antennae 6- or 7-segmented; ostioles present, but often only poorly developed; hind coxae with a group of translucent pores; cerarii on end of abdomen only, 1 to 3 pairs; circuli present or absent, if present, up to 3 small oval circuli developed; anal ring with 6 setae, latter twice as long as diameter of ring; and lobes poorly developed; minute simple pores scattered on both surfaces in at least *B. takahashii;* trilocular pores evenly distributed on both surfaces; quinquelocular pores absent; multilocular pores form a band along body margin of both derms, and in transverse bands on abdominal sternites; oral collar ducts on both surfaces, often of two sizes; distributed similarly to multilocular pores and often mixed with those, their collar flange-shaped and occupies half or nearly half of duct length; short setae distributed on both derms.

Distribution. Trans-Palearctic; with 4 species included; 1 in CE.

Biology. On grasses, often under leaf sheaths. Biology not known. *B. takahashii* (McKenzie, 1964) was observed to damage turf.

Note. The genus is related to *Trionymus* and *Kiritshenkella* Borchsenius, but differs from latter principally in the trilocular pores being more evenly distributed, while in *Kiritshenkella* these are concentrated in the midline of both surfaces.

18. **Balanococcus boratynskii** Williams – Sea-shore Mealybug

Syn. & Lit.: *Balanococcus boratynskii* Williams,* 1962:15; McKenzie, 1964:219; Danzig, 1968:499, 1980a:186; Danzig & Ivanova, 1976:76; Koteja, 1986a (in print).

Adult female elongate, almost parallel-sided, dark colored ca 3.5 mm long; antennae 6-segmented 260–310 μ long; labium conical, 70–75 μ long; ostioles poorly developed, with 1–3 trilocular pores on each lip; legs slender ca 490–560 long, hind coxae with a group of translucent pores; cerarii 2 pairs, anal lobe cerarii with 2 conical setae, 5–6 trilocular pores and 1 auxiliary seta; C_{17} with only 1 conical seta and 1 or 2 trilocular pores; 1–3 circuli present, these small and oval, posterior smallest; anal ring with 2 rows of pores and 6 setae about twice length of ring diameter; anal lobes moderately developed; trilocular pores evenly distributed on both surfaces; multilocular pores form a marginal band on both surfaces, and a transverse row on sternite 5, also bands on sternites 6–8 of abdomen; oral collar ducts of one size, short, their collar flange occupies slightly less than half the total length of duct; ducts mixed with multilocular pores on venter and on dorsal margin, also sparsely scattered on most of the rest of dorsum; short and slender setae distributed on both surfaces, those on venter of head and on terminal abdominal sternites somewhat longer than others (Williams, 1962).

Distribution. England, Hu, It, Po, and USSR (European parts and Far East). Second instars examined by us from Germany and determined by Lindinger as *Rhizoecus halophilus* probably belong to this species.

Biology. It was collected from leaf sheaths of *Deschampsia caespitosa* in England and on *Festuca* sp. in USSR. Females noted in August on sandy seashore.

Note. It is possible that the records of *R. halophilus* crawlers invading a home in Germany (Weidner, 1970) refer to this species. McKenzie (1964) noted close relationship between *B. boratynskii* and *B. takahashii,* but the latter species lacks the dorsal marginal band of multilocular pores present in *B. boratynskii.*

Genus **Boreococcus** Danzig

Syn. & Lit.: *Boreococcus* Danzig, 1960:117, 1964:630; Miller & McKenzie, 1970:439; Tereznikova, 1975:164; Kozár & Danzig, 1976:67; Kosztarab & Kozár, 1978:34.

Type-species: *Boreococcus ingricus* Danzig, 1960. Genus monotypic.

Females covered with fine white waxy powder. Adult female elongate oval; antennae 9-segmented; 4 ostioles present; legs short; 4–5 pairs of cerarii; circulus present; anal ring with 2–3 rows of pores and 6 setae; anal lobes poorly developed; trilocular pores absent; numerous multilocular pores on both surfaces, pores larger on body margin and on terminal sternites, smaller in median area; oral collar tubular ducts on both derms; setae short and thin on dorsum, longer and more numerous on venter.

Distribution. Eastern and Central Europe.

Biology. Feed on *Carex* spp.

Note. Danzig (1960) suggested its close relationship to *Annulococcus* James, an African genus. Miller and McKenzie (1970) found similarities with *Heterococcus* Ferris, but noted the following distinguishing characters: the multilocular pores are not present on both surfaces in *Heterococcus,* also the quinquelocular pores are distributed on both surfaces in latter but are not in *Boreococcus.*

19. **Boreococcus ingricus** Danzig – Sedge Mealybug
Fig. 54

Syn. & Lit.: *Boreococcus ingricus* Danzig, 1960:117, 1964:630; Tereznikova, 1975:166; Kosztarab & Kozár, 1978:35; Koteja, 1986a (in print).

Adult females forming loose white cottony ovisacs. Adult female light pink, ca 2.5 mm long, 1 mm wide; antennae ca 250 μ long; stylets short, ca 1.5 times length of labium; all 4 ostioles present; hind coxae without translucent pores, claws with small denticle; only 4–5 pairs of cerarii, each with 2 thick spines and a group of quinquelocular and multilocular pores; 1 circulus, small, circular, diameter 43 μ; anal ring almost circular with setae ca 108 μ long; few quinquelocular pores on dorsum; three types of multilocular pores, each with 8–17 pores, smaller ones mostly in median area of venter, larger ones on dorsum and ventral submargin; star-shaped pores present singly in the cerarian groups and on body margin, 1 or 2 near spiracles; oral collar tubular ducts on both surfaces.

Distribution. In CE: Hu, Po, and USSR (Ukraine and Leningrad Region). A rare species.

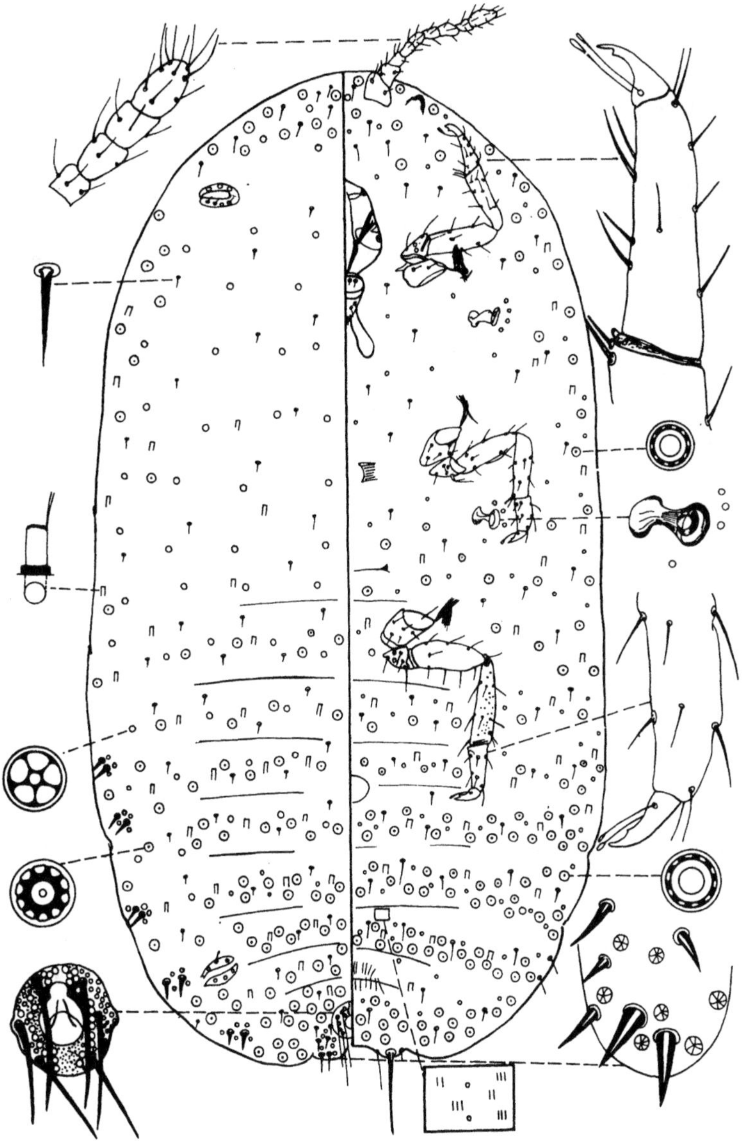

Fig. 54. *Boreococcus ingricus* Danzig, female (original)

Biology. A mesophil species. It was found at the base of leaf sheaths of *Carex pilosa* and *Carex* sp. Adult females laying eggs were observed in the fall. Their yellow eggs overwintered in cottony white ovisacs (Danzig, 1960). In Poland 3rd instars, young females and adult males were collected in August.

Genus **Brevennia** Goux

Syn. & Lit.: *Brevennia*, subgenus of *Ripersia* Goux, 1940:58; Morrison 1945:41; Borchsenius, 1949:270. Miller,* 1975:47. *Heterococcus* (in part), Miller & McKenzie, 1970:439; Koteja et al., 1978:502.

Type-species: *Brevennia tetrapora* Goux, 1940.

Female covered with white wax powder. Adult female elongate oval, derm weakly sclerotized, segmentation distinct; antennae 6- or 7-segmented (1 species with 8 or 9 segments); ostioles, on abdomen only (also on thorax in 1 species); claws without denticle (present in 1 species); tarsal digitules with acute apices; spiracles with associated trilocular pores; at least 1 pair of cerarii present; circulus absent; anal ring with 2 pore rows and 6 setae; anal lobes poorly developed, each with an apical seta; trilocular pores in small number, in anal lobe cerarii and near spiracles (absent in 1 species); quinquelocular pores on both surfaces, more numerous than other disc pores; multilocular pores along body margin and in transverse rows and bands on terminal abdominal segments: oral collar tubular ducts on both surfaces, more than twice as long as wide; short setae distributed on dorsum, longer hairlike setae on venter. All instars of *B. rehi* (Lindinger) described by Miller (1975).

Distribution. In the Palearctic and Ethiopian Regions with 8 species; 2 species in CE.

Biology. On grasses.

Note. This genus can be separated from the related *Heterococcus* as follows: in the latter the swirled trilocular pores are absent, antennae normally are 8 or 9-segmented; usually they possess two pairs of ostioles.

Key to species of *Brevennia*

Cerarii 2 pairs; claws without denticle; adult female ca 1.6 mm long; antennae 6-segmented; known from France only .... *tetrapora*, p. 79

Cerarii 3 or 4 pairs; claws with a small denticle; adult female 1.0–2.75 mm long; antennae 6- or 7-segmented; known from other parts of Europe than France ... . *pulveraria*, p. 78

20. **Brevennia pulveraria** (Newstead) – Bluegrass Mealybug

Syn. & Lit.: *Ripersia pulveraria* Newstead, 1892:145. *Dactylopius pulverarius*, Newstead, 1903:174. *Trionymus pulverarius*, Goux, 1941b:31. *Heterococcus pulverarius*, Williams, 1961:673, 1962:31; McKenzie, 1967: 191; Miller & McKenzie,* 1970:445; Koteja, 1974a:99; Ter-Grigorian, 1973:198; Tereznikova, 1975:193; Dziedzicka, 1977:66; Kosztarab & Kozár, 1978:44. *Brevennia krishtali* Tereznikova, 1962:122. *B. pulveraria*, Miller, 1975:49; Koteja et al., 1978:503.

Females and eggs covered by white wax powder. Adult female elongate oval, light yellow; according to Miller and McKenzie (1970) 1–2.7 mm long, 0.9–1.2 wide; antennae 6 or 7-segmented; ostioles on abdomen only; legs robust, dorsum of hind femora and tibiae with translucent pores, tarsal digitules setose, claws with a small denticle; cerarii 3 or 4 pairs, anal lobe cerarii each with 2 conical setae, 1 or 2 trilocular pores, a small group of quinquelocular pores and 1 or 2 auxiliary setae; circulus absent;

anal ring circular with 2 rows of pores and 6 setae; latter ca twice as long as diameter of anal ring; anal lobes poorly developed; a few simple pores scattered on both derms; trilocular pores few, mostly in median area of dorsum of prosoma, a few near spiracles on venter; quinquelocular pores of two sizes, a few distributed on both surfaces; multilocular pores on margin and on last few abdominal segments, on venter scattered on entire surface, also forming transverse rows on abdominal segments 6 to 9, in small clusters near legs and in submarginal areas on rest of body; oral collar tubular ducts elongate, a few in marginal area of dorsum, on venter associated with multilocular pores in bands and clusters; body setae short on dorsum, elongate, hairlike on venter.

Distribution. In CE: Hu, Po, and USSR; also known from England and USA.

Biology. Found between stem and leaf sheaths of *Agropyron* sp. and *Agrostis vulgaris,* also on roots of *Dactylis*. Biology not known.

Natural enemies. Parasites: *Ericydnus longicornis* and *E. sypilus* (Trjapitzin, 1978).

Note. It is often confused with *Heterococcus nudus,* but the latter has 8- or 9-segmented antennae, slender legs and no trilocular pores (Miller and McKenzie, 1970).

21. **Brevennia tetrapora** Goux – Goux's Mealybug

Syn. & Lit.: *Brevennia tetrapora* Goux, 1940:58; Tereznikova, 1962:122; Miller, 1975:47. *Heterococcus tetraporus,* Miller & McKenzie, 1970:447; Kosztarab & Kozár, 1978:44.

Adult female elongate-oval, pink, ca 1.6 mm long and 0.95 mm wide; antennae 6-segmented, ca 170 μ long; labium ca 50 μ long, 60 μ wide at base; legs robust, with or without translucent pores on hind tibiae, tarsal claw digitules setose, claws without denticle; spiracles well-developed, apparently with associated trilocular pores (Goux found some small peculiar pores probably triloculars, but structure not verified); cerarii with 2 conical setae restricted to last 2 abdominal segments; circulus absent; anal ring normal for the genus: trilocular pores probably present at least near spiracles; quinquelocular pores on both surfaces; multilocular pores along lateral margins and in transverse rows or bands on abdominal segments 5–9 of both surfaces; elongate oral collar ducts on both surfaces (Goux, 1940). First instar described by Goux (1940).

Distribution. In France only. According to Koteja et al. (1978) the record on the occurrence of this species in Poland refers to *B. pulveraria*.

Biology. In leaf sheaths of grass, *Agrostis* sp. Biology unknown.

Genus **Chaetococcus** Maskell

Syn. & Lit.: *Chaetococcus* Maskell 1898:249; Morrison & Morrison, 1922:55; Goux, 1937:94 (considered as a subgenus); Borchsenius, 1949:314; Williams, 1958:206; Danzig, 1964:625; Yang & Kosztarab, 1967:3; Ter-Grigorian, 1973:227; Tereznikova, 1975:168; Kosztarab & Kozár, 1978:36. *Antonina,* Zimmerman, 1948:149.

Type-species: *Sphaerococcus bambusae* Maskell, 1893.

Female partly or completely enclosed in a white ovisac. Derm of adult female heavily sclerotized, especially at posterior end of abdomen and on margin, oval, tapering toward apex of abdomen; antennae reduced to unsegmented stubs;

mouthparts often close to center of venter; dorsal ostioles, cerarii, legs and ventral circuli absent; spiracles large, often enclosed in a sclerotized frame or deep pocket; anal ring with 6 setae, at end of anal tube; apical setae absent or short and thick; normally simple, trilocular, and often multilocular disc pores on dorsum and venter; sievelike pores and/or ducts forming a large cluster just posterior of hind spiracles; oral collar tubular ducts with dome-shaped cap on dorsum and often on ventral margin; setae stout, short and rare, most on body margin or submargin. Adult male described by Afifi and Kosztarab (1967); nymphs by Morrison and Morrison (1922), Šulc (1944a), Borchsenius (1949) and Yang and Kosztarab (1967).

Distribution. World-wide with 3 species; 2 in CE.

Biology. On Gramineae, including bamboo, reed and grasses; common under leaf sheaths at the nodes. Probably only 1 generation per year in CE, but more in the tropics. Females develop through 4 instars, males through 5 (Yang & Kosztarab, 1967).

Key to species of *Chaetococcus*

Adult females over 3.0 mm long; multilocular pores present on both surfaces; most tubular ducts small, slender, over 1 ½ times as long as wide, located on venter only; on *Phragmites communis* .... *phragmitis,* p. 80

Adult females less than 3.0 mm long; multilocular pores absent; most tubular ducts large, but short and wide, less than 1 ½ times as long as wide, located on dorsum, few along margin of venter; on *Festuca* spp. .... *sulci,* p. 82

22. **Chaetococcus phragmitis** (Marchal) – Legless Reed Mealybug
Figs 37, 55, 56

Syn. & Lit.: *Antonina phragmitis* Marchal, 1909:871; La Face,* 1921:254; Kiritchenko, 1928:113; Balachowsky, 1932:199; Goux, 1935b:94; Afifi & Kosztarab, 1967:21, 37; Yang & Kosztarab, 1967:28, 61, 68; Gomez-Menor, 1968:557. *A. waterstoni* Newstead, 1920: 119. *C. phragmitis* Borchsenius, 1949:318; Parker, 1960:170; Danzig, 1964:626; Ter-Grigorian, 1973:227; Tereznikova, 1975:171; Kosztarab & Kozár, 1978:37.

Females incompletely covered by white waxy secretion. Adult females flat, elongate oval, yellowish, 3.0–6.6 mm long, 1.8–3.4 mm wide (larger when parasitized), last abdominal segment protruding and lobed; antennae as unsegmented cylindrical stubs, each with ca 9 setae; mouthparts on mesothorax, stylet loop short, only 2–3 times length of labium; spiracles with sclerotized anterior bar, surrounded by trilocular pores; anal ring wide with a wide band of pores, located in a small cavity at end of anal tube; anal ring hairs 60–125 μ long, extend well beyond apex of abdomen; no apical setae; few trilocular pores scattered on both derms; multilocular pores form a marginal band on both derms with segmental clusters on abdomen; minute ducts with sievelike pores form a large cluster posterior of each hind spiracle; oral collar tubular ducts scattered on venter; only few short setae distributed on derm, most along body margin; larger stout setae on anal lobe margin. All instars described by La Face (1921), adult male (fig. 37) by Afifi and Kosztarab (1967), first instar by Yang and Kosztarab (1967).

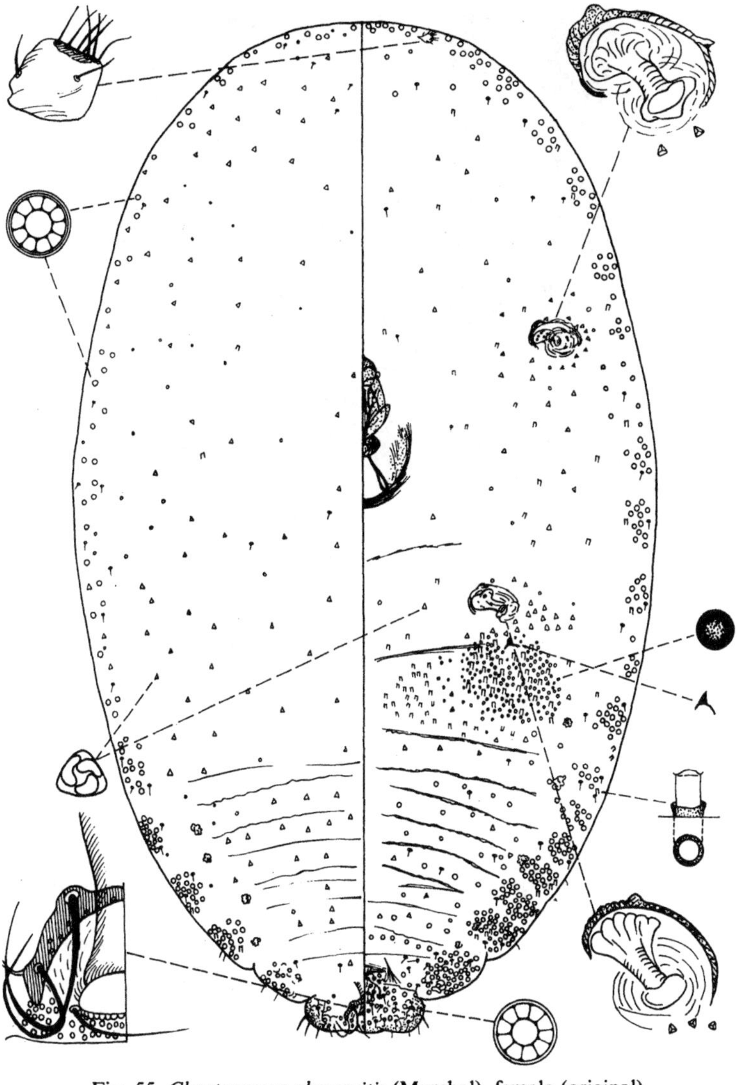

Fig. 55. *Chaetococcus phragmitis* (Marchal), female (original)

Distribution. Circum-Mediterranean. In CE: Au, Fr, Ge (?), Hu, It, USSR and Yu. Also known from Greece, Spain, Middle East, North Africa and USA. It is often intercepted in U.S. Plant Quarantine in reed products originating from Europe. It was found in reed sent from Germany and Poland. It was also recently found (Nakahara, 1975) in New Jersey, U.S.A., on common reed. Based on distribution records of its specific parasites, this species is probably distributed in more countries in CE.

Biology. Apparently feeds only on reed, *Phragmites communis*. The record from Egypt from *Arundo donax*, is questionable. It infests the leaf sheaths near the nodes on

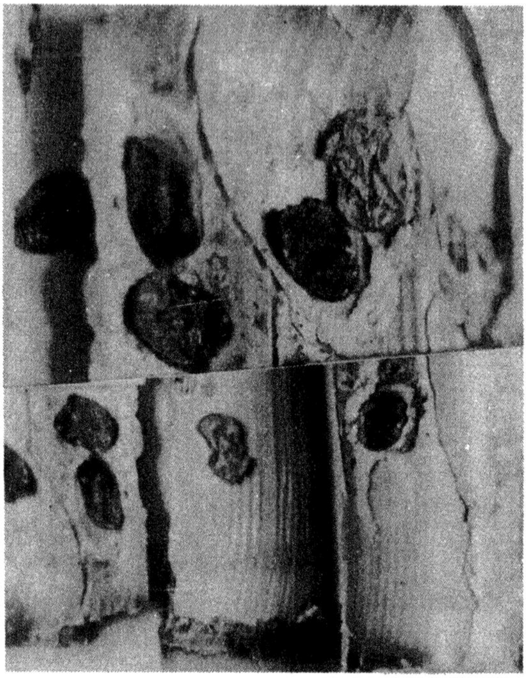

Fig. 56. *Chaetococcus phragmitis* (Marchal), old females on reed (Photo M. Kosztarab)

stem, usually up from 2nd or 3rd node above the water level. Only first instars found in the spring, thus we assume that the nymphs overwintered; they crawl from dried reed to new shoots. Only first instars have legs. Females develop by mid-July in Hungary and start laying eggs; these apparently hatch soon. Probably 1 yearly generation in CE, but La Face (1921) reported 3 generations from Italy.

Natural enemies. Parasites: *Asitus phragmitis, Boucekiella depressa, Parasyrpophagus lindus, Platencyrtus parkeri* and *Platyrhopus meridionalis; Protyndarichus balatonicus* (Kosztarab, 1959; Parker, 1960; Trjapitzin, 1978). Predators: Not certainly known. Parker (1960) observed some Lepidoptera larvae in colonies of this species.

23. **Chaetococcus sulci** (Green) – Legless Fescue Mealybug

Syn. & Lit.: *Antonina sulci* Green, 1934:510; Schmutterer, 1955:101; Danzig & Kozár, 1974:11. *C. sulci,* Goux, 1937a:94 (in subgenus *Chaetococcus*); Borchsenius, 1949:316; Reyne, 1953:235; Schmutterer, 1955b:101; Danzig, 1964: 626; Tereznikova, 1975:171; Kosztarab & Kozár, 1978:38; Koteja & Zak-Ogaza, 1983:475. *Antoninella sulci,* Šulc,* 1944a:148.

Females pealike, enclosed in a thin white ovisac. Adult females ovoid with protruding sclerotized anal lobes, sand-colored, 1.6–2.4 mm long, 1.5–2.1 mm wide; antennae as unsegmented stubs (Green, 1934, noted 2 segments) with 6–8 terminal setae; stylet loop short, extends to center of body; anterior legs absent, middle and posterior legs present as small stubs with claw; spiracles large, surrounded by a sclerotized frame and some trilocular pores; anal ring with a wide band of pores, its setae extend beyond body margin for about half their length; each apical seta short and thick; multilocular pores absent; dorsum with scattered simple and trilocular pores, venter with triloculars, also a group of minute ducts with sievelike pores posterior of hind spiracles; these ducts smaller than trilocular pores; short broad oral collar tubular ducts with dome-shaped caps on dorsum and along ventral margin; few short setae scattered on both derms, larger setae along body margin and on posterior abdominal segments of both derms; lateral ends of vulva with fingerlike apodemes when on slide. Immature stages described by Šulc (1944a), 1st instar by Borchsenius (1949).

Distribution. In CE: Cz, Fr, Ge, Hu, Po and USSR.

Biology. It is common in warm steppe habitats on root crown and at the base of stems of *Festuca* spp. Adult females were collected from the beginning of summer through late fall. Biology unknown. Apparently a biparental species (Šulc, 1944a).

Genus **Chnaurococcus** Ferris

Syn. & Lit.: *Chnaurococcus* Ferris, 1950:40; Williams, 1962:17; McKenzie, 1967:83; Ter-Grigorian, 1973:77; Tereznikova, 1975:173; Kosztarab & Kozár, 1978:38.

Type-species: *Ripersia villosa* Ehrhorn, 1899.

Adult female broad oval, often at full maturity very rotund; antennae 6- or 7-segmented; dorsal ostioles present; legs somewhat reduced, claw denticle absent, hind coxae or rarely also mid-coxae with translucent pores; cerarii confined to anal lobes; circulus present or absent; anal ring at apex of abdomen, circular, with 6 setae, each about twice as long as diameter of anal ring; anal lobes at most slightly protruding; trilocular pores evenly distributed on both surfaces; quinquelocular pores absent; multilocular pores few, mostly in midregion of abdominal venter; oral collar tubular ducts rarely of two sizes, present or absent on dorsum, present on venter of abdomen, rarely in submargin of prosoma; setae distributed on both surfaces.

Distribution. Most are known from the Holarctic Region, 1 from Colombia, and 1 from Madagascar. Of the 9 described species, 3 are in CE.

Biology. Infesting herbaceous and woody plants; often found on roots, bulbs and in ant nests.

Note. This genus is somewhat related to *Cryptoripersia* Cockerell, but in the latter the dorsal ostioles are absent and the anal ring is located at some distance from the apex of the abdomen.

Key to species of *Chnaurococcus*

1. Circulus present; with oral collar tubular ducts on both body surfaces; body grey . . . . *danzigae*, p. 84
   Circulus absent; oral collar tubular ducts restricted to abdominal sternites; body not grey 2

2. Each cerarius with 16–20 trilocular pores, 3–4 slender auxiliary setae and 2 conical setae; oral collar tubular ducts scattered, never in marginal groups on terminal abdominal sternites; antennae less than 200 μ long; body dark pink to violet .... *parvus*, p. 84

Each cerarius with 6–7 trilocular pores, 1 slender auxiliary seta and 2 conical setae; oral collar tubular ducts forming groups on margin of terminal abdominal sternites; antennae more than 200 μ long; body reddish-brown .... *subterraneus*, p. 86

24. **Chnaurococcus danzigae** Kozár & Kosztarab – Danzig's Mealybug
Fig. 57

Syn. & Lit.: *Chnaurococcus danzigae* Kozár & Kosztarab, 1976:300; Kosztarab & Kozár, 1978:38.

Adult female oval, tapering toward prosoma, gray, ca 3.0–4.0 mm long, 1.7–2.2 mm wide at third abdominal segment; antennae 6- or 7-segmented, 255–265 μ long; distance between antennal bases ca. 350 μ; stylet loop normally twice as long as labium; legs relatively short, hind coxa with 50–70 translucent pores; each anal lobe cerarius with 2 thick conical setae, 3 slender auxiliary setae and 14–20 trilocular pores; circulus small oval, ca 25 μ in diameter, not always detected; anal ring diameter 70–75 μ; with 2 rows of pores, setae 60–80 μ long; anal lobes not protruding; apical setae ca 150 μ long; trilocular pores numerous on both surfaces; multilocular pores forming transverse rows of 16–30 pores each on abdominal venter, also in groups on lateral margins of abdominal venter and dorsum; a few oral collar ducts on venter and on lateral margins of abdominal dorsum, some mixed among multilocular pores; body setae very similar on both surfaces.

Distribution. It is known only from Hungary.

Biology. This species was collected from root crowns of *Carex* sp. Adult females, eggs and first instars were present in July; males are unknown.

Note. *C. danzigae* is related to *C. villosa* (Ehrhorn), but in the latter the circulus is as large as anal ring; each cerarius has fewer than 10 trilocular pores; and the tubular ducts are scattered on the entire dorsum.

25. **Chnaurococcus parvus** (Borchsenius) – Russian Root Mealybug

Syn. & Lit.: *Pseudococcus parvus* Borchsenius 1949:158. *Ripersia corynephori* Signoret (Kiritchenko 1940a:124, a misidentification). *Trionymus parvus*, Matesova, 1968a:105. *C. parvus*, Ter-Grigorian, 1973:77; Tereznikova, 1975:173; Kozár & Kosztarab, 1976:302; Kosztarab & Kozár, 1978:39.

Adult female covered with white wax powder and a few short wax threads on abdominal apex; body oval, from dark pink to violet, ca 2.8 mm long, 1.7 mm wide; antenna 6-, rarely 7-segmented, ca 273 μ long; beak conical, stylet loop extending to anterior spiracles; ostioles well developed; hind coxae with a few translucent pores; anal lobe cerarii each with 2 conical setae, 3–4 stout auxiliary setae, and 16–20 trilocular pores; circulus absent; anal ring oval, with 2 rows of pores and 6 setae, each ca 60–70 μ long; anal lobes poorly developed; numerous trilocular pores distributed on

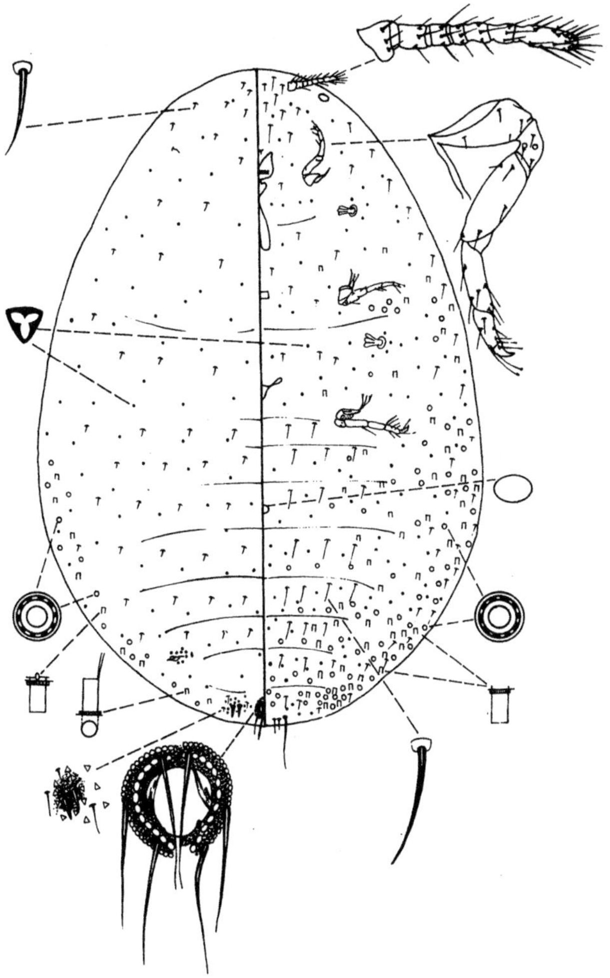

Fig. 57. *Chnaurococcus danzigae* Kozár & Kosztarab, female (modified after Kozár and Kosztarab, 1976)

both surfaces; multilocular pores in transverse rows on venter of terminal 3–4 abdominal segments; oral collar ducts few, forming transverse rows on abdominal sternites; dorsal body setae thick and short; those on venter more slender. First and second instars described by Ter-Grigorian (1973).

Distribution. A rare species. It is known from the USSR only, including the Transcarpathian Region, Armenian SSR, and Kazakhstan.

Biology. Hosts include a variety of herbaceous plants in the genera: *Agropyron, Centaurea, Festuca, Filipendula, Koeleria, Oxytropis, Phragmites, Poa, Potentilla, Stipa, Trifolium* and *Verbascum*. It was found in Armenia at up to 3,500 m elevation. Adult females were collected from May through October. Eggs and nymphs were present during most of the summer. Second instars were collected in October, thus we are assuming that N_1 and N_2 stages are overwintering (Ter-Grigorian, 1973). Males unknown. It was found in ant nests of *Lasius alienus* (Matesova, 1968a).

26. **Chnaurococcus subterraneus** (Newstead) – Root Mealybug

Syn. & Lit.: *Ripersia subterranea* Newstead, 1893:79. *R. tomlinii* Newstead, 1903:187 (in part). *R. formicarii* Newstead, 1907:5. *C. subterraneus*, Williams, 1962:17; Koteja & Zak-Ogaza, 1969:360; Ossiannilsson, 1971:53; Koteja, 1974a:100; Kozár & Kosztarab, 1976:302; Kozarzhevskaya & Reitzel, 1977:379; Kosztarab & Kozár, 1978:39.

Adult female broadly oval, reddish-brown, 1.5–2.5 mm long, 1.0–1.5 mm wide; antennae 6-segmented, according to Williams (1962) 235–260 μ long; labium 3-segmented, elongate conical, ca 112 μ long, with 14 setae (Koteja, 1974a), stylet loop slightly longer than labium; ostiole lips each with 4–5 trilocular pores and an occasional seta; anterior legs slightly longer than antennae, posterior longer, 2 very short digitules on each claw; each cerarius on anal lobes with 2 short conical setae, 1 slender seta and 6–7 trilocular pores; circulus absent; anal ring with 2 rows of pores, setae only slightly longer than diameter of ring; anal lobes moderately developed, apical setae ca twice as long as anal ring setae; trilocular pores numerous, distributed on both surfaces; multilocular pores of one size, forming single transverse rows on abdominal venter; oral collar tubular ducts of one size, forming transverse single rows and small submarginal groups on abdominal venter; body setae short and rather thick.

Distribution. A rare species. Known from Po in CE, also from Denmark, England, Scotland and Sweden.

Biology. Collected from roots of grasses, *Festuca, Nardus* and *Poa,* often in ants' nests. Adult females were noted from July to September, nymphs during fall. Males are unknown. Associated ants, *Formica flava, Lasius alienus* and *L. niger*. Lives in xerothermic rocky areas (Koteja and Zak-Ogaza, 1969).

Genus **Chorizococcus** McKenzie

Syn. & Lit.: *Chorizococcus* McKenzie, 1960:692; 1964:219, 1967:86; Williams, 1962:19; Miller & McKenzie, 1971:569; Kozár, 1983:89.

Type-species: *Chorizococcus wilkeyi* McKenzie, 1960.

Adult female elongate oval to broadly oval; antennae 6- to 8-segmented; dorsal ostioles well developed; legs variable, from small and stout to long and slender; hind coxa often with group of translucent pores at its base, claws with or without denticles; cerarii 0 to 4 pairs, normally 3 or 4 pairs on terminal abdominal segments; anal lobe cerarius with 2 conical spines, a few trilocular pores and 1 or more slender auxiliary setae; circulus present or absent; anal ring with pores and 6 setae, each seta about twice as long as diameter of anal ring; anal lobe usually poorly developed; trilocular pores usually evenly distributed on both surfaces; multilocular pores form transverse rows or bands on terminal abdominal sclerites; tubular ducts of two types, oral rim ducts always present somewhere on body, and a few oral collar ducts normally present, except in *C. irishii* (Cockerell); dorsal body setae small and slender, ventral setae usually longer than those on dorsum.

Distribution. World-wide with about 38 species; (25 in North America); 2 in CE.

Biology. On roots, leaves and stem of a wide variety of herbaceous plants; at least one species known from woody plants only (*Juniperus*); a number of species known on grasses and on *Agave*.

Key to species of *Chorizococcus*

Circulus rotund, small, not divided; 1–12 multilocular pores at each anterior leg; 7 oral rim ducts on abdominal tergite VIII; stylet loop twice as long as labium .... *rostellum*, p. 87

Circulus oval, large and divided; no multilocular pores near anterior legs; 3 to 5 oral rim ducts on abdominal tergite VIII; stylet loop shorter than labium .... *viktorina*, p. 89

27. **Chorizococcus rostellum** (Lobdell) – Bermuda-Grass Mealybug

Fig. 58

Syn. & Lit.: *Trionymus rostellum* Lobdell, 1930:219. *T. vallis* Ferris, 1950:275. *C. rostellum* McKenzie,* 1967:113; De Lotto, 1973:35.

Female covered with creamy-white wax. Adult female elongate-oval, ca 1.4–3.7 mm long, 0.5–1.7 mm wide; antennae 8-segmented, rarely 7-segmented; labium as long as wide at base, stylet loop twice as long as labium; legs slender, hind coxae with a group of translucent pores; usually only anal lobe cerarii developed, each with 2 short conical spines, 2–3 slender auxiliary setae, and without apparent concentration of trilocular pores; penultimate and antepenultimate cerarii sometimes represented by 2 long slender setae; circulus rotund and small; anal ring with 6 setae and 2 rows of pores; anal lobes weakly developed; trilocular pores distributed on both surfaces; multilocular pores form transverse bands on abdominal sternites 5–8, some also on thorax, none on head; oral rim ducts form transverse rows on most abdominal tergites, and on dorsal and ventral submargins of prosoma; oral collar ducts present in multilocular pore bands on abdominal sternites; most body setae small, slender, on venter slightly longer than on dorsum.

Distribution. A common species in North America. De Lotto (1973) reported it from Bari, Italy; recently, Kozár collected it in Hungary.

Biology. It is found between the leaf sheaths and stem, also on roots of various grasses and other herbaceous plants, also on *Platanus* and *Yucca* in N America. Specimens from Italy are from *Cynodon*, those from Hungary are from *Agropyron*.

Natural enemy. Parasite: *Anagyrus swezeyi* (Beardsley, 1969).

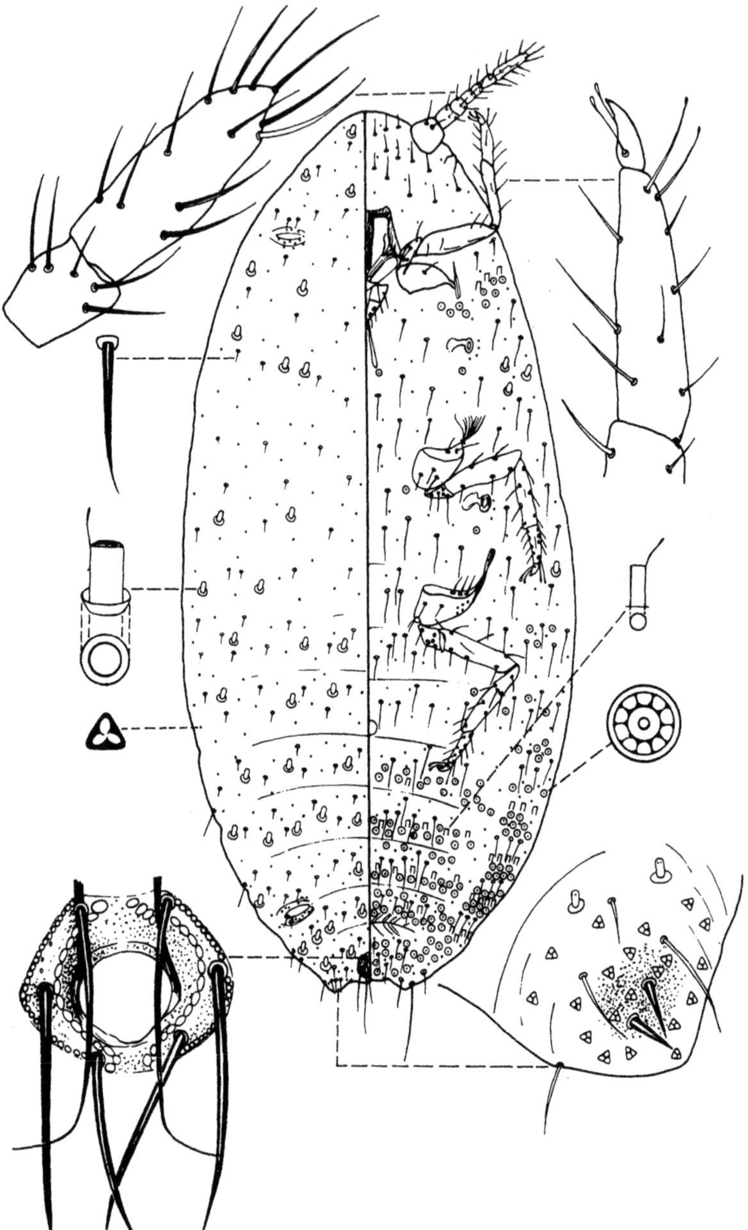

Fig. 58. *Chorizococcus rostellum* (Hoke), female (original)

28. **Chorizococcus viktorina** Kozár – Viktorin Grass Mealybug

Syn. & Lit.: *Chorizococcus viktorina* Kozár, 1983:89.

Female pink when live. Adult female elongate oval, ca 2.0–3.5 mm long, 1.0–1.5 mm wide; antennae 8-segmented, terminal segment longest; labium appears 2-segmented; stylet loop shorter than labium; legs robust, each hind coxa with 10–40 translucent pores; only anal lobe cerarii developed, these on sclerotized plate, each with 3–4 slender setae and 20–30 trilocular pores; circulus large, oval and divided; anal ring with two rows of pores, inner row with large oval pores, outer row with small circular pores; anal lobes poorly developed; trilocular pores numerous, on both body surfaces; multilocular pores few, forming 3–4 transverse rows on terminal abdominal sternites; oral rim ducts few, only 5 on abdominal tergite VIII; oral collar ducts form transverse bands on last few abdominal sternites, few scattered on venter of thorax; dorsal ostioles well-developed with trilocular pores and short setae; body setae hairlike, shorter on dorsum than on venter.

Distribution. Known from Hungary and Greece.

Biology. Inside leaf sheaths of *Chrysopogon gryllus* grass.

Genus **Coccidohystrix** Lindinger

Syn. & Lit.: *Coccidohystrix* Lindinger, 1943:218; De Lotto, 1969:2; Kosztarab & Kozár, 1978:39. *Echinococcus* Balachowsky, 1936:157 (a homonym). *Centrococcus* Borchsenius 1948:953, 1949:307; Danzig, 1964:631; Ter-Grigorian, 1973: 224; Tereznikova, 1975:166.

Type-species: *Echinococcus echinatus* Balachowsky, 1936.

Female body completely covered by thick, compact, feltlike ovisac. Adult female elongate oval, body segments distinct, dorsum normally with transverse and longitudinal rows of round or oval sclerotized plates with a group of thick spines; antenna 7–9-segmented; eyes prominent, with basal cone well developed; labium conelike; ostioles and circuli absent; legs well developed, tibia much longer than tarsus; spiracles wide, without disc pores in atrium; anal ring oval, with 2–3 rows of pores and 6 setae; each anal lobe with apical, subapical and other shorter setae; 3 types of disc pores: trilocular, quinquelocular and multilocular; all tubular ducts of invaginated type, often of 2 sizes; dorsal setae short, conelike; ventral setae hairlike.

Distribution. In the Palearctic, Ethiopian and Oriental Regions, with 5 species; in CE 1 species.

Biology. On roots, stems, twigs and leaves of woody and herbaceous plants, belonging to the genera *Artemisia, Hibiscus, Sida* and *Solanum.*

Note. The presence of dorsal sclerotized plates with conical setae brings this genus close to Putoini, but in the latter the ostioles are developed. A study of adult males will assist with the proper phylogenetic placement of this genus.

29. **Coccidohystrix artemisiae** (Kiritchenko) – Wormwood Mealybug
Fig. 59

Syn. & Lit.: *Echinococcus artemisiae* Kiritchenko, 1937:395. *Centrococcus artemisiae*, Borchsenius, 1949:308; Danzig, 1964:631; Matesova, 1968a:113; Tereznikova, 1975:168. *C. artemisiae*, Kosztarab & Kozár, 1978:40.

Ovisac snow-white, 3–4.3 mm long, and 1.5–3 mm wide. Adult female according to Borchsenius (1949) oval, yellowish-green, ca 2.5 mm long, 1.7 mm wide; antenna slender, 8-segmented, 2nd segment longest; stylet loop 1.5–2 times as long as labium; hind coxae with translucent pores; claw with small denticle; each spiracle with 3–5 associated trilocular and a few quinquelocular pores; anal ring setae 160 μ long; each anal lobe wide, with apical seta shorter than anal ring setae, 1 subapical seta, ca 65 μ long, and 2 short setae; numerous trilocular pores on dorsum, few on abdominal sternites; few quinquelocular pores, these on venter of prosoma and on abdominal sternites 2–6; multilocular pores form rows and bands on most abdominal sternites; invaginated tubular ducts form wide bands on dorsum, on venter form a marginal band and 1–2 transverse rows on middle of abdominal sternites; dorsal setae on sclerotized plates, each plate with 2–17 such stout conical setae, and 3–5 translucent pores; venter with hairlike setae.

Distribution. Known only from USSR, in the vicinity of Odessa and in Kazakhstan.

Biology. A steppe inhabiting xerophilous species living on leaves of *Artemisia* spp. Matesova (1968a) found the females in egg sacs during June in Kazakhstan. Biology unknown.

Genus **Coccura** Šulc

Syn. & Lit.: *Coccura* Šulc, 1908:64; Borchsenius, 1949:299; Danzig, 1964:631; 1980a:135; Ter-Grigorian, 1973:220; Tereznikova, 1975:173; Kosztarab & Kozár, 1978:40. *Tetrura* (preoccupied) Lichtenstein 1882:124.

Type-species: *Coccus comari* Künow, 1880.

Female covered with fine white wax powder; egg sac small, saucer-like. Adult female broadly oval; antennae 8- or 9-segmented; eyes prominent, with basal cone; labium conical; dorsal ostioles small; legs very small in comparison to body, claw with denticle; cerarii 18 pairs, usually each with 2–3 thick setae and a few trilocular pores; circuli 3 or 4, large, oval; anal ring oval, with pores and 6 setae; anal lobes only slightly developed; disc pores of 3 types: trilocular pores on both body surfaces, quinquelocular pores scattered on venter only, multilocular pores rarely absent, when present on abdominal sternites; tubular ducts of 2-types: invaginated and oral collar type, usually both on each body surface; short body setae on dorsum, longer hairlike setae on venter.

Distribution. Palearctic with 3 or 4 species; 1 in CE.

Biology. On woody and herbaceous plants. Biology of only *C. comari* was studied.

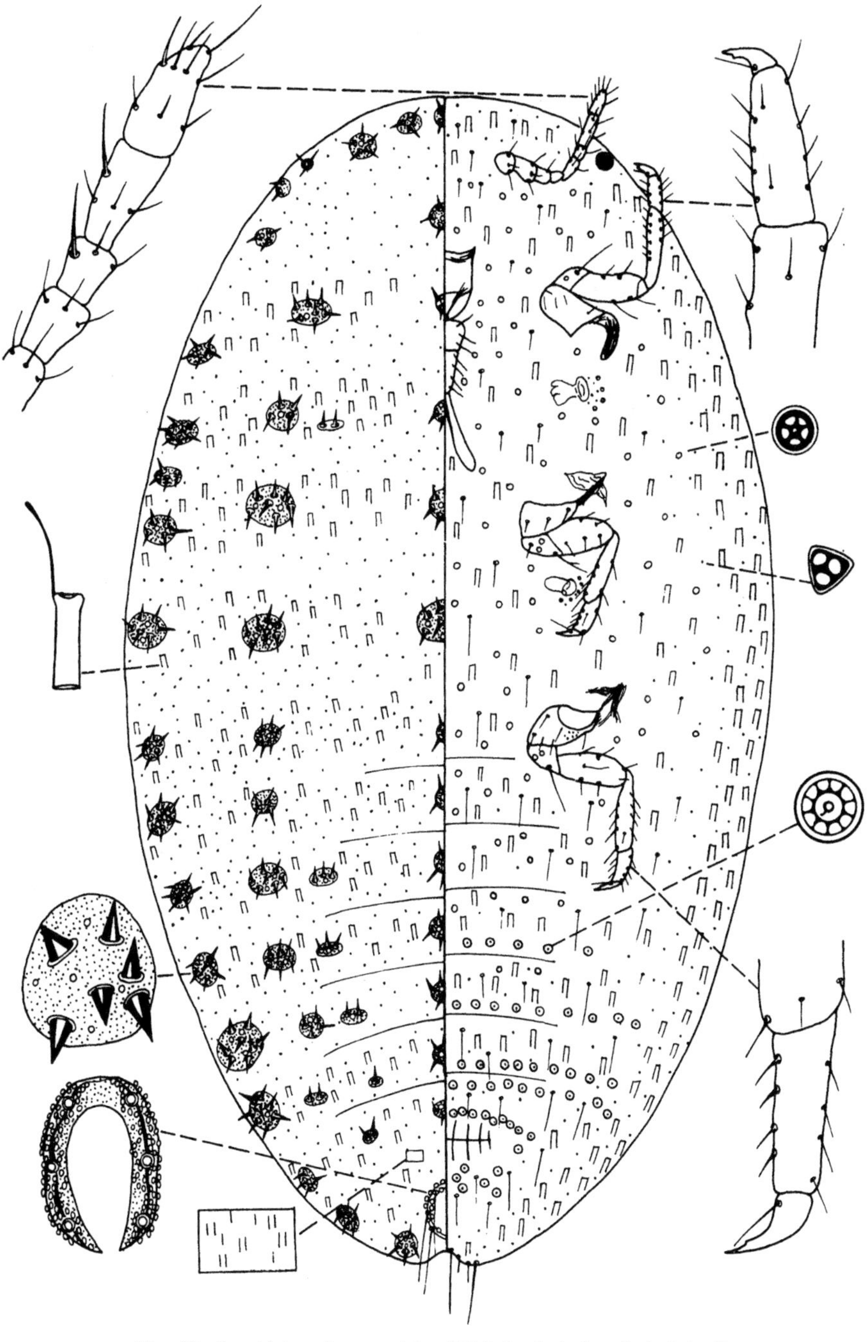

Fig. 59. *Coccidohystrix artemisiae* (Kiritchenko), female (original)

30. **Coccura comari** (Künow) – Künow's Mealybug
Figs 6, 60

Syn. & Lit.: *Coccus comari* Künow, 1880:46. *Phenacoccus comari*, *Schmutterer, 1952b:400. *Coccura comari*, Borchsenius, 1949:300; Danzig, 1964:631; Matesova, 1968a:112; Koteja, 1974a:98; Ter-Grigorian, 1973:220;* Tereznikova, 1975:175; Dziedzicka, 1977:65; Koteja, Zak-Ogaza & El-Nabawi, 1978:502; Kosztarab & Kozár, 1978: 40. *Tetrura rubi* Lichtenstein, 1882:124.

Egg sac white, saucer-like, exposing only dorsum of female. Adult female broadly oval, 3.3–7.5 mm long, globular, light brown or light gray to violet, covered with white wax powder; antennae 9- or 8-segmented, 370–400 μ long; labium almost as wide as long, 110 μ long, with 15 pairs of setae; ostioles present; legs 500, 560, 580 μ long respectively; spiracles with few associated trilocular pores; 18 pairs of cerarii, C_{18} on small circular sclerotized plate with 3–4 spines and 8–10 trilocular pores; 3 large oval circuli present; anal ring with 2 outer rows of pores and 1 inner row, anal ring setae ca 142 μ long; apical setae about as long as anal ring setae; trilocular pores few on both surfaces; quadrilocular pores sometimes near vulva and on terminal abdominal sternites; quinquelocular pores scattered on venter of prosoma and on anterior abdominal sternites; multilocular pores in transverse rows on terminal sternites of abdomen; oral collar and invaginated tubular ducts distributed on both body surfaces; body setae short on dorsum, of different lengths on venter. First instars described by Schmutterer (1952b).

Distribution. In CE is known from: Au, Cz, Fr, Ge, Hu (recent finding by Kozár), Po, USSR (from Ukraine to Middle Asia) and Yu.

Biology. Usually found on *Rubus*, but nymphs collected from *Alchemilla, Fragaria* and roots of *Sedum*. The females were found on root crowns of *Comarum, Potentilla, Rosa* and *Sanquisorba*. Schmutterer (1952b) noted egg laying to start in mid-June, nymphs hatch in a few hours, apparently second instars overwinter. It has one generation per year. Males are unknown. Occasionally reaches high density. Associated ant *Lasius alienus* (Matesova, 1968a).

Natural enemies. Parasites: *Anagyrus coccurae* (Trjapitzin, 1978); *Cerapterocerus mirabilis* (Sugonjaev, 1962); *Cheiloneurus* sp., *Pseudaphycus* sp. (Matesova, 1968a).

Genus **Dysmicoccus** Ferris

Syn. & Lit.: *Dysmicoccus* Ferris, 1950:53; McKenzie, 1960:706; 1967:156; Miller & McKenzie, 1973:503; Kosztarab & Kozár, 1978:40; Danzig, 1980a:169. *Ferrisicoccus* Ezzat & McConnell, 1956:31. *Kaicoccus* Takahashi, 1958:5.

Type-species: *Dactylopius brevipes* Cockerell, 1893.

Female body covered with white powdery waxy secretion; usually with long lateral wax filaments; normally producing a filamentous ovisac. Adult females oval, of various colors; antennae normally 8-segmented, seldom 7-segmented; labium elongate conical; with 4 dorsal ostioles; legs usually short, but fully developed, claws without denticle; normally with 15–17 pairs of cerarii (seldom fewer), cerarius normally with more than 2 conical setae, and usually with slender auxiliary setae and a cluster of

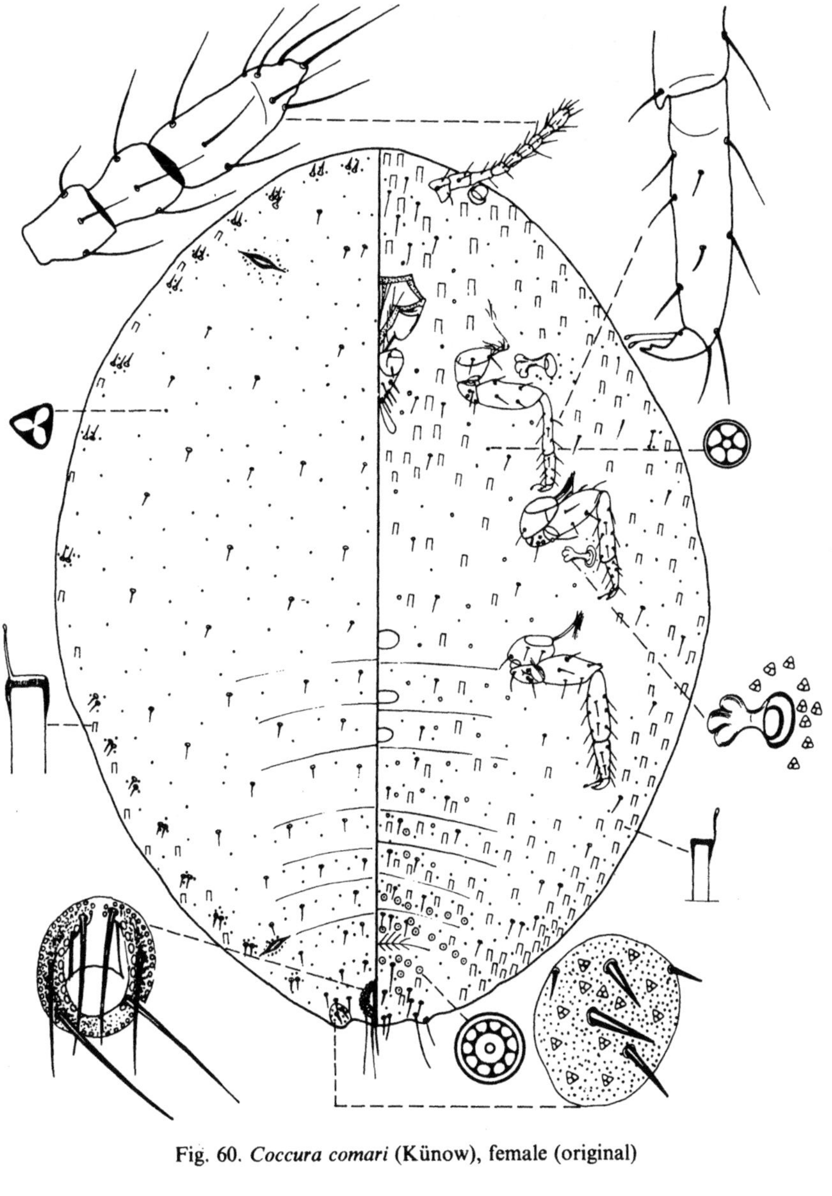

Fig. 60. *Coccura comari* (Künow), female (original)

trilocular pores; circulus present or absent; anal ring usually large with pores and 6 setae, each ca twice as long as diameter of anal ring; minute circular pores often on dorsum; trilocular pores distributed on both surfaces; at least a few multilocular pores form transverse rows on venter; oral collar tubular ducts at least on abdominal sternites; oral rim ducts absent; body setae of various shapes and sizes.

Distribution. World-wide, with several dozen species; ca 30 in North America; only 1 in CE.

Biology. Found on a variety of woody and herbaceous plants and can occur on all parts of vegetation. Many species infest grasses. Yearly generations vary from 1 to 7. Most are oviparous, some ovoviviparous. Except for a few species, biologies are poorly known.

Note. The genus needs a revision. Some species are difficult to separate from others. See Reviewer's remark under *Trionymus*.

31. **Dysmicoccus walkeri** (Newstead) – Walker's Mealybug
Fig. 61

Syn. & Lit.: *Dactylopius walkeri* Newstead, 1891:164. *Pseudococcus rectus* Borchsenius, 1949:136. *P. walkeri*, Schmutterer, 1952b:338. *D. walkeri*, Williams, *1962:21; Danzig, 1972a:331, 1980a:171; Koteja, 1974a:98; Dziedzicka, 1977:67; Kosztarab & Kozár, 1978:41.

Female covered by white wax powder; margin with cottony appendages, those on body apex longer and stouter than others. Adult female elongate oval, brown, 3.2–4.5 mm long, ca 1.2 mm wide; antennae normally 8-segmented, (7 in our specimens from Yugoslavia); according to Williams (1962) 410–515 μ long; labium 117–130 μ long, with 14 setae; ostioles with 3–4 trilocular pores and 2–3 setae on each lip; legs large, 840, 900, 1010 μ long respectively; 15–17 pairs of cerarii, anal lobe cerarii each with 2 conical setae, ca 10 auxiliary setae and many trilocular pores, all borne on a sclerotized area larger than anal ring; circulus small, oval, not divided; anal ring with 2 pore rows, 6 setae, ca. twice as long as diameter of ring; anal lobes poorly developed; trilocular pores evenly distributed on both body surfaces; a few multilocular pores on penultimate tergite, numerous and forming transverse bands on abdominal sternites, also a few on venter of prosoma; oral collar tubular ducts of two sizes on both body surfaces, smaller ones only on posterior segments of both surfaces; numerous slender body setae on both body surfaces.

Distribution. Trans-Palearctic. In CE: known from Fr, Ge, It, Po, USSR (to Far East), and Yu (Kozár's recent collection). Also known from Great Britain and Mongolia. A rare species.

Biology. Its hosts are various herbaceous plants: *Agropyron, Agrostis, Brachypodium, Calamagrostis, Calluna, Carex, Dactylis, Elymus, Holcus, Molinia, Piptochaetium* and *Poa*. On leaves and in leaf sheaths. Probably the eggs overwinter. Second instars were found during first half of June and third instars at end of June. Adult females appeared during second half of July and at beginning of August. Apparently eggs are laid in September (Schmutterer, 1952b). Males unknown.

Genus **Euripersia** Borchsenius

Syn. & Lit.: *Euripersia* Borchsenius, 1948b:955; 1949:179; Williams, 1962:24; Ter-Grigorian, 1973:94; Tereznikova, 1975:175; Kosztarab & Kozár, 1978:41; Danzig, *1971:372; 1980a:140.

Type-species: *Euripersia amnicola* Borchsenius, 1948.

Body of adult female covered by white powdery wax; egg sac globose or oval, compact, white. Adult female oval, strongly convex; antennae 6–7-segmented; eyes

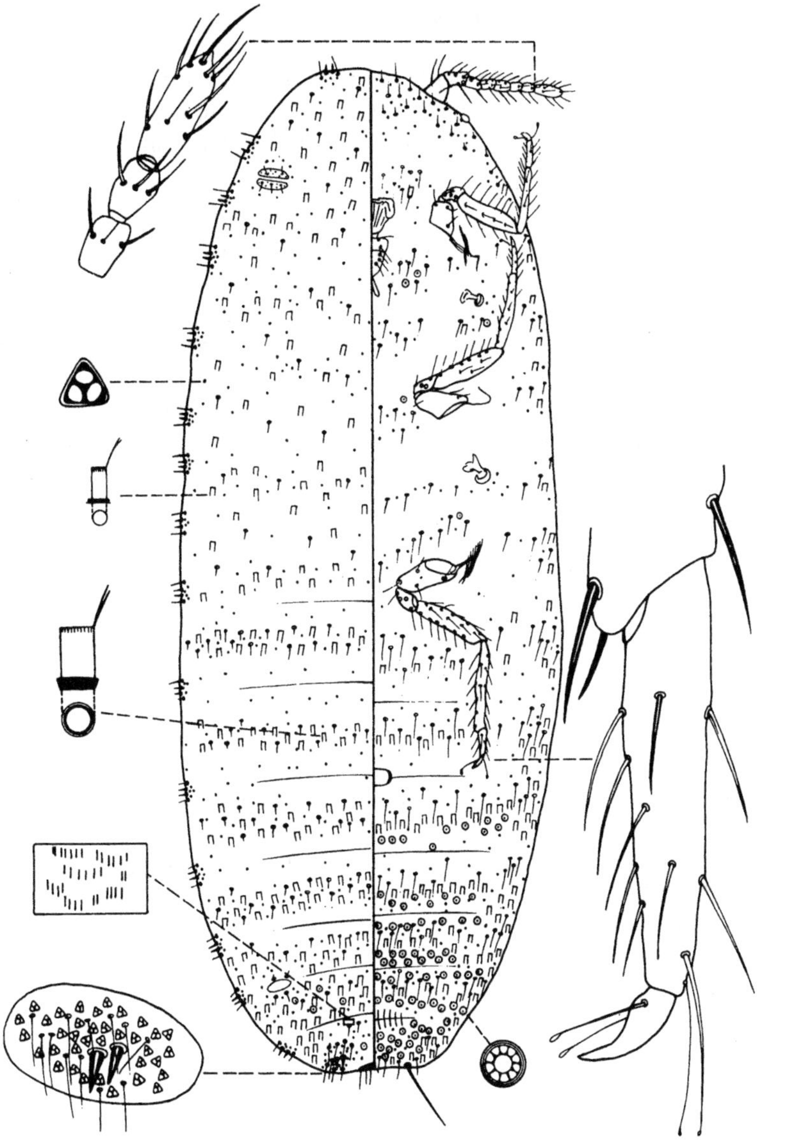

Fig. 61. *Dysmicoccus walkeri* (Newstead), female (original)

normal; labium broadly conical; ostioles present; legs small, claw normally with denticle, rarely absent, hind coxae rarely with translucent pores; at most 4 pairs of cerarii; circuli 1–3 or absent; anal ring with 2 rows of pores and 6 setae; anal lobe broad, with stout apical seta; trilocular pores on both surfaces; quinquelocular pores usually present on venter of prosoma; multilocular pores on abdominal sternites; oral collar ducts sometimes on venter only; body setae of various sizes, normally longer on venter.

First instar described by Borchsenius, 1949:180; three nymphal stages by Ter-Grigorian (1973).

Distribution. Palearctic Region, with 7 species; in CE 2 species.

Biology. Six of the 7 species feed on roots of Gramineae, in steppe regions of Eurasia. With up to 3 generations a year; viviparous or oviparous. Males unknown.

Key to species of *Euripersia*

Circuli absent; anal ring setae as long as diameter of ring; oral collar tubular ducts on venter only; quinquelocular pores forming groups near bases of coxae and antennae . . . . *europea,* p. 96

Circuli (2–3) present; anal ring setae over twice as long as diameter of anal ring; oral collar tubular ducts on both surfaces; quinquelocular pores few, only in the area of mouthparts . . . . *tomlini,* p. 97

32. **Euripersia europea** (Newstead) – Grass Mealybug

Syn. & Lit.: *Ripersia europea* Newstead, 1897b:167; Schmutterer, 1952b:391. *R. wuenni* Reyne, 1953:235. *E. brevispina* Borchsenius & Ter-Grigorian, 1956:17; Ter-Grigorian, 1973:99; Kosztarab & Kozár, 1978:42. *E. europea,* Williams,* 1962:24; Koteja & Zak-Ogaza, 1969:351; Danzig, 1971b:373; Tereznikova, 1975:178; Kosztarab & Kozár, 1978:42.

Female body densely covered with white powdery wax. Adult female oval, yellowish-pink or pink, ca 2.5 mm long; antenna 6-segmented, 225–274 µ long; labium 90–113 µ long, with 30 setae, stylet loop as long as labium; each ostiole lip with ca 4 trilocular pores; claw with or without denticle; 2–4 pairs of cerarii, each anal lobe with 2 slender setae and 5–15 trilocular pores; circuli absent; anal ring setae slightly longer than ring diameter; anal lobes poorly developed; trilocular pores on entire surfaces; quinquelocular pores in groups on venter, near bases of antennae and at coxae; multilocular pores on abdominal sternites 7–10; oral collar ducts on entire surface; ventral setae slender, longer than those on dorsum. Two nymphal stages described by Ter-Grigorian (1973).

Distribution. A common species. In CE: Ge, Hu, Po, USSR (from Ukraine to Armenia); also known from England, Holland, and Mongolia.

Biology. Common on roots of grasses and on other herbaceous plants, often found under rocks and visited by ants. On *Agropyron, Agrostis, Corynephorus, Deschampsia, Festuca, Poa, Secale, Thymus, Trifolium* and *Zerna.* In Armenia present at 3,000 m elevation (Ter-Grigorian, 1973). In Germany overwinters as N_2, N_3; egg laying females observed from end of May through early June, each laid 28–78 eggs; eggs hatched in a few hours; second generation females appeared in July, third generation females in September (Schmutterer, 1952b). According to Ter-Grigorian (1973) this species (*"E. brevispina"*) does not lay eggs, but gives birth to young. Males unknown. Associated ants: *Lasius flavus, L. niger* (Williams, 1962).

Natural enemy. Parasite: *Dusmetia euripersiae* (Ter-Grigorian, 1973).

Note. There are some morphological differences between the samples described as *E. europea* and *E. brevispina* (Kosztarab and Kozár, 1978); their taxonomic status needs to be further investigated, especially in the light of the biological differences (lay eggs or give birth to young) noted by Schmutterer (1952b) and Ter-Grigorian (1973).

33. **Euripersia tomlini** (Newstead) – Tomlin's Mealybug
Fig. 62

Syn. & Lit.: *Ripersia tomlinii* Newstead, 1892:146. *R. exul* Green, 1924:46. *R. mesnili* Balachowsky, 1934b:67; *Pseudococcus bufo* Kiritchenko, 1936:149. *Phenacoccus agropyri* Borchsenius 1937:52. *Phenacoccopsis agropyri* Borchsenius, 1948:954, 1949:200. *Phenacoccopsis nasonovi* Borchsenius, 1949:201; *Phenacoccopsis bufo* Borchsenius, 1949:202. *E. tomlini*, Williams,* 1962:26; Danzig, 1971:373, Koteja, 1972:567, 1974a:99; Tereznikova, 1975:180; Kosztarab & Kozár, 1978:41; Kozarzhevskaya & Reitzel, 1975:15. *Phenacoccopsis tomlini*, Kozár 1984: 127.

Female ovisac compact, globose or ovate, white, diameter ca 5 mm. Adult female broadly oval, about 2.6–3.5 mm long; antennae 6–7-segmented, according to Williams (1962) 230–260 μ long; stylet loop slightly longer than labium; each ostiole lip with ca 3 setae and rarely with 1 small seta; legs 435–540 μ long, hind tibia almost 1½ times as long as tarsus, claw with denticle; 2 pairs of cerarii, on anal lobes each with 2 small conical setae and ca 5 trilocular pores; 2–3 circuli, circulus on fourth segment much smaller than others, rarely absent; anal ring with 3 rows of pores, setae over twice as long as diameter of ring; anal lobes poorly developed; each lobe with 1 long apical and a shorter subapical seta; trilocular pores on both surfaces; a few quinquelocular pores in area of mouthparts; multilocular pores forming single transverse rows on most abdominal sternites; oral collar ducts scattered on both surfaces; setae on dorsum sparse, those on venter longer than those on dorsum. First instar described by Tereznikova (1975).

Distribution. A Palearctic species. In CE: Fr, Ge, Hu, Po, USSR (Transcarpathian Region to Eastern Siberia); also known from Denmark, England, Holland and Mongolia. Rarely a mass infestation.

Biology. On roots of a variety of herbaceous plants, especially grasses: *Agropyron, Arrhenatherum, Artemisia, Atropis, Avenastrum, Corynephorus, Dactylis, Deschampsia, Elymus, Festuca, Galium, Koeleria, Phleum, Poa* and *Stipa*. It was found at up to 3000 m elevation in Armenia (Ter-Grigorian, 1973.) In Germany it has 1 generation a year; N_2 overwinters; females develop by June and start laying an average of 450 eggs per female from mid-July (Schmutterer, 1952b). Associated ant: *Lasius flavus* (Kozarzhevskaya and Reitzel, 1975).

Natural enemy. Parasite: *Ericydnus longicornis* (Schmutterer, 1955c).

Note. Henriksen (1921) places *E. tomlini* as a synonym of *Ripersia corynephori* Signoret. Based on the detailed description of *R. corynephori* by Reyne (1951), the two species are distinct.

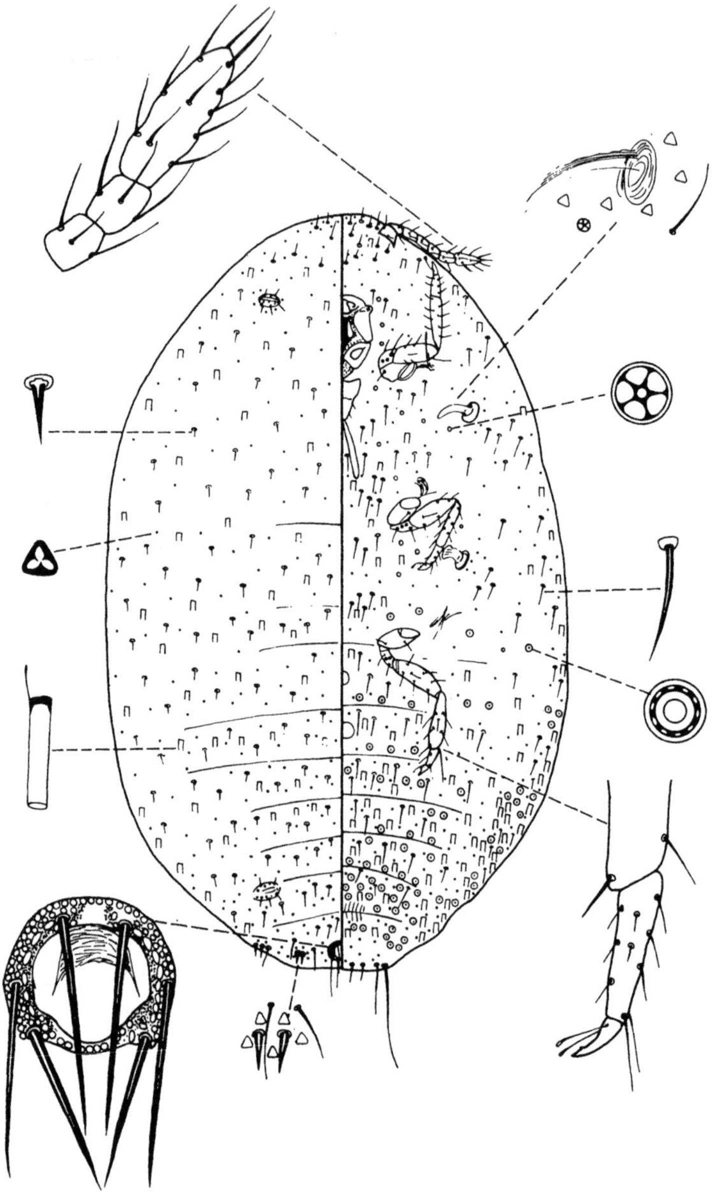

Fig. 62. *Euripersia tomlini* (Newstead), female (original)

Genus **Heliococcus** Šulc

Syn. & Lit.: *Heliococcus* Šulc, 1912:39; Borchsenius, 1949:271; Ferris, 1950:96; Balachowsky, 1953e:236; Goux, 1953:104; Williams, 1962:28; Danzig, 1964:630; 1980a:145; Matile-Ferrero 1970:176; Ter-Grigorian, 1973:200; Miller,* 1974:177; Bazarov, *1974a:78, *1974b:97; Koteja, 1974a:98; Tereznikova, 1975:182; Kosztarab & Kozár, 1978:42.

Type-species: *Heliococcus bohemicus* Šulc, 1912.

Adult female oval; antennae normally 9-segmented; labium elongate cone-shaped with 30 setae; with 2 pairs of ostioles; hind legs with or without translucent pores, each claw with a large denticle and a pair of capitate digitules; cerarii 2–18 developed, normally 17 or 18 pairs; with 2 or 3 conical setae in each cerarius; normally 1 circulus; anal ring with 3–5 rows of pores and 6 setae; anal lobes large, protruding, usually sclerotized; small simple disc pores scattered on both surfaces; trilocular pores on entire derm, except medial and mediolateral areas of venter; quinquelocular pores normally restricted to venter and confined to medial and mediolateral areas; multilocular pores, when present, normally restricted to vulvar area of venter; crateriform tubular ducts with orifices borne at apex of sclerotized, projecting tubercles, present, which gives the genus special characteristics; oral collar tubular ducts usually present, restricted to venter. Nymphal stages described by Schmutterer (1952b), Goux (1931b, 1953) and Miller (1974); adult male by Goux (1931b).

Distribution. Almost exclusive to the Holarctic Region, except 1 questionable species (*phaseoli* Laing) known from the Ethiopian Region; a total of 47 species and subspecies, of these 39 in the Palearctic, in the CE 6 species.

Biology. On both woody and herbaceous plants, including lichens and mosses, usually on roots, stems, trunk, twigs, and leaves. Biology poorly known. Schmutterer (1952b) and Zahradnik (1972) reported one yearly generation for two species; nymphs overwintered; reproduction was ovoviviparous. Males in most species unknown.

Key to Species of *Heliococcus*

1. With only 1 type (large) of crateriform duct; without cerarii on prosoma (including *H. nivearum austriacus*), ... *nivearum*, p. 102
   With 2 to 4 types of crateriform ducts; with cerarii on entire body margin. 2
2. Crateriform ducts of 2 sizes 3
   Crateriform ducts of 3 or 4 sizes 5
3. Each large crateriform duct surrounded by a group of 35–60 small crateriform ducts .... *cydoniae*, p. 100
   Large crateriform ducts not surrounded by a group of 35–60 small crateriform ducts 4
4. Anal lobes, cerarii and bases of large crateriform ducts sclerotized; multilocular pores present on abdominal sternites .... *danzigae*, p. 102
   Anal lobes, cerarii and bases of large crateriform ducts not sclerotized; multilocular pores absent .... *radicicola*, p. 103

5 (2). Small crateriform ducts with 1 (rarely 2) spinule; antenna 1/5 to 1/8 the length of body; on herbaceous plants only .... *sulci*, p. 103
Small crateriform ducts without spinules; antenna less than 1/5 the length of body; on trees, shrubs and woody vines only .... *bohemicus*, p. 100

34. **Heliococcus bohemicus** Šulc – Bohemian Mealybug
Fig. 63

Syn. & Lit.: *Heliococcus bohemicus* Šulc, 1912b:40; Goux, 1934:170; Borchsenius, 1949:275; Schmutterer, 1955b:101; Danzig, 1964:631, 1980a:151; Zahradnik, 1972:399; Ter-Grigorian, 1973:203; Danzig & Kozár 1974:11; Bazarov, 1974b:101; Tereznikova, 1975:184; Kosztarab & Kozár, 1978:43; Bognár & Vinis, 1979:15; Kawecki, 1985:15. *H. stacyos,* Kiritchenko, 1931:313. *H. hystrix* Kiritchenko, 1932:261.

Body of adult female with protruding glassy wax filaments, giving it a porcupine-like appearance. Adult female oval, pink, up to 3.5 mm long, 2 mm wide; antenna 9-segmented, ca 740 μ long; stylet loop 2–2½ times longer than labium; hind coxae without translucent pores; with 18 pairs of cerarii, C_{18} with 2 conical setae and 8–11 trilocular pores; 1 circulus, elongate oval; anal ring with 2, rarely 3 rows of pores on outer margin and 1 row on inner margin with setae ca 200 μ long; anal lobes protruding, apical seta ca 245–330 μ long; trilocular pores on entire dorsum, on venter of abdominal sternites and in a wide marginal band on prosoma; quinquelocular pores in median area of venter; multilocular pores in transverse rows on abdominal sternites 6 to 8; according to Borchsenius (1949) crateriform ducts of 3 sizes; large ones, ca 40 μ long, 8–9 μ wide in median and marginal longitudinal band on dorsum, middle-sized ducts, with 1–2 spinules, ca 33 μ long, 5.5 μ wide, in pairs along middle line of abdominal tergites 7 and 8, small ducts without spinules, ca 31–36 μ long and 2.5 μ wide form a wide band along margin on both surfaces of body, also in a group on frons, and in transverse rows on thorax and on tergites of abdomen; oral collar ducts singly or in transverse rows on abdominal sternites.

Distribution. Trans-Palearctic. In CE: Cz, Fr, Ge, Hu, Po, USSR (European part to Far East), and Yu. Also known from Spain and Sweden. Rare.

Biology. On trunk and branches of *Aesculus, Calluna, Caragana, Catalpa, Pinus, Platanus, Populus, Quercus, Robinia,* and *Vitis.* Occasionally a pest on grape vines. Reproduces ovoviviparously in June and July–August in Poland. No records on males. Biology poorly known.

Natural enemy. Parasite: *Leptomastix histrio* (Zahradnik, 1972).

35. **Heliococcus cydoniae** Borchsenius – Quince Mealybug

Syn. & Lit.: *Heliococcus cydoniae* Borchsenius, 1949:272; Danzig, 1964:630; Ter-Grigorian, 1973:201; Bazarov, 1974b:97; Tereznikova, 1975:183; Kosztarab & Kozár, 1978:43.

Body of female with protruding thin glassy wax filaments of rainbow color. Adult female oval, yellowish pink, up to 4 mm long, 2.7 mm wide; antennae 9-segmented, ca 776 μ long; labium elongate conical, stylet loop as long as labium; hind coxae with a few translucent pores; 18 pairs of cerarii, each C_{18} with 2 slender setae, ca 30–33 μ long,

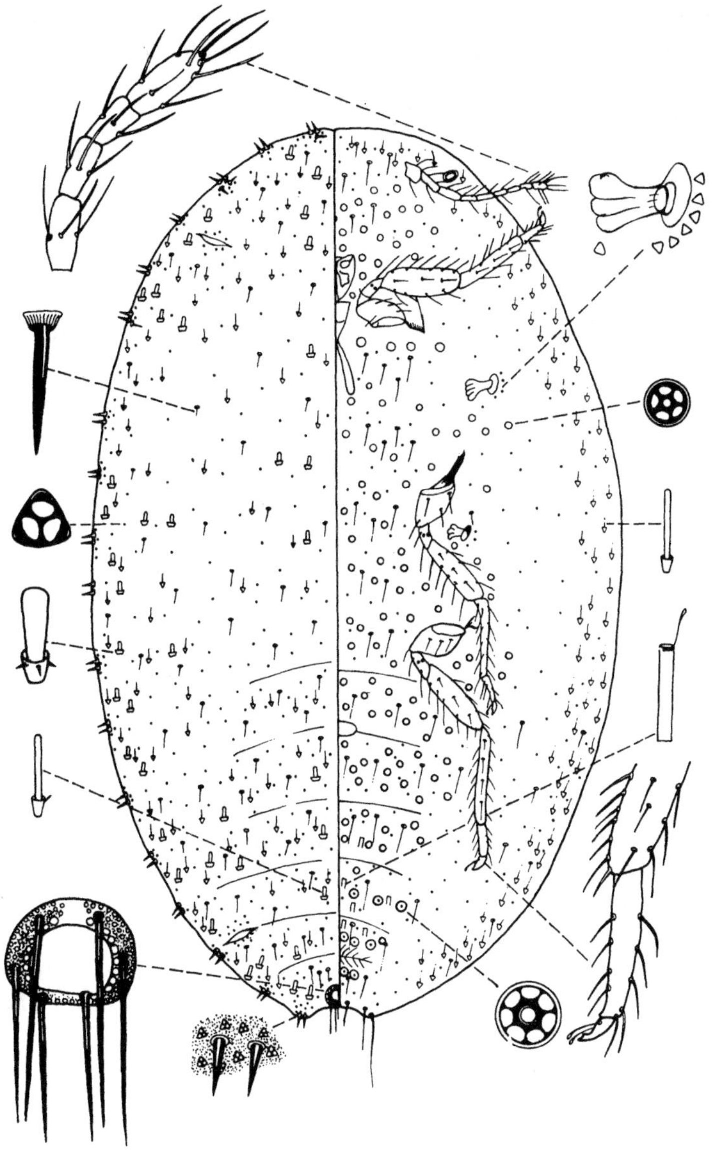

Fig. 63. *Helicoccus bohemicus* Šulc, female (original)

and 12–14 trilocular pores; anal ring with 3, rarely 4 outer rows of pores and 1 row on inner margin, its setae 200–225 μ long, twice as long as ring diameter; apical setae of anal lobes ca 280 μ long; trilocular pores on both surfaces; quinquelocular pores on all sternites; multilocular pores in groups around vulva; large crateriform ducts, ca 44 μ long, with 4–5 (seldom 3–7) spinules each, forming 5 longitudinal bands on dorsum; small crateriform ducts, ca 28 μ long, without spinules, ca 35–60, grouped around each large crateriform duct on dorsum, and in large groups on margin of venter; few oral collar ducts on venter; body setae of various lengths and thicknesses on both surfaces.

Distribution. Known only from Hu and USSR (including the Transcarpathian Region). Infestations rarely in masses (Danzig, 1972c).

Biology. A polyphagous species on roots and branches of herbaceous and woody plants, such as: *Agropyron, Cydonia, Genista, Lathyrus, Rosa* and *Trifolium*. Adult females were collected from June through September (Ter-Grigorian, 1973). Biology and males unknown.

36. **Heliococcus danzigae** Bazarov – Bazarov's Mealybug

Syn. & Lit.: *Heliococcus danzigae* Bazarov, *1974a:81; Danzig, 1977a:199, 1978:9, 1980a:150; Koteja et al., 1978:502.

Adult female oval, according to Bazarov (1974a) 2.6–3.5 mm long, ca 1.6 mm wide; antenna 9-segmented; stylet loop slightly longer than labium; legs large, hind tibiae with translucent pores; 18 pairs of cerarii on sclerotized areas, C_{18} with 2 conical setae and 12–14 trilocular pores; circulus narrowly oval; anal ring with 2–3 outer rows and 1 inner row of pores, its setae ca 140 μ long; anal lobes with sclerotized tip, apical seta ca 270 μ long; trilocular pores evenly distributed on dorsum, form a marginal band on venter and occur singly on abdominal sternites 5–8; quinquelocular pores on almost entire venter; multilocular pores form transverse rows on abdominal sternites 6 and 7, and 2 small groups on sternite 8; large crateriform ducts, 46 μ long, 7–9 μ wide, with 2–3, rarely 4 spinules on sclerotized platelets form a single or double interrupted row along dorsal margin; small crateriform ducts with or without spinules, 40 μ long, 2 μ wide, distributed on almost entire body surface; oral collar tubular ducts form transverse rows on sternites 3–7, a few on venter of prosoma, also a few on abdominal sternites; body setae, up to 11 μ long on entire dorsum and along ventral margin.

Distribution. Known from Hu, Po, and USSR (Leningrad area and Far East), also Mongolia.

Biology. A polyphagous species feeding on leaves of herbaceous plants and on the bark of woody plants, including *Bergenia, Carex, Empetrum, Hieracium, Larix, Rosa, Rubus, Solidago* and *Ulmus*. Biology unknown.

37. **Heliococcus nivearum** Balachowsky – Snow Mealybug

Syn. & Lit.: *Heliococcus nivearum* Balachowsky, 1953e:238, 1953b:295; 1953a:93; Zahradnik 1957:5; Tereznikova, 1959b:796; Miller, 1974:178. *Heliococcus nivearum austriacus* Balachowsky, 1953e:240.

Adult female oval, pinkish, ca 1.5 mm long, 0.9 mm wide; antenna 9-segmented, ca 130 μ long; labium elongate conical; 4 ostioles distinct; legs slender, tibia longer than tarsus, claws with a small denticle; spiracles with few associated trilocular pores; cerarii on sclerotized plates, each with 2 conical setae and ca 6 trilocular pores, except C_{18}, which has 3 setae and more trilocular pores, cerarii absent on prosoma; circulus small rotund; anal ring setae about twice as long as diameter of anal ring; anal lobes protruding, with long apical seta, and 3–4 large crateriform ducts on dorsal surface; trilocular pores uniformly distributed on entire dorsum, few on venter, these near spiracles; only in *H. n. austriacus* forming 2 transverse rows along mesothoracic suture; quinquelocular pores on dorsum, others unevenly distributed on most of venter; multilocular pores few, only in the perivulvar area; only large crateriform ducts, a total of 80–100, these ca 25 μ long, with 2–3 lateral spinules on dorsum forming a marginal row, with few scattered in median area and along submargin, on venter few on margin of prosoma; oral collar tubular ducts entirely absent.

Distribution. Rare. Collected only in the French and Austrian Alps at 3,410 and 3,600 m elevations.

Biology. On *Androsace, Saxifraga,* also lichens and mosses. Adult females collected in July and August. Biology unknown.

Note. The subspecies *H. n. austriacus* of Austria is separated from *H. nivearum* of France by the presence of quinquelocular pores on dorsum in the first one.

38. **Heliococcus radicicola** Goux – Pink Root Mealybug

Syn. & Lit.: *Heliococcus radicicola* Goux, 1931b:113; Borchsenius, *1949:279; Schmutterer, *1952b:402; Ossiannilsson, 1959:195; Danzig, 1964:631, 1971:388; Zak-Ogaza & Koteja, 1964:420 (misidentification); Ter-Grigorian, 1973:206; Bazarov, 1974b:77; Koteja, 1974a:98 (misidentification); Tereznikova, 1975:184; Kosztarab & Kozár, 1978:43; Koteja et al., 1978:501.

Adult female when fully developed, oval, pink, 2–2.6 mm long, 1.2–1.4 mm wide; antenna 9-segmented, ca 420 μ long; stylet loop reaches line of anterior spiracles; hind coxae without translucent pores; 18 pairs of cerarii, C_{18} with 2 conical setae 11–16 μ long and 8–10 trilocular pores; circulus large, oval; anal ring with 2 inner and 1 outer pore row, setae ca 130 μ long; anal lobes protruding, apical setae ca 200–215 μ long; trilocular pores scattered on entire dorsum, and in a band along ventral margin; quinquelocular pores in median area of venter; without multilocular pores; large crateriform ducts with 1 or 2 spinules form longitudinal rows on dorsum, small ones without spinules on both body surfaces; oral collar tubular ducts singly on abdominal sternite 7 or absent; body setae hairlike. First and second instars and adult male described by Goux (1931b), first instars also by Schmutterer (1952b).

Distribution. Palearctic. A rare species. In CE: Fr and USSR (Crimea and Armenia); also reported from Sweden. An earlier report on its occurrence in Poland has been discounted by Koteja et al. (1978).

Biology. On roots and stems of a variety of herbaceous plants: *Dianthus, Potentilla, Poterium, Rumex, Sanguisorba, Taraxacum, Teucrium,* and *Thymus.* Adult females were collected in July, first and second instars during August; reproduction ovoviviparous (Schmutterer, 1952b).

39. **Heliococcus sulci** Goux – Šulc's Mealybug

Syn. & Lit.: *Heliococcus sulcii* Goux, 1934:167; Borchsenius, 1949:283; Reyne, 1953:235; Danzig, 1964:819; Koteja, 1974a:98; Tereznikova, 1975:188; Kosztarab & Kozár, 1978:43. *H. sulci,* Bazarov, 1974b:109. *H. slavonicus* Borchsenius & Tereznikova, 1959b:491. Bazarov, 1974b:112; Koteja et al., 1978:502.

Adult female oval, up to 2.8 mm long, 1.5 mm wide; antennae 9-segmented, ca 350 μ long; eyes large, dome-shaped; stylet loop reaches line between coxae of middle legs; legs very large, hind coxae without translucent pores; 18 pairs of cerarii, C_{18} with 2 conical setae, ca 18 μ long, and 8–11 trilocular pores; 1 oval circulus; anal ring with 2 partly double rows of pores on outer and 1 row on inner margin, its setae about 1 ½ times as long as diameter of anal ring, 110–130 μ long; anal lobes protruding, semicircular, apical setae ca 265 μ long; trilocular pores numerous on dorsum, along ventral margin, few in mid-line of venter; quinquelocular pores few, located on venter in middle area of prosoma and on last 2 abdominal sternites; multilocular pores form short transverse rows on abdominal sternites 6 to 8; crateriform ducts of 3 sizes: large ducts, ca 33 μ long, 6 μ wide, with 2–3 spinules, on anal lobes only; middle-sized ducts ca 33 μ long, 4 μ wide with 2 spinules form transverse and marginal rows on dorsum; small ducts, ca 31 μ long and 3 μ wide with 1 (rarely 2) spinule, interspersed with middle-sized ducts in the transverse rows on dorsum and form a row along ventral margin, and groups at bases of mid-and hind coxae; oral collar ducts in transverse rows on abdominal sternites 3 to 7; dermal setae stout, of various lengths.

Distribution. A rare species. In CE: Fr, Ge, Hu, It, Po, and USSR.

Biology. On roots and stems of steppe inhabiting herbaceous plants, such as: *Artemisia, Dianthus, Euphorbia, Galium, Genista, Ranunculus, Sedum, Thymus* and *Trifolium*. Adult females collected during June–August in Ukraine (Tereznikova, 1975). Biology unknown.

Genus **Heterococcus** Ferris

Syn. & Lit.: *Heterococcus* Ferris, 1918:65, 1953:382; Morrison, 1945:38; Borchsenius, 1949:266; *Williams 1961:671, 1962:29; Miller and McKenzie, 1970:438; Ter-Grigorian, 1973:198; Tereznikova, 1975:188; Miller, 1975:1; Kosztarab & Kozár, 1978:44; Danzig, 1980a:143.

Type-species: *Heterococcus arenae* Ferris, 1918.

Adult female elongate oval; antennae with 8 or 9 segments, rarely with 7, but with 6 in *abludens* Borchsenius; all 4 ostioles developed (absent in 1 species); legs slender, usually hind femur and tibia, rarely coxa, with translucent pores, claw with denticle; at least 1 pair of cerarii developed; circulus usually absent, rarely developed; anal ring with 2 pore rows and 6 setae; anal lobes poorly developed; minute simple pores usually present; swirled trilocular pores absent; quinquelocular pores on dorsum and venter, smaller than multilocular pores and more numerous than other pores combined; multilocular pores usually on abdomen, seldom on head; oral collar ducts on dorsum and venter often of two sizes, at least twice as long as wide; setae on dorsum short and thick, on venter long, hairlike. Adult male and immature stages described by Miller and McKenzie (1970) and Miller (1975).

Distribution. In the Holarctic and Ethiopian Regions with about 15 species; 2 in CE.

Biology. All species known are from Gramineae; feeding on leaf sheaths, rarely on roots.

Note. It is closely related to *Brevennia*. See distinguishing characters at that genus. Comparison with other related genera found in Miller and McKenzie (1970).

Key to Species of *Heterococcus*

Multilocular disc pores present on dorsum; cerarii 5–8 pairs; anal lobe cerarii not on basal sclerotization; hind coxae without translucent pores; oral collar ducts of one size; anal ring setae normally ca 1 ½ times as long as diameter of ring .... *nudus*, p. 105

Multilocular disc pores absent on dorsum; cerarii 1–4 pairs; anal lobe cerarii often on basal sclerotization; hind coxae with many translucent pores; oral collar ducts of two sizes; anal ring setae not longer than diameter of anal ring ... *tritici*, p. 105

40. **Heterococcus nudus** (Green) – Naked Grass-Mealybug
Fig. 64

Syn. & Lit.: *Phenacoccus nudus* Green, 1926:172. *Heterococcus borchsenii* Morrison, 1945:48; Koteja & Zak-Ogaza, 1966:317. *H. graminicola* Morrison, 1945:48. *H. occidentalis* Morrison, 1945:53. *H. variabilis* Schmutterer, 1958:18. *H. pulverarius* Newstead 1892:145 (in part), Williams, 1961:673; McKenzie, 1967:191; Danzig, 1968:502. *H. nudus* Miller & McKenzie, 1970:443; Miller, 1975:11; Tereznikova, 1975:190; Koteja et al., 1978:503: Kosztarab & Kozár, 1978:45; Danzig, 1980a:143.

Female covered with a white wax bloom; venter producing a loose filamentous ovisac; eggs yellow. Adult female very elongate oval, almost parallel-sided, light yellow, according to Miller (1975) 1.2–3.3 mm long, 0.4–1.6 mm wide; antennae with 8 or 9 segments, rarely with 7, 248–318 μ long; hind femur and tibia with numerous translucent pores; cerarii 5 to 8 pairs; circulus normally absent; anal ring setae usually 1 ½ times as long as diameter of ring; few simple discoidal pores scattered on both surfaces of prosoma; quinquelocular pores of 1 or 2 sizes distributed on entire derm; multilocular pores in a marginal band on both surfaces, and in transverse rows or bands on both surfaces of abdomen; oral collar ducts of 1 size, normally absent from median area of dorsum; forming transverse rows on terminal 4th, 5th or 6th abdominal sternites; body setae short on dorsum, long hairlike on venter. Immature stages of both sexes and adult male described by Miller (1975).

Distribution. Holarctic. In CE: Ge, Hu, It, Po and USSR. It is known from England and USA (Miller, 1975).

Biology. A mesophilous species. It is oligophagous; known from *Agropyron, Agrostis, Alopecurus, Arrhenatherum, Briza, Bromus, Elymus, Festuca, Hordeum, Lolium, Phleum, Poa, Setaria* and *Sieglingia*. It infests the leaf sheaths and the crown of grasses. Tereznikova (1975) observed nymphs overwintering; adult females and eggs were found during the summer months.

Note. This species is closely related to *H. abludens* Borchsenius but the latter species has only 6-segmented antennae, and the multilocular disc pores are absent from the prosoma.

41. **Heterococcus tritici** (Kiritchenko) – Wheat Mealybug

Syn. & Lit.: *Trionymus tritici* Kiritchenko 1932:136. *Heterococcus confertus* Borchsenius, 1949:267; Miller & McKenzie, 1970:442. *H. tritici*, Borchsenius, 1949:268; Miller & McKenzie, 1970:447; Tereznikova, 1975:190: Miller. 1975:16; Kosztarab & Kozár, 1978:45.

Female covered with waxy powder; ovisac snow white. Adult female elongate oval, pale-pink; according to Miller (1975) 2.4–4.0 mm long, 0.8–1.4 mm wide; antennae normally 9-segmented, rarely 8, 270–358 μ long; hind coxae enlarged and with many translucent pores, fewer pores on hind femur and tibia; cerarii 1 to 4 pairs; anal lobe cerarii often on basal sclerotization; anal ring setae slightly shorter than diameter of anal ring; minute simple pores scattered over dorsum, rare on venter; numerous quinquelocular pores of one-size abundant on both derms; multilocular pores absent

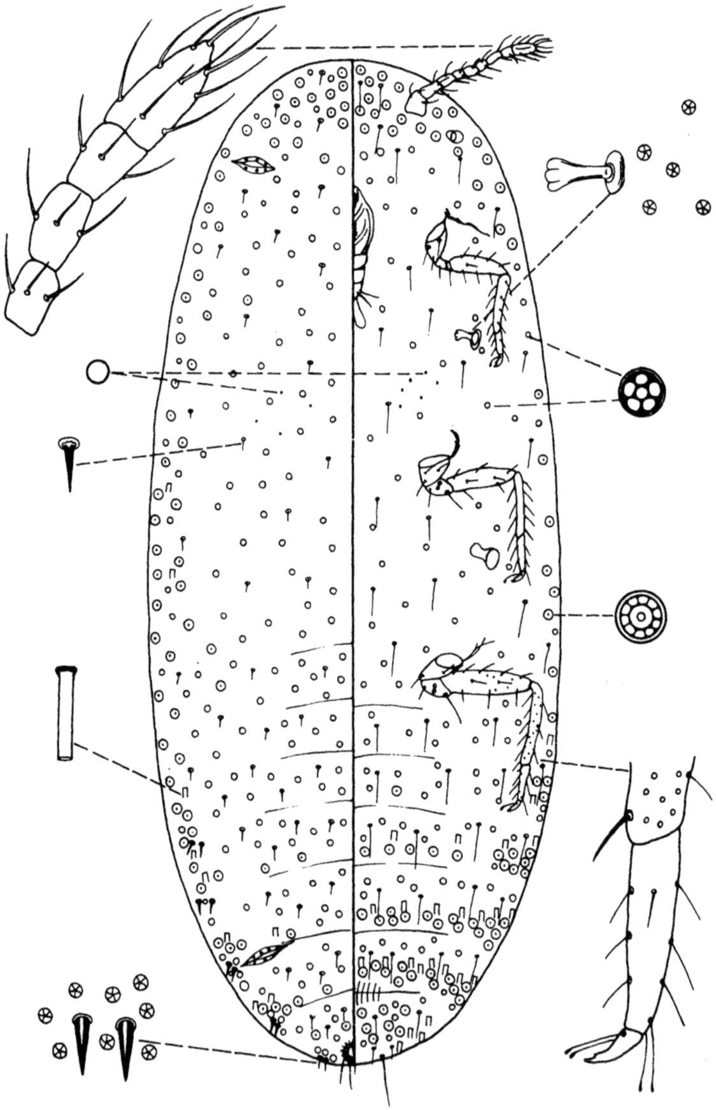

Fig. 64. *Heterococcus nudus* (Green), female (original)

on dorsum except 1 or 2 7-locular pores on body margin, on venter abundant over entire surface; oral collar ducts of two sizes on both surfaces, few on dorsum, more on ventral submargin and on abdominal sternites; dorsal setae short, bristle-like, on venter long, hair-like.

Distribution. In USSR only, Ukraine, Tadzhik SSR and Turkmen SSR.

Biology. Infesting Gramineae; *Agropyron, Bromus, Secale* and *Triticum*. Occurring in leaf sheaths. Biology unknown.

Genus **Longicoccus** Danzig

Syn. & Lit.: *Longicoccus* Danzig, 1975b:49. *Mirococcus* (in part) Borchsenius, 1949:181; Koteja, 1971a:3.

Type-species: *Mirococcus clarus* Borchsenius, 1949.

Adult female elongate oval, body segmentation distinct. Antennae 7- to 9-segmented; eyes distinct with short basal cone; 2 pairs of ostioles, their lips with or without trilocular pores; hind legs with translucent pores; claw with denticle; cerarii absent or only C_{17} and/or C_{18} developed; circulus absent; anal ring with 6 setae and few pores; anal lobes small but distinct; minute simple pores present or absent; trilocular pores evenly distributed on both surfaces of body; quinquelocular pores absent; multilocular pores on posterior sternites; oral collar tubular ducts on both body surfaces; body setae short, needle-like or slender on dorsum, elongate hairlike on venter.

Distribution. Palearctic Region with 7 species; in CE 2 species.

Biology. On roots and leaf sheaths of steppe grasses.

Key to species of *Longicoccus*

Antennae 9-segmented; dorsal tubular ducts in transverse rows on most segments; hind coxae with a few small indistinct translucent pores .. *psammophilus*, p. 108

Antennae 7-segmented; dorsal tubular ducts on last abdominal tergite only; hind coxae without translucent pores .... *festucae*, p. 107

42. **Longicoccus festucae** (Koteja) – Koteja's Mealybug

Syn. & Lit.: *Mirococcus festucae* Koteja, *1971a:14, 1974a:98; Koteja & Liniowska, 1976:662; Kosztarab & Kozár, 1978:48. *Longicoccus festucae*, Danzig, 1975b:49.

Adult female elongate oval, according to Koteja (1971a), up to 1.5 mm long, 0.5 mm wide; antenna 7-segmented, ca 200 μ long; eye diameter ca 30 μ; labium 60–66 μ long with 30 setae, stylet loop ca 200 μ long; each spiracle with ca 10 associated trilocular pores; legs small, hind legs 310 μ long, claw digitules setiform, shorter than claw, claw with a small denticle; small, indistinct semitranslucent loculi few on hind femur and tibia; circulus absent; anal ring pentagonal, with a single row of pores; anal lobes with small dorsal areas of sclerotization, and reduced C_{18}; each cerarius with 2 stout setae, 1 hairlike seta and ca 5 trilocular pores; simple disc pores and trilocular pores on both body surfaces; multilocular pores relatively few on last 5 abdominal sternites; oral collar tubular ducts few, on dorsum only on last abdominal tergite, on venter in small groups along body margin and a few in transverse rows on last 4 sternites; dorsal setae up to 15 μ long, ventral setae hairlike, up to 57 μ long.

Distribution. Known only from Hu and Southern Poland.

Biology. Under the leaf sheaths of *Festuca pallens*. Adult females and nymphs present in September. Biology unknown.

43. **Longicoccus psammophilus** (Koteja) – Sand-loving Mealybug
Fig. 65

Syn. & Lit.: *Mirococcus psammophilus* Koteja, *1971a:8, 1974a:98; Koteja & Liniowska, 1976:662; Kosztarab & Kozár, 1978:47. *Longicoccus psammophilus*, Danzig, 1975b:49.

Adult female elongate, nearly parallel-sided, according to Koteja (1971a), yellowish-brown, covered by fine wax powder, 3.5–4.8 mm long, 1–1.4 mm wide; antennae 9-segmented, 310–370 μ long; diameter of eyes ca 28 μ; labium 68–76 μ long with 30 setae, stylet loop ca 160 μ long; claw with large denticle; few very small translucent pores on leg segments; cerarii indistinct; circulus absent; anal ring circular, diameter ca 50 μ, with 2 rows of pores; anal lobes not sclerotized, C_{18} with 2–3 stout setae and ca 8 trilocular pores, apical seta ca 110 μ long; small simple disc pores, trilocular pores and oral collar ducts on both body surfaces; multilocular pores in transverse bands on terminal abdominal sternites; dorsal setae up to 25 μ long, ventral setae up to 95 μ long.

Distribution. Known only from Baltic shore of Poland, and from Hungary, if the record by Zak-Ogaza (1966) on *Mirococcus clarus* applies to this species (Koteja, 1971a).

Biology. Living under the leaf sheaths of *Deschampsia, Festuca,* and other grasses on the sand dunes near the Baltic Sea Coast, where adults are present by the first half of September (Koteja, 1971a).

Genus **Metadenopus** Šulc

Syn. & Lit.: *Metadenopus* Šulc, 1933:1; Lindinger, 1935:122; Goux, 1941b:41; Borchsenius, 1949:185; Tereznikova, 1975:193; Kosztarab & Kozár, 1978:46. *Longisomus* Kiritchenko, 1931:392 (a nomen nudum).

Type-species: *Metadenopus festucae* Šulc, 1933.

Adult female oval to elongate oval; antennae with 6 thick segments; only posterior ostioles developed; some hind leg segments with translucent pores, claws without denticle; cerarii absent; circuli present or absent; anal ring setae longer than diameter of anal ring; anal lobes reduced, with 1 to 3 long apical setae; trilocular and quinquelocular pores absent; multilocular pores on both body surfaces; tubular ducts of one kind, few; only short setae on body, except on anal lobes.

Distribution. Palearctic, with 2 species; in CE 1 species.

Biology. On roots and leaf sheaths of grasses.

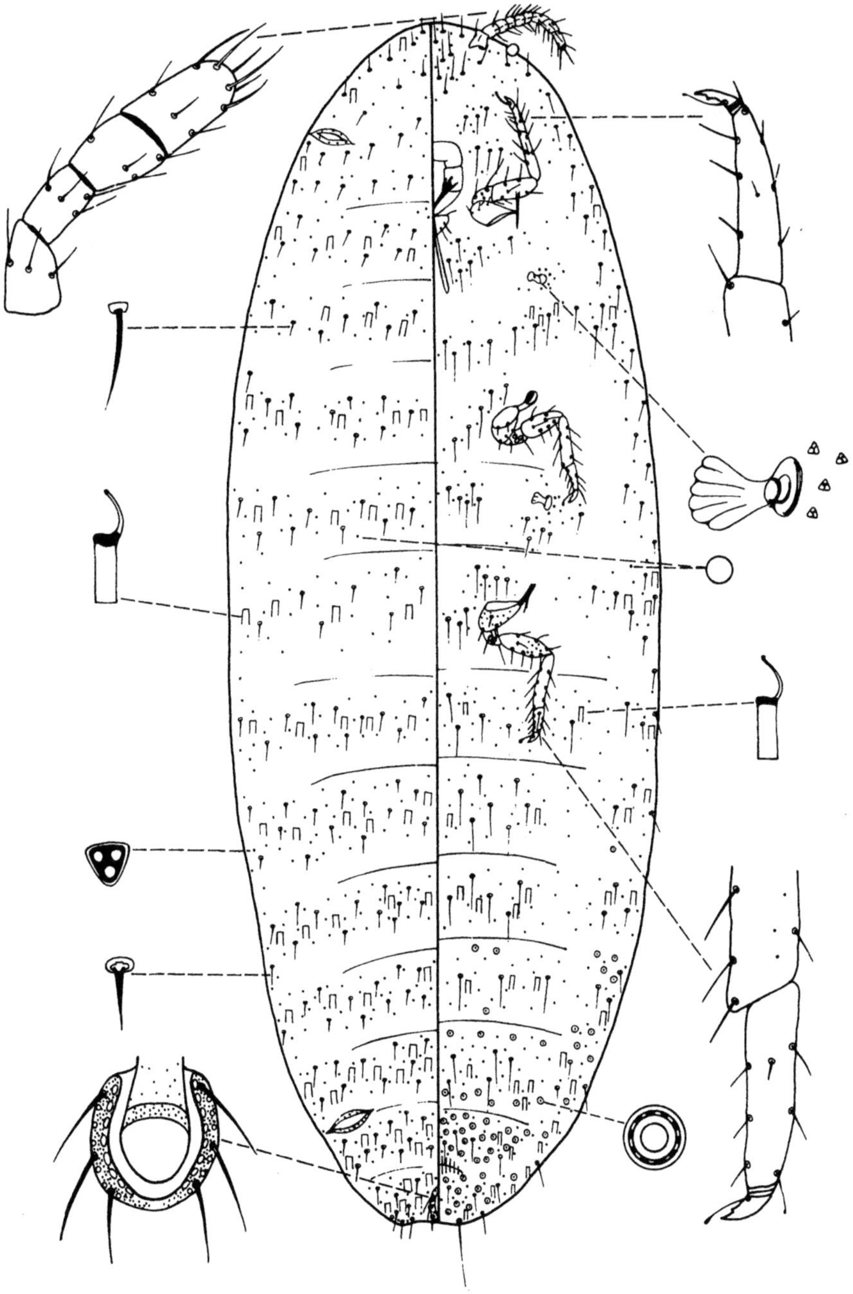

Fig. 65. *Longicoccus psammophilus* (Koteja), female (original)

44. **Metadenopus festucae** Šulc – Long Mealybug
Fig. 66

Syn. & Lit.: *Metadenopus festucae* Šulc, 1933:2; Lindinger, 1935:122; Borchsenius, 1949:186; Danzig, 1964:626, 1972:332; Koteja & Zak-Ogaza, 1969:360; Koteja, 1972:568; Tereznikova, •1975:194; Kosztarab & Kozár, 1978:46. *Longisomus festucae* Kiritchenko, 1931:312 (a nomen nudum); Goux, 1937c:93. *Metadenopus festuceli* Goux, 1941b:43.

Female covered by white powdery wax; ovisac of loose wax filaments. Adult female elongate oval, almost parallel-sided, pinkish, up to 2 mm long, 0.45–0.6 mm wide; antennae, ca 130 μ long; labium ca 50 μ long with 28 setae; stylet loop reaches line between mid-coxae; hind femur swollen, this and tibia with translucent pores; circuli absent; anal ring circular with 6 setae, and with a partly double row of sparsely spaced pores, setae 60–70 μ long; 3 apical setae, each 60–80 μ long; multilocular pores form a wide band on ventral margin, scattered on rest of body; oral collar tubular ducts on ventral margin and on last 3 sternites. Nymphs and adult male described by Šulc (1933).

Distribution. A rare species. In CE: Cz, Fr, Hu, Po and USSR (Ukraine), also known from Mongolia.

Biology. On roots and in leaf sheaths of grasses: *Elymus* and *Festuca*. Usually on warm rocky slopes. Šulc (1933) reported copulation in Moravia during mid-August, females overwintered and layed eggs following June, these hatched in July.

Natural enemies. Parasite: *Chamaemyia juncorum* reared from undetermined *Metadenopus* species (Papp, 1979). Šulc (1933) also observed predation by undetermined Diptera larvae during August 1932.

Genus **Mirococcopsis** Borchsenius

Syn. & Lit.: *Mirococcopsis* Borchsenius, 1948c:582, 1949:168; Danzig, 1964:627; Ter-Grigorian, 1973:86; Tereznikova, 1975:194; Bazarov & Nurmamatov, 1975:94; Kosztarab & Kozár, 1978:46.

Type-species: *Mirococcopsis rubidus* Borchsenius, 1948.

Adult female oval or elongate-oval; body covered by white wax powder; antennae 6- to 8- (rarely 9-) segmented; eyes present; labium wide at base; posterior ostioles developed, anteriors present or absent; legs well-developed, hind coxae normally with translucent pores; tibia longer than tarsus, claw denticle absent; spiracles rarely with associated trilocular pores; cerarii absent; circulus present or absent; anal ring with single or double row of pores and 6 setae; anal lobes wide with 1–2 long apical setae and 3 or more associated shorter setae; trilocular pores, multilocular pores and 1 type of oral collar duct present; body setae of variable size.

Distribution. Trans-Palearctic, with 6 species; in CE 3 species.

Biology. Primarily on roots and in leaf sheaths of grasses, but also on other herbaceous plants.

Note. It appears from the original description that this genus might be identical to *Volvicoccus* Goux (1945), but since no material was available for study on the latter, their status will need to be investigated in the future.

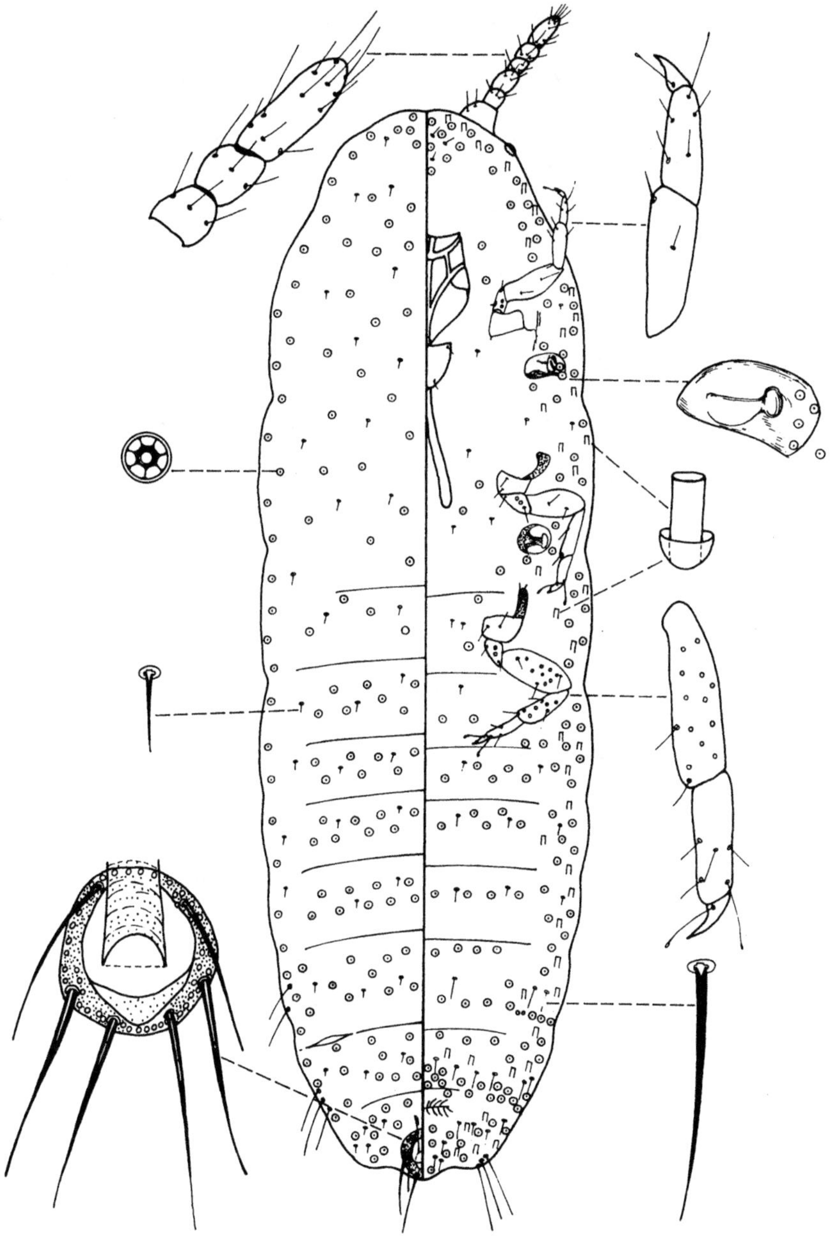

Fig. 66. *Metadenopus festucae* Šulc, female (original)

Key to species of *Mirococcopsis*

1. Antennae 8-, rarely 9-segmented; spiracles with associated trilocular pores .... *nagyi,* p. 112
   Antennae 6- or 7-segmented; spiracles without associated trilocular pores . 2
2. Antennae 7-segmented; anal ring setae ca 114 μ long; multilocular pores on head at bases of antennae .... *stipae,* p. 114
   Antennae 6-segmented; anal ring setae ca 50 μ long; no multilocular pores on head .... *elongatus,* p. 112

45. **Mirococcopsis elongatus** Borchsenius – Elongate Mealybug

Syn. & Lit.: *Mirococcopsis elongatus* Borchsenius, 1948c:583, 1949:171; Danzig, 1964:627.

Adult female elongate-oval, pinkish-violet, up to 2.1 mm long, 0.7 mm wide; antennae stout, 6-segmented; labium as long as wide, stylet loop reaches line between anterior spiracles; lips of dorsal ostioles without setae; leg segments thick, hind coxae with numerous translucent pores; circulus oval or quadrangular, small; anal ring broad hexagonal, with 1 row of large oval, sparsely spaced pores and 3 pairs of smaller pores on outer margin, setae ca 50 μ long; each anal lobe with 2 dorsal setae, ca 25 μ long, and 3–4 shorter ventral setae; trilocular pores distributed on both body surfaces; multilocular pores located singly on abdominal tergites 6 and 7 and on sternite 5 of abdomen, but forming 1–2 sparse transverse rows on abdominal sternites 6–8; oral collar ducts located singly on margin of abdominal segments 2 to 8; body hairs few, short and slender.

Distribution. Known only from Hungary and USSR (Chkalovskaja District and Aktjubinsk).

Biology. In leaf sheaths of *Stipa capillata.* Oviposition and hatching observed during first half of August in Hungary. Biology unknown.

46. **Mirococcopsis nagyi** Kozár – Nagy's Mealybug
Fig. 67

Syn. & Lit.: *Mirococcopsis nagyi* Kozár, 1981:315.

Adult female elongate oval, almost parallel-sided, 3–6 mm long, 1.5–2 mm wide; antennae 8-segmented, 220 μ long; stylet loop ca 60 μ long; ostioles reduced, anterior pair hardly detectable; legs ca 330 μ long, leg segments with few, short and thin setae; each spiracle with 7–10 associated trilocular pores; in area of C_{18} two hairlike setae and 3 trilocular pores; circulus absent; anal ring oval, ca 50 μ in diameter, with setae ca 50 μ long and inner row of large oval pores and some circular pores in outer row; anal lobes not developed; trilocular pores numerous on both body surfaces; multilocular pores form 3 or 4 transverse rows and bands on terminal abdominal sternites and a band along margin of abdomen; oral collar tubular ducts of 2 sizes, numerous on venter, but only on last few tergites of abdomen; body setae short and thin.

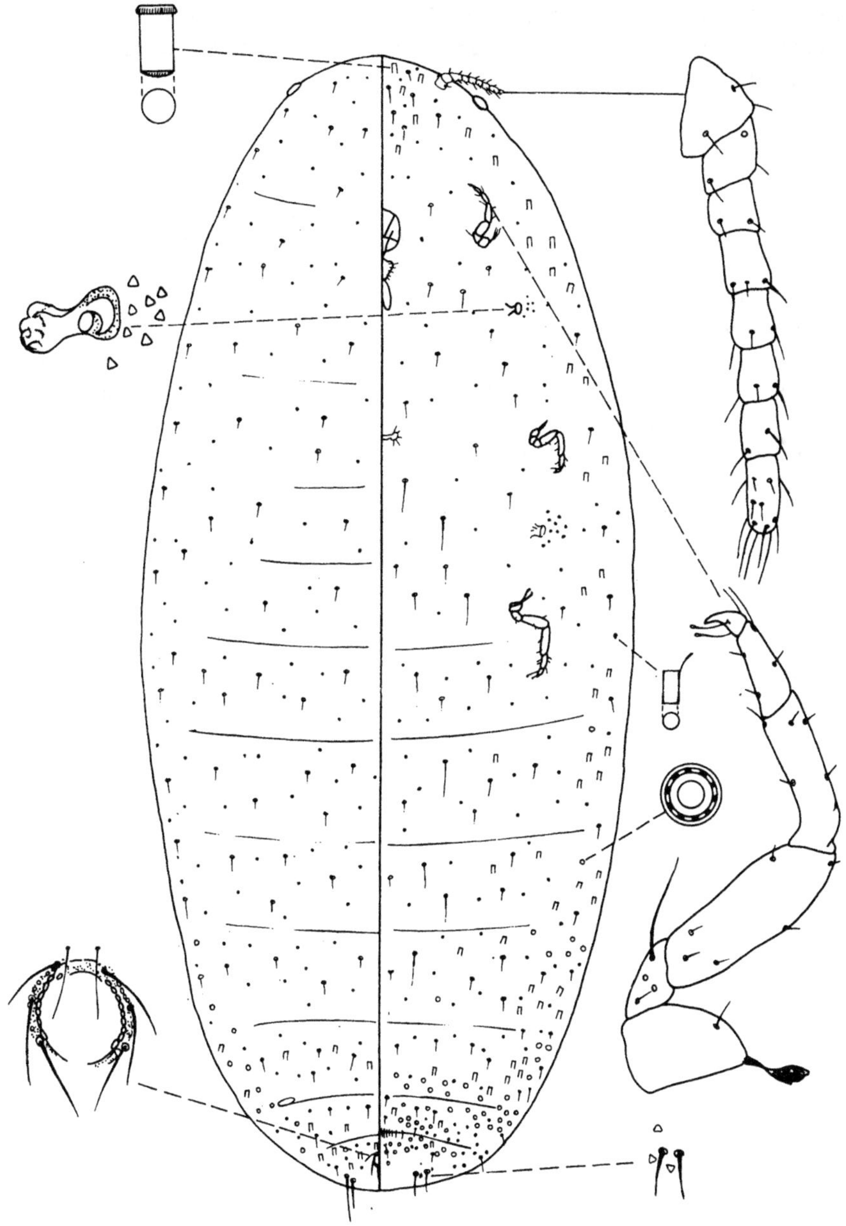

Fig. 67. *Mirococcopsis nagyi* Kozár, female (original)

Distribution. Known only from NW Hungary (Vértes Mtn.).

Biology. Found in leaf sheaths of *Stipa pennata*. Adult females collected during mid-June. Biology unknown.

47. **Mirococcopsis stipae** Borchsenius – Needlegrass Mealybug

Syn. & Lit.: *Mirococcopsis stipae* Borchsenius 1949:179; Ter-Grigorian, 1973:86; Koteja, 1974a:101; Tereznikova, 1975:196; Koteja & Liniowska, 1976:663; Kozár et al., 1977:71, 1979:130; Kosztarab & Kozár, 1978:46.

Adult female elongate oval, almost parallel-sided, according to Borchsenius (1949) up to 4.4 mm long and 1 mm wide; antenna elbowed, 7-segmented, ca 235 μ long, apical segment longest, 3rd and 4th segment shortest; labium 66 μ long with 28 setae, stylet loop reaches line between anterior spiracles; legs relatively small and stout, hind coxae with many translucent pores, tibia ca 1/3 longer than tarsus; circulus quadrangular; anal ring circular with a partly double row of oval pores, its setae thick, ca 114 μ long; anal lobes reduced, apical seta ca 150 μ long, with 2–3 associated shorter setae; numerous trilocular pores on both body surfaces; multilocular pores distributed singly along ventral margin, in a transverse row on 6th sternite and in transverse bands on sternites 7 and 8 of abdomen, (Ter-Grigorian [1973] noted some in a marginal band on 3 terminal tergites of abdomen); oral collar ducts occur together with multilocular pores; body setae short and rare.

Distribution. Known only from Bu, Hu, and USSR (from Ukraine to Armenia).

Biology. A rare steppe-inhabiting xerophilous species. In leaf sheaths and rarely on roots of *Stipa* spp. It lays eggs at end of June in Ukraine (Tereznikova, 1975). Biology unknown.

Note. Based on Goux's (1945) original description of *Volvicoccus volvifer*, the latter species is probably conspecific with *M. stipae*. Since Goux's material is not available for study, recollected topotype specimens could clarify the taxonomic status of the two species.

Genus **Mirococcus** Borchsenius

Syn. & Lit.: *Mirococcus* Borchsenius, 1947:141, 1949:181; Danzig, 1964:627; Koteja, 1971a:3; Ter-Grigorian, 1973:102; Tereznikova, 1975:197; Kosztarab & Kozár, 1978:47. *Aegyptococcus* Ezzat, 1963:192.

Type-species: *Phenacoccus inermis* Hall, 1925.

Adult female elongate oval or ovoid, body segmentation distinct; antennae 7–9-segmented; eyes with short basal cone; labium with 30 setae; 2 pairs of ostioles, lips with few trilocular pores; claw with denticle; spiracles with few associated trilocular pores; cerarii absent; circulus present or absent; anal ring with 1 row or sometimes with a second incomplete row of pores and 6 setae, setae about as long as diameter of ring; anal lobe variously developed, without sclerotization; minute simple and trilocular pores on both body surfaces; quinquelocular pores sometimes present on venter; multilocular pores and oral collar ducts on both surfaces; body setae on dorsum short, hairlike, longer and more numerous on venter.

Distribution. In the Palearctic and Ethiopian Regions with 5 species; in CE 1 species.

Biology. On roots of dicotyledons and on roots and under leaf sheaths of grasses.

48. **Mirococcus inermis** (Hall) – Harmless Mealybug
Fig. 68

Syn. & Lit.: *Phenacoccus inermis* Hall, 1925:7. *Mirococcus inermis* Borchsenius, 1949:181; Williams 1958b:4; Danzig, 1964:627, 1975:49; Bazarov, 1968:70; Koteja, *1971a:6, 1974a:98; Ter-Grigorian, 1973:102; Tereznikova, 1975:197; Kosztarab & Kozár, 1978:47. *Aegyptococcus inermis* Ezzat, 1963:192, 1965:2.

Female covered by white powdery wax, but during oviposition enclosed in a loose ovisac. Adult female yellowish, up to 3.5 mm long and 2.6 mm wide; antennae 9-segmented, ca 326 μ long; labium 115–140 μ long; sytlet loop almost reaching mid-coxae; legs relatively small, hind legs longest, without translucent pores; C_{18} reduced to 2–3 setae; circulus oval, large; anal ring oval, with 2 rows of pores, setae 30–40 μ long; apical setae ca 130 μ long; quinquelocular pores mostly on venter of prosoma; multilocular pores form transverse rows on dorsum and bands on venter; oral collar ducts form transverse rows on both body surfaces; most body setae 22–29 μ long.

Distribution. In the USSR known from Ukraine to Central Asia, also in Greece, North Africa, including Sudan.

Biology. On roots of different dicotyledons including: *Alhagi, Amaranthus, Arastation, Atriplex, Chenopodium, Cleome, Colutea, Cousinia, Cressa, Euphorbia, Fagonia, Frankenia, Hordeum, Neuraga, Nitraria, Polycarpaea, Spinacia, Xanthium* and *Zygophyllum*. Bazarov (1968) collected it at 1800 m in Tadzhikistan. Adult females and nymphs present during May in Armenia (Ter-Grigorian, 1973).

Genus **Peliococcopsis** Borchsenius

Syn. & Lit.: *Peliococcopsis* Borchsenius, 1948b:954, 1949:262; Danzig, 1964:630; Ter-Grigorian, 1973:196; Tereznikova, 1975:198; Kosztarab & Kozár, 1978:50.

Type-species.: *Phenacoccus caucasicus* Borchsenius, 1939.

Adult female covered by white powdery wax, body margins with wax threads. Adult female elongate oval; antennae 9-segmented, all segments slender, apical longest; labium conical, stylet loop about 1 ½ to 2 times as long as labium; both pairs of ostioles distinct; hind legs without translucent pores; tibia longer than tarsus, claw with small denticle; cerarii complete or reduced in number; circulus absent; anal ring with 1 or 2 outer rows and 1 inner row of pores; anal lobes distinct with well developed apical and subapical setae; trilocular, quinquelocular and multilocular pores on dorsal and ventral surfaces; oral collar tubular ducts of two sizes.

Distribution. In Central Palearctic. Europe, as far east as Armenia in USSR, with 2 species; 1 in CE.

Biology. On stems and under leaf sheaths of grasses.

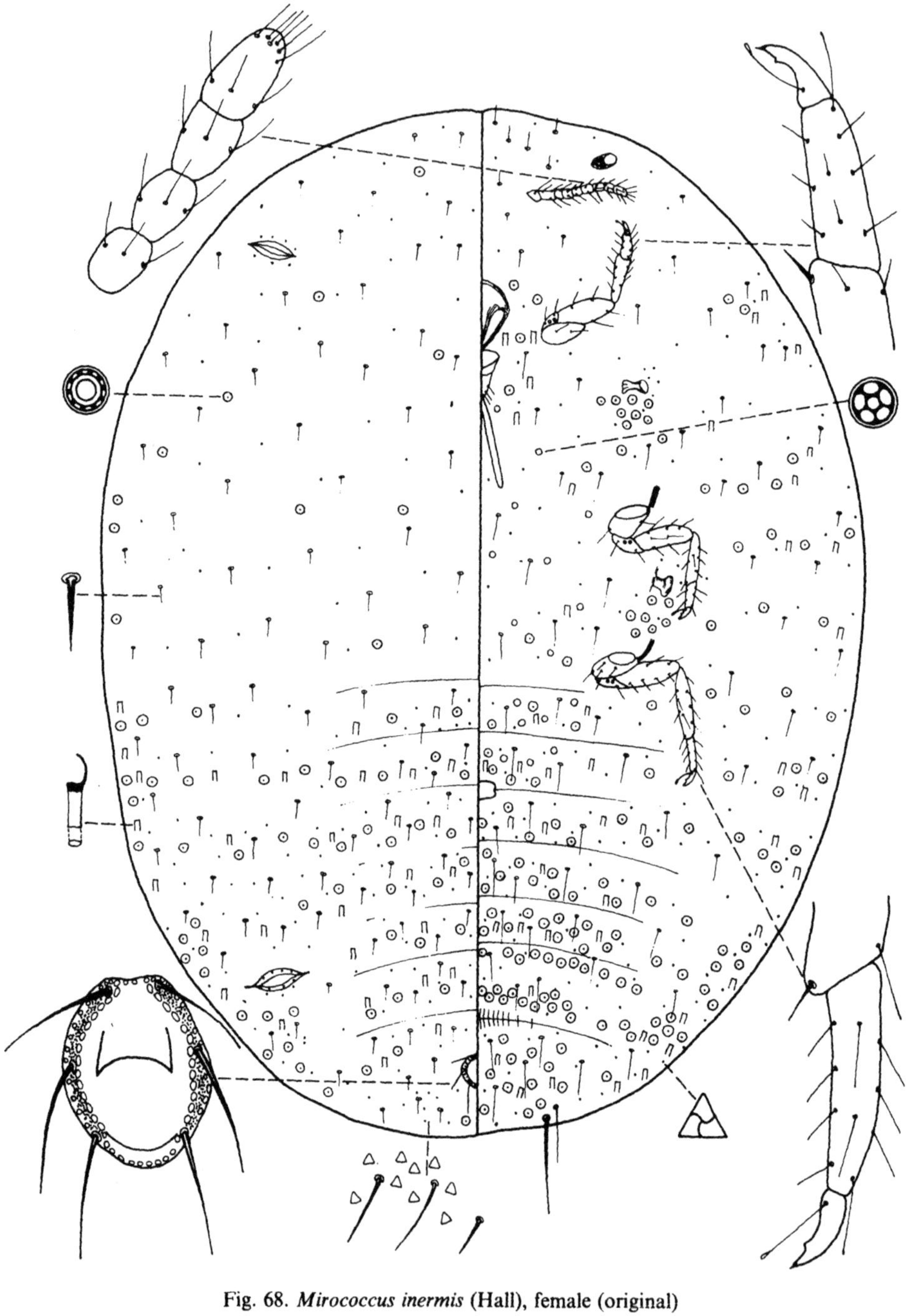

Fig. 68. *Mirococcus inermis* (Hall), female (original)

49. **Peliococcopsis parviceraria** (Goux) – Small Cerarian Mealybug
Fig. 69

Syn. & Lit.: *Phenacoccus parvicerarius* Goux, *1937b:223. *Peliococcopsis parvicerarius*, Danzig, *1960:178, 1964:630; Tereznikova, 1975:200; Koteja et al., 1978:502. *P. parviceraria*, Kosztarab & Kozár, 1978:50.

Adult female oval, light yellow; according to Danzig (1960) 2.5 mm long, 1.5 mm wide; antennae 9-segmented, 250–290 μ long; stylet loop about 1½ times as long as labium; lips of dorsal ostioles with trilocular pores; tarsus about half as long as tibia; only a few cerarii developed, C_{16-18} distinct, each with 2 setae and 2–5 trilocular pores; anal ring circular with 1 inner and 2 outer rows of pores, setae ca 86 μ long; anal lobes distinct, apical seta ca 142 μ long, trilocular pores sparse, these near spiracles, on lips of ostioles, in the cerarii and few along dorsal margin; quinquelocular pores numerous on both body surfaces; multilocular pores form groups of 3–4 (rarely 2 or 5–6) around tubular ducts on dorsum and along ventral margin, also some in 2 transverse rows on sternites 5–7 and in a band on sternite 8 of abdomen; small tubular ducts primarily distributed within the multilocular pore groups, large ducts form transverse rows on abdominal sternites 5–7, and occur singly along margin of sternites 5–8; dorsal setae small and numerous, ventral setae slender, hairlike. Nymphs and adult male described by Goux (1937b).

Distribution. In CE: Fr, Po, USSR (Ukraine and Leningrad area). A rare species.

Biology. Found inside leaf sheaths and at stem bases of *Corynephorus, Cynodon* and *Poa*. Females present in July and August. Biology unknown.

Genus **Peliococcus** Borchsenius

Syn. & Lit.: *Peliococcus* Borchsenius, 1948b:954, 1949:241; Ferris, 1950:115; Williams, 1962:31; DeLotto, 1964:364; Danzig, 1964:629, 1980a:118; Ter-Grigorian, 1973:175; Tereznikova, 1975:200; Kosztarab & Kozár, 1978:51. *Spinococcus*, Danzig, 1980a:118.

Type-species: *Phenacoccus chersonensis* Kiritchenko, 1936.

Adult female oval; basal eye cone distinct; antennae normally 9-segmented, rarely 8- or 7-segmented; labium conical; all four ostioles well developed; legs normal, tibia much longer than tarsus, claw with a denticle; usually 18 pairs of cerarii present; cerarii in a few species located on membranous prominences of body margin; normally with 1 circulus, this rarely absent; anal ring oval, with 1–2 outer and 1 inner row of pores and 6 setae; anal lobes normally well developed, with apical and subapical setae; trilocular pores scattered on both surfaces; quinquelocular pores only on venter, seldom absent; multilocular pores normally form small clusters and surround at least one minute tubular duct; others form transverse rows on both surfaces, seldom occurring singly; oral collar ducts of two sizes; dorsal setae short, those on venter slender, hairlike.

Distribution. Most species are present in the Holarctic Region, few in S Africa, with a total of about 27 species; in CE with 3 species.

Biology. Primarily on roots and foliage of herbaceous plants; rarely on leaves and branches of woody plants.

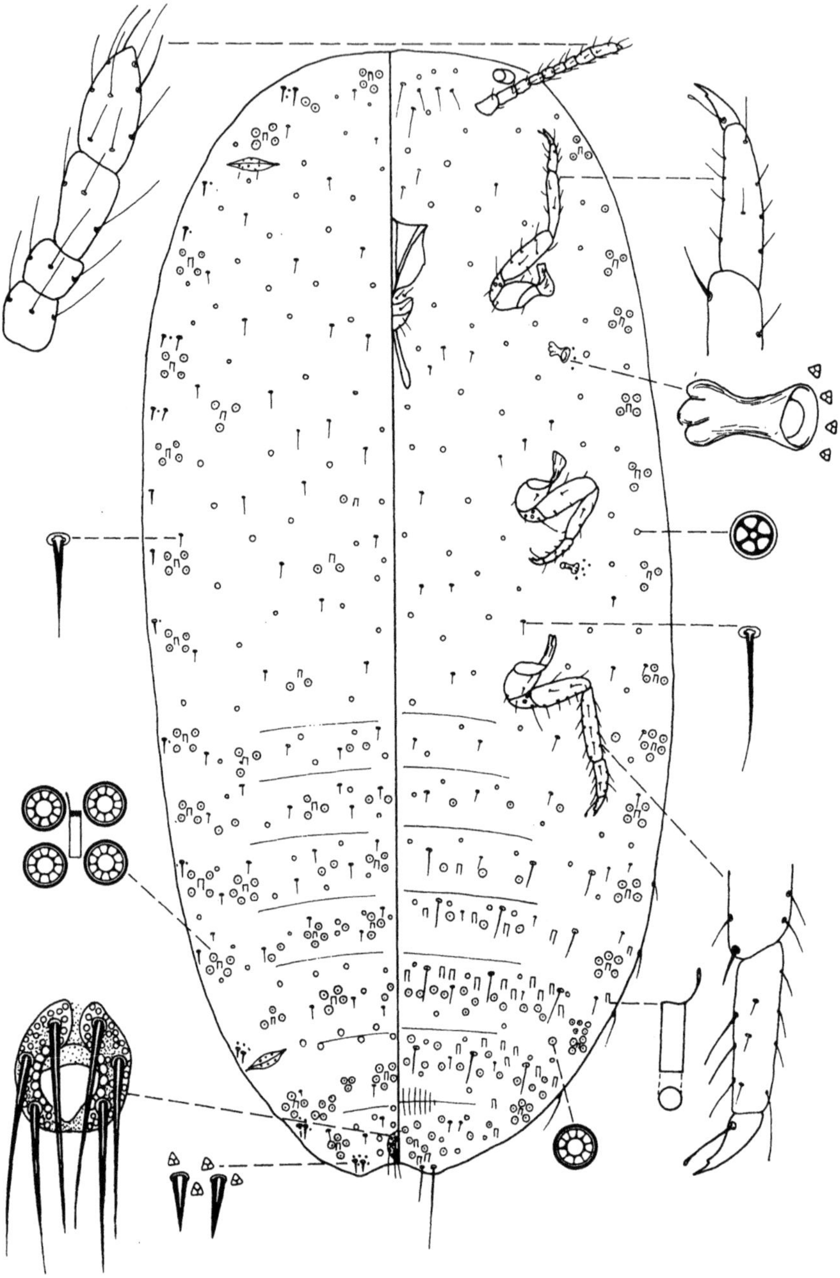

Fig. 69. *Peliococcopsis parviceraria* (Goux), female (original)

Note. Danzig (1980a) reduced the genus *Spinococcus* to synonym of *Peliococcus*. We consider these genera as distinct. For further discussion see Reviewer's remark under *Spinococcus*.

Key to species of *Peliococcus*

1. Multilocular pores on both body surfaces form clusters of 3–6, with a minute tubular duct in center; C_{18} with 4–6 large thick setae and 13–16 trilocular pores .... *balteatus*, p. 119
   Multilocular pores occur singly or in pairs forming transverse rows or bands on both body surfaces; C_{18} with 2–3 large thick setae and 4–8 trilocular pores 2
2. Antennae 8-, rarely 7-segmented; quinquelocular pores few, around labium; cerarii in 13–18 pairs .... *slavonicus* p. 121
   Antennae 9-segmented; quinquelocular pores entirely absent; cerarii in 3–12 pairs .... *perfidiosus*, p. 121

50. **Peliococcus balteatus** (Green) – Girdled Mealybug
Fig. 70

Syn. & Lit.: *Phenacoccus balteatus* Green, 1928a:21; *Peliococcus balteatus*, Borchsenius, 1949:244; Schmutterer, 1956a:65; Zahradnik, 1956:51; Williams, *1962:32; Danzig, 1964:629, 1980a:122; Koteja, 1972:569; Tereznikova, *1975:201; Kosztarab & Kozár, 1978:51.

Body of adult female covered with white powdery wax secretion, margins with a fringe of tassels. Adult female elongate oval, pale yellow, up to 3.5 mm long, and 1.2–1.8 mm wide; antennae 9- (rarely 8-) segmented, 450–490 μ long; labium elongate, stylet loop slightly longer than labium; ostioles with 2–4 setae and a few trilocular pores on each lip; coxae without translucent pores; 18 pairs of cerarii, C_{18} with 4–6 thick setae, and 13–16 trilocular pores; circulus oval; anal ring with partly double outer row and 1 inner row of pores, setae 100–120 μ long; anal lobes moderately developed, apical setae 200–220 μ long; trilocular pores sparse on both body surfaces; quinquelocular pores in middle of prosoma and forming transverse rows on sternites 2–7 of abdomen; multilocular pores in clusters of 3–6 with a minute tubular duct in center, clusters form segmental transverse rows on dorsum, and on sternites 3–5 of abdomen, simple pores form transverse bands on abdominal sternites 6–8, often with clusters along margin; large oral collar ducts in transverse rows on sternites 6–8; small ducts in middle of multilocular pore clusters; body setae of various lengths.

Distribution. A rare species. In CE: Cz, Ge, Po, USSR (areas of Leningrad, Crimea and Far East), also known from UK, Sweden.

Biology. On leaves and in leaf sheaths of grasses: *Agrostis, Arrhenatherum, Deschampsia, Festuca, Melica* and *Sieglingia*. Adult females collected in August–September. Biology and males unknown.

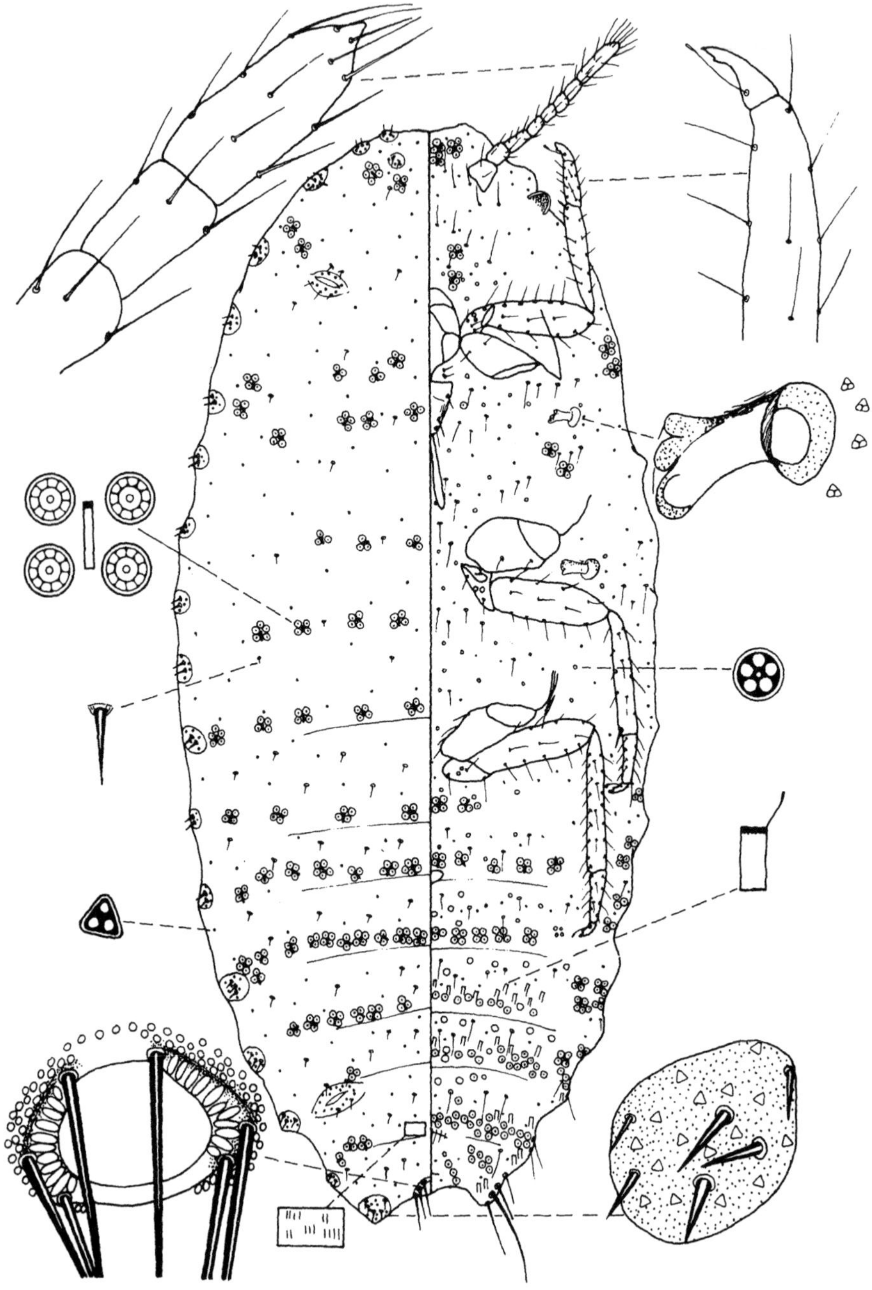

Fig. 70. *Peliococcus balteatus* Green, female (original)

51. **Peliococcus perfidiosus** Borchsenius – Malicious Mealybug

Syn. & Lit.: *Phenacoccus latipes* Kiritchenko, 1931:313. *Peliococcus perfidiosus* Borchsenius, 1949:257; Danzig, 1964:630; Tereznikova, *1975:209; Kosztarab & Kozár, 1978:52.

Adult female, according to Borchsenius (1949), oval, light green, up to 3 mm long and 1.8 mm wide; antennae 9-segmented, 2nd and 3rd segments longest; stylet loop reaches line between mid-coxae, tibia twice as long as tarsus; 3–12 pairs of cerarii, C_{18} with 2 slender setae, 1–2 shorter setae and 4–8 trilocular pores; circulus large, oval; anal ring oval, with 1 outer and 1 inner pore row, setae 73–100 μ long; anal lobes wide, apical setae 145 μ long; trilocular pores numerous on both body surfaces; quinquelocular pores absent; multilocular pores arranged singly or in pairs in 4 bands on dorsum of prosoma and in transverse rows on abdominal tergites, on venter arranged in groups on prosoma, also in transverse rows on sternites 2–3 and in bands on sternites 4–8 of abdomen; tubular ducts of 2 sizes on dorsum and on venter only a single size similar to the small size on dorsum; the ducts form transverse rows on both surfaces of abdominal segments; setae on dorsum short and slender, setae on venter longer.

Distribution. A rare species. Known from the USSR (Ukraine, Crimea, and N Caucasus) and Greece.

Biology. On roots and tubers of herbaceous plants: *Achillea, Althea, Artemisia, Jurinea, Malva, Melilotus, Nicotiana, Salvia, Scorzonera, Solanum* and *Verbascum*. It could become a pest when feeding on roots of tobacco and on potato tubers. Biology and males unknown.

52. **Peliococcus slavonicus** (Laing) – Slavonic Mealybug

Syn. & Lit.: *Phenacoccus latipes* var. *slavonicus* Laing, 1929:475. *Peliococcus slavonicus*, Borchsenius, *1949:255; Danzig, 1972:193; Tereznikova, 1975:206; Kosztarab & Kozár, 1978:51.

Adult female, according to Borchsenius (1949), oval, up to 2.5 mm long, 1.2 mm wide; antennae 8-, rarely 7-segmented, ca 310 μ long, apical segment longest; sometimes segments 3 and 4 fused; labium elongate, stylet loop reaches line between mid-coxae; legs thick, hind coxae with a few translucent pores; 13–18 pairs of cerarii, C_{18} with 2–3 slender setae and 6–7 trilocular pores; circulus large, oval; anal ring with partly double outer pore row and 1 inner pore row, setae ca 95 μ long; trilocular pores numerous on both body surfaces; quinquelocular pores present around labium only; multilocular pores singly or in pairs form transverse rows on both body surfaces; oral collar tubular ducts of two sizes, stouter ducts on both derms, slender ducts on venter only.

Distribution. A rare species. Known from Ukraine in the USSR, in the Poltava district and from the area of Adlera near the Black Sea coast.

Biology. Collected from roots of *Hordeum "distychum"*. Biology and males unknown.

Note. There are some doubts about the validity of this species, since Laing's (1929) original description was very brief, and neither he nor Borchsenius (1949) illustrated the species, also because we have so far not studied type material held in the British Museum (Nat. Hist.).

Genus **Phenacoccus** Cockerell

Syn. & Lit.: *Phenacoccus* Cockerell, 1893:318; Kiritchenko, 1940a:187; Borchsenius, 1948b:953, 1949:208; Ferris, 1950:120; Williams, 1962:32; Danzig, 1964:629, 1980:122; McKenzie, 1967:213; Ter-Grigorian, 1973:134; Tereznikova, 1975:210; Kosztarab & Kozár, 1978:52. *Paroudablis* Cockerell, 1900:87. *Peukinococcus* Šulc, 1944b:2.

Type-species: *Pseudococcus aceris* Signoret, 1875.

Adult female oval; antennae normally 9-segmented, rarely 8 or less; 2 pairs of ostioles, ostiole lips normally with trilocular pores and setae; legs well developed, claw with denticle or tooth on plantar surface; spiracles without disc pores; normally 18 pairs of cerarii, occasionally these reduced to 1 or 2 pairs; circuli present (1–4) or absent; anal ring oval, apically located, with 1 or 2 rows of pores and 6 setae; anal lobes usually well developed, with a long apical seta and 1 shorter sub-apical seta; disc pores of 1 to 3 types: triloculars, normally on both body surfaces, quinqueloculars, present on venter or absent, multilocular pores, normally on venter, rarely on dorsum; oral collar ducts of 1 or 2 sizes usually present; oral rim ducts rare, never in CE species; dorsal setae short, thick, often lanceolate, ventral setae hairlike, elongate. All immature stages and adult males described for two species by Šulc (1943, 1944).

Distribution. World-wide; at least 100 species have been assigned to this genus, some erroneously. There are about 50 species in the Palearctic Region; in CE 11 species.

Biology. On a variety of woody and herbaceous plants, from roots to leaves. Life cycles vary by species. One species, *P. dearnessi* King in North America, is known to make a series of migrations on its host: to feeding sites, overwintering sites, molting sites and oviposition sites.

Note. There is a wide range in variation of morphological characters within this genus; variations are often host-induced within the same species.

Key to species of *Phenacoccus*

1. Tubular ducts on dorsum normally twice as wide as ducts on venter 2
   Tubular ducts on both body surfaces of uniform size, or less than 1 ½ times as wide on dorsum as on venter; ducts rarely absent on dorsum 4
2. Multilocular pores form transverse rows on both dorsum and venter; C_1–C_3 not borne on sclerotized platelets; living on needles of conifers *Abies* and *Picea* . . . . *piceae*, p. 131
   Multilocular pores on venter only; C_1–C_3 borne on slightly sclerotized platelets; living on grasses 3
3. Circulus large, hourglass-like; 18–20 trilocular pores on C_{18}; antennae more than 530 μ long . . . . *evelinae*, p. 126
   Circulus absent; 10–15 trilocular pores on C_{18}; antennae less than 530 μ long . . . . *interruptus*, p. 130

4 (1). Multilocular pores form transverse rows on both dorsum and venter.....5
Multilocular pores on venter only.....6

5. Dorsal multilocular pores form transverse rows only on sternites 5 and 6 of abdomen; 4–7 trilocular pores in each C_{18}; apical segment of antenna longest ..... *pumilus*, p. 132
Dorsal multilocular pores form transverse bands on entire surface; 7–14 trilocular pores in each C_{18}; second segment of antenna longest ..... *hordei*, p. 129

6 (4). With 18 pairs of cerarii; circulus present; quinquelocular pores present .. 7
With 1 or 2 pairs of cerarii; circulus absent; quinquelocular pores absent .9

7. First 4 pairs of cerarii borne on sclerotized platelets; C_{18} with only 6–8 trilocular pores; on grasses only ..... *avenae*, p. 125
First 4 pairs of cerarii not borne on sclerotized platelets; C_{18} with 16 or more trilocular pores; on deciduous trees only 8

8. With 2 or more circuli; dorsal tubular ducts present on prosoma; antennae about 660 μ long ..... *aceris*, p. 123
With 1 circulus; dorsal tubular ducts absent from prosoma; antennae about 440 μ long..... *mespili*, p. 130

9 (6). With 1 pair of cerarii; 7–8 trilocular pores in each C_{18} .... *incertus*, p. 129
With 2 pairs of cerarii; 4–6 trilocular pores in each C_{18} 10

10. Tubular ducts absent from dorsum of prosoma; anal ring setae 48–60 μ long; multilocular pores present along ventral margin of most sternites of abdomen ..... *phenacoccoides*, p. 131
Tubular ducts present on dorsum of prosoma; anal ring setae 95–105 μ long; multilocular pores present in transverse rows on 3 terminal abdominal sternites only ..... *bicerarius*, p. 126

53. **Phenacoccus aceris** (Signoret) – Polyphagous Tree Mealybug
Fig. 71

Syn. & Lit.: *Pseudococcus aceris* Signoret, 1875:329. *P. quercus* Douglas 1888:154. *P. ulicis* Douglas, 1888a:88. *P. ulmi* Douglas, 1888b:124. *P. socius* Newstead, 1892:144. *P. pergandei* Cockerell, 1896:16. *Dactylopius vagabundus* Schilling, 1897:248 (in part). *Phenacoccus polyphagus* Borchsenius, 1949:213. These were only the more common synonym names. *Phenacoccus aceris*, Marchal, 1908:239; Rau, 1942:118; Šulc, *1943:1; Borchsenius, 1949:212; Ferris, *1950:126; Schmutterer, 1952b:396; Williams, 1962:34; Danzig, 1964:629, 1980a:124; Panis, 1969b:383; Zahradnik, 1972:398; *Ter-Grigorian, 1973:136; Koteja, 1974a:97; Tereznikova, 1975:214; Kosztarab & Kozár, 1978:55.

Ovisac well developed, long, woolly and white. Female with fine wax powder on derm and with 18 pairs of wax extensions on body margins. Adult female oval with well-developed anal lobes, greenish yellow, up to 4.3 mm long, 2.5 mm wide; antennae normally 9-segmented, ca 660 μ long; labium ca 200 μ long with 30 setae, stylet loop reaches line between mid-coxae; hind coxae without translucent pores, claws with small but distinct denticle; 18 pairs of cerarii, C_{18} with 2 conical setae, 3 or more smaller setae and 16–20 trilocular pores, other cerarii with 2 setae, except those on head with 3 or 4; circuli 2, rarely more, second much larger than first; anal ring with 2½ outer rows

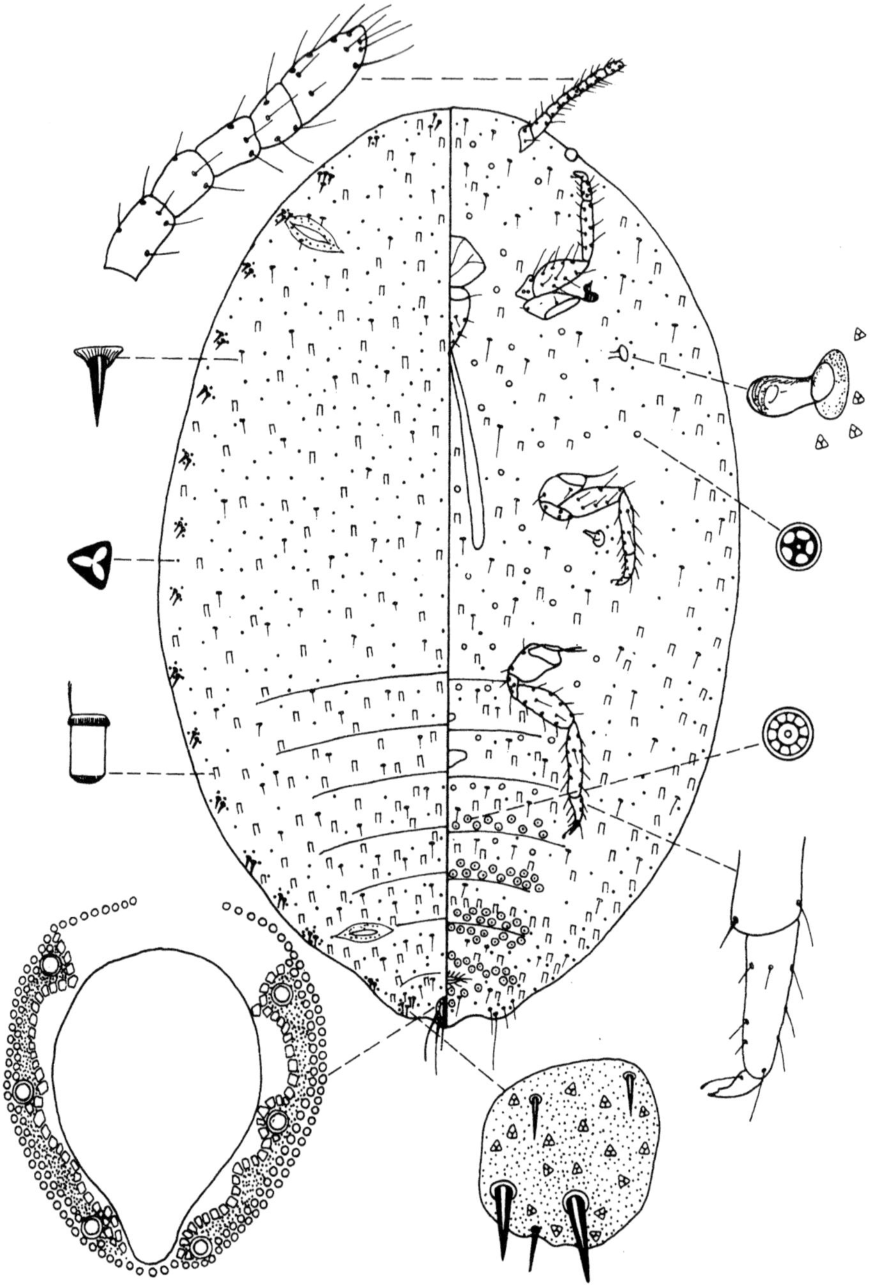

Fig. 71. *Phenacoccus aceris* (Signoret), female (original)

and 1 inner row of pores, setae 167–195 μ long; anal lobes wide, apical seta ca 300 μ long; trilocular pores scattered on both body surfaces; quinquelocular pores on venter located in middle of prosoma and on abdominal sternites 1–6; numerous multilocular pores in mid-region of abdominal venter forming transverse rows and bands; oral collar tubular ducts scattered on both body surfaces but only few on head; body setae of various lengths, some on venter of mid-abdomen up to 142 μ long. Nymphal instars and adult male described in detail by Šulc (1943); nymphs by Ter-Grigorian (1973).

Distribution. A common holarctic species. In CE: Au, Bu, Cz, Fr, Ge, Hu, It, Po, Ro, Sw, USSR and Yu. It is also known from UK, Japan, Canada and USA.

Biology. A polyphagous tree-infesting species. Host genera: *Acer, Actinidia, Aesculus, Alnus, Aralia, Betula, Buxus, Carpinus, Castanea, Cornus, Corylus, Cotoneaster, Crataegus, Cytisus, Euonymus, Fagus, Fraxinus, Juglans, Laburnum, Lonicera, Malus, Mespilus, Myrica, Morus, Pavia, Prunus, Pyrus, Quercus, Ribes, Salix, Sorbaria, Sorbus, Spiraea, Thelycrania, Tilia, Ulex, Ulmus,* and *Vitis.* Occasionally with mass infestations on ornamental and/or fruit trees it becomes a pest. According to Schmutterer (1952b), the species overwinters as third instar female nymphs and male pupae; adults appear by mid-April in S Germany, lay eggs (776–3605 per female) in cottony egg sacs in bark crevices from end of April; eggs hatch in about 30 days after laying; nymphs feed on leaves and return for overwintering to bark; third instar nymphs develop by the second half of September. It has 1 generation a year in CE.

Natural enemies. Parasites: *Allotropa utilis* (found most effective in biological control of this species by Bartlett, in Clausen, 1978), *A. mecrida, Anagyrus schoenherri, Aphycus apicalis, Hyperimerus pusillus, Leptomastidea bifasciata, Microterys chalcostomus, M. trjapitzini* (Panis, 1969b; Zahradnik, 1972; Trjapitzin, 1978; Kozár and Sugonjaev, 1979; Podsiadlo, 1981b). Predators: *Porcellio scaber, Nephus quadrimaculatus, Piocoris luridus, Leucopis silesiaca* (Schmutterer, 1952b; Schmutterer et al., 1957).

Associated ants. *Formica rufa pratensis, Lasius fuliginosus, L. brunneus, L. emarginatus, L. niger niger, Myrmica laevinodis* (Schmutterer, 1952b).

54. **Phenacoccus avenae** Borchsenius – Oat Mealybug

Syn. & Lit.: *Phenacoccus avenae* Borchsenius, 1949:217, Ter-Grigorian, *1973:168.

Adult female oval, almost parallel-sided, up to 2.5 mm long and 1 mm wide; antennae 9-segmented, ca 412 μ long, second segment longest; stylet loop only slightly longer than labium; legs thin, ca 554 μ long, tibia ca 2 ½ times as long as tarsus; 18 pairs of cerarii, C_{18} with 2 conical setae and 6–8 trilocular pores; circulus small oval; anal ring oval with 1 outer row and 1 inner row of pores (Ter-Grigorian [1973] found 2 outer pore rows), setae 98 μ long; anal lobes conical, apical seta 180 μ long; trilocular pores numerous on both body surfaces; quinquelocular pores in middle of ventral side of prosoma, form transverse rows on abdominal sternites 2 and 3, and found singly on sternites 4 and 5 of abdomen; multilocular pores form transverse rows and bands on sternites of abdomen; oral collar tubular ducts of two sizes: larger ones form sparse transverse rows on dorsum, smaller ones form transverse rows on abdominal sternites and a few found along ventral margin of prosoma; small setae 5–8 μ long, scattered on entire dorsum; ventral setae longer, 20–67 μ long. Third instar described by Ter-Grigorian (1973).

Distribution. Known only from USSR (Armenia), and recently found in Hungary.

Biology. Adult females were collected from *Avena* and *Poa* during May and June in Armenia, and from inside leaf sheaths of *Poa* during August in Hungary. Biology, and males unknown.

55. **Phenacoccus bicerarius** Borchsenius – Bicerarian Mealybug

Syn. & Lit.: *Phenacoccus bicerarius* Borchsenius, 1949:225; Matesova, 1968a:110; Koteja, 1971b:321, 1974a:98; Ter-Grigorian, 1973:170; Bazarov & Nurmamatov, 1975:54; Kosztarab & Kozár, 1978:55.

Adult female elongate oval, nearly parallel-sided, 2 mm long, 0.7 mm wide; antennae 9-segmented, ca 380 μ long, segments slender, apical segment longest; labium 61 μ long, with 30 setae (Koteja, 1974a), stylet loop as long as labium; legs slender, ca 560 μ long; tibia twice as long as tarsus; 2 pairs of cerarii, C_{17} and C_{18} on circular sclerotized platelets, each with 2 conical setae and 4–6 trilocular pores, circulus absent; anal ring oval, with setae 95–105 μ long; anal lobes produced, apical seta 160–195 μ long; trilocular pores numerous on both surfaces; quinquelocular pores absent; multilocular pores form transverse rows and bands on 3 terminal abdominal sternites; oral collar tubular ducts somewhat wider on dorsum than those on ventral abdominal sternites; dorsal setae 8–14 μ long, on venter up to 100 μ long.

Distribution. Known only from Poland and USSR (Armenia, Tadzhikistan, Kazakhstan).

Biology. Found in leafsheaths of grasses: *Corynephorus, Elymus* and *Sorghum.* Adult females were collected in June and July. Biology and males unknown.

56. **Phenacoccus evelinae** (Tereznikova) – Evelyna's Mealybug
Figs 72, 73

Syn. & Lit.: *Paroudablis graminis* Tereznikova, (preoccupied) 1968:472; Ter-Grigorian, 1973:174; Kosztarab & Kozár, 1978:50. *Phenacoccus evelinae* Tereznikova, 1975:211.

Female body covered with fine wax powder. Adult female elongate oval, nearly parallel-sided; lead colored, up to 4.3 mm long and 1.6 mm wide; antennae with 9 thin segments, ca 540 μ long, second segment longest; labium 124 μ long 99 μ wide, stylet loop shorter than labium; leg segments thin, tibia at least twice as long as tarsus; 8 or rarely 9–10 pairs of cerarii, last 3 pairs situated on sclerotized platelets; circulus wide, large, hourglass-like; anal ring oval, with 2 outer and 1 inner pore rows, setae ca 143 μ long; apical setae ca 277 μ long; numerous trilocular pores on dorsum, on venter some laterad of spiracles; quinquelocular pores common on venter of prosoma and on abdominal sternites; multilocular pores form transverse rows and bands on last 5 abdominal sternites; oral collar ducts of 2 sizes; wider ducts form transverse rows on

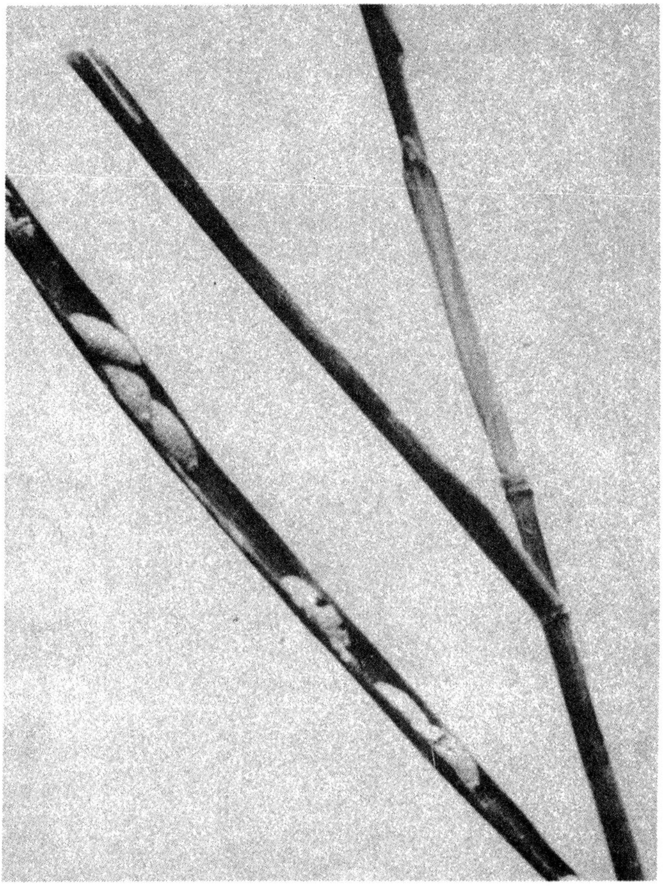

Fig. 72. *Phenacoccus evelinae* (Tereznikova), female ovisacs on leaf of grass (Photo F. Kozár)

dorsum and groups along body margin, thinner ducts form transverse rows on prosomal sternites and on anterior abdominal sternites and bands on posterior sternites; dorsal setae appear narrowed at base; ventral setae of various sizes.

Distribution. Common in warm habitats on roadsides. In CE: Hu, Po and USSR (Ukraine and Armenia).

Biology. On upper leaf surface, inside leaf sheaths and rarely on roots of *Setaria, Triticum* and other grasses. Females move to thatch during oviposition and prepare a very long (up to 10 mm) ovisac. Adult females present during June in Hungary. Males unknown.

Natural enemy. Predator: *Leucopis szepligeti* (Kozár and Sugonjaev, 1979).

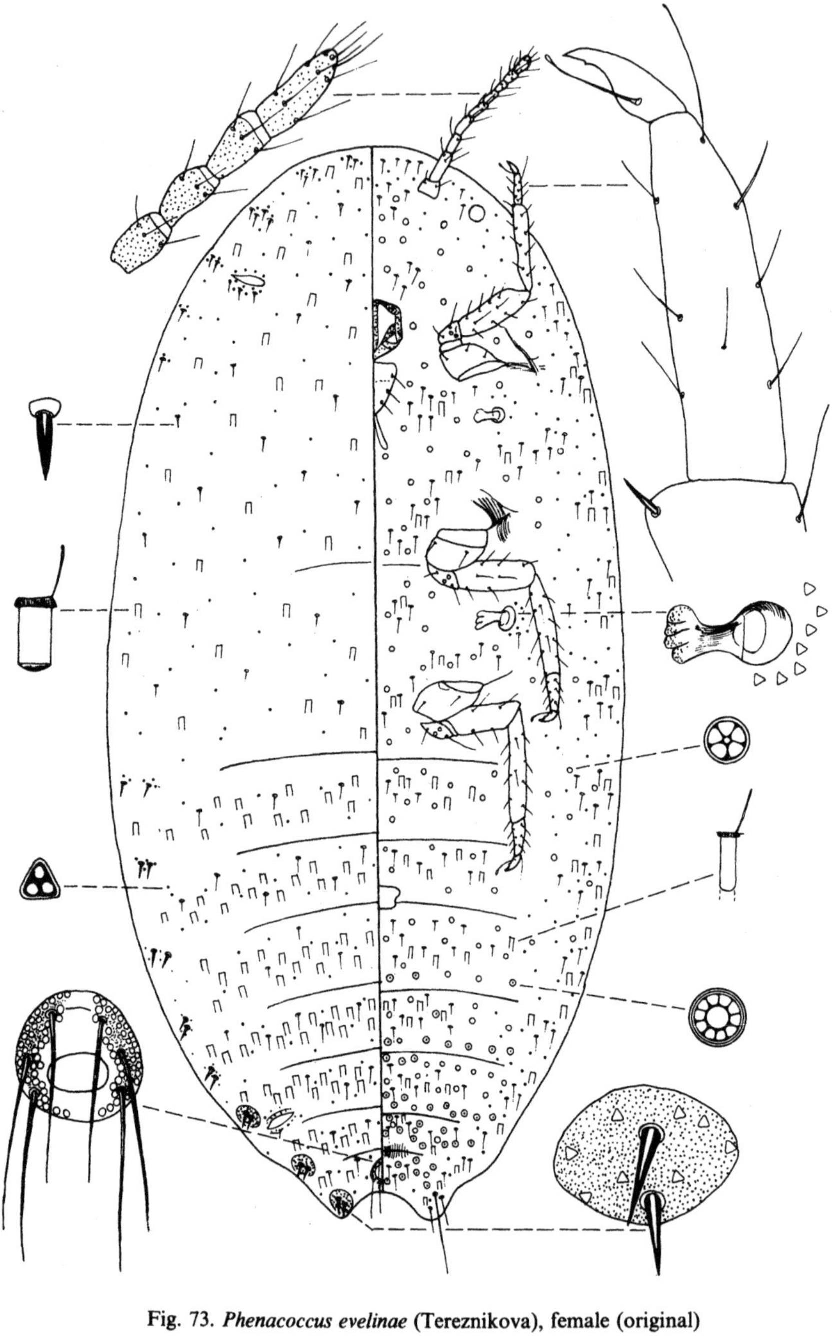

Fig. 73. *Phenacoccus evelinae* (Tereznikova), female (original)

57. **Phenacoccus hordei** (Lindeman) – Barley Mealybug

Syn. & Lit.: *Westwoodia hordei* Lindeman, 1886:6; *Pseudococcus graminis* Reuter (preoccupied). *P. elongatus* Reuter, 1904:66, 251. *Phenacoccus cholodkovskyi* Marchal, 1908:245; Schmutterer, *1952b:397. *P. hordei,* Borchsenius, 1949:231; Danzig, 1964:629; Wetzel, 1971:297; Ter-Grigorian, 1973:157; *Tereznikova, 1975:222; Kosztarab & Kozár, 1978:53.

Adult female oval, pale yellow or pink, up to 3.3 mm long and 1.5 mm wide; antennae 9- (rarely 8-) segmented, second segment longest; labium about as long as wide, stylet loop 1 ½–2 times as long as labium; hind coxae with few translucent pores; 16–18 cerarii, C_{18} with 2–3 thick conical and 2–3 small setae and 7–14 trilocular pores; circulus absent; anal ring with 1 inner and 2 outer rows of pores, setae 85–115 μ long; anal lobes not protruding, wide, apical seta 180–195 μ long; trilocular pores numerous on both body surfaces; quinquelocular pores found singly around clypeolabral shield; multilocular pores numerous, form transverse rows and bands on both body surfaces; oral collar tubular ducts on both surfaces; dorsal setae slender, ventral setae of various lengths.

Distribution. Fairly common. In CE: Fr, Ge, Hu, It, Po and USSR. Also known from Finland and Holland.

Biology. On roots of plants, especially grasses: *Agropyron, Arrhenantherum, Bromus, Festuca, Hordeum, Phleum, Poa, Secale, Triticum, Vicia, Zozimia.* Rarely a minor pest. Can produce up to 3 generations per year. According to Schmutterer (1952b), they overwinter as second, rarely as third instars; adults develop by end of April, or beginning of May; start laying eggs an average of 159 per female by mid-May; eggs hatch in 10–14 days and develop into adults by end of July, they lay eggs from beginning of August, if weather permits, the nymphs could develop into adults by second half of September and produce the overwintering generation. It frequents warm, dry, sunny slopes.

Natural enemies. Parasite: *Coccidencyrtus phenacocci.* (Schmutterer, 1955c). Predator: Mite, *Tyrophagus sachsi* (Schmutterer, 1952b).

58. **Phenacoccus incertus** (Kiritchenko) – Uncertain Mealybug

Syn. & Lit.: *Ripersia incerta* Kiritchenko, 1940b:124. *Phenacoccus incertus,* Borchsenius, 1949:224; Tereznikova, *1975:219; Kosztarab & Kozár, 1978:55; *Caulococcus incertus,* Danzig, 1964:629.

Female covered with fine white wax powder. Adult female elongate-oval, nearly parallel-sided, pink to yellow, 1.2–1.6 mm long and ca 0.5 wide; antennae 9-segmented; stylet loop as long as labium; tibia about twice as long as tarsus; of cerarii only C_{18} developed, each with 2 conical setae, and 7–8 trilocular pores; circulus absent; anal ring almost circular on outer side with a few pores around setal bases and with a loose inner row of pores, setae ca 28 μ long, shorter than diameter of anal ring; anal lobes distinct, apical setae ca 56 μ long; trilocular pores numerous on both body surfaces; no quinquelocular pores; multilocular pores form transverse rows and bands on abdominal sternites 5–8; oral collar ducts somewhat larger on dorsum and on margin of venter, smaller ducts form transverse rows on abdominal sternites; dorsal setae up to

14 μ long, those on venter longer and form transverse bands on abdominal sternites. Third instar described by Tereznikova (1975).

Distribution. Known only from USSR, in Ukraine, at Odessa and Crimea.

Biology. A steppe inhabiting xerophilous species. On roots of grasses: *Festuca* and *Stipa*. Life cycle and males unknown.

59. **Phenacoccus interruptus** Green – Green's Mealybug

Syn. & Lit.: *Phenacoccus interruptus* Green, 1923:215; Williams, *1962:35; Tereznikova, 1975:219; Danzig 1980:131. *Paroudablis interruptus,* Borchsenius, 1949:240; Danzig, 1964:629; Zak-Ogaza & Koteja, 1964:423; Podsiadlo & Komosinska, 1976:88; Kosztarab & Kozár, 1978:50. *Phenacoccus assimilis* Kiritchenko, 1940a:188 (nomen nudum).

Female covered with white wax powder. Adult female elongate oval, nearly parallel-sided, ca 3.5–4.3 mm long and 1.2–1.6 mm wide; antennae slender, 9-segmented, 380–525 μ long; labium 90–120 μ long, with 30 setae; legs well-developed, claw with a distinct denticle, tarsi less than half as long as tibiae; 9 pairs of cerarii, 5 pairs at anterior and 4 pairs at posterior ends of body, each borne on a slightly sclerotized platelet, C_{18} with 2 thick conical setae, 2 small setae and 10–15 trilocular pores; circulus absent; anal ring with 2 outer and 1 inner pore rows, its setae about 160 μ long; anal lobes distinct, apical setae ca 309 μ, 3 times as long as anal ring setae; trilocular pores evenly distributed on both surfaces, except median area of venter; quinquelocular pores only in median area of venter; multilocular pores form transverse rows and bands on abdominal sternites 5–8; oral collar tubular ducts of 2 sizes: large ducts scattered on dorsum of prosoma and form transverse rows on abdominal tergites, small ducts in transverse rows on abdominal sternites, few on margin of thorax; body setae of uniform small size on dorsum, of various lengths on venter.

Distribution. A common species. In CE: Fr, Hu, Po, USSR (Ukraine, Armenia, Kazakhstan, Siberia); also known from England, Greece, Korea and Mongolia.

Biology. A steppe inhabiting species in CE. On grasses: *Agropyron, Arundinella, Elymus, Festuca* and *Sieglingia*. Females collected in June. Biology and males unknown.

Note. Matesova (1968a) reported two aggressive ant species (undetermined) that protect these mealybugs in Kazakhstan.

60. **Phenacoccus mespili** (Signoret) – Apple Mealybug

Syn. & Lit.: *Pseudococcus mespili* Signoret, 1875:333. *Phenacoccus mespili,* Borchsenius, 1949:220; Schmutterer, et al., 1957:439; Danzig, 1964:629; Ter-Grigorian, *1973:140; Bazarov & Nurmamatov, 1975:54; Tereznikova, *1975:217; Kosztarab & Kozár, 1978:55.

Female body covered with a thin layer of white powdery wax. Adult female oval, pink, up to 5.2 mm long and 2.5 mm wide; antennae 9-segmented, ca 440 μ long; stylet loop reaches line between midlegs; legs slender, ca 514 μ long, tibia almost twice as long as tarsus; 18 pairs of cerarii, C_{18} with 2–3 large, 1–2 small setae and 16–19 trilocular pores; 1 circulus; anal ring with 2 outer and 1 inner row of pores; setae 132–150 μ long; anal lobes wide, apical seta 214–300 μ long; trilocular pores numerous on both body surfaces; quinquelocular pores scattered on venter of prosoma and on sternites 1–6; multilocular pores form transverse rows on sternites 4–6 and bands on sternites 7 and 8 of abdomen; oral collar tubular ducts on dorsum only on abdominal tergites 6 and 7, scattered on venter, predominant in a submarginal band; short setae scattered on entire dorsum, slender setae form transverse rows on venter. First and second instars described by Ter-Grigorian (1973).

Distribution. Reported from a number of European countries, but probably often confused with *P. aceris*, especially since there are no voucher specimens left behind. In USSR is known from its European part and from Central Asia.

Biology. Lives on bark of trunk and branches of: *Armeniaca, Fraxinus, Malus, Mespilus, Persica, Prunus,* and *Pyrus.* Occasionally a pest on *Malus* and *Fraxinus.* Males and biology unknown.

Natural enemies. Parasites: *Aphycus hadzibejliae, Pseudaphycus phenacocci* (Jasnosh, 1967; Trjapitzin, 1978).

61. **Phenacoccus phenacoccoides** (Kiritchenko) – Pink Grass-Mealybug

Syn. & Lit.: *Trionymus phenacoccoides* Kiritchenko, 1932:136. *P. phenacoccoides,* Borchsenius, 1949:228; Matesova, 1968:110; Tereznikova, *1975:222; Kosztarab & Kozár, 1978:54. *Caulococcus phenacoccoides,* Danzig, 1964:629.

Ovisac porous, snow-white (Kiritchenko, 1932). Female derm covered by fine white wax powder. Adult female elongate oval, pale-pink, 2.7–3.5 mm long, 0.8–1.4 mm wide; antennae 9-segmented, first and last segment longest; stylet loop reaches line between anterior spiracles; hind coxae with few translucent pores; only last 2 pairs of cerarii developed, these situated on sclerotized platelets, C_{18} with 2 conical setae and 5–6 trilocular pores; circuli absent; anal ring circular, setae 48–60 μ long; anal lobes hardly noticeable, apical seta 95–135 μ long; numerous trilocular pores on both body surfaces; quinquelocular pores absent; multilocular pores form transverse rows on 4 terminal abdominal sternites, few also along abdominal ventral margin; oral collar ducts present on terminal segments of abdomen on both surfaces, also a few on venter of prosoma; dorsal setae small and thin, those on venter more slender.

Distribution. Known only from Hu, USSR (from Transcarpathian Region to Siberia) and Mongolia.

Biology. It is a steppe-inhabiting mesophilous species. On stems under leaf sheaths of grasses in the genera: *Agropyron, Cleistogenes, Cynodon, Elymus, Elytrigia, Phleum* and *Poa.* Adults were collected in the vicinity of Odessa from May through October (Kiritchenko, 1932). Males and biology unknown.

62. **Phenacoccus piceae** Löw – Spruce Mealybug

Syn. & Lit.: *Boisduvalia piceae* Löw, 1883a:267. *Peukinococcus piceae* Šulc, *1944b:2. *Paroudablis piceae,* Borchsenius, 1949:238; Danzig, 1964:629; Zahradnik, 1972:397; Koteja, 1974a:98; Kosztarab & Kozár, 1978:48. *Phenacoccus piceae,* Marchal, 1908:243; Schmutterer, 1952b:396; Tereznikova, 1975:214.

Female body covered with white wax powder. Ovisac white, cottony. Adult female oval, pink, up to 2.5 mm long and 1.7 mm wide; antennae 9-segmented, 2nd and apical segments longest; labium 88 μ long, as wide as long, with 28–30 setae, stylet loop nearly reaches line between mid-coxae; legs slender and long, tibia more than twice as long as tarsus; 6–7 pairs of cerarii, normally 4 pairs on head and 3 pairs on apex of abdomen, C_{18} with 2 conical setae, 1–2 short setae and 15–20 trilocular pores; circulus oval; anal ring with 1 inner and 2 outer rows of pores, setae ca 155 μ long; anal lobes moderately developed, apical seta ca 165 μ long; trilocular pores scattered almost on entire body surface; quinquelocular pores on venter of prosoma and on abdominal sternites 2–6; multilocular pores form transverse rows on dorsum, also on almost entire venter; oral collar tubular ducts of 2 sizes: large ducts form transverse rows on dorsum, small ducts form transverse rows on sternites of abdomen and scattered on venter of prosoma; dorsal setae small and slender, ventral setae of various lengths. All life stages described by Šulc (1944b).

Distribution. A common species. In CE: Au, Bu, Cz, Fr, Ge, Hu, Po, Ro, Sw, USSR, and Yu: also reported from Mongolia.

Biology. On needles of *Abies* and *Picea*. All life stages commonly found throughout the season. It develops 1–3 yearly generations, depending on climatic conditions. Lays ca 150 eggs per female; females overwinter, mostly as third instars, males as pupae; often under bark flakes or in old galls produced by the aphid *Sachiphantes abietis* (L.). Often found in the mountains at high elevations. Occasionally a pest on physiologically weakened ornamental spruce trees (Schmutterer, 1952b; Zahradnik, 1972).

Natural enemies. Parasites: *Aphycus sumavicus, Allotropa mecrida, Anagyrus schmuttereri, Eunotus nigriclavis, Microterys subcupratus, Pseudaphycus austriacus, Tetracnemoidea piceae* and *Tetracnemus diversicornis* (Schmutterer, 1955c; Sugonjaev, 1962; Trjapitzin, 1978). Predator: *Exochomus quadripustulatus* (Schmutterer, 1952b).

63. **Phenacoccus pumilus** Kiritchenko – Dwarf Mealybug

Syn. & Lit.: *Phenacoccus pumilus* Kiritchenko, 1935:134; Borchsenius, 1949:233; Danzig, 1964:629; Matesova, 1968a:110; Ter-Grigorian, •1973:159; Tereznikova, •1975:225; Bazarov & Nurmamatov, 1975:54; Kosztarab & Kozár, 1978:53. *P. latus* Kiritchenkó, 1935:132 (see Borchsenius, 1949:234).

Female body covered with fine white wax powder. Adult female oval, pink, 1–3.7 mm long, 0.8–2.3 mm wide; antennae 9-segmented, ca 360 μ long, apical segment longest; stylet loop almost reaches line between mid-coxae; legs stout, ca 440 μ long, hind coxae without translucent pores; 10–18 pairs of cerarii, C_{18} with 2 slender setae, 1–2 shorter setae, and 4–7 trilocular pores; circulus absent; anal ring oval, with 1 outer row and 1 inner row of pores, setae 75–85 μ long; anal lobes wide, apical setae 145–163 μ long; trilocular pores numerous on both body surfaces; quinquelocular pores occasionally present in area of mouthparts and at bases of coxae; multilocular pores on dorsum form transverse rows on abdominal tergites 5 and 6, on venter rarely near bases of mid- and hind legs, also arranged in transverse rows on abdominal sternites 3–5 and in bands on sternites 6–8; oral collar ducts form transverse rows on most abdominal sternites and tergites; short thick setae on dorsum, on venter more slender and sparse. Three nymphal instars described by Ter-Grigorian (1973).

Distribution. Known only from USSR, Ukraine to Kazakhstan and Tadzhikistan.

Biology. A steppe inhabiting xerophilous species. On roots of a variety of dicotyledonous herbaceous plants mainly in the genera: *Achillea, Ajuga, Alhagi, Alyssum, Amaranthus, Artemisia, Beta, Chenopodium, Convolvulus, Cucumis, Daucus, Dianthus, Galium, Inula, Isatis, Lallemantia, Lepidium, Linaria, Medicago, Melilotus, Onobrychis, Plantago, Scabiosa, Scutellaria, Sonchus, Sysimbrium, Taraxacum, Teucrium, Thymus, Trigonella, Tragopogon, Vicia, Xeranthemum,* and *Ziziphora.* Occasionally a pest. Egg-laying females and first instars collected in Armenia from May through October (Ter-Grigorian, 1973). Biology and males unknown.

Genus **Polystomophora** Borchsenius

Syn. & Lit.: *Polystomophora* Borchsenius, 1948b:955; 1949:184; Danzig, 1964:627; Ter-Grigorian, 1973:109; Tereznikova, 1975:229; Kosztarab & Kozár, 1978:56; Vinis & Kozár, 1981:209.

Type-species: *Phenacoccus ostiaplurimus* Kiritchenko, 1940.

Female covered by white powdery wax. Adult female oval; antenna 9-segmented; labium large and wide at base; with 1 or 2 pairs of ostioles; legs thick, hind coxae with or without translucent pores, claw with denticle; cerarii absent or only C_{18} developed, this with 2 long thick setae and several trilocular pores; 1–5 oval circuli; anal ring with 2 rows of pores and 6 setae; anal lobes poorly developed, each with a long apical seta; with 3 types of disc pores; oral collar tubular ducts form transverse rows on both body surfaces; body setae short and slender.

Distribution. In the Palearctic Region with 2 species; 1 in CE.

Biology. On roots and leaves (grasses) or herbaceous plants (*P. orientalis* Matesova) and on trunk and branches of deciduous trees (*P. ostiaplurima* (Kiritchenko)).

64. **Polystomophora ostiaplurima** (Kiritchenko) – Maple Mealybug
Fig. 74

Syn. & Lit.: *Phenacoccus ostiaplurimus* Kiritchenko, 1940b:188. *P. ostiaplurima* Borchsenius, 1948b:955, 1949:184; Danzig, 1964:627; Zahradnik, 1972:399; Ter-Grigorian, 1973:109; Tereznikova, 1975:229; Kosztarab & Kozár, 1978:56; Vinis & Kozár,* 1981:209.

Adult female oval, with distinct body segmentation, light pink, up to 3 mm long, 1.2–1.8 mm wide; antennae of 8 or 9 thick segments; stylet loop as long as body or longer; only posterior ostioles developed; hind coxae with a few translucent pores; cerarii absent; with 5 oval circuli, first smallest; anal ring setae 60–65 μ long; apical setae 124–136 μ long; trilocular pores scattered on dorsum and on abdominal sternites, also in a marginal band on venter; quinquelocular pores occur singly on abdominal tergites and scattered on entire venter; multilocular pores form sparse transverse rows on entire dorsum and bands on terminal abdominal sternites; oral collar ducts common on both derms; body setae few, short, slender on both surfaces. Adult male and all nymphal stages described by Vinis and Kozár (1981).

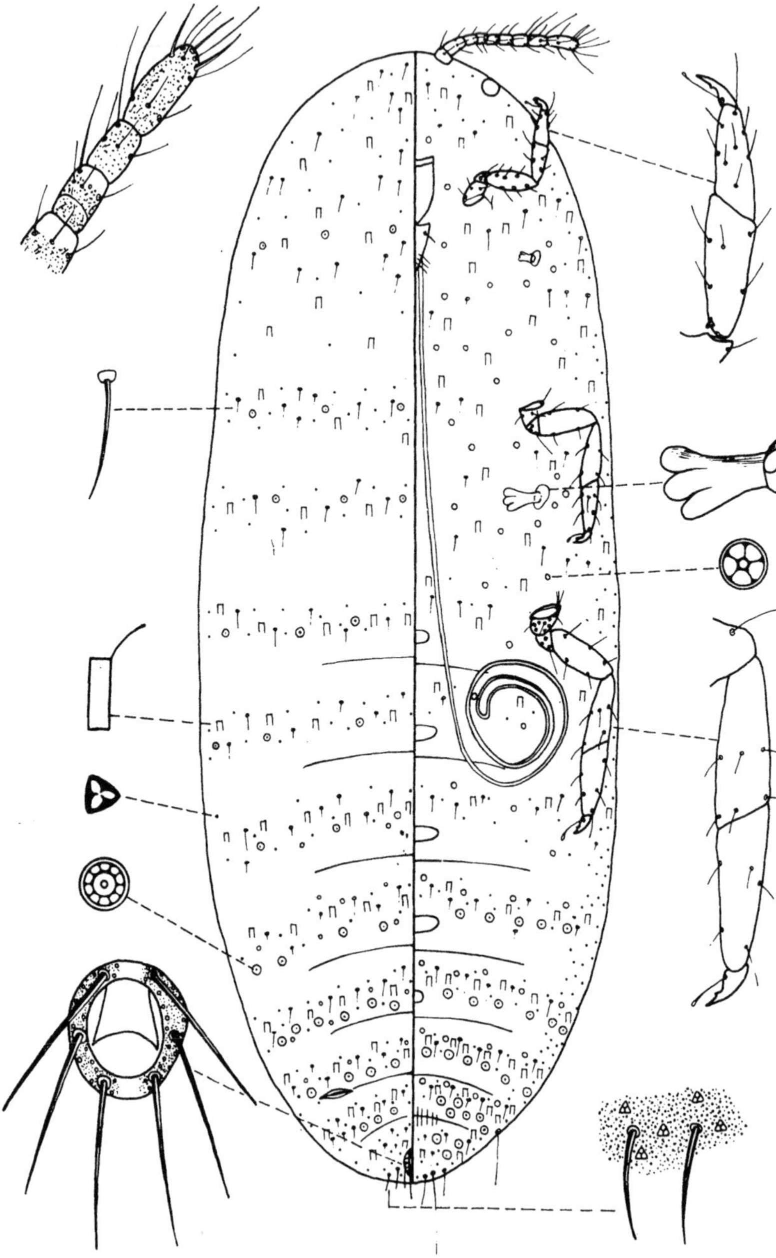

Fig. 74. *Polystomophora ostiaplurima* (Kiritchenko), female (original)

Distribution. Rare. Known only from Hu (Budapest) and USSR (areas of Kiev and Odessa, in Ukraine).

Biology. On trunks and branches of *Acer* and *Aesculus*. Often becomes a pest on roadside trees. First and second instars overwinter in bark crevices; adults appear from early April; start laying eggs (60–120 per female) from beginning of May through end of June; eggs hatch in 12–16 days, nymphs mature by beginning July. Eggs of second generation hatch in August and September. The two yearly generations overlap. Three male morphs occur, but all males of first generation are neotenic (Vinis and Kozár, 1981).

Natural enemy. Parasite: *Pseudaphycus hungaricus* (Kosztarab, 1956b).

Genus **Puto** Signoret

Syn. & Lit.: *Putonia* Signoret 1875:375 (preoccupied). *Puto* Signoret, 1875:428; Šulc, 1898:18; Ferris, 1950:190; Balachowsky, 1953c:301; Reyne, 1954:291; McKenzie, 1967:326; Koteja, 1974b:292; Tereznikova, 1975:238; Danzig, 1980a:110. *Ceroputo* Šulc, 1898:1; Borchsenius, 1949:285; Balachowsky, 1953:301; Reyne, 1954:291; Afifi, 1968:155; Ter-Grigorian, 1973:213; Kosztarab & Kozár, 1978:36. *Macrocerococcus* Leonardi, 1907:151; Borchsenius, 1948d:31; 1949:289; Schmutterer, 1952b:383; Balachowsky, 1953c:301; Ter-Grigorian, 1973:215; Kosztarab & Kozár, 1978:45.

Type-species: *Putonia antennata* Signoret, 1875.

Live females normally covered with an irregular mass of white waxy secretion, also short and wide wax filaments usually attached laterally and longer filaments arise from anal lobes; ovisac seldom formed. Adult females broadly oval, up to 5.0 mm long; antennae normally 9-segmented, seldom 8-; eyes often protruding; labium with 15–17 pairs of setae; dorsal ostioles present; legs well developed, trochanter with 3 sensoria on each surface, claw with a large subapical denticle, rarely with basal denticles (no claw denticles in *P. janetscheki*); cerarii 17–26 pairs on margin, but some in median area, each on a sclerotized plate with 6–20 or rarely more spines, mixed with trilocular pores, seldom with tubular ducts; circulus present or absent, when present, usually large, broadly oval, anal ring normally with 6 setae, rarely 8 or even more; each anal lobe usually with 3 to 7 setae; trilocular pores large, on both surfaces, characteristically constructed; quinquelocular pores in *P. pilosellae* only; multilocular pores absent from dorsum, present or absent on venter; oral collar tubular ducts elongate, seldom on dorsum, only few on sternites of abdomen, and on prosoma, occasionally larger oral-collar ducts on dorsum; dorsal setae short and thick, resembling setae in cerarii, often borne on a sclerotized tubercle, on venter elongate hairlike. Adult males described by Šulc (1898), Schmutterer (1952b), Reyne (1954), Beardsley (1962), and Afifi (1968); immatures by Šulc (1898).

Distribution. Primarily Holarctic, with ca 44 species world-wide, only few in the Neotropical, Oriental and Australian Regions. In CE 4 species.

Biology. On herbaceous and woody plants; often found at high altitudes under alpine conditions. Most species with 1 generation a year, but some with 1 generation every 2–4 years. Usually ovoviviparous. At least 3 species known to be of economic importance.

Note. Zahradnik (1959) and Ossiannilsson (1959) listed *Ceroputo balachowskyi*

Zahradnik & Schmutterer, but without a description; it is a nomen nudum, and all specimens were destroyed in mailing.

Reviewer's remark. The genera *Puto, Ceroputo* and *Macrocerococcus* are considered as distinct or synonymous by various authors. On the other hand, the North American species of *Puto* are fairly different from the Palearctic ones (Reyne, 1954: Nur, private communication) which are close to *Phenacoccus*. The American *Puto* exhibits several primitive characteristics which are unique among the Coccoidea (e. g. the XX–XO chromosome system) and has been elevated to family rank equivalent to Pseudococcidae (Beardsley, 1969a), while the Palearctic males *(Ceroputo pilosellae)* are considered to be closer to such genera as *Nairobia, Planococcus*, etc. than to *Puto* (Afifi, 1968). For further discussion see Koteja (1974b), Nur (1980), Miller (1984).

Key to species of *Puto*

1. Marginal cerarii 20–26 pairs; with 6 longitudinal rows of median cerarii; oral collar tubular ducts on venter of head only; usually with few quadrilocular pores on venter .... *superbus*, p. 139
   Marginal cerarii 17 or 18 pairs; with 0–4 longitudinal rows of median cerarii; oral collar tubular ducts not only on head; without quadrilocular pores on venter 2
2. Quinquelocular pores present on venter of prosoma; claw digitules thick and with large apical knob; with 18 pairs of marginal cerarii .... *pilosellae*, p. 137
   Quinquelocular pores absent from venter; claw digitules slender and pointed at apex; with 17 pairs of marginal cerarii 3
3. Claws without regular denticle; claws without spine-like basal denticles; anal ring with only 6 large setae; with 2 submedian and 2 submarginal longitudinal rows of accessory cerarii; feed on a variety of host plants, including *Juniperus* .... *janetscheki*, p. 137
   Claws with regular denticle; each claw with 2 spine-like basal denticles; anal ring with 6 large and 2–4 smaller accessory setae; without longitudinal rows of accessory cerarii in submedian and submarginal areas; feed on conifers only .... *antennatus*, p. 136

65. **Puto antennatus** Signoret – Conifer Mealybug

Syn. & Lit.: *Putonia antennata* Signoret, 1875:341. *Puto antennata* Šulc, 1898:18, table 2; Lindinger, 1909:225, 1912:258; Schimitchek, 1939:118; Reyne, *1954:291; Zahradnik, 1956:50, 1972:397, 1974:141; Afifi, 1968:10. *Ceroputo antennatus*, Balachowsky, 1932b:63.

Dorsum of female covered by white powdery wax; with small wax processes along body margin. Adult female oval, dark grey; according to Reyne (1954) 3.0–4.0 mm long. 2.0–2.5 mm wide; antennae 9-segmented, dark brown, ca 0.90–0.96 mm long; eyes sclerotized, protruding; stylet loop extends to line of midlegs; legs dark brown, with many setae; claw with large subapical denticle and 2 small spinelike basal denticles, claw digitules only slightly knobbed; cerarii, apparently 17 pairs along dorsal margin, anal lobe cerarii with 26–33 trilocular pores and 12–16 spines; circulus large with wrinkled rim; anal ring with 3–4 rows of pores, and 6 large setae, each 220–270 μ long, and 2–4 smaller accessory setae; similar trilocular pores on both body surfaces; multilocular pores with 8–10 or 12 loculi, most numerous on venter around vulva, on midventer mixed with trilocular pores; oral collar tubular ducts rare, most at end of body; dorsal setae on abdomen in transverse rows, and in bands near apex, ventral setae hairlike. Adult male and immature stages described by Reyne (1954).

Distribution. A rare species. In CE: Au, Cz, Fr, Ge and Sw.

Biology. Living at high elevations, e. g. 1,600–1,900 m in the Bavarian Alps; young nymphs feed on needles of conifers in the genera: *Abies, Picea,* and *Pinus;* they migrate to bark crevices. According to Schmutterer, as cited by Reyne (1954), these insects probably cannot feed on bark, but both sexes complete their development in bark crevices of the trunk and older branches. Adult males were observed during the last week of July; nymphs often crawl into empty cocoons of males. We still do not know where the eggs are laid or details on their life cycle.

66. **Puto janetscheki** Balachowsky – Alpine Mealybug

Syn. & Lit.: *Puto janetscheki* Balachowsky, *1953b:295; Zahradnik, 1956:50; 1957:5, 1974:141.

Adult female oval, ca 5.0–5.2 mm long; antennae slender, 9-segmented; eyes on a thick stalk; dorsal ostioles well-developed; claws without denticles; cerarii 16 or 17 dorsal marginal pairs, 10–12 submarginal pairs and ca 10–13 poorly developed submedian pairs; anal lobe cerarii each with ca 15–16 thick setae and 42–48 trilocular pores; circulus large; anal ring with 2–3 rows of pores and 6 setae; each anal lobe with a group of 5–7 long setae; trilocular pores distributed in large number on dorsum and venter; multilocular pores scattered on entire venter; oral collar tubular ducts on venter of two sizes, more slender on head region; short, wide on abdominal sternites 1–9; dorsal body setae short and thick, slender, hairlike on venter.

Distribution. In CE: Cz and Fr. Rare.

Biology. Found on roots and lower parts of *Alchemilla, Juniperus, Luzula, Poa* and *Vaccinium.* Apparently polyphagous. In Alpine habitats at elevations between 1,600–2,760 m. Biology unknown.

67. **Puto pilosellae** (Šulc) – Hairy Mealybug
Fig. 75

Syn. & Lit.: *Ceroputo pilosellae* Šulc, 1898:2; Kawecki, 1948:3; Borchsenius, 1949:286; Balachowsky, 1953c:301; Reyne, 1954:291; Koteja & Zak-Ogaza, 1966:317; 1969:359; Mitic-Muzina, 1968:303; Afifi, 1968:155; Koteja, 1974a:97; Zahradnik, 1974:142; Dziedzicka, 1977:65; Kosztarab & Kozár, 1978:36. Kozár et al., 1979:130. *Puto pilosellae,* Danzig & Kozár, 1974:11; Tereznikova, 1975:243; Danzig, 1980a:113.

Females covered with white wax plates, giving them the appearance of Ortheziidae. No egg sacs formed. Adult female oval, 1.7–3.5 mm long, 1.0–2.0 mm wide, tapering toward both ends, 1.7–3.5 mm long, 1.0–2.0 mm wide, distinctly segmented; antennae slender, 8- or 9-segmented; eyes with prominent basal cone; labium broad conical ca

105 μ long, with 15 pairs of setae, stylet loop slightly longer than labium; legs thick 940–1010 μ long, with many long spines, tibia twice as long as tarsus, claw with broad base and large denticle, claw digitules thick, with knobs, exceed claw by a quarter of its length; marginal cerarii 18 pairs, also 1 pair on top of head, each on a sclerotized plate with numerous thick setae and trilocular pores; circulus small, oval; anal ring with 2 rows of pores and 6 setae; anal lobes protruding; trilocular pores numerous on both

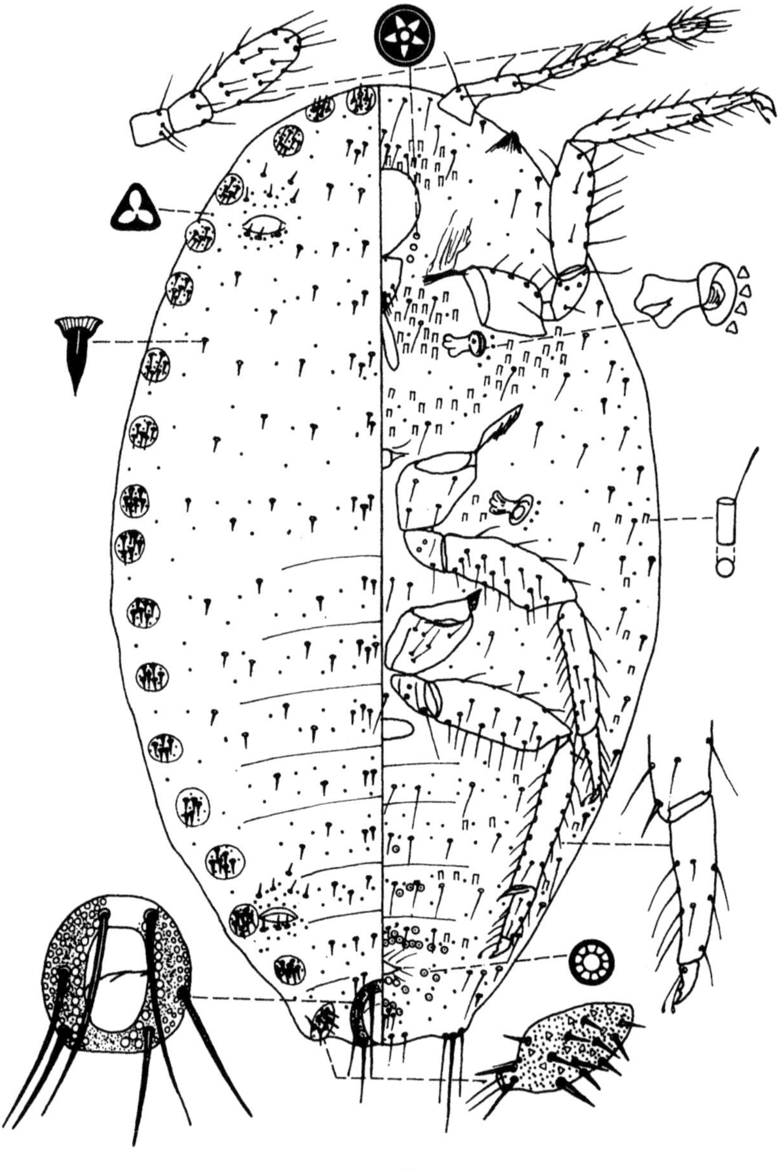

Fig. 75. *Puto pilosellae* Šulc, female (original)

surfaces; quinquelocular pores scattered on venter of prosóma; multilocular pores form short transverse rows on abdominal sternites 5 to 8; oral collar tubular ducts numerous on venter; conical setae form transverse rows on dorsum, slender setae on venter. Adult male described by Afifi (1968), first and second instars by Šulc (1898); also first instar by Tereznikova (1975).

Distribution. A common Trans-Palearctic species. In CE: Bu, Cz, Fr, Ge, Hu, Po, Sw, USSR (Ukraine, Tadzhikistan, Kazakhstan and Far East), Yu; also in Mongolia.

Biology. Polyphagous. On underside of leaves of plants of genera: *Achillea, Betonica, Carex, Centaurea, Dactylis, Fragaria, Galium, Geranium, Hierarcium Leontodon, Plantago, Taraxacum* and especially on *Thymus*. In heavy populations it became a pest of strawberries in Serbia (Yugoslavia) where second instars overwintered, adults matured by end of April, mated by second half of May; first instars of first generation emerged at end of May, beginning June; each entire life cycle lasted an average of 37 to 55 days (except for the overwintering generation), but only 31 days at constant temperature of 30°C; 2–4 yearly generations developed; ovoviviparous, adult males lived 1–2 days (Mitic–Muzina, 1968).

Natural enemy. Parasite: *Moraviella inexpectata* (Trjapitzin, 1978).

68. **Puto superbus** (Leonardi) – Superb Mealybug

Fig. 76

Syn. & Lit.: *Macrocerococcus superbus* Leonardi, 1907:152; Borchsenius, 1949:296; Schmutterer, 1952b:383; Reyne, 1954:319; Ter-Grigorian, 1973:218; Koteja, 1974a:97; Dziedzicka, 1977:64; Kosztarab & Kozár, 1978:45; Kozár et al., 1979:129. *Ceroputo volynicus* Nassonov, 1908b:471. *Ceroputo superbus*, Balachowsky, 1932b:63. *Puto superbus*, Balachowsky, 1953c:304; Tereznikova, 1975:241.

Body of female covered with 8 rows of white wax plates, resembling Ortheziidae. Adult female oval, 3,5–4.5 mm long, 2–2.5 mm wide; antennae 9-segmented, ca 1.45 mm long; eyes protruding dome-shaped; labium ca 250 μ long, with 16–17 pairs of setae, stylet loop extending halfway between anterior and posterior spiracles; ostioles well developed; legs robust, 1.5–1.8 mm long, claw with large denticle, claw digitules pointed, longer than claws; marginal cerarii 20–26 pairs, 6 more longitudinal rows of cerarii on most of dorsum, marginal cerarii each with 6–18 conical setae and 6–20 trilocular pores; circulus ca 300 μ wide, 70 μ long; anal ring with 6 setae, with 3–4 rows of pores, setae ca 335 μ long; trilocular pores numerous on both surfaces; a few quadrilocular pores on venter; most multilocular pores in middle of venter; oral collar tubular ducts narrow and long, grouped on head between mouthparts and antennae; trilocular pores at base of dorsal spines; venter with many long setae, these form bands on abdominal sternites. Adult male and first instar described by Leonardi (1907), male by Schmutterer (1952b) and Tereznikova (1975).

Distribution. Widely distributed in the Mediterranean area. In CE: Bu, Fr, Ge, Hu, It, USSR and Yu. Also known from Greece, Corsica, Sardinia, Algeria and Morocco.

Biology. Polyphagous, known to feed on stems and leaves of a large variety of herbaceous plants including 28 plant genera. In CE most common on grasses. Nymphs and eggs overwinter, adult females were observed developing by end of May, the wingless males were observed during lst third of June in S Germany by Schmutterer (1952b). Probably ovoviviparous. Prefers warm but shady areas in CE.

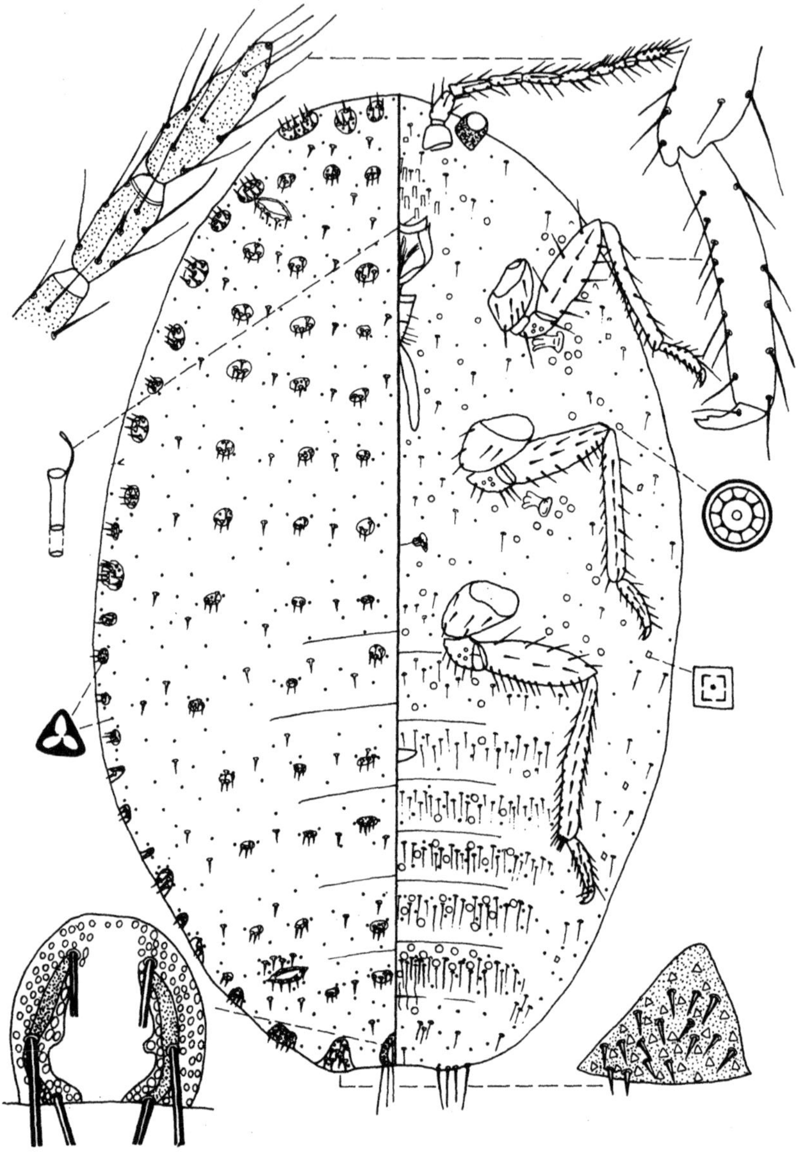

Fig. 76. *Puto superbus* (Leonardi), female (original)

Genus **Rhizoecus** Künckel d'Herculais

Syn. & Lit.: *Rhizoecus* Künckel d'Herculais, 1878:163; Borchsenius, 1949:174; Ferris, 1953:426; Borchsenius & Tereznikova, 1959a:322; Williams, 1962:41; DeLotto, 1964:381; McKenzie, 1967:370; Ter-Grigorian, 1973:89; Tereznikova, 1975:246; Hambleton, 1976:1; Kosztarab & Kozár, 1978:56; Danzig, 1980a:196.

Type-species: *Rhizoecus falcifer* Künckel d'Herculais, 1878.

Adult female elongate oval to broadly oval; usually with an irregular cephalic plate anterior of mouthparts; antennae geniculate, 5- or 6-segmented, apical segment with thick sensory setae, a straight one at apex and 2–3 on sides bent inwards; eyes present or absent; labium normally elongate with 20–28 setae; ostioles usually distinct, strongly sclerotized, seldom inconspicuous or absent; claws elongate without denticle, claw digitules usually short, setose, seldom longer than claw; cerarii absent in all CE species, normally replaced by 1 or 2 slender setae; 1 circulus or rarely 2–6; anal ring usually distinct with big oval pores and 6 setae; anal lobes seldom distinct, each with 3 or more apical setae; bitubular and/or tritubular ducts normally present; trilocular pores on both surfaces; quinquelocular pores and multilocular pores present or absent; tubular ducts of oral collar type; body setae of variable lengths, but mostly short.

Distribution. World-wide, with ca 89 species, but so far most (64) have been collected in the New World; in CE 5 species.

Biology. Most species are polyphagous; they infest rootlets of plants belonging to over 62 families; only few are monophagous. Several species are injurious, especially to tropical economic plants, also to nursery and greenhouse plants in the temperate regions. Species, especially in the tropics and greenhouses, develop in 1 ½ to 4 months, thus they can produce several generations each year. Males known in a few species. Reproduction oviparous or ovoviviparous; eggs laid in ovisacs, often in 1–3 batches.

Key to species of *Rhizoecus*

1. Both bitubular and tritubular pores absent; hind coxae with few large translucent pores; multilocular pores form segmental transverse rows almost on entire body surface .... *poltavae*, p. 145
   - Bitubular or tritubular pores present; apparently without translucent pores on hind coxae; multilocular pores restricted in distribution to certain areas of body or absent 2
2. With bitubular pores 3
   - With tritubular pores 4
3. Ostioles small, hardly noticeable; few multilocular pores on venter of prosoma; few oral rim ducts on both body surfaces .... *caesii*, p. 142
   - Ostioles normal, inner margin of lips sclerotized, each with ca 2 setae and 2–3 trilocular pores; no multilocular pores on venter of prosoma, all on 3 terminal sternites of abdomen; few oral collar ducts on both body surfaces .... *halophilus*, p. 144

4(2). Multilocular pores entirely absent; body 1.3–1.8 mm long; antennae 165–170 μ long .... *albidus*, p. 142
   - Multilocular pores present on 3 terminal sternites of abdomen; body ca 0.85 mm long; antennae ca 110 μ long .... *franconiae*, p. 144

69. **Rhizoecus albidus** Goux – White Root-Mealybug

Fig. 77

Syn. & Lit.: *Rhizoecus albidus* Goux, 1936:40, 1943a:53; Schmutterer, 1952b:394; Williams, *1962:41; Matesova, 1968a:107; Tereznikova, 1975:248; Kosztarab & Kozár, 1978:57; Danzig, 1980a:196. *R. uniporus* Borchsenius & Tereznikova, 1959a:323; Ter-Grigorian, 1973:92.

Adult female elongate oval, with posterior end rounded, white, 1.3–1.8 mm long, 0.5–0.8 mm wide; cephalic plate quadrate to circular with 4–6 marginal setae; antennae 6-segmented, according to Williams (1962) 165–170 μ long; ostioles distinct, each lip, except lower lip of posterior ostiole, with 2–4 setae and 4–7 trilocular pores; claw digitules as long as claw, or longer; circulus distinctly sclerotized, truncate-conical; anal ring setae twice as long as diameter of ring; apical setae about same length as anal ring setae; tritubular pores sparse on both surfaces, on dorsum 1–3 pores per segment, on venter fewer; trilocular pores on both surfaces; multilocular pores absent; oral collar ducts very small, sclerotized, sparse on both body surfaces; body setae short and slender, except longer ones on body margin. Adult male described by Schmutterer (1952b).

Distribution. A rare Trans-Palearctic species. In CE: Fr, Ge, Hu, USSR (Ukraine to Far East); also known from England and Greece.

Biology: A steppe-inhabiting mesophilous species. On rootlets of grasses: *Agrostis, Arrhenatherum, Corynephorus, Deschampsia, Festuca* and *Holcus*. According to Schmutterer (1952b), except first instar, all other stages overwinter. First generation adults developed in S. Germany by end of May, females were ovoviviparous; first instars appeared by end of July, and after three months developed into adults by first part of August; by end of August and by beginning of September the second and third instars developed; in October the adults appeared and their progeny overwintered. Ants associated: *Lasius flavus, L. niger, L. n. alienus* (Schmutterer, 1952b).

Natural enemies. Parasite: *Anomalicornia tenuicornis* (Trjapitzin, 1978).

Note. The species discussed under *R. uniporus* Borchsenius & Tereznikova, by Ter-Grigorian (1973) differs in its morphology, host (*Veronica*), and ecology (subalpine), from other population samples.

70. **Rhizoecus caesii** Schmutterer – Carnation Root Mealybug

Syn. & Lit.: *Rhizoecus caesii* Schmutterer, 1956b:516.

Adult female, according to Schmutterer (1956b), elongate oval, white, sparsely covered with powdery wax, ca 1.35 mm long, 0.65 mm wide; antennae 6-segmented, ca 150 μ long, apical and first segments longest, apical segment with 3 thick, lightly inward bent sensory setae, preapical segment with 1 similar seta; eyes present; labium elongate conical, ca 80 μ long, and 40 μ wide at base; ostioles small and hardly noticeable; legs relatively small, tibiae longer than tarsi, each hind tibia with 2 light spots on dorsal side, claws elongate and narrow, hind legs ca 250 μ long; circulus small, diameter ca 10 μ, located between abdominal segments 3 and 4; anal ring with outer row of small

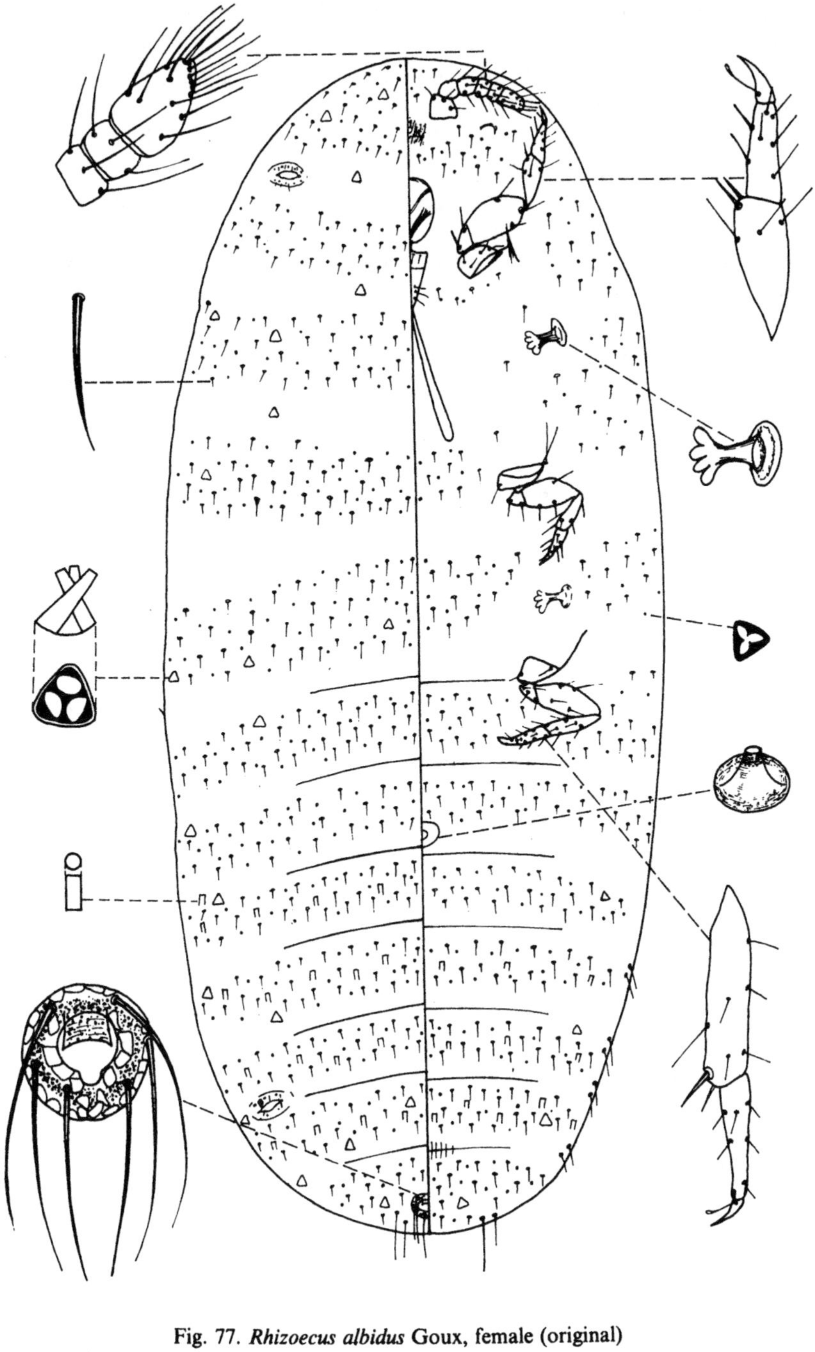

Fig. 77. *Rhizoecus albidus* Goux, female (original)

elongate light cells, inner row of wider darker cells; anal lobes indistinct, each with 3 setae; bitubular pores common on both surfaces; trilocular pores most common pore on both body surfaces; most multilocular pores with 10 loculi, located on terminal abdominal segments of venter, but not on dorsum of terminal segments, few single pores on prosoma of venter; oral rim ducts, ca 5 μ long, sparse, but present on both surfaces; apparently ducts and bitubular pores more numerous on dorsum than on venter; body setae of various sizes, largest setae on anal lobes, on margin of terminal abdominal segments and on head.

Distribution. Known only from western part of Germany near Bad Münster.

Biology. Collected from rootlets of *Dianthus caesius,* on the southern slope of a rocky cliff. Biology and males unknown.

Note. Due to the presence of bitubular pores, not tritubular ones, this species should probably be included in the genus *Ripersiella* Tinsley.

71. **Rhizoecus franconiae** Schmutterer – Franconian Root Mealybug

Syn. & Lit.: *Rhizoecus franconiae* Schmutterer, *1956b:519; Danzig, 1971b:209.

Adult female, according to Schmutterer (1956b), oval, white, sparsely covered with powdery wax, ca 0.85 mm long and 0.55 mm wide; antennae 6-segmented, ca 110 μ long, apical and first segments longest, apical segment with 3 thick lightly and inwardly bent sensory setae, preapical segment with 1 similar seta; labium ca 55 μ long, 38 μ wide at base; ostioles small, with few setae and trilocular pores on the lips; legs small, each hind tibia with 2 large, light, circular spots on dorsum, hind legs ca 200 μ long; circulus ca 6–7 μ in diameter, located between abdominal segments 3 and 4; anal ring outer pore row with small narrow, parallel-sided cells, middle pore row of large wide cells, inner pore of somewhat smaller and darker cells; anal lobes only slightly developed, each with 2 apical setae; tritubular pores on venter and dorsum not in definite longitudinal and transverse rows, absent from terminal segments of venter; trilocular pores numerous on both body surfaces; most multilocular pores with 10 loculi, in moderate numbers on venter of 3 terminal abdominal segments, none on dorsum; oral collar tubular ducts, ca 5 μ long, located singly on terminal abdominal segments also on body margin, including prosomal region of both body surfaces; tritubular and trilocular pores more numerous on dorsum than on venter; body setae few and relatively small, longest setae on terminal abdominal segments and on head.

Distribution. Known only from S Germany, near Erlangen and Berchtesgaden.

Biology. On rootlets of *Achillea, Corynephorus, Hieracium* and *Tanacetum.* Females during oviposition found inside white feltlike ovisac. Biology and males unknown.

72. **Rhizoecus halophilus** (Hardy) – Hardy Root Mealybug

Syn. & Lit.: *Coccus halophilus* Hardy, 1868:136. *R. halophilus,* Lindinger, 1938a:23; Williams, *1962:47; Koteja, 1972:569, 1974a:101; Tereznikova, 1975:251; Zahradnik, 1977:122; Kosztarab & Kozár, 1978:57. *R. pratensis,* Borchsenius & Tereznikova 1959a:322. *Ripersia halophila,* Newstead, 1903:192.

Female covered with white wax powder, with 2 wax filaments at caudal end. Adult female oval, opaque white, 1.5–1.9 mm long, 1.3 mm wide; cephalic plate of irregular shape, bordered by ca 8 setae; antennae 6-segmented, ca 155–170 μ long, apical segment longest and with at least 2 thick sensory setae; eyes minute; labium 68–75 μ long, with 28 setae; stylet loop reaches line between mid-coxae; ostioles with inner lip margin sclerotized, each lip with ca 2 setae and 2–3 trilocular pores; legs normally developed, claw digitules slightly longer than claw and with very small apical knob; circulus small, sclerotized, truncate conical, its center with small circles; anal ring with 2½ pore rows, its setae twice as long as ring diameter; anal lobes poorly developed, each with 3 setae, about as long as anal ring setae; bitubular pores sparse, on dorsum about 8 on each body margin, and 2–3 on midline, none on venter (except in specimens from Poland); few trilocular pores on each body surface; multilocular pores totaling 30–50, on 3 terminal sternites of abdomen; oral collar tubular ducts few on both body surfaces; body setae generally short, except those on margin of abdomen.

Distribution. A rare species. In CE: Cz, Ge, Hu, Po, USSR (Ukraine). Also known from UK.

Biology. On roots of various plants: *Armeria, Artemisia, Calluna, Erica, Hieracium, Ligusticum, Radiola,* also on undetermined grasses. Biology and males unknown.

Note. Weidner (1970) reported nymphs of this species invading a home with thatched roof in Sylt, Germany. We believe that it probably was *Balanococcus* sp., not a *Rhizoecus*. This species apparently belongs to *Ripersiella.*

73. **Rhizoecus poltavae** Laing – Laing's Root Mealybug

Syn. & Lit.: *Rhizoecus poltavae* Laing, 1929:469; Kiritchenko, 1931:312; Goux, 1943:47; Borchsenius, 1949:117; Tereznikova, *1975:251; Kosztarab & Kozár, 1978:57. *Radicoccus poltavae,* Hambleton, 1946:50.

Adult female broadly oval, ca 2.1 mm long, 1.3 mm wide; antennae 6-segmented, apical segment twice as long as any other segment; eyes apparently absent; stylet loop twice longer than labium; legs of thick segments, hind coxae with few large translucent pores, each hind tibia with 2 large subcircular spots on dorsal outer margin; circulus small, round; anal ring circular with 2–3 rows of elongate cells, setae ca 60 μ long; anal lobes not projecting, each with 3 apical setae, ca 60 μ long; bi- and tritubular pores apparently absent; trilocular pores very sparsely scattered on both body surfaces; multilocular pores form segmental transverse rows on both body surfaces, with double rows on sternites 7 and 8 of abdomen; few oral collar tubular ducts along body margin of both surfaces; body hairs slender, few on entire derm.

Distribution. Known only from Hu, USSR, Poltava area in the Ukraine.

Biology. Biology, host, adult males and immature stages unknown.

Note. Due to the absence of bi- or tritubular pores, this species probably should not be included in *Rhizoecus* or *Radicoccus* Hambleton. A new genus might need to be erected for its inclusion after the examination of the type material, not seen by us at present.

Genus **Rhodania** Goux

Syn. & Lit.: *Rhodania* Goux, 1935:291; 1940:57; Borchsenius, 1949:187; Gomez-Menor, 1956a:615; Danzig, 1964:626; Ter-Grigorian, 1973:113; Tereznikova, 1975:254; Kosztarab & Kozár, 1978:58.

Type species: *Rhodania porifera* Goux, 1935.

Females covered with white powdery wax. Adult female circular to elongate-ovoid; anal lobes not developed; antennae 6- or 7-segmented, segments stout; dorsal ostioles absent; legs thick, very small; spiracles with associated disc pores; cerarii and circuli absent; anal ring wide with 6 setae and 1 or 2 rows of pores; anal lobes poorly developed; trilocular pores absent; numerous quinquelocular pores on both body surfaces; without multilocular pores; oral collar tubular ducts short, on both body surfaces; body hairs short and slender. Nymphs described by Goux (1935) and Schmutterer (1952c).

Distribution. In the Palearctic Region with 3 species, 2 in CE.

Biology. Feed on roots and at base of leaves of grasses in the genera: *Agrostis, Corynephorus* and *Festuca.*

Key to species of *Rhodania*

Antennae 7-segmented, ca 180 μ long; hind tibiae thickened; body circular .... *porifera,* p. 147

Antennae 6-segmented, ca 140 μ long; hind tibiae normal; body elongate-oval .... *occulta,* p. 146

74. **Rhodania occulta** Schmutterer – Schmutterer's Mealybug

Syn. & Lit.: *Rhodania occulta* Schmutterer, *1952c:555; Koteja & Zak-Ogaza, 1966:319; Koteja, 1974a:99; Koteja & Liniowska, 1976:662; Kosztarab & Kozár, 1978:58.

Adult female, according to Schmutterer (1952c), elongate oval, light yellow, ca 1.9 mm long, 0.95 mm wide, body segmentation well visible; antenna 6-segmented, ca 140 μ long; eyes small but distinct, near antennal base; labium about as long as wide, ca 52 μ long, with 28 setae; leg segments normal; spiracles large, oval, each atrium with 4–8 quinquelocular pores; anal ring with ca 36 pores and 6 long slender setae; numerous quinquelocular pores and oral collar ducts on both surfaces of body; body setae short, and sparse on both surfaces; 1 small seta in place of C_{18}. First instars described by Schmutterer (1952c).

Distribution. Known in CE from Ge and Po; it was also reported from Holland.

Biology. According to Schmutterer (1952c) it has one yearly generation; adult females overwinter; lay eggs during second half of April–first half of May; eggs hatch in about 2 weeks after egg laying; adults develop by September. Feed underside leaf blades, at base of stem on *Agrostis* and *Corynephorus.*

75. **Rhodania porifera** Goux – Pore-bearing Mealybug
Fig. 78

Syn. & Lit.: *Rhodania porifera* Goux, 1935:291, 1936:35; Borchsenius, 1949:187; Schmutterer, 1952b:92, 393; Danzig, 1964:626; Ter-Grigorian, 1973:113; Koteja, 1974a:99; Tereznikova, *1975:254; Kosztarab & Kozár, 1978:58.

Adult female nearly circular, pinkish, 1.7–2.5 mm long and 1.2–2.0 mm wide; antennae 7-segmented, ca 180 μ long; labium wide at base, ca 64 μ long, with 28 setae, stylet loop nearly reaches mesothoracic coxae; hind tibia thickened, with many translucent pores; each spiracle with 2–4 associated quinquelocular pores; anal ring with 16–20 pores and with 6 spinelike setae, each 14–23 μ long; anal tube well developed; apical seta 70–90 μ long; quinquelocular pores and short oral collar ducts on both body surfaces; body setae slender, ca. 9 μ long; 2 setae in place of C_{18}. Adult males, 1st and 2nd instars described by Goux (1935), also 1st instars by Schmutterer (1952b), 2nd instars by Ter-Grigorian (1973).

Distribution. A Trans-Palearctic species. In CE: Fr, Ge, Po and USSR (from Transcarpathian Region to Armenia); also known from Mongolia.

Biology. A steppe-inhabiting xerophilous species that tolerates warm conditions. Living on roots of *Festuca* and other grasses. Sometimes a pest on pastures (Ter-Grigorian, 1973). Adult females in Germany (Schmutterer, 1952b) overwinter, start laying eggs (52–148 per female) from mid-May, these hatch in 10–14 days, from first part of June; adults appear in August. Males observed in S France (Goux, 1935). Schmutterer (1952b) found the ants *Lasius flavus* and *L. niger alienus* visiting *R. porifera* colonies. In Poland young females and males collected with Berlese funnel in early July.

Genus **Ritsemia** Lichtenstein

Syn. & Lit.: *Ritsemia* Lichtenstein, 1879:387, 1882b:329; Lindinger, 1912:350; Kozár & Vinis, 1980:97. *Ripersia* (in part), Fernald, 1903:119. *Physococcus* Hadzibejli, 1958:903; Ter-Grigorian, 1973:111.

Type-species: *Ritsemia pupifera* Lichtenstein, 1879.

Adult female oval, dorsum convex, sclerotized; antennae 7–9-segmented; basal eye cone well-developed; ostioles absent; legs thick, sclerotized, claw with denticle; spiracles with associated disc pores, with 3-, 4-, 5-loculi; only C_{18} developed, but reduced; with 5 oval circuli; anal ring circular with 2 rows of pores and 6 setae; anal lobes poorly developed; with 4 kinds of disc pores on both body surfaces; tubular ducts absent; body setae form transverse rows on both dorsum and on abdominal sternites.

More information at *R. pupifera,* the only included species.

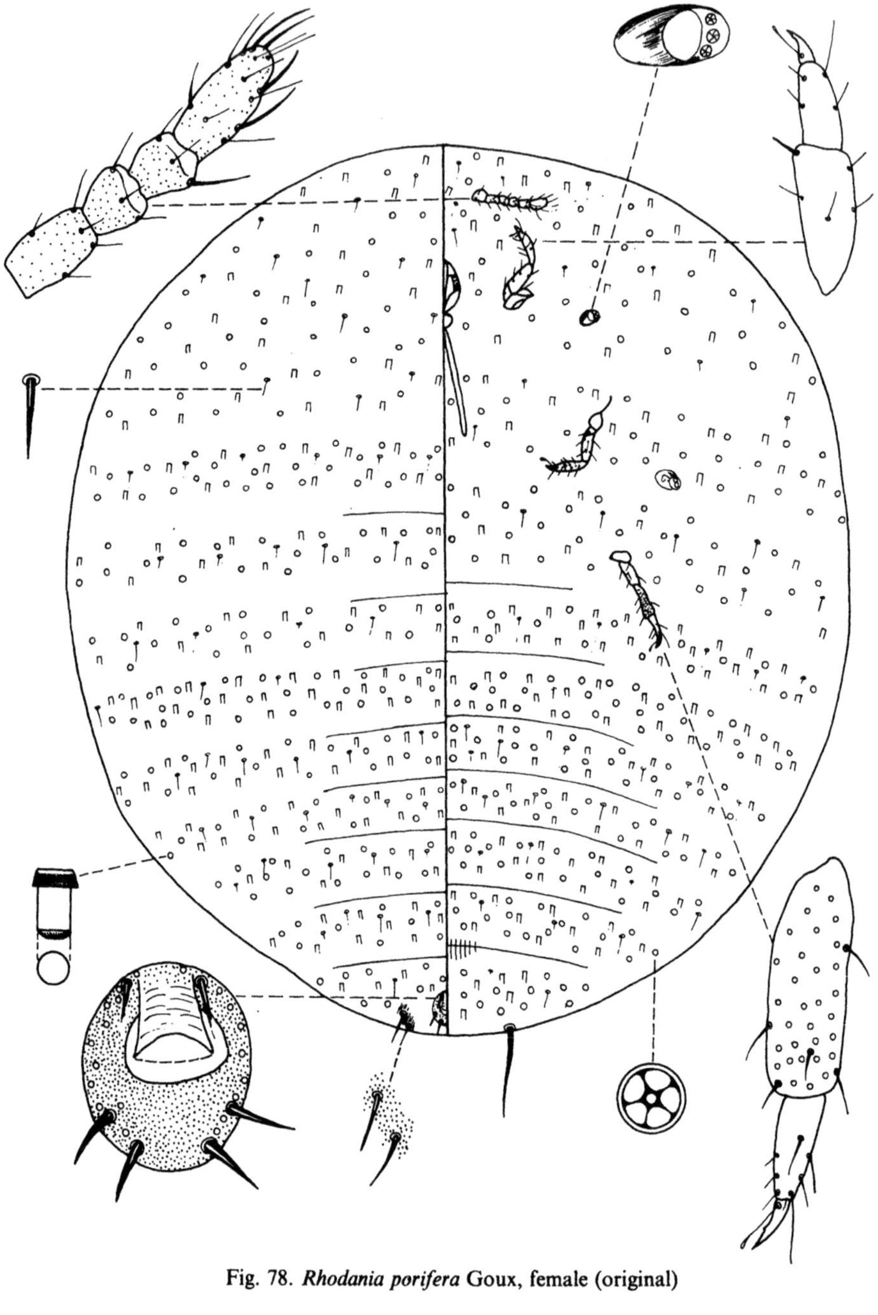

Fig. 78. *Rhodania porifera* Goux, female (original)

76. **Ritsemia pupifera** Lichtenstein – Elm Bark Scale
Fig. 79

Syn. & Lit.: *Ritsemia pupifera* Lichtenstein, 1879:387; Lindinger, 1912:350, 1949:212; Kozár & Vinis, 1980:97. *Physococcus nanus* Hadzibejli, 1958:903; Ter-Grigorian, 1973:112.

Adult female raspberry-red colored, 0.45–3 mm long, 0.3–1.6 mm wide, resembles a *Kermes* species; antenna 240 μ long; labium ca 62μ long and wide, stylet loop longer than body; eye diameter 11 μ; claw digitules much longer than claw; each C_{18} reduced, with 2 slender setae, each 23 μ long, and 2–3 trilocular pores; diameter of circuli 26–36 μ; anal ring setae 43 μ long; apical setae 132 μ long; disc pores with 3-, 4-, 5-, and 6-loculi on both body surfaces; setae shorter on dorsum than on venter. Adult male described by Hadzibejli (1958), Kozár and Vinis (1980); immature stages by Hadzibejli (1958), Ter-Grigorian (1973), Kozár and Vinis (1980).

Distribution. A rare species. Known only from Fr, Hu, and USSR (Caucasus).

Biology. Lives in bark crevices of elm, *Ulmus* spp.; it has 1 yearly generation; second instars overwinter in the bark crevices; adults emerge during April–May; mass male emergence observed on April 24, 1979, in Budapest; females lay eggs (counted 50–60 embryos per female); eggs hatch from end of June until end of August (Kozár and Vinis, 1980).

Genus **Saccharicoccus** Ferris

Syn. & Lit.: *Saccharicoccus* Ferris, 1950:216; Williams, *1962:50, *1970:171; Kosztarab & Kozár, 1978:58.

Type-species: *Dactylopius sacchari* Cockerell, 1895.

Adult female oval or elongate oval, antennae 7-, rarely 8-segmented; labium conical with 28 setae; ostioles poorly or well developed, lips with 0–2 setae and 2–3 trilocular pores; claws without denticle, hind coxae with a few translucent pores in type-species; cerarii on anal lobes only, each with 2 short setae and 7–11 trilocular pores; circulus usually large, hourglass-shaped, longer than wide; 6 anal ring setae, these twice as long as diameter of anal ring; anal lobes moderately to well developed, each with 1 long apical seta; trilocular and multilocular pores present on both body surfaces; large number of irregularly shaped simple pores around hind coxae; tubular ducts of 1 or 2 sizes, all oral collar type, these either few on venter of abdomen or numerous on both surfaces; body setae short, except on margin of terminal abdominal segments.

Distribution. World-wide with 2 species; 1 in CE.

Biology. The biology of *S. sacchari* (Cockerell) was treated in detail by Uichanco and Villanueva (1932), also by Beardsley (1962b). Both species are known from grasses. *S. sacchari* is a pest on *Saccharum officinarum,* but also infests rice, sorghum, and other cultivated plants; this species develops a new generation every 20 to 40 days in S. Asia, females lay ca 200 eggs.

Note. This genus is close to *Trionymus*. For discussion see Reviewer's remark under *Trionymus*.

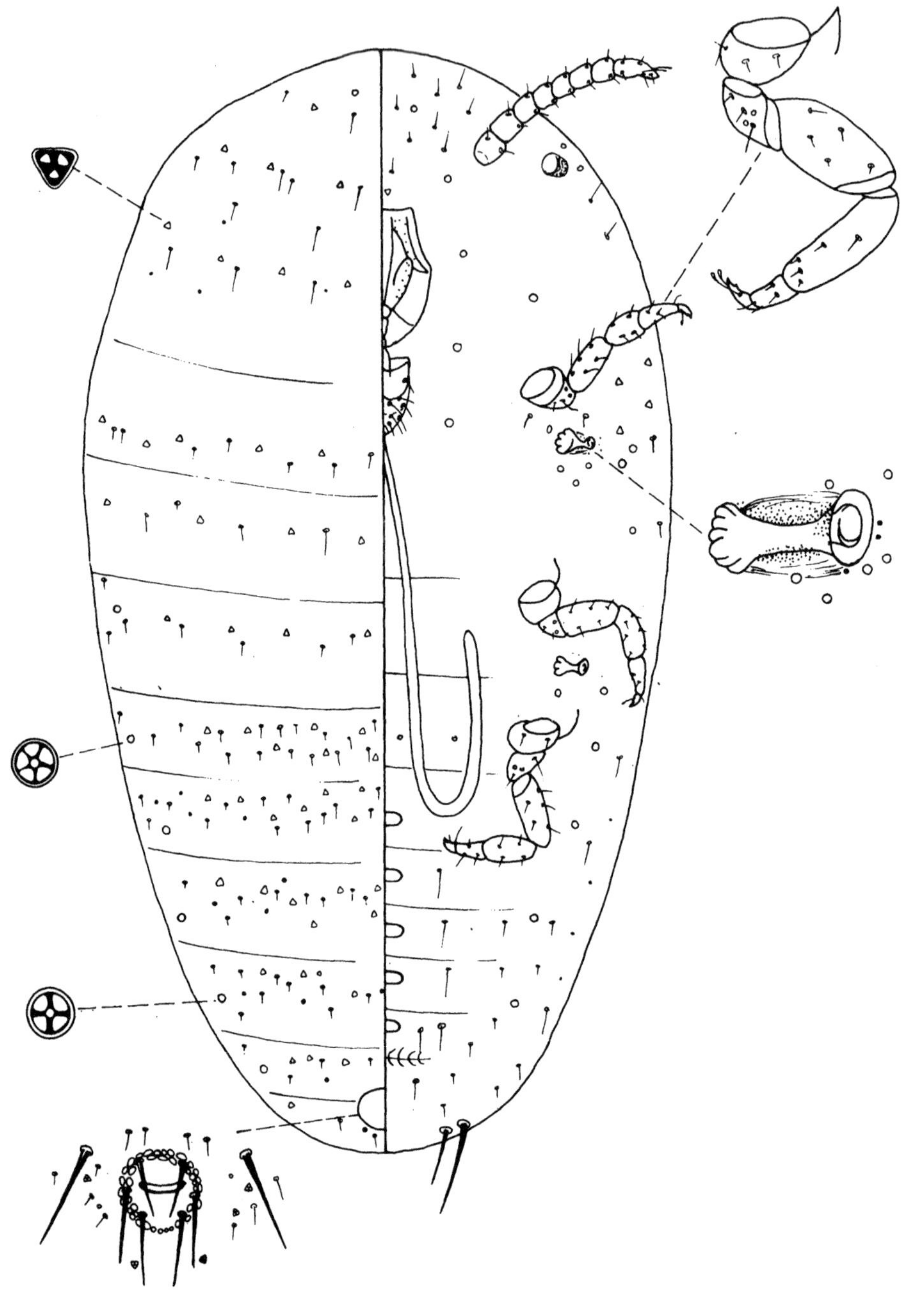

Fig. 79. *Ritsemia pupifera* Lichtenstein, female (modified after Kozár and Vinis, 1980)

77. **Saccharicoccus penium** Williams – Williams' Grass Mealybug
Fig. 80

Syn. & Lit.: *Saccharicoccus penium* Williams, *1962:50; Koteja, 1969:4; Danzig, 1968:499; Koteja & Zak-Ogaza, 1969:361; Kosztarab & Kozár, 1978:59. *Trionymus penium*, Danzig, 1980a:183. *T. isfarensis* Danzig, 1983:518.

Adult female elongate, almost parallel-sided, according to Williams (1962), up to 3.5 mm long; antennae 7-segmented, ca 240–250 μ long, terminal segment longest; labium 70–90 μ long; stylet loop 3–4 times as long as labium; ostioles poorly developed, lips without setae and each with 2–3 trilocular pores; legs slender; each C_{18} with 2 short conical setae, 1 long slender seta, and 7–9 trilocular pores; circulus between abdominal segments 4 and 5, elongate hourglass-like, rarely round, oval, small and occasionally absent; anal ring circular; anal lobes moderately developed; apical setae as long as anal ring setae; minute simple pores of different shape and size around hind legs; trilocular pores evenly distributed on both body surfaces, more sparse on venter; multilocular pores forming transverse rows on posterior ends of tergites and sternites 5–8 of abdomen, scattered on rest of dorsum and venter, except more numerous on sternites 9–10 of abdomen; oral collar tubular ducts of 2 types, smaller and larger, each on both body surfaces; body setae short and slender, longer on venter.

Distribution. In CE: Po and USSR (Leningrad Province to Far East); also known from England.

Biology. Found at base of grasses and in leafsheaths: *Agropyron, Agrostis, Beckmannia, Festuca* and *Poa*. Nymphs, egg laying and parasitized females were collected in Poland during August–September. Biology and males unknown.

Genus **Spilococcus** Ferris

Syn. & Lit.: *Spilococcus* Ferris, 1950:219; Williams, 1962:52; McKenzie, *1967:410; Danzig, 1980a:158.

Type-species: *Dactylopius gutierreziae* Cockerell, 1896.

Females normally covered with wax powder and form ovisacs. Adult female elongate to broadly oval; antennae 7- or 8-segmented; ostioles usually well developed and lips with setae and trilocular pores; legs range from small and stout to long and slender, hind coxae and tibiae often with translucent pores, claws with or without denticle. Ferris (1950) described 0–17 pairs of cerarii, while McKenzie (1967) limited the genus to 6–17 pairs of cerarii, (we noted 7 pairs in type-species); circulus present or absent; anal ring at posterior apex of abdomen, with 6 setae, these about twice as long as diameter of ring; anal lobes never prominent; trilocular pores on both body surfaces; multilocular pores always on venter of abdomen, rarely extending to prosoma; oral collar and oral rim ducts of various sizes usually present on both body surfaces; body setae short and slender on dorsum, on venter most are longer, especially on abdomen.

Distribution. World-wide with over 30 species, only 1 in CE.

Biology. On woody and herbaceous plants, from roots to foliage. Little is known about their biology.

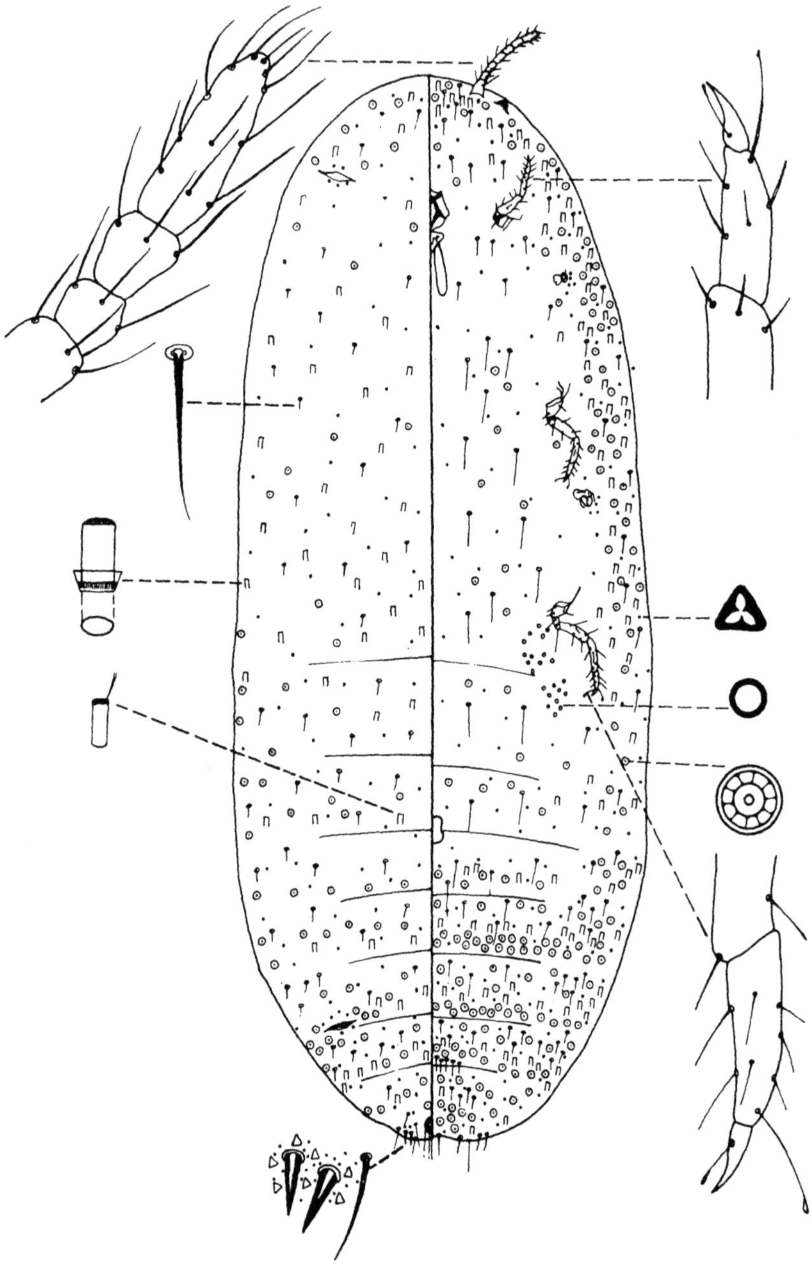

Fig. 80. *Saccharicoccus penium* Williams, female (original)

78. **Spilococcus nanae** Schmutterer – Birch Mealybug

Syn. & Lit.: *Spilococcus nanae* Schmutterer, 1957:70; Bachmaier, 1963:553, 1965:142; Danzig, 1968:499; Zahradnik, 1972:400.

Female lightly covered with wax powder. Ovisacs of loose wax filaments. Adult female, according to Schmutterer (1957), oval, reddish-violet, ca 1.5 mm long, 1 mm wide; antennae ca 300 μ long, lightly elbowed, 7-segmented, apical segment as long as any other 2 segments; eyes fairly large; labium triangular; ostioles poorly developed, lips with a few trilocular pores and setae; legs normally developed, femur and tibia of similar length, tibia about twice as long as tarsus, hind coxae with a few translucent pores, hind leg ca 450 μ long; anterior abdominal segments and prosoma without cerarii, apparently with at least 4 pairs of cerarii well recognizable, C_{18} with 2 relatively thick conical setae and ca 15 trilocular pores; circulus almost oval, ca 35 μ long and 20 μ wide, between sternites 3 and 4 of abdomen; anal ring with 6 setae each, ca 100 μ long; anal tube distinct, well sclerotized; anal lobes poorly developed, apical setae ca 140 μ long; trilocular pores numerous on both body surfaces; multilocular pores few, only on terminal sternites of abdomen; oral collar and oral rim ducts on both body surfaces, except anal lobes; body setae of various sizes, larger setae on head and terminal sternites of abdomen, on dorsum these shorter. First instar described by Schmutterer (1957).

Distribution. Known only from S Germany and USSR (Northern Karelia). In Germany in high moors, at 700–1,000 m elevation.

Biology. Under the bud scales and on twigs of *Betula nana*. Reproduction biparental. It has 1 generation a year. Females lay about 20 eggs in egg sacs during beginning of July. Both sexes overwinter as N_2, seldom as N_3 in females.

Natural enemies. Parasites: *Allotropa mecrida, Leptomastidea bifasciata, Leptomastix brevipennis* and *Hungariella spilococci* (Bachmaier, 1965).

Note. If we accept McKenzie's (1967) interpretation of the genus, this species would be included in the genus *Chorizococcus* Ferris.

Genus **Spinococcus** Borchsenius

Syn. & Lit.: *Spinococcus* Borchsenius; Kiritchenko, 1931:314 (a nomen nudum); Borchsenius, 1948b:953, 1949:203; Danzig, 1960:178, 1964:628; Williams, 1962:56; Ter-Grigorian, 1973:127; Tereznikova, 1975:257; Kosztarab & Kozár, 1978:60. *Synacanthococcus*, Balachowsky, 1953:282. *Peliococcus*, Danzig, 1980a:118.

Type-species: *Acanthococcus marrubii* Kiritchenko, 1936.

Female covered with white powdery wax. Ovisac woolly, white. Adult female oval; antennae 8- or 9-segmented; eyes oval, on small basal cone, protruding; labium conical; ostioles well-developed; legs normal, claw with distinct denticle; 17–18 pairs of cerarii, each normally on elevated sclerotized tubercle, with 2 conical setae and 2–10 trilocular pores; circulus present or absent; anal ring oval, with 1 inner and 1 or 2 outer rows of pores and 6 setae; anal lobes distinct with apical and subapical setae; trilocular pores on both body surfaces; quinquelocular pores on venter, but sometimes absent; multilocular pores form transverse rows and bands on sternites of abdomen, but often on both body surfaces; oral collar tubular ducts on entire body, often of 2 sizes; dorsal body setae conical, similar to cerarian setae, form segmental transverse rows and usually a longitudinal row on mid-line of body; most ventral setae slender, hairlike.

Distribution. Trans-Palearctic, except 1 species in S Africa. About 8 species are recognized at present, but some may belong to other genera. Only 3 species in CE.

Biology. On a variety of herbaceous and woody plants. Apparently the biology was studied for only *S. calluneti* Lindinger by Schmutterer (1952b). For records see that species.

Reviewers' remark. The diagnostic characters of *Spinococcus*, as currently understood, are: cerarii consisting of 2 closely arranged setae and some trilocular pores, and the dorsal setae are similar to cerarian setae, born on a sclerotized tubercle and associated with trilocular pores. The type species of *Peliococcus*, *P. chersonensis*, and other species, share characters of both these genera. In the type species of *Spinococcus*, *S. marrubii*, the multilocular pores are not characteristically arranged, but are in *S. calluneti* and some other species. For these reasons Danzig (1980a) reduced *Spinococcus* to a synonym of *Peliococcus*, but she suggested that *Peliococcus balteatus* and some others are not related with the *Peliococcus–Spinococcus* complex and a new genus will have to be established for them.

Key to species of *Spinococcus*

1. Multilocular pores on both body surfaces, forming transverse rows on dorsum; circulus present 2
   Multilocular pores restricted to venter of terminal abdominal segments; circulus absent . . . . *marrubii*, p. 155
2. Circulus oval and large; C_{18} with 1 trilocular pore, not on sclerotized plate; second segment of antennae longest . . . . *morrisoni*, p. 157
   Circulus round and small; C_{18} with 10–12 trilocular pores on large sclerotized plate; apical segment of antennae longest . . . . *calluneti*, p. 154

79. **Spinococcus calluneti** (Lindinger) – Heather Mealybug

Syn. & Lit.: *Pseudococcus calluneti* Lindinger, 1912:90; Schmutterer, 1952b:390. *Parapedronia calluneti*, Balachowsky, 1953:230. *S. calluneti*, Zahradnik, 1959b:537; Danzig, 1960:176, 1964:628; Williams, *1962:56; Koteja, 1974a:99; Tereznikova, 1975:259; Kosztarab & Kozár, 1978:60.

Adult female oval, 1.5–2.5 mm long, 0.7–1. mm wide; antennae with 8 segments rarely with 7 or 9, 245–324 μ long; labium ca 83 μ long with 30 setae, stylet loop twice as long as labium; inner margins of ostiole lips sclerotized, each lip normally with 3 trilocular pores; claws stout, with enlarged base of plantar surface, and with a distinct denticle; 18 pairs of cerarii, C_{18} on a large sclerotized plate, with 2 conical setae, 1–2 small setae and 10–12 trilocular pores; dorsal cerarii on midline, each with 2 conical setae and 3–4 trilocular pores; circulus small, circular, on segment 4; anal ring with 2 pore rows, and setae twice as long as diameter of ring; anal lobes moderately developed, apical setae 120–138 μ long; trilocular pores evenly distributed on dorsum; few on venter, these form a marginal band; quinquelocular pores in median area of venter; multilocular pores form transverse segmental rows on both body surfaces, except on venter of prosoma; oral collar ducts on dorsum interspersed with multilocular pores, on venter of 2 sizes, larger ones distributed similarly to those on dorsum, smaller type mostly in median area of abdomen; dorsal setae short, conical, similar to cerarian setae, form transverse rows; ventral setae slender, of various lengths. Adult male described by Schmutterer (1952b).

Distribution. A common species. In CE: Cz, Ge, Po and USSR. Also known from England, Denmark and Sweden.

Biology. On roots, underground stems and branches of *Calluna, Empetrum, Erica, Fragaria, Oxycoccus, Ramischia* and *Vaccinium*. Eggs overwinter in Germany (Schmutterer, 1952b), and hatch during April; adults develop by end of May and first half of June; most eggs are laid from end of June through mid-July, 18–68 eggs per female, these hatch from mid-July; adults of second generation develop by end of September. Associated ants: *Formica fusca fusca, Lasius niger alienus, Tetramorium caespitum* (Schmutterer, 1952b).

Natural enemies. Parasites: *Allotropa mecrida, Anagyrietta pantherina, Leptomastix epona* (Trjapitzin, 1978), *L. histrio* (Schmutterer, 1955c). Predator: *Novius testaceus* (Schmutterer, 1952b).

80. **Spinococcus marrubii** (Kiritchenko) – Hoarhound Mealybug
Fig. 81

Syn. & Lit.: *Acanthococcus marrubii* Kiritchenko, 1935:156. *Phenacoccus multispinosus* Borchsenius, 1937:55. *Spinococcus marrubii*, Borchsenius, 1949:204; Danzig, 1964:629; Matesova, 1968:109; Ter-Grigorian, 1973:128; *Tereznikova, 1975:257; Kozár et al., 1979:130.

Ovisac cottony, female covered with white wax powder. Adult female oval, pink, 2.6–5.0 mm long and 1.0–2.5 mm wide; antennae 9-segmented, ca 490 μ long, 2nd segment longest; stylet loop reaches line between mid-coxae; legs large and thick, ca 580–640 μ long; 18 pairs of cerarii along body margin and 9–10 pairs along midline of body, each with 2 conical setae and trilocular pores; circulus absent; anal ring with 1 inner and 2 outer rows of pores, setae 125–150 μ long; anal lobes conical, apical setae 170–195 μ long; trilocular pores numerous on terminal tergites of abdomen and along body margin, fewer on ventral surface; quinquelocular pores on middle of prosomal venter and on sternites 1–5 of abdomen; a few multilocular pores form transverse rows and bands on terminal abdominal sternites; oral collar tubular ducts numerous, distributed similarly to trilocular pores; dorsal conical setae form sparse transverse rows, with 1 trilocular pore located near base of each seta, other smaller setae also present on dorsum; ventral setae more numerous, of different lengths. Second instar described by Ter-Grigorian (1973).

Distribution. Known from Bu, Hu, and USSR (Ukraine, Caucasus and Kazakhstan).

Biology. A steppe inhabiting species. On roots of *Marrubium, Thymus,* and grasses. In Hungary was found at the base of *Euphorbia* stems, and egg laying females were present from June through August. Males unknown.

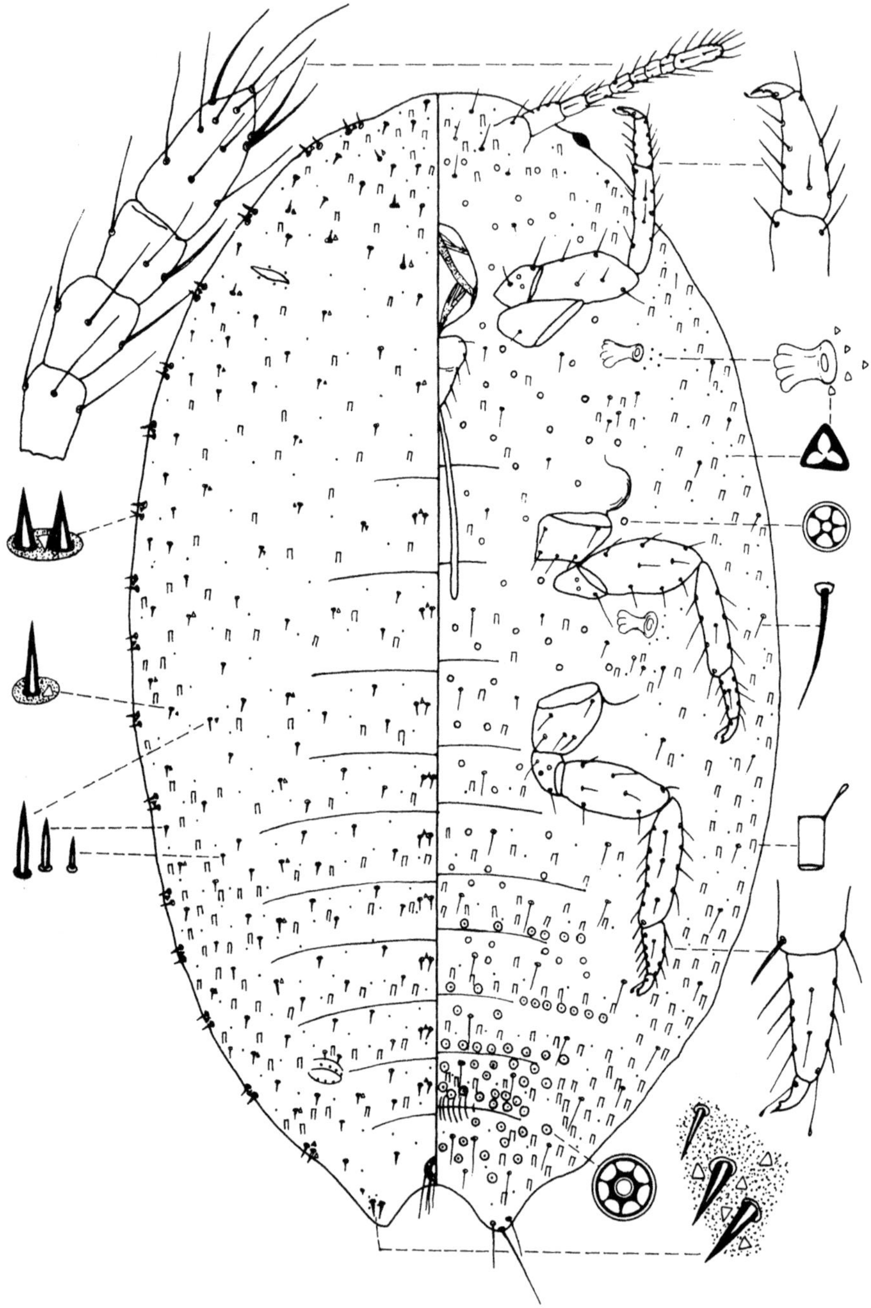

Fig. 81. *Spinococcus marrubii* (Kiritchenko), female (original)

81. **Spinococcus morrisoni** (Kiritchenko) – Morrison's Mealybug

Syn. & Lit.: *Phenacoccus morrisoni* Kiritchenko, 1935:141. Danzig, 1964:629; Zahradnik, 1972:399; Ter-Grigorian, 1973:130; Tereznikova, *1975:261; Kosztarab & Kozár, 1978:60.

Adult female elongate oval, gray or pink, up to 5 mm long, 2.5 mm wide; antennae 9-segmented, ca 770 μ long, 2nd segment longest; labium ca 93 μ long and wide, stylet loop reaches line between hind coxae; legs ca 950 μ long, hind tibia 2.5 times as long as tarsus; 18 pairs of cerarii, each cerarius on an elevated tubercle, with 2 slender conelike setae and 1 trilocular pore; a row of dorsal cerarii on midline of body; circulus large and oval; anal ring oval with 2 outer and 1 inner pore rows, setae 195–220 μ long; each anal lobe with large oval sclerotized plate, apical seta 225–385 μ long; trilocular pores numerous on entire body, except prosomal venter; quinquelocular pores on middle of prosomal venter and form transverse rows on abdominal sternites 2–6; multilocular pores fewer on dorsum, form transverse rows and bands on both abdominal surfaces and groups on entire body margin; oral collar ducts few on dorsum, form transverse rows and groups on abdominal sternites and on body margin; most of the conical dorsal setae on tubercles, each with a trilocular pore near base. Third instar described by Ter-Grigorian (1973).

Distribution. Known from USSR (European Part), and recently found in Hungary.

Biology. On trunks and branches of *Carpinus* and *Rosa*. Females collected from May through November. Overwinters in Ukraine as eggs (Tereznikova, 1975) in Armenia as N_3. Reproduction biparental.

Genus **Trionymus** Berg

Syn. & Lit.: *Trionymus* Berg, 1899:78; Goux, 1941b:31; Ferris, 1950:251, 1953:482; Ezzat, 1962:67; Williams, 1962:58; DeLotto, 1964:386; McKenzie, 1967:459; Ter-Grigorian, 1973:61; Miller & McKenzie, 1973:530; Tereznikova, 1975:261; Kosztarab & Kozár, 1978:61; Danzig, 1980a:178. *Westwoodia* Signoret 1875:337 (preoccupied). *Signoretia* Kraatz, 1888a:176 (preoccupied). *Bergrothia* Kraatz, 1888b:360 (preoccupied). *Bergrothiela* Reitter, 1898:54 (preoccupied). *Pseudococcus* Borchsenius, 1949:90.

Type-species: *Westwoodia perrisii* Signoret, 1875.

Adult female usually elongate-oval; antennae 6–8-segmented; 4 dorsal ostioles (rarely 2) present; legs well-developed, hind legs often with translucent pores; claws without denticle; with fewer than 6 pairs of cerarii (except in *multivorus,* where sometimes up to 10 pairs present), C_{18} with 2 conical setae, often sclerotized; 1 circulus normally present, rarely more, small, undivided; anal ring oval or circular; anal lobes not prominent; trilocular pores on both body surfaces; quinquelocular pores absent; multilocular pores always on venter, often present on dorsum; oral collar tubular ducts on both body surfaces, often of two sizes, seldom (e. g. *newsteadi*) ducts with deep collar and a flange-shaped membranous ring, resembling oral rim ducts; only cerarian setae thick, others on body surface usually slender, hairlike. Adult male described by Afifi (1968).

Distribution. Probably of world-wide distribution, with many species erroneously assigned to it. There are at least 15 species in CE.

Biology. Most species feed on grass leaves; usually under leafsheaths, stems or roots. Only *T. newsteadi* (Green) is known from trees. Generally the nymphs overwinter; the number of yearly generations varies, the species in CE having 1–3 generations; egg production varies, usually around 100 per female.

Reviewer's remark. The concept of *Trionymus* in temperate regions of Palearctic is based on the type species –*T. perrisii*– which is elongate, feeds on grasses, lives within leaf sheaths and bears few cerarii. In other regions, particularly in the Nearctic, species assigned to *Trionymus* are broadly oval and feed on woody plants, being very close to the members of *Dysmicoccus*, except that the latter have more than 5 or 7 pairs of cerarii. However, in the Palearctic, elongate grass-infesting (living on leaves) species with a greater number of cerarii are also assigned in *Dysmicoccus*. It is thus clear that the *Dysmicoccus-Trionymus* complex includes several (perhaps 3) groups of species and that the number of cerarii is of little value in recognizing natural units. On the other hand, some genera (*Ferrisicoccus*, *Kaicoccus*) have been recently synonymized with *Dysmicoccus* (Danzig 1980a).

Several members of *Trionymus* (e.g., *T. singularis*, *T. radicum*, etc.) have short tubular ducts arranged together with multilocular pores, in broad submarginal bands; these features bring them very close to *Balanococcus*. Also *Saccharicoccus penium* is close to *Trionymus* (transferred to this genus by Danzig, 1980a). The species of *Neotrionymus* practically cannot be distinguished from members of *Trionymus*. Apparently, *Trionymus* includes a number of morphologically similar, and inadequately described species; thus the whole group needs a thorough revision.

Key to species of *Trionymus*

1. Multilocular pores on venter only except some on head and abdomen (*T. singularis*) and few on dorsum (*T. thulensis*) 2
   Multilocular pores on both body surfaces 5
2. With abnormal tubular ducts, these with deep collar and a flange-shaped membranous ring; only posterior ostioles developed; on beech trees only .... *newsteadi*, p. 164
   With normal oral collar tubular ducts; both anterior and posterior ostioles developed; on herbaceous plants, most on grasses 3
3. Translucent pores present on hind coxae, femora and tibiae; anal lobe cerarii on large sclerotized plates, each with many trilocular pores and over 10 auxiliary hairlike setae .... *placatus*, p. 166
   Translucent pores on hind coxae only; anal lobe cerarii not on sclerotized plates, each with fewer than 10 trilocular pores and fewer than 9 auxiliary hairlike setae 4
4. Only anal cerarii developed, these normally with 3–6 trilocular pores; antennae ca 200–230 μ long .... *singularis*, p. 168
   Two pairs of cerarii developed, anal lobe cerarii each normally with 7–9 trilocular pores; antennae ca 380 μ long .... *thulensis*, p. 168

5 (1). Hind coxae without translucent pores 6
   Hind coxae with translucent pores 7

6. 2 pairs of cerarii; each anal lobe cerarius with 2 conical setae, 5–6 auxiliary hairlike setae and 13–15 trilocular pores; antennae ca 310 μ long .... *elymi*, p.161
   3–10 pairs of cerarii; each anal lobe cerarius with 2 conical setae, 7–12 auxiliary hairlike setae and ca 40 trilocular pores; antennae ca 397 μ long .... *multivorus*, p. 163

7(5). Antennae 7-, rarely 6-segmented 8
Antennae 8-segmented 10
8. Oral collar ducts with flange-shaped collars occupying close to half length of ducts; 2 pairs of cerarii developed; antennae over 300 μ long .... *radicum*, p. 167
Oral collar ducts with flange-shaped collars never occupying close to half length of ducts; normally only 1 pair of cerarii developed (sometimes 2 pairs in *isfarensis*); antennae less than 300 μ long 9
9. Anal lobe cerarii on minute sclerotized plates with 2 conelike setae, 3–5 hairlike setae and 6–7 trilocular pores .... *isfarensis*, p. 162
Anal lobe cerarii not on sclerotized plates, with 2 conelike setae, 3–4 hairlike setae and 15–20 trilocular pores .... *phragmitis*, p. 166
10 (7). With only anal lobe cerarii developed .... *hamberdi*, p. 161
With 2, rarely 3, pairs of cerarii developed 11
11. Circulus large, oval, divided by a distinct intersegmental fold .... *tomlini*, p. 169
Circulus small (except in *luzensis*), oval or round, never divided by intersegmental fold 12
12. With fewer than 50 trilocular pores in each anal lobe cerarius; antennae less than 400 μ long 13
With more than 50 trilocular pores in each anal lobe cerarius; antennae more than 400 μ long 14
13. Anal ring horseshoe-shaped with setae ca 43–72 μ long; with 8–12 trilocular pores in each anal lobe cerarius; no multilocular pores on dorsum of prosoma (fig. 82)............................... *aberrans*, p. 159
Anal ring subcircular with setae ca 140 μ long; with 26–45 trilocular pores in each anal lobe cerarius; with some multilocular pores on dorsum of prosoma .... *perrisii*, p. 164
14 (12). Each lip of anterior ostioles with 6–18 trilocular pores; smaller type of tubular ducts present on both surfaces of prosoma; with 14–21 hairlike setae in each anal lobe cerarius................................ *luzensis*, p. 162
Each lip of anterior ostioles with 4–6 trilocular pores; smaller type of tubular ducts on abdominal segments only; with 9–12 hairlike setae in each anal lobe cerarius .... *phalaridis*, p. 165

82. **Trionymus aberrans** Goux – Aberrant Mealybug
Fig. 82

Syn. & Lit.: *Trionymus aberrans* Goux, 1938:166; Koteja & Zak-Ogaza, 1966:318; Matesova, 1968a:103; Danzig, 1972c:197; Ter-Grigorian, 1973:69; Tereznikova, 1975:265; Kozár et al., 1977:71; Kosztarab & Kozár, 1978:62. *T. pulverarius* Kiritchenko, 1928:3. *Pseudococcus aberrans*, Borchsenius, 1949:149.

Female covered with white powdery wax; abdominal apex with 2 white wax filaments. Adult female elongate-oval, almost parallel-sided, 3–4.5 mm long, 1.2–1. 6 mm wide; raspberry-red colored; antennae 8-segmented, ca 346–408 μ long, apical segment longest; labium ca 75 μ long, 93 μ wide, stylet loop reaches line between

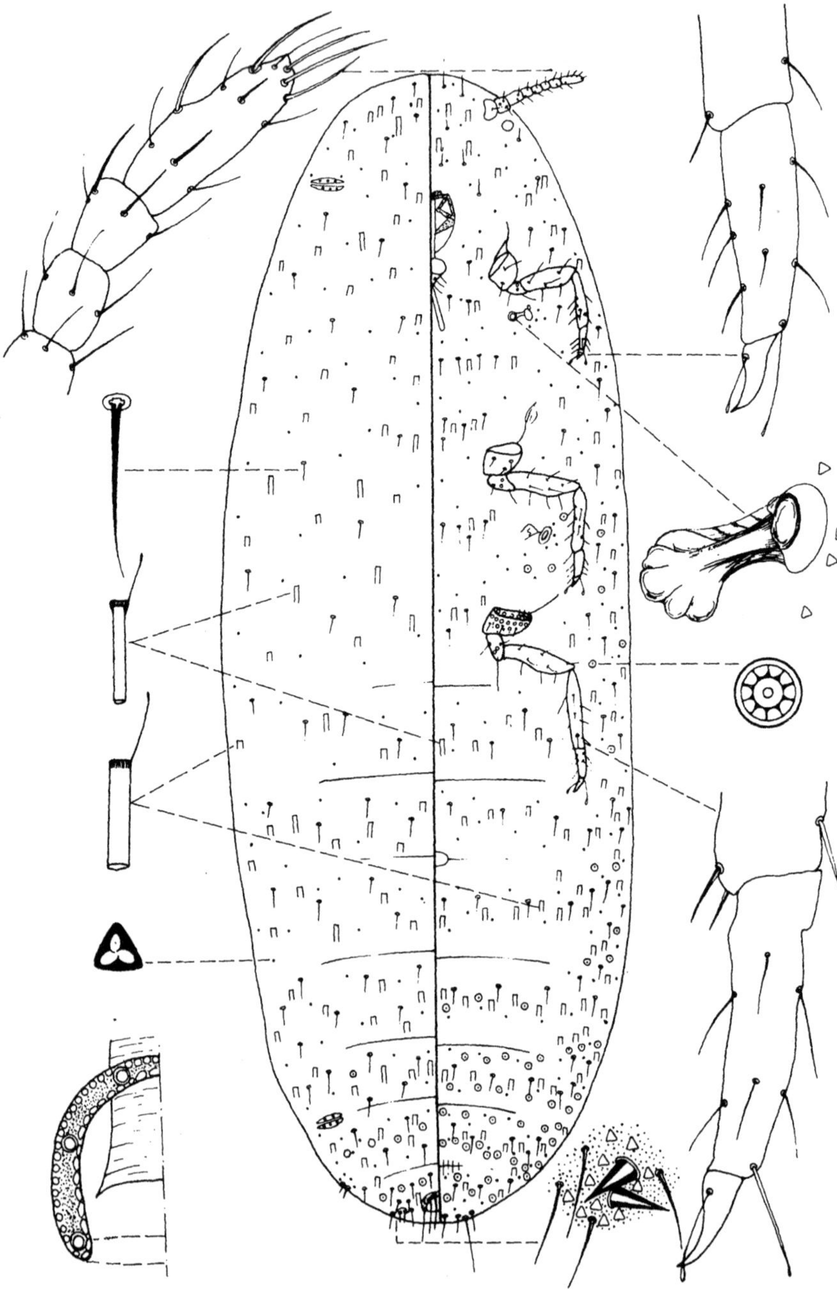

Fig. 82. *Trionymus aberrans* Goux, female (original)

anterior spiracles; hind coxae with translucent pores; 2 pairs of cerarii, C_{18} with 2 conical setae and 8–12 trilocular pores; circulus rarely present, absent in Hungarian specimens; anal ring horseshoe-shaped, with 2 rows of pores, setae 70–93 μ long; anal lobes not developed, apical seta 160–186 μ long; trilocular pores common on both body surfaces; multilocular pores forming transverse rows on both sides of terminal abdominal segments, more numerous on venter but few on prosomal venter; oral collar ducts of 2 sizes, forming transverse rows on both body surfaces; body setae relatively few. First instar described by Goux (1938).

Distribution. A common species. In CE: Bu, Fr, Ge, Hu, Po, USSR (Ukraine to Middle Asia); also known from England (?) (Ter-Grigorian, 1973).

Biology. In leaf sheaths of a variety of monocotyledons : *Agropyron, Avena, Briza, Festuca, Hordeum, Lolium, Poa, Stipa* and *Triticum*. Reproduction biparental; egg laying females present in June and October.

Natural enemies. Parasites: *Cheiloneurus elegans* (Schmutterer, 1955c), *Parochthiphlia coronata* (Papp, 1979), *Rhopus parvulus* (Trjapitzin, 1978).

83. **Trionymus elymi** (Borchsenius) – Borchsenius's Mealybug

Syn. & Lit.: *Pseudococcus elymi* Borchsenius, 1949:146. *Trionymus elymi*, Matesova, 1968a:103; Tereznikova, 1975:267; Kosztarab & Kozár, 1978:63.

Egg sac oval, white, cottony. Adult female oval, up to 2.5 mm long, 1.7 mm wide; antennae 7–8-segmented, ca 310 μ long, apical segment longest; stylet loop almost reaches line between mid-legs, hind coxae without translucent pores; 2 pairs of cerarii, C_{18} with 2 conical setae, 5–6 long hairlike setae and 13–15 trilocular pores; circulus absent; anal ring oval, with 2 pore rows, setae ca 110 μ long; anal lobes convex; trilocular pores numerous on both body surfaces; multilocular pores form transverse rows and bands on both sides of abdomen, often a few around hind coxae; oral collar ducts of 2 sizes, forming transverse rows and bands that widen toward body margin on both surfaces; body setae few, most short.

Distribution. Only known from the European part of USSR, and from Kazakhstan.

Biology. A xerophilous species. On roots of herbaceous plants: *Achillea, Artemisia, Cirsium, Elymus, Festuca, Inula, Potentilla, Stipa, Tanacetum,* and *Ziziphora.* In Kazakhstan females develop by June and start laying eggs in July (Matesova, 1968a). Biology and males unknown. Associated ants in Kazakhstan: *Lasius alienus, Tetramorium caespitum semilaeve* (Matesova, 1968a).

84. **Trionymus hamberdi** (Borchsenius) – Hamberd's Mealybug

Syn. & Lit.: *Pseudococcus hamberdi* Borchsenius, 1949:154. *Trionymus hamberdi,* Ter-Grigorian, 1973:70; Kosztarab & Kozár, 1978:61.

Adult female elongate oval, pink, up to 4 mm long; antennae 8-segmented, ca 360 μ long, apical segment longest; legs slender, ca 490 μ long, hind coxae with a few translucent pores; only 1 pair of cerarii, C_{18} with 2 slender conical setae, 3–6 hairlike setae and 13–16 trilocular pores; circulus small, round; anal ring with an outer row of circular pores and an inner row of oval pores, setae 100–125 μ long; anal lobes wide, hardly detectable, apical setae 186 μ long; trilocular pores numerous on both body surfaces; multilocular pores scattered on both body surfaces, more numerous along body margin and on terminal abdominal sternites; oral collar ducts present on entire dorsum and on terminal sternites of abdomen; body setae short and few.

Distribution. Known only from Armenia in USSR and from Hungary.

Biology. On leaves and in leaf sheaths of *Agropyron, Hordeum* and *Phragmites*. Biology and males unknown.

85. **Trionymus isfarensis** (Borchsenius) – Tadzhik Mealybug

Syn. & Lit.: *Pseudococcus isfarensis* Borchsenius, *1949:152; Matesova, 1955:201; Koteja & Zak-Ogaza, 1979:672; *Saccharicoccus penium*, Danzig, 1983:518.

Adult female according to Borchsenius (1949), elongate oval, nearly parallel-sided, pink, up to 5 mm long and 2 mm wide; antennae 7-segmented, ca 260 μ long, apical segment longest; hind coxae with a few translucent pores; 7–8 trilocular pores forming a semi-circle in front of each spiracle; 1–2 pairs of cerarii, C_{18} on a small oval sclerotized plate, with 2 conical setae, 3–5 hairlike setae and 6–7 trilocular pores; 2–3 circuli, one between segments 3 and 4 large with sides notched, others small and nearly circular; anal ring almost circular, with sharply projecting posterior end, setae 110–160 μ long; anal lobes broad, only slightly projecting, apical seta ca 140 μ long; trilocular pores numerous on both body surfaces; multilocular pores also numerous on both body surfaces, forming transverse rows and bands, also a band along body margin; oral collar ducts associated with multilocular pore bands; body hairs stout, numerous on terminal abdominal segments, sparse on other parts of body.

Distribution. Recently reported from Poland (Koteja and Zak-Ogaza, 1979), known also from USSR in Tadzhikistan.

Biology. On stems of *Calamagrostis* and *Phragmites*. Biology and males unknown.

Note. This species is difficult to separate from *T. phragmitis* and may be identical with it.

86. **Trionymus luzensis** Komosinska – Komosinska's Mealybug

Syn. & Lit.: *Trionymus luzensis* Komosinska, 1980:257.

Ovisac cylindrical, white 7–12 mm long, 3–4 mm wide. Adult female (description extracted from Komosinska, 1980) oval, dark violet, 4–5.1 mm long, 1.9–2.8 mm wide; antennae 8-segmented, rarely asymmetrically 9-segmented, 434–616 μ long, apical segment longest; labium ca 145 μ long, 131 μ wide with 26 setae, stylet loop ca 252 μ long; each ostiole lip with 6–25 trilocular pores and 0–1 setae; hind coxae with small translucent pores; 2 pairs of cerarii, C_{18} on a large sclerotized plate, with 2 conical setae, 14–21 hairlike setae and 73–98 trilocular pores; circulus oval, 52–93 μ long, 124–176 μ wide; anal ring with 2 inner and 2 outer rows of pores, setae slightly longer than ring diameter; anal lobes convex, only slightly protruding; apical setae slightly longer than anal ring setae; simple disc pores on both body surfaces sparse; trilocular pores numerous, evenly distributed on both body surfaces; multilocular pores ca 80 on venter, these scattered on prosoma, forming transverse rows and bands on abdominal sternites, on dorsum fewer and scattered; oral collar ducts of 2 sizes scattered on both body surfaces; body setae hairlike. First instar also described in detail by Komosinska (1980).

Distribution. Known only from Poland.

Biology. On leaves of grasses: *Calamagrostis* and *Deschampsia,* but ovisacs also found on fallen dry leaves of deciduous trees. Egg laying females present during first part of September.

87. **Trionymus multivorus** (Kiritchenko) – Polyphagous Mealybug

Syn. & Lit.: *Pseudococcus multivorus* Kiritchenko, 1935:151; Borchsenius, 1949:138. *P. mendosus* Kiritchenko, 1935:153. *P. saxatilis* Borchsenius 1937b:182 (nomen nudum). *P. marrubium* Borchsenius, 1937b:184 (nomen nudum). *P. taraxaci* Borchsenius, 1937b:184 (nomen nudum). *Trionymus multivorus,* Matesova, 1968a:104, 1971:26; Ter-Grigorian, 1973:63; Danzig & Kozár, 1974:11; Tereznikova, 1975:263; Kozár et al., 1977:71; Kosztarab & Kozár, 1978:62. *Dysmicoccus multivorus,* Koteja & Zak-Ogaza, 1979:672.

Female covered with fine white wax powder, with waxy tassels along abdominal margins. Adult female according to Borchsenius (1949), oval, pink to violet, depending on age, 1.5–4.5 mm long and 0.7–3 mm wide; antennae 8-segmented, ca 402 μ long; length of stylet loop varies, rarely reaches line between hind spiracles; hind coxae without translucent pores, hind tibia at least twice as long as tarsus; cerarii 3–10 pairs, C_{18} on small circular sclerotized plate, with 2 conical setae, 7–12 hairlike setae and 26–40 trilocular pores; circulus absent; anal ring setae 86–130 μ long; anal lobes broad, apical seta 160–176 μ long; trilocular pores numerous on both surfaces; multilocular pores on both sides of terminal abdominal segments form transverse rows; oral collar ducts of 2 sizes common on both body surfaces; body setae slender, 36–84 μ long. First instar described by Borchsenius (1949), Tereznikova (1975), this and second instar by Ter-Grigorian (1973).

Distribution. A common species. Known from Hu, It, Po and USSR (European parts to Middle Asia).

Biology. On stems, leaves, but most often on roots and root crowns of a variety of herbaceous plants, primarily dicotyledons belonging to at least 35 plant genera. Overwintering in Hungary as first instars on roots of lucerne (alfalfa), the females developed by the end of May and oviposition was in white egg sacs on root crown of lucerne; second generation adults developed by end of August. High population produced damage on lucerne (Kozár, 1973). The damage was associated with plant virus infestation, and the mealybugs may serve as vectors of the virus disease. Associated ant: *Lasius alienus* (Matesova, 1968a).

Natural enemies. Parasites: *Anagyrus diversicornis*, *Leptomastix flava* and *Pezaphycus obenbergeri* (Trjapitzin, 1978).

Note. This species apparently belongs to *Dysmicoccus.*

88. **Trionymus newsteadi** (Green) – Beech Mealybug

Syn. & Lit.: *Pseudococcus newsteadi* Green, 1917:265; Borchsenius, 1949:157; Schmutterer, *1952b:388; Zahradnik, 1956:50. *Trionymus newsteadi*, Williams, *1962:62; Koteja, 1964a:99, 1969:5; Koteja & Zak-Ogaza, 1966:319; Zahradnik, 1972:399; Ter-Grigorian, 1973:72; Tereznikova, 1975:270; Kosztarab & Kozár, 1978:64; Vinis, 1981:207.

Female venter thinly, dorsum thickly covered with wax powder, terminal abdominal segments with short waxy tassels. Adult female oval, pink to purplish grey, up to 3 mm long, ca 1.5 mm wide; antennae 8-segmented, 340–464 μ long; labium 105–132 μ long, with 26 setae, stylet loop reaches line between mid-coxae; only posterior ostioles developed, each lip with 3–4 trilocular pores; hind coxae and tibiae with many translucent pores; 3–5 pairs of cerarii, C_{18} with 2 conical setae, 2–4 slender auxiliary setae and 6–15 trilocular pores, without sclerotized plate; circulus large, ca 124 μ long and 156 μ wide, often notched laterally, posterior margin normally shorter than anterior; anal ring with 2 rows of pores, setae 124–155 μ long; anal lobes moderately developed, apical setae 164–190 μ long; trilocular pores distributed in moderate numbers on both body surfaces; multilocular pores on venter only, forming transverse rows and groups on terminal abdominal sternites, a few scattered on prosoma; oral collar ducts of 2 sizes on venter, small ones on median area of terminal sternites, seldom in submargin, large ones form transverse rows and groups on both surfaces, ducts with deep collar and a flange-shaped membranous ring, resembling oral rim ducts; body setae of various lengths, mostly slender. Adult male described by Afifi (1968).

Distribution. In CE: Cz, Ge, Po and USSR (Ukraine and Caucasus). Collected recently in Hungary and Yugoslavia. Also known from England and Holland.

Biology. Nymphs on underside of leaves and on young shoots, adults on woody parts, trunk and branches of *Fagus*. Adult females found from May to September; third instars overwinter, and transform into adults which start laying eggs from mid-May, 86–213 per female, these hatch in about 25 days, from mid-May to end of June; the first instars feed on underside of leaves and molt from end of July through August (Schmutterer, 1952b). Reproduction biparental.

Natural enemy. Parasite: *Doliphoceras pseudococci* (Trjapitzin, 1978).

89. **Trionymus perrisii** (Signoret) – Perris' Grass Mealybug

Syn. & Lit.: *Westwoodia perrisii* Signoret, 1875:337. *Dactylopius hibernicus* Newstead, 1895a:167. *D. pulverarius* Newstead, 1903:174 (in part). *Pseudococcus perrisi*, Borchsenius, 1949:142. *Trionymus phalaridis*, Tereznikova, 1975:265. *T. perrisii*, Marchal, 1909a:248; Goux, 1941b:37; Schmutterer, *1952b:404; Williams, *1962:63; Koteja, 1969:5; Koteja & Zak-Ogaza, 1969:361; Ter-Grigorian, 1973:66; Tereznikova, 1975:265; Kozár et al., 1977:71; Kosztarab & Kozár, 1978:63; Kozár et al., 1979:130; Danzig, 1980a:178. *T. pulverarius*, Zahradnik, 1977:122.

Female covered with fine dustlike wax powder, abdominal apex with 3–4 pairs of short tassels. Adult female elongate oval, nearly parallel-sided, pink, 3–6 long, 1.3–1.6 mm wide; antennae 8-segmented, 320–371 μ long; labium ca 92 μ long, stylet loop 250–280 μ long; ostioles moderately developed, each lip with 4–6 trilocular pores, seldom with setae; hind coxae with translucent pores, hind tarsus about half as long as tibia, cerarii 2–3 pairs, normally 2 in CE, C_{18} on large sclerotized plate larger in area than anal ring, with 2 conical setae, 9–13 hairlike setae and 26–45 trilocular pores; circulus 0–5, normally 1, oval, ca 50 μ long, 26 μ wide; anal ring subcircular with 2 pore rows, setae 118–140 μ long; twice as long as diameter of ring; anal lobes moderately developed, apical setae 160–190 μ long; trilocular pores numerous, evenly distributed on both surfaces; multilocular pores forming transverse rows on both surfaces of abdomen, some scattered on both sides of prosoma; oral collar ducts of two sizes, forming transverse rows and bands and marginal group on both surfaces; body setae slender, of various sizes.

Distribution. A common Trans-Palearctic species. In CE: Bu, Cz, Fr, Ge, Hu, It, Po, Ro and USSR (European Part to Far East) and Yu. Also known from Holland, Ireland, UK, Korea and Mongolia.

Biology. A mesophilous species. On stems and leaf sheaths of grasses: *Agropyron, Agrostis, Brachypodium, Bromus, Calamagrostis, Corynephorus, Dactylis, Deschampsia, Festuca, Leontodon, Poa,* and *Sieglingia.* Reproduction biparental. Second, rarely third instars, overwinter in S. Germany; adults develop by May, each lays an average of 100–150 eggs; first instars appear at end of May and develop into adults by mid-July, lay eggs during second half of July; their nymphs overwinter. Schmutterer (1952b) and Koteja (1969) found 2 yearly generations. Associated ants: *Lasius niger alienus, Solenopsis fugax, Tetramorium caespitum* (Schmutterer, 1952b).

Natural enemies. Parasites: *Allotropa mecrida, Mayridia myrlea* (Trjapitzin, 1978). Predator: *Novius testaceus.*

Reviewer's remark. *Trionymus perrisii* is the most common representative of the grass-infesting *Trionymus* in the Palearctic, and is also very variable and/or represents a combination of several species. On the other hand, some species (*T. phalaridis, T. dactylis,* etc.) are practically indistinguishable when compared with a larger collection of *T. perrisii* (see also Williams, 1962).

90. **Trionymus phalaridis** Green – Canarygrass Mealybug

Syn. & Lit.: *Pseudococcus phalaridis* Green, 1925a:37. *Trionymus californicus* Kiritchenko, 1931:312. *T. festucae* Borchsenius, 1937:184. *P. phalaridis,* Borchsenius, 1949:140. *T. phalaridis,* Williams, *1962:67; Kosztarab & Kozár, 1978:63. *T. perrisii,* Tereznikova, 1975:265.

Female covered with fine wax powder, abdominal apex with 4 short waxy tassels. Adult female elongate oval, nearly parallel-sided, reddish-purple to dark grey, up to 6.5 mm long, 1.6 mm wide; antennae 8-, rarely 9-segmented, according to Williams (1962), 450–495 μ long; labium conical, ca 100 μ long, stylet loop twice as long as labium; ostioles distinct, without setae, each anterior ostiole lip with ca 4–6, each posterior ostiole lip with 10–15 trilocular pores; legs slender, hind coxae with few translucent pores, tarsus about 1/3 as long as tibia; hind spiracles much larger than anteriors; 2 pairs of cerarii, C_{18} on sclerotized plate, with 2 conical setae, 9–12 hairlike setae and ca 72 trilocular pores; circulus small, oval; anal ring with 2 pore rows, setae ca 135–140 μ long, about twice as long as diameter of ring; anal lobes distinct, apical seta ca 155 μ long; trilocular pores numerous, evenly distributed on both surfaces; multilocular pores scattered on both surfaces of prosoma, forming transverse rows on both terminal abdominal segments with large concentration on last 2 sternites of abdomen; oral collar ducts of 2 sizes, smaller type in median area of abdomen on both surfaces, larger type more abundant and forming transverse rows and marginal groups on most abdominal sternites and tergites, scattered on both surfaces of prosoma; body setae slender of various sizes.

Distribution. Known from England and USSR (Ukraine and Caucasus).

Biology. On leaves, leaf sheaths and stems of grasses: *Bromus, Calamagrostis, Elymus, Festuca, Phalaris* and *Stipa*. Adult females collected in August and September in England. Green (1925a) found 408 eggs in 1 ovisac. Biology and males unknown.

Natural enemy. Predator: *Chamaemyia polystigma* larvae were found as very effective predators of this species (Green, 1925a).

91. **Trionymus phragmitis** (Hall) – Hall's Reed Mealybug

Syn. & Lit.: *Ripersia phragmitis* Hall, 1923:10; Green, 1931:101; Kiritchenko, 1935:132. *Pseudococcus phragmitis*, Borchsenius, 1949:153. *Trionymus phragmitis*, Ezzat, *1962:79, 167; Williams, 1970:181; Tereznikova, *1975:269; Kosztarab & Kozár, 1978:61. *T. polyporus*, Kiritchenko, 1930:312 (misidentification).

Adult female oval, according to Ezzat (1962) 3.4–7.0 mm long, 1.7–3.8 mm wide; antennae 6- or 7-segmented, when 6-segmented, ca 220 μ long, segment 3 with some degree of articulation; labium about as long as wide, at base, 112 μ; ostioles inconspicuous, anterior lips of thoracic spiracles without pores, each posterior lip with 1–3 trilocular pores, each lip of abdominal ostiole with ca 4 trilocular pores; translucent pores on hind coxae and few on hind and mid-tibiae; only C_{18} developed, this with 2 conical setae, 3–4 hairlike setae and ca 15–20 trilocular pores; 1–3 oval circuli; anal ring subcircular, diameter ca 90 μ with partly double outer row of pores and 1 inner pore row, setae 115–155 μ long; anal lobes broad, convex, apical setae 125–190 μ long; trilocular pores distributed on both body surfaces; multilocular pores numerous on dorsum and on abdominal sternites where they form transverse bands, also a wide submarginal band present on venter; oral collar ducts of 2 sizes on both surfaces of body; body setae relatively short and few.

Distribution. A rare species. Known only from USSR (Ukraine and Herson Region), Egypt and Tunis.

Biology. Between leaf sheaths and stems of *Arundo, Hierochloe* and *Phragmites*. Biology and males unknown.

92. **Trionymus placatus** (Borchsenius) – Pleasant Grass Mealybug

Syn. & Lit.: *Pseudococcus placatus* Borchsenius, *1949:158. *Trionymus placatus*, Ivanova, 1973:24; Tereznikova, 1975:270; Kosztarab & Kozár, 1978:64.

Adult female elongate oval, nearly parallel-sided, pink, according to Borchsenius (1949) ca 4 mm long, 1.6 mm wide; antennae 8-segmented, ca 420 μ long, all segments slender; legs very long, hind coxae, femora and tibiae with translucent pores; 2 pairs of cerarii, C_{18} on large sclerotized plate, with 2 stout conical setae, 11–13 hairlike setae and numerous trilocular pores; circulus large, oval; anal ring oval, setae ca 160 μ long; anal lobes distinct, broad, with 1 long and 3–4 short setae on venter of each lobe; trilocular pores numerous on both body surfaces; multilocular pores restricted to venter, located singly in middle of prosoma, and forming transverse rows and bands on abdominal sternites 4–8; tubular ducts of 2 sizes on dorsum, large types form groups along body margin and transverse rows on thorax and on abdominal tergites 1–7 with 1 long seta at base of ducts, small type of duct mixed in rows with large ones predominantly on abdominal tergites; on venter a median size duct located singly on abdominal sternites 3 and 4 and forming transverse rows on sternites 5–8; body hairs slender.

Distribution. Known only from USSR (Ukraine in Hmelnick and Kaliningrad Regions).

Biology. Tereznikova (1975) reported it from foliage of grasses. Biology and males unknown.

93. **Trionymus radicum** (Newstead) – Thrift Root Mealybug

Syn. & Lit.: *Dactylopius radicum* Newstead, 1895b:235. *D. hibernicus*, Newstead 1903:172 (in part, misidentification). *Ripersia donisthorpei* Newstead, *In* Donisthorpe, 1907:4. *Trionymus radicum*, Williams, *1962:69; Koteja, 1969:7, 1971b:322; Kosztarab & Kozár, 1978:64.

Female during oviposition completely enclosed in a loose cottony egg sac.

Adult female elongate.oval, pale reddish-pink, ca 3.5 mm long; antennae 7-segmented, according to Williams (1962) 305–335 μ long; labium conical, 70–75 μ long; ostioles poorly developed, each lip with 4–7 trilocular pores; legs slender, tarsi much shorter than tibiae; hind coxae with few translucent pores; 2 pairs of cerarii, C_{18} with 2 conical setae, with sclerotized area between the bases, and with 1 hairlike seta, and 6–7 trilocular pores; circulus oval, very small; anal ring with 2 rows of pores, setae longer than diameter of ring; anal lobes moderately produced, apical seta longer than anal ring setae; trilocular pores evenly distributed on both body surfaces; multilocular pores very few, scattered on dorsum, on venter numerous on 2 terminal abdominal segments, forming transverse rows on sternites 5–8, scattered on rest of venter; oral collar ducts of 2 sizes, a larger type on dorsum forming transverse rows on entire body except on head, on venter a larger type forming groups along body margin, few in median area of segments 5–8, smaller ducts primarily in median area of abdominal venter, all ducts with flange-shaped collar that occupies almost half total length of duct; body setae slender, those on venter longer.

Distribution. In CE known only from Poland, Hungary and Yu. Also reported from England, Wales and Holland.

Biology. Feeds on roots of *Armeria* and on roots and at base of *Poa* and other grasses. Adults present in May–June, August–September with probably 2 yearly generations. Biology and males unknown. Sometimes found in nest of ants, *Ponera contracta* (Williams, 1962). Newstead (1895) stated that nymphs were collected at 10–12 cm below surface, especially on plants in rock fissures, in sunny exposures.

94. **Trionymus singularis** Schmutterer – Solitary Mealybug

Syn. & Lit.: *Trionymus singularis* Schmutterer, 1952c:560; Zahradnik, 1977:122; Koteja & Zak-Ogaza, 1979:672.

Female during oviposition produces a loose cottony ovisac attached to end of body. Adult female (description extracted from Schmutterer, 1952c) after egg laying, elongate-oval, flesh colored or orange-red, ca 3.2 mm long, ca 1.2 mm wide; antennae 6 or 7-segmented, ca 200 to 230 μ long; labium slightly longer than wide, with 26 setae; each ostiole with many trilocular pores on lips; legs well-developed, hind coxae enlarged, with many translucent pores on underside; each spiracle opening surrounded by a group of 8–11 trilocular pores; only C_{18} developed, each with 2 conical setae, and 3–6 trilocular pores; circulus very small, circular, between abdominal segments 3 and 4; anal ring circular with outer row of small circular pores and inner row of larger irregularly shaped pores, setae ca 1½ times as long as diameter of ring; anal lobes indistinct, apical setae as long as anal ring setae; trilocular pores numerous on both body surfaces, fewer on 3 terminal sternites of abdomen; multilocular pores especially numerous on 3 terminal sternites of abdomen, on margin of other abdominal and thoracic segments, absent or few on dorsum; tubular ducts atypical, numerous on 3–4 terminal abdominal segments and on ventral margin of other abdominal and thoracic segments, rarely a few on head; ducts also numerous on 3 terminal abdominal segments of dorsum, less frequent on margin of anterior half of abdomen; average number of setae on both surfaces relatively small.

Distribution. Known in CE from: Cz, Ge and Po, and recently Kozár found it in USSR (Moscow District).

Biology. Schmutterer (1952c) collected adult females at end of May from a dry hillside, from under leaf sheaths of grasses, *Agrostis, Festuca* and *Poa*, also from nest of ants, *Lasius niger*. Some females were found parasitized by chalcid larvae.

95. **Trionymus thulensis** Green – Northern Mealybug

Syn. & Lit.: *Trionymus thulensis* Green, 1931:265; Williams, *1962:69; Koteja, 1969:7; Kosztarab & Kozár, 1978:64.

Adult female elongate oval, up to 4 mm long, 1.2–2 mm wide, antennae 7- or 8-segmented, 380–495 μ long; labium conical, ca 110 μ long, stylet loop less than twice length of labium; ostioles moderately developed, with ca 6 trilocular pores on each lip; legs slender, hind coxae with a few translucent pores; each spiracular opening surrounded by a few trilocular pores; 2 pairs of cerarii, C_{18} with 2 conical setae (only between their bases slightly sclerotized), 1 hairlike seta and 7–9 trilocular pores; circulus round and small; anal ring setae twice as long as diameter of ring; anal lobes only slightly produced, apical seta longer than anal ring setae; trilocular pores numerous, evenly distributed on both body surfaces; multilocular pores on venter only (except in some Polish specimens also on dorsum), forming transverse rows and bands on 5 terminal sternites of abdomen, very few on other parts of venter; oral collar ducts of 2 sizes on venter, larger type scattered on prosomal margin, and mixed with smaller ducts forming transverse rows on abdominal sternites; larger ducts only on dorsum, these form transverse rows on thoracic and abdominal tergites, with only few scattered on head; short slender setae on dorsum, longer body setae on venter.

Distribution. Known in CE from: Fr, Hu, Po; also known from England, Wales and Iceland.

Biology. On stems, leaf sheaths and roots of grasses: *Agropyron, Agrostis, Deschampsia, Festuca, Holcus, Phragmites, Trisetum* and *Triticum*. Adult females present from May through September. Biology and males unknown.

96. **Trionymus tomlini** Green – Guernsey Grass Mealybug

Syn. & Lit.: *Pseudococcus (Trionymus) tomlini* Green, 1925b:525; *T. tomlini,* Williams, *1962:72; Koteja, 1969:7; Koteja & Zak-Ogaza, 1969:362; Podsiadlo & Komosinska, 1976:88; Kosztarab & Kozár, 1978:63.

Adult female elongate oval, slightly tapering towards apex of abdomen, 3.4–4 mm long; antennae 8-segmented, 412–425 μ long; labium conical, 95–113 μ long; stylet loop 3–4 times length of labium; ostioles distinct, each lip with 6–8 trilocular pores, normally without setae; legs long and slender, hind coxae with a few small translucent pores, hind tibiae about twice as long as tarsi; spiracular openings with trilocular pores in a semi-circle; 2 pairs of cerarii, C_{18} on oval sclerotized plate with 2 conical setae, 8–9 hairlike setae and 35–42 trilocular pores; circulus oval, large, normally with distinct intersegmental fold; anal ring setae ca twice as long as diameter of ring; anal lobes distinct, apical setae much longer than anal ring setae, ca 220 μ long; numerous trilocular pores evenly distributed on both body surfaces; multilocular pores on dorsum absent on 2 terminal abdominal segments, only a few on thorax, also sparse on anterior abdominal tergites, on venter more numerous, especially on posterior half of abdomen where in median area they form transverse segmental bands, and larger groups along body margin; oral collar tubular ducts of 2 sizes on both surfaces, forming transverse rows and bands and marginal groups; body setae slender, numerous, on both body surfaces.

Distribution. A rare species. Known in CE from: Hu and Po; also from the Channel Islands, England.

Biology. Collected from leaves of herbaceous plants, especially from beneath leaf sheaths of grasses: *Agropyron, Brachypodium, Bromus, Festuca, Hypochoeris,* and *Psamma*. Females found during August–September. Biology and males unknown.

FAMILY **COCCIDAE** – SOFT SCALES, COCCIDS

Syn. & Lit.: Coccides Fallén, 1814:23. Coccinae, Fernald, 1903:127. Lecaninos, Gomez-Menor, 1937:253. Lecaniinae, Newstead, 1903:5; Leonardi, 1920:276; Silvestri 1920:501; Balachowsky, 1948:255; Schmutterer, 1952b:544. Coccidae, Steinweden, 1929:197; Šulc, 1932:1; Costa Lima, 1942:246; Zimmerman, 1948:291; Bodenheimer, 1953:92; Borchsenius, 1957:1; Canard, 1958:185, 1968:91; Řehaček, 1960:1; Kawecki, 1962a:1; Danzig, 1967:829, 1980a:244; Giliomee, 1967:1; Ali, 1971:14; Williams and Kosztarab, 1972:1; Koteja, 1974b:304, 1979a:585; Dziedzicka, 1977:38; Kosztarab & Kozár, 1978:89; Kawai, 1980:138; Nakahara 1981:283; Tereznikova, 1981:92; Kosztarab, 1982:466.

Fortunately, members of this family were studied in Europe recently more than in any other parts of the world. Among the European workers who stand out are Šulc, Kawecki, Borchsenius, Schmutterer, Řehaček, Canard, Koteja and Tereznikova. Our present work was made easier by utilizing their comprehensive research publications. Still, there are the genera of *Eulecanium* and *Parthenolecanium* (especially the latter), where generic revisions are badly needed. The classification of Coccidae is difficult because of the wide range of variation within many species. A number of them exhibit host-induced morphological variations, e.g. *P. corni*; thus the reason for the many synonym names in some species. Most of the early original species descriptions of Coccidae were based on postreproductive females, where the dorsum was heavily sclerotized, and the specimens deformed, therefore, these descriptions are useless for modern taxonomic work. Only young, non-sclerotized adults, preferably fixed in hot 70% ethyl alcohol, are worth slide-mounting and useful for study in this group. Females develop through three or four instars (e.g. *Sphaerolecanium prunastri*), males through five. As in other families of Coccinea, adult males, except for a few studies (e.g. Giliomee, 1967), and immature stages have been neglected, therefore identification keys are based on adult female characters, as are our following species descriptions.

Ovisac. In most Eriopeltinae ovisac completely encloses females, except in *Psilococcus,* while in *Scythia* it is feltlike and includes nymphal exuviae; egg laying female shrinks and after oviposition drops from ovisac; rarely without an ovisac, e.g. *Parafairmairia,* where female is found with eggs in a translucent wax test; in Pulvinariini the ovisac is cottony, located behind or under the body; while in Coccini no ovisac is formed.

Postreproductive females. Heavily sclerotized, strongly convex, (especially in Coccini), rarely margins bent under, e.g. *Rhodococcus,* usually brown and wrinkled.

Teneral females. With membranous derm, usually subcircular or elliptical in shape, with an anal cleft at posterior end of body that extends to a pair of anal plates; anal plates absent in *Physokermes. Venter.* Antennae usually of 5 to 9 slender segments, 3rd segment normally longest; segments reduced to 1–4 in *Scythia,* or stout in *Lecanopsis* and *Psilococcus*; labium hemispherical, indistinctly 2-segmented, with 5 pairs of setae; legs normally well-developed, 5-segmented, but short in *Eriopeltis,* and only 1–2-segmented and stublike in *Scythia,* tibio-tarsal articulation heavily sclerotized in some groups; claw digitules often thick and with very large apical knob, e.g. *Luzulaspis*; claw with or without denticle; spiracles might be located in a funnel-shaped depression and very large; spiracular setae rarely absent, e.g. *Physokermes, Vittacoccus, Eriopeltis,* when present, usually 3 or 2 in each group terminating in spiracular pore bands, their shape and size are of taxonomic importance in separating genera and species; spiracular pore bands of disc pores with 5-, rarely 3–8 loculi; marginal setae in most species normally hairlike or spinelike, slender, and pointed, in a few variable, rarely with fringed tip, e.g. *Chloropulvinaria,* or chisel-shaped with notched tip, in *Lichtensia viburni*; width of intervals between marginal setae in relation to their length is often taxonomically significant; submarginal setae arranged in an even or irregular row; body setae of varying number, shape and size, normally in the medial area hairlike, interantennal setae longest; simple disc pores rarely present, e.g. in *Eriopeltis festucae,* 1–2 at base of each marginal seta; dark-rimmed pores sometimes on surface, e.g. *Psilococcus*; multilocular pores with 7–10 loculi in the anal area, often form transverse bands, rarely a few on thorax near coxae or at spiracles; microducts normally present, usually scattered on entire surface, but more numerous in submarginal area; tubular ducts common, of various sizes and filaments. *Dorsum.* Body setae usually short and slender, scattered on entire surface, in *Eriopeltis* these large and truncate; submarginal tubercles rarely present, e.g. *Parthenolecanium persicae*; simple disc pores usually randomly scattered on entire surface; bilocular pores sometimes also present; discoidal pores with sievelike openings form a medial longitudinal band, or a group anterior of anal plates; minute ducts rarely on entire surface; tubular ducts often of various sizes, common on entire surface; anal ring with translucent pores and normally with 6–10 setae, rarely ring reduced, without pores and setae, e.g. *Rhodococcus*; anal plates vary in shape, usually triangular with setae, especially apical setae, that are of taxonomic significance.

Immature stages. Described only in a few species. Literature sources for these and males are given at species descriptions. **First instars,** or crawlers, elongate-oval, ca 0.4–0.7 mm long, 0.2–0.4 wide with an anal cleft. *Venter.* Antennae slender, 6-segmented, (rarely 5); stylet loop sometimes extends beyond end of body; legs well-developed, claws slender, normally with very small denticle; spiracular pore row of quinquelocular pores, these rarely with 3 to 7 loculi, such pores may occur near spiracles, or on submargin of thorax; spiracular setae normally 3, rarely 2 or 4 in each group, these blunt, or pointed, equal or unequal in size; marginal setae short and thick, or long and slender; body setae of various sizes. *Dorsum.* Bilocular pores often in a longitudinal medial double row, some rarely scattered on head; 2 trilocular disc pores on frontal margin of head; anal ring with a row of small pores and 6 setae; anal plates elongate, lobe-like, each with a long, thick, apical seta. Sexes normally not distinguishable in this stage. **Second instars,** females and males after molting oval, ca 1–1.6 mm long, 0.3–0.6 mm wide, with an anal cleft. *Venter.* Antennae usually thick, 6-segmented in females and 7-segmented in males, legs variable; claw stout, often with a very small denticle; more spiracular pores present than in first instar, these may form a marginal row in males; spiracular setae often similar to marginal setae; latter stout, conical in most, marginal setae as in adult females; body setae as in first instar; dark-rimmed disc pores sometimes present; multilocular pores absent from abdomen. *Dorsum.* Rarely with submarginal bicylindrical duct tubercles; body setae short, spine-like, 2, rarely 3–4 per segment; usually some minute disc pores on body margin; very small bilocular pores normally present; tubular ducts that produce wax test, present in male nymphs and form rows on body margin, and in submedial area; anal ring with pores and 6 setae; anal plate triangular, without long apical seta. **Third instar female,** (not an adult), when present, very similar to adult female, but smaller and with fewer (5–9) antennal segments, also fewer setae, disc pores and tubular ducts, latter rarely absent, multilocular pores rarely absent from abdominal venter, vulva not developed. **Third instar male**, or prepupa, with eyes, but reduced legs and antennae; wing pads short and stublike; found under wax cover produced by second instar. **Fourth instar male,** or pupa, with eyes; antennae and legs fully developed; wings elongate padlike; some disc pores normally present near anterior spiracles; found under wax cover.

Fifth instar male or adult. Elongate, with distinct circular head, this well-separated from thorax; 2–5 pairs of simple eyes and pair of lateral ocelli; postoccipital ridge always absent (except *Parafairmairia*); antennae filiform, 10-segmented, with many setae; forewings with 2 veins, hind wings reduced to hamulohalteres, or absent; when present, each hamulohalter with 1–4 apically hooked setae; scutal area clear; legs long, with many setae, 1-segmented tarsus and single claw; abdomen parallel-sided, tapering to posterior end, with 8 pregenital and 1 genital segment; with glandular pouch and caudal lobes; penial sheath long, incomplete. For more detailed information on adult males see Giliomee (1967) and Koteja (1970).

Male test or "puparium". Produced by second stage nymph. Encloses third and fourth instar males and adults until their emergence; test oval, made of very thin glassy semitranslucent wax, flat, often divided with sutures into a central and several lateral plates, rounded in front and with sloping margin that is glued to the surface. For more information, see Šulc (1932) and Koteja (1966a). Many Coccidae species can be easily identified through their characteristic male test, therefore more attention should be paid to their taxonomic utilization in the future.

Distribution. This is the third largest coccoid family; world-wide with ca 100 genera and over 1,000 species. Twenty genera and 51 species are known in CE to date. The following 9 introduced species are known only from greenhouses and house plants, thus they are not treated here: *Ceroplastes japonicus* Green, *C. rusci* (Linnaeus), *Coccus hesperidum* Linnaeus, *C. pseudohesperidum* (Cockerell), *Eucalymnatus tessellatus* (Signoret), *Parasaissetia nigra* (Nietner), *Pulvinariella mesembryanthemi* (Vallot), *Saissetia coffeae* (Walker), *S. oleae* (Bernard). For further information on some of the above listed introduced species consult the publication of Zahradnik (1968) and the additional sources listed by him.

Biology. Members of the Coccidae feed on a variety of plants belonging to different plant orders. Some are polyphagous, but most oligophagous or monophagous. There is a certain degree of host specificity among the genera. For example, *Parthenolecanium* species infest woody plants. While members of *Eriopeltis, Luzulaspis, Psilococcus* and *Scythia* live on herbaceous plants. All *Luzulaspis* feed only on species of *Carex, Juncus* and *Luzula*. Species of *Physokermes* and *Nemolecanium* feed exclusively on conifers, latter on *Abies* only. Often they are also specific to a feeding site. Members of *Lecanopsis* and *Rhizopulvinaria* are found in CE only on roots and on the root crowns of plants. All stages of *Sphaerolecanium prunastri* develop on woody parts of plants, while in *Parthenolecanium corni* the first instars feed on leaves, other stages on woody parts of trees and shrubs. Species of *Eriopeltis* develop on leaves of grasses while *Psilococcus* inside leaf sheaths.

Life cycles have been studied only in a few species. Biologies vary by genera and species. Some overwinter as follows: species of *Luzulaspis, Palaeolecanium* and *Psilococcus* in egg stage; *Eulecanium, Parthenolecanium, Sphaerolecanium* and *Physokermes* as second instars; while *Pulvinaria* and *Rhizopulvinaria* as adult females. Most species have one generation per year, but some are known under warm climates to develop a second or even third generation, e.g. *Parthenolecanium corni* and *P. persicae*. The life cycle of most species, with one generation per year and which overwinter as second instars on trees, follows the general pattern given below. The first instars or crawlers move on the plant surface from females or ovisacs to a new site for settling, usually on leaves where they feed and molt. Crawlers are not protected by a wax covering this time, therefore, they are very susceptible to pesticides. This is the time to apply chemical controls if needed. Before the leaves fall off, the second instars crawl to the woody parts of trees and shrubs for overwintering. The males produce a glassy plate-like wax covering in this stage while females are protected by a thin wax coat. After molting into third instar, males become motionless as prepupae and pupae under their translucent wax test from where they emerge as winged adults. Females that have a third nymphal stage develop further into adults, after molting for the last time. Adults usually mate in the late spring or early summer. In the majority of the species the females are large, thus they generally produce 1,000–2,000 eggs, but in some over 8,000. A few species are known to produce live young. Reproduction is almost exclusively biparental in Central Europe but some, *Parthenolecanium, Pulvinaria*, and others, are known to reproduce both parthenogenetically and sexually.

Coccidae produce honeydew, and as such their colonies are normally attended by ants, also are often visited by other honeydew seeking organisms. Koteja (1981) studied honeydew production in some coccids. Their honeydew usually drips to the plant leaves and on it sooty mold develops. Latter produces a black coating on the leaves thus making the spotting of infested trees possible from some distance. The mold inhibits photosynthesis and reduces the aesthetic qualities of the infested trees and shrubs. Damage to the infested trees is primarily produced with their sucking of plant juices. Among the species of major economic importance in CE are: *Eulecanium tiliae, Parthenolecanium corni, P. persicae, Pulvinaria betulae, P. vitis*, and *Sphaerolecanium prunastri*. These are all polyphagous species that attack various fruit and ornamental trees.

Natural enemies are commonly found in the colonies, those are listed under each species. Most important and frequent natural enemies include hymenopterous parasites, mainly chalcid wasps, that were used for biological control, fungus weevils, coccinellid beetles, Neuroptera, Cecidomyiidae, some Lepidoptera larvae, birds, also some entomogenic fungi, eg. *Cephalosporium lecanii,* common in tropical areas. Records on biological control of three species discussed here are given by Bartlett (1978).

Reviewer's remark. The enlarged frontal part of head, obscurely 2-segmented labium, anal cleft, anal lobes, and the glassy male puparium well define the Coccidae. However, their classification has not yet been elaborated. Steinweden (1929) suggested relationships of some genera; Bodenheimer (1953) and Borchsenius (1957) proposed comprehensive classification based on adult male features which appeared to be convergencies in many instances, thus the system appears artificial. Giliomee (1967) based his new concept on male morphology, but unfortunately, only about 20 species were available to him.

Without considering the large number of genera (some very peculiar), that are distributed all over the world, the European Coccidae may be divided, according to Giliomee (1967), into 3 subfamilies (groups of genera in the original paper): the Eriopeltinae, Coccinae, and Eulecaniinae.

The Eriopeltinae, partly covering the Borchsenius' Filippinae, are well characterized by the elongate female body, 2, or no spiracular setae; also dorso-ventrally flattened head in male, narrow wings, box-shaped (with perpendicular lateral walls) male puparium and the monocotyledon host plant (mainly Poaceae and Cyperaceae). Nine genera are included in this subfamily (7 in CE)–*Exaeretopus, Poaspis, Hadzibejliaspis, Luzulaspis, Vittacoccus, Eriopeltis, Scythia, Psilococcus* and *Lecanopsis*; but not *Filippia, Parafairmairia*, etc.

The Coccinae are characterized by the constricted apex of antenna, presence (in a number of species) of dorsal submarginal tubercles, 3 spiracular setae, with the median the longest; anterodorsal part of male head strongly bulged, absence of halteres, etc. However, the egg-protecting systems are extremely diverse, convergent with those in the Eulecaniinae. Based on this convergency, Borchsenius (1957) recognized the

Coccinae, with Pulvinarini and Coccini, and Ceroplastinae. In CE the Coccinae, s. Giliomee, are represented by *Parthenolecanium, Pulvinaria, Chloropulvinaria,* etc., and in southern Europe also by *Coccus, Saissetia, Eucalymnatus, Cereplastes,* and others.

The Eulecaniinae are a diverse group that provisionally include the remaining genera. Some of them represent the most primitive forms among the Coccidae; with numerous simple eyes, broad wings, halteres, etc. in the males.

Key to genera of Coccidae

(Morphological characters given are of teneral females, unless stated otherwise)

1. Teneral females at least twice as long as wide (except in *Lecanopsis,* which has no spiracular setae, and *Lichtensia,* in which most marginal setae chisel-shaped); dorsum and venter of postreproductive females membranous or only weakly sclerotized, rarely anterior and posterior ends, also body margins, more strongly sclerotized; adult females completely enclosed in ovisac, except in *Psilococcus,* rarely without ovisac, then with translucent test (e. g. *Parafairmairia*); on monocotyledons (Gramineae, Cyperaceae, Juncaceae and Iridaceae), except *Lichtensia* (subfamily Eriopeltinae) .. 2

   Teneral females less than twice as long as wide, normally less than 1 ½ times as long as wide; dorsum of postreproductive females thick and sclerotized in varying degrees, venter membranous; adult females not enclosed in ovisac, if cottony ovisac present (e.g. Pulvinariini), located behind or under body of female; on gymnosperms, dicotyledons, especially woody plants .... (subfamily Coccinae) 10

2. Legs small, reduced, of short segments, or stublike; large conical truncate setae on body margin and often on entire dorsum; anal cleft very short 3

   Legs well-developed; setae on body margin or on dorsum not conical, truncate, usually well-developed and pointed; anal cleft moderately long, except in *Lecanopsis* and *Parafairmairia* 4

3. Antennae 6–9-segmented; legs short, but clearly divided into segments; ovisac fluffy, does not include nymphal exuviae (figs 7, 84) .. *Eriopeltis,* p. 179

   Antennae stublike, 1–4-segmented; legs stublike, 1–2-segmented; ovisac dense, feltlike, includes nymphal exuviae divided in 2 parts (figs 107, 108)...... .... *Scythia,* p. 250

4 (2). Spiracles in funnel-shaped depressions, with disc pores on their walls; legs and antennae of stout segments, often reduced; spiracular setae absent; on underground parts of grasses or in leaf sheaths of *Carex.* 5

   Spiracles on body surface, not in funnel-shaped depressions; legs and antennae of slender segments, well-developed; normally with spiracular setae, except in *Vittacoccus;* not on underground parts of plants 6

5. Spiracles very large; body broadly oval; without dark-rimmed disc pores on venter; disc pores with 3–6 loculi, do not form submarginal band on venter; ovisac encloses egg-laying female; on underground parts of Gramineae (fig. 87) .... *Lecanopsis,* p. 192

   Spiracles relatively small, ca 37 or 46 μ long; body elongate, parallel-sided; with many dark-rimmed disc pores on venter; disc pores with 5, rarely 3, 4 or 6 loculi form a submarginal band on venter; ovisac encloses eggs only; in leaf sheaths of *Carex* (fig. 102) .... *Psilococcus*, p. 236

6 (4). Marginal setae chisel-shaped, notched at tip; tubular ducts form a submarginal band on both surfaces; on woody and herbaceous plants, not on monocotyledons (fig. 88) .... *Lichtensia*, p. 197

Marginal setae pointed, never chisel-shaped and notched at tip; tubular ducts scattered on both surfaces, not forming a submarginal band (except in *Parafairmairia*); on monocotyledons only 7

7. Dorsal body setae absent, except on margin; with dark-rimmed pores on venter; anal cleft short; ovisac absent, test of female of translucent wax sheaths (figs 94, 95) .... *Parafairmairia*, p. 213

Dorsal body setae present; without dark-rimmed pores on venter, except in *Vittacoccus*; anal cleft moderately long; ovisac present, encloses female. 8

8. Spiracular setae absent; with wide marginal band of quinquelocular pores on dorsum; dark-rimmed disc pores in a ventral submarginal band from spiracles to anal lobes (fig. 110).................. *Vittacoccus*, p. 257

Spiracular setae present, usually 2 in each group; without marginal band of quinquelocular pores on dorsum; dark-rimmed disc pores absent from venter 9

9. With 35–70 interantennal setae; claw digitules slender, diameter of apical knob smaller than middle of claw; anal ring with 8 setae; on Gramineae .... *Poaspis*, p. 235

With 9–31 interantennal setae; claw digitules thick, diameter of apical knob wider than middle of claw or of same width; anal ring with 6 setae, rarely 8 in *pieninica*; on Cyperaceae and Juncaceae (fig. 89) ..... *Luzulaspis*, p. 199

10 (1). Tibia and tarsus usually freely articulated; tibio-tarsal articulation in most heavily sclerotized; dorsum of postreproductive females strongly sclerotized, moderately convex, rarely convex, hemispherical; female forms white cottony ovisac behind or under her body (tribe Pulvinariini) 11

Tibia and tarsus normally not freely articulated; tibio-tarsal articulation in most not sclerotized; dorsum of postreproductive females almost always strongly sclerotized, strongly convex, often hemispherical or almost hemispherical, rarely moderately convex or flat, in latter cases ventral tubular ducts absent or not in a submarginal band; female does not form ovisac (tribe Coccini) 14

11. With hairlike marginal setae, simple or fringed 12

With spinelike marginal setae, simple 13

12. Marginal setae with sharp-pointed tips, never expanded or fringed at tip (fig. 104) .... *Pulvinaria*, p. 238

All or many marginal setae expanded and fringed at tip (in *floccifera* they number only 2 to 30) (fig. 83) .... *Chloropulvinaria*, p. 177

13 (11). Ventral tubular ducts large, on dorsum much smaller, latter rarely absent; spiracular setae in each group of about same size; on roots and root crowns of perennial herbaceous or semi-woody plants in dry habitats (fig. 105) . ... *Rhizopulvinaria*, p. 244

Large submarginal tubular ducts on venter and dorsum about the same size; medial spiracular seta in each group larger than laterals; on leaves and branches of ericaceous shrubs in moist habitats (fig. 99) .... *Phyllostroma*, p. 227

14 (10). Anal plates, also marginal and spiracular setae absent; postreproductive females nearly spherical, budlike or kidney-shaped; only on conifers, e.g. *Picea, Abies* (fig. 101) .... *Physokermes*, p. 230

Anal plates, also marginal and spiracular setae present; postreproductive females only rarely spherical (*Sphaerolecanium, Rhodococcus*), budlike or kidney-shaped; most on deciduous trees, few on conifers 15

15. Tubular ducts absent; numerous dorsal discoidal pores and hairlike setae form a wide medial longitudinal band from head to anal plates; postreproductive females dark brown, almost black; on Rosaceae, common on stone-fruit trees (fig. 109) .... *Sphaerolecanium*, p. 254

Tubular ducts present; dorsal discoidal pores scattered, or medial longitudinal band without numerous setae; postreproductive females often brown, but not almost black; on a variety of host trees 16

16. Spiracles small, diameter of peritreme 2–3, rarely 1 ½ times shorter than coxa; antennae usually long, 6–9-segmented, 222–504 μ long 17

Spiracles large, diameter of peritreme almost as large or exceeds length of coxa; antennae moderately short in proportion to body length, 5–8-segmented, 190–320 μ long 19

17. Marginal setae normally longer or of same length as medial spiracular setae; discoidal pores form a narrow medial longitudinal band on dorsum; postreproductive female with 2, rarely 1 pair of dorsal tubercles; on *Rosaceae* and *Juglans* (fig. 93) .... *Palaeolecanium*, p. 210

Marginal setae shorter than medial spiracular setae; discoidal pores if arranged in a longitudinal band, then the band short and located just anterior of anal plates; dorsal tubercles absent in postreproductive females 18

18. Marginal setae poorly developed, short and slender, form an indistinct row, intervals between them (60–200 μ) exceed many times their length; only on firs, *Abies* .... *Nemolecanium*, p. 208

Marginal setae well-developed, form 1 or 2 distinct rows, intervals between them normally do not exceed 2–3 times their length; on other host trees than firs (fig. 97) .... *Parthenolecanium*, p. 217

19 (16). With 3 spiracular setae at least in two groups; anal ring well-developed, each with 6–10 large setae, and 2 rows of pores; postreproductive females oval to circular, convex; body margins not bent under; on various trees and shrubs (fig. 85) .... *Eulecanium*, p. 184

With 1 or 2 spiracular setae in each group; anal ring small, narrow, without pores, often without setae, or latter poorly developed; postreproductive females rarely spherical; with margins bent under; on *Rosa* or *Spiraea* (fig. 106) .... *Rhodococcus*, p. 246

Genus **Chloropulvinaria** Borchsenius

Syn. & Lit.: *Chloropulvinaria* Borchsenius, 1952:299, 1957:205; Canard, 1965:424; Řehaček, 1960:38; Ali, 1971:45.

Type-species: *Coccus floccifera* Westwood, 1870.

Ovisac moderately convex, white, protrudes from behind body or under abdomen. Postreproductive female oval, moderately convex, dorsum slightly sclerotized, venter membranous, marginal fold sharply indicated. Teneral female often greenish, yellowish or brownish, with yellowish or brownish bands and spots when live, up to 5 mm long. *Venter.* Antennae 8-, rarely 5- to 7-segmented; claw digitules stout, strongly expanded at tip; spiracular pore bands developed with quinquelocular pores; spiracular setae normally 3, rarely 4–6 in each group, median seta much longer than laterals; marginal setae often with forked, or fringed tips; body setae of various sizes; multilocular pores on sternites; tubular ducts numerous, especially in a submarginal band. *Dorsum.* Submarginal tubercles usually developed; small spine-like setae scattered on entire surface; discoidal pores present medially anterior of anal plates; tubular ducts of 1 or 2 sizes on entire surface; anal ring with 8, rarely 6 or 10 setae and double row of pores; anal plates triangular.

Distribution. World-wide with ca 14 species; 1 in CE.

Biology. On trees and shrubs where N_1 feeds on leaves and after molting usually moves to twigs for maturation; adult female migrates to leaves, produces ovisac and lays eggs. One generation per year and biparental reproduction in species studied.

97. **Chloropulvinaria floccifera** (Westwood) – Cushion Scale, Cottony Camellia Scale
Fig. 83

Syn. & Lit.: *Coccus flocciferus* Westwood, 1870:308. *Pulvinaria floccifera,* Leonardi, 1920:333; Balachowsky & Mesnil, 1936:1602; Gomez-Menor, 1937:284; Steinweden, 1946:6; Takahashi, 1955a:153; Schmutterer, 1955b:100; Ezzat & Hussein, 1969:417; Williams & Kosztarab, 1972:135. *Pulvinaria camellicola* Signoret, 1873:32. *P. brassiae* Cockerell, 1895:135. *P. phaiae* Lull, 1899:237. *Chloropulvinaria floccifera,* Borchsenius, 1952:300, 1957; 205; Řehaček, 1960:39; Ali, 1971:46; Tereznikova, 1981:132.

Ovisac elongate, straight or curved, white. Postreproductive female oval or elongate-oval, slightly convex; somewhat transversely ridged, dark brown. Teneral female oval, or elongate-oval, cream to tan, mottled with brown; 3.1–4.6 mm long, 2.1–3.3 mm wide. *Venter.* Antennae 8-segmented, seldom 7, 438–474 μ long; labium 85 μ long and 109 μ wide (Koteja, 1974a); legs well-developed 715–884 μ long, with tibio-tarsal articulation, claws without denticle; spiracular pore bands 3–4 pores wide, each with 50–70 quinquelocular pores; spiracular setae 3 in each group, median seta 2 to 3 times length of lateral setae; marginal setae 44–76 μ long, acute, bifid or fringed, most curved; the number of bifid or fringed setae vary from 2 to 30 (Steinweden, 1946); submarginal setae 19–32 μ long, acute, straight or slightly curved near tip; body setae scattered on entire surface 8–19 μ long; multilocular pores 6–8 μ in diameter, most with 7 loculi, form a cluster around vulvar area and transverse bands or rows on abdominal sternites; microducts with square-shaped orifice, scattered on entire surface; tubular

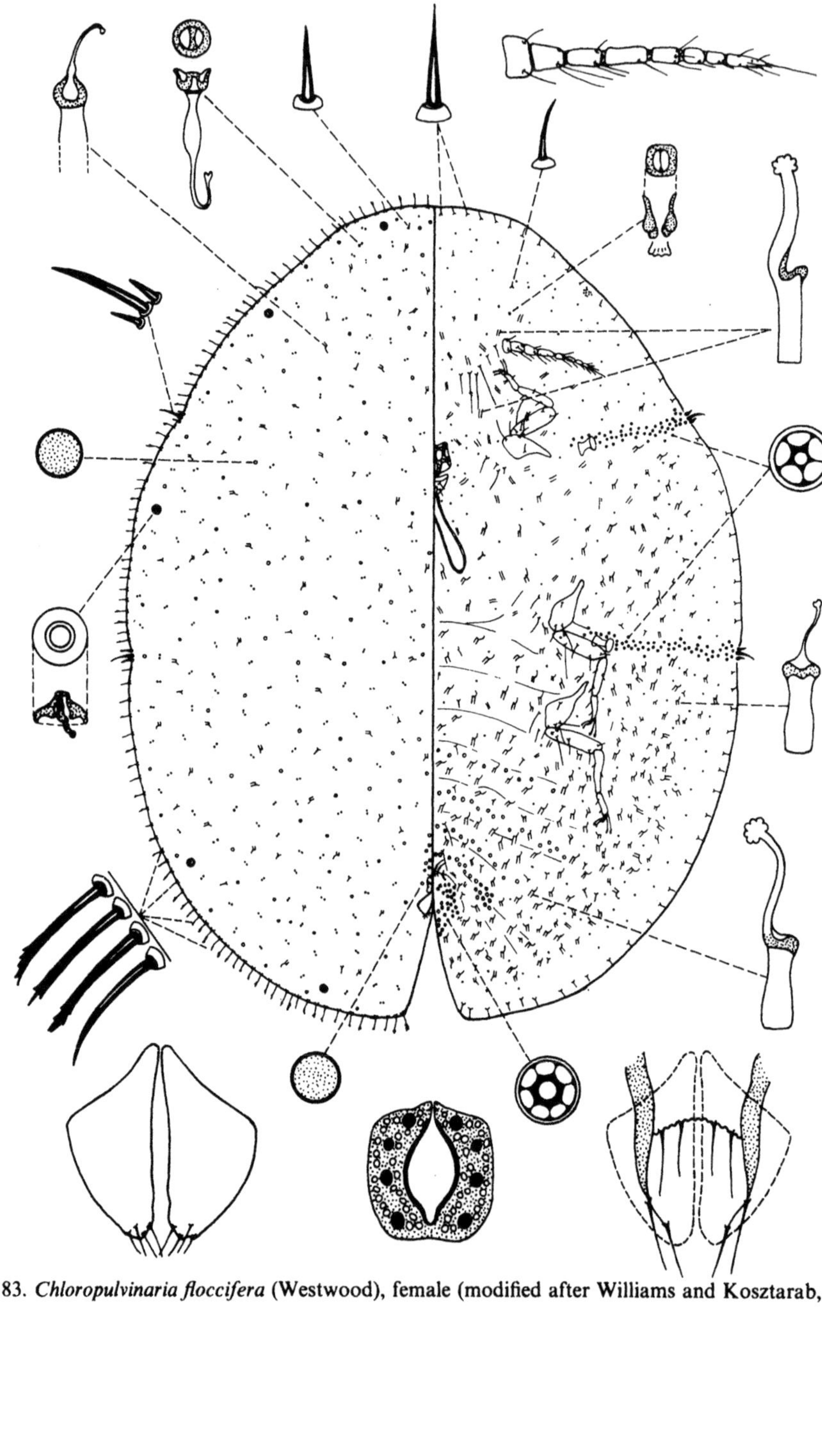

Fig. 83. *Chloropulvinaria floccifera* (Westwood), female (modified after Williams and Kosztarab, 1972)

ducts of 3 types, most with flowery tipped filaments, few ducts on head, more densely distributed toward posterior end. *Dorsum.* Submarginal bicylindrical tubercles 4–11 around entire body; body setae 8–12 μ long, scattered on entire surface; bilocular pores interspersed with simple disc pores and scattered over dorsum; ca 25 discoidal pores, 2.5–3.7 μ in diameter, in median area anterior of anal plates; tubular ducts with sclerotized rim, 12 μ long, 2.5 μ wide with slender filament ca 11 μ long, scattered on entire surface; anal ring rectangular, ca 62 μ long, 57 μ wide with 8 setae and 2 rows of pores; anal plates triangular, 125–234 μ long, 63–88 μ wide, each with 4 apical and 2 subapical setae. Male test and N_1 described by Řehaček (1960).

Distribution. A common Mediterranean species with almost world-wide distribution. It is found principally in greenhouses, e.g. Po, rarely outdoors in CE: Cz, Fr, Ge, Hu, It, Ro, Sw, USSR and Yu; also in Denmark, England, Holland, Portugal, Spain, and Sweden. Also on all other continents and on most large islands.

Biology. A polyphagous species, known from at least 25 plant genera. Occurs on leaves and twigs. Often found on *Camellia,* where it may cause injury, also was reported as a pest from some countries, especially from those with tropical or subtropical climate. One generation per year. In Czechoslovakia N_2 overwinter, adults develop by April, and females start laying eggs in May; eggs hatch at end of June and N_1 settle on leaves, N_2 migrate to twigs in fall where they overwinter and mature. Males are present (Řehaček, 1960). Canard (1965) noted 1100 eggs per female.

Natural enemies. Parasite: *Coccophagus lycimnia.* Predators: *Chilocorus bipustulatus, C. renipustulatus, Cryptolaemus montrouzieri, Exochomus flavipes, E. quadripustulatus, Hyperaspis campestris, Leucopis lusoria, Rhizobius litura* (Borchsenius, 1957; Tereznikova, 1981). Fungi associated: *Cephalosporium lecanii, Fusarium juruanum* (Schmutterer et al.,.1957).

Genus **Eriopeltis** Signoret

Syn. & Lit.: *Eriopeltis* Signoret 1872a:429; Newstead, 1903:21; Leonardi, 1920:352; Steinweden, 1929:235; Šulc, 1941:3, 16; Schmutterer, 1952b:555; Borchsenius, 1956:397, 1957:140; Kawecki, 1957:198; Řehaček, 1960:28; Morrison & Morrison, 1966:70; Williams & Kosztarab, 1972:62; Koteja, 1974b:306; Danzig, 1975d:808, 1980a:251; Kosztarab & Kozár, 1978:95; Tereznikova, 1981:99.

Type-species: *Coccus festucae* Fonscolombe, 1834.

Ovisac elongate-oval, feltlike, fluffy, cottony, white or grayish-white, up to 15 mm long; encloses egg-laying female completely. Postreproductive female oval or elongate-oval, moderately convex, with heavily sclerotized derm; up to 10 mm long. Teneral female with short anal cleft. *Venter.* Antennae 6 to 9-segmented, most segments short and thick; legs small, tibia and tarsus sometimes fused, claws short, wide at base; spiracles large; spiracular setae absent; spiracular pore band often only partly developed, pores quinquelocular; marginal setae short, blunt, conical; multilocular pores form wide transverse bands on abdominal sternites; tubular ducts of 2 sizes, dense in a wide submarginal band, sparse in median area. *Dorsum.* Submarginal tubercles absent; surface usually densely covered with thick truncate setae, discoidal pores in a narrow longitudinal median band from anal plates to head, where discoidal pores replaced setae; normally minute simple disc pores and bilocular pores with heavily sclerotized rim in an irregular dense band around body margin, sparsely on rest of dorsum; tubular ducts scattered on entire surface; anal ring with pores and 6–10 setae; anal plates triangular, all angles bluntly pointed.

Distribution. Holarctic, with 6 species; in CE 3 species.

Biology. On leaves of grasses and *Luzula(?)*; 1, rarely 2, generations per year.

Key to species of *Eriopeltis*

1. Large truncate dorsal setae only on anterior and posterior end of body; other areas covered with small spinelike setae .... *lichtensteini*, p. 182
   Large truncate dorsal setae on most of surface, except for a narrow medial longitudinal band 2
2. At least 30 setae on each lateral margin; 1–3 disc pores at base of majority of setae; submarginal setae on dorsal lateral margins of body close to each other (fig. 84); on broad-leaf grasses .... *festucae*, p. 180
   Less than 10 setae on each lateral margin; no disc pores at base of setae; submarginal setae on dorsal lateral margins of body far apart; on narrow-leaf grasses .... *stammeri*, p. 183

98. **Eriopeltis festucae** (Fonscolombe) – Cottony Grass Scale
Figs 7, 84

Syn. & Lit.: *Coccus festucae* Fonscolombe, 1834:216. *Eriopeltis festucae*, Patch, 1905:169; Borchsenius, 1956a:398, 1957:154; Řehaček, 1960:32; Williams & Kosztarab, *1972:63; Danzig, 1975d:809, *1980a:252; Kosztarab & Kozár, 1978:97; Tereznikova, 1981:100. *Eriopeltis brachypodi* Giard, 1894:199 (nomen nudum). *E. agropyri* Borchsenius, 1956a:399. *E. araxis* Borchsenius, 1956a:401. *E. caucasicus* Borchsenius, 1956a:402. *E. desertus* Borchsenius, 1956a:403. *E. eversmanni* Borchsenius, 1956a:403. *E. ferganensis* Borchsenius, 1956a:404. *E. hamberdiensis* Borchsenius, 1956a:405. *E. maximus* Borchsenius, 1956a:408. *E. phragmitidis* Borchsenius, 1956a:408. *E. pratensis* Borchsenius, 1956a:411. *E. rasinae* Borchsenius, 1956a:411. *E. stammeri*, Borchsenius, 1956a:419 (not Schmutterer, 1952b). *E. zolotarevae* Borchsenius, 1956a:416. We deleted a number of early literature citations because the early records were unreliable and could not be authenticated.

Ovisac feltlike, white or greyish-white, up to 15 mm long. Female after egg laying drops from ovisac through an opening at anterior end. Teneral female 3–9 mm long, 1.5–4 mm wide, anal cleft twice as long as anal plates. *Venter*. Antennae 7- or 8-, rarely 6- or 9-segmented, 150–243 μ long; labium 85 μ long, 119 μ wide (Koteja, 1974a); claws with broad base; each spiracular furrow with 28–57 quinquelocular pores; marginal setae form a row, with over 30 setae on each side and with 1–2 simple disc pores at base of each seta; multilocular pores numerous in vulvar area, in transverse bands on 2–3 preceding abdominal sternites; tubular ducts on entire venter. *Dorsum*. Truncate setae variable, often at base twice as wide as their tip; arrangement of minute simple disc pores, bilocular pores, discoidal pores and tubular ducts as given for the genus; anal ring with 3 irregular rows of pores and 8 setae, 2 of these shorter and more slender than others. Adult male described by Giliomee (1967), male test and N_1 by Green (1917), Řehaček (1960); nymphs by Borchsenius (1957) and Leonardi (1920).

Distribution. Because of the nomenclatural uncertainties in the pre-1975 literature (Danzig, 1975d), the actual distribution is not known. It appears to be common in the steppe and forested areas of the Holarctic Region and based on literature records in CE is present in: Au, Bu, Cz, Fr, Ge, Hu, It, Po, Ro, Sw, USSR and Yu, also known from Denmark, England, Finland, Holland, Portugal, Spain, Sweden, Middle East, Mongolia and USA.

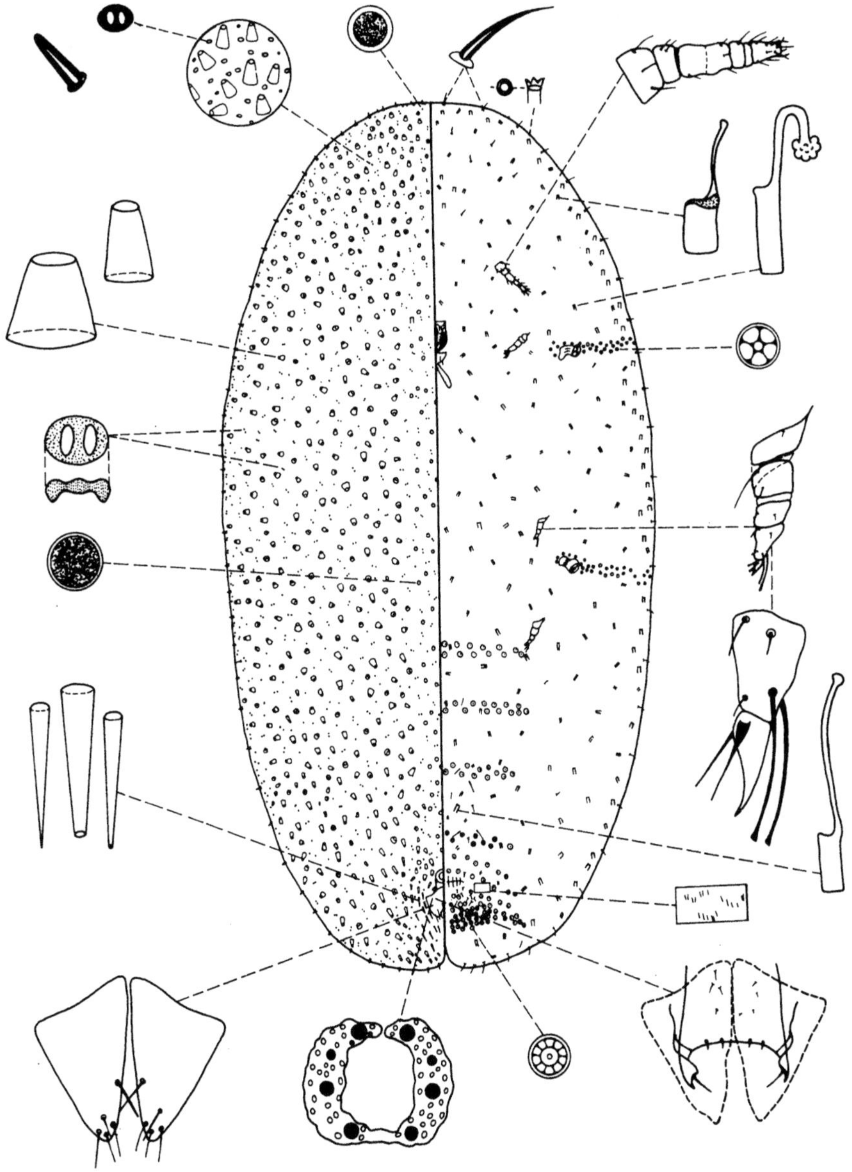

Fig. 84. *Eriopeltis festucae* (Fonscolombe), female (modified after Williams and Kosztarab, 1972)

Biology. On broad-leaved grasses: *Agropyron, Agrostis, Aneurolepidium, Arrhenatherum, Arundinella, Brachypodium, Bromus, Calamagrostis, Festuca* (broad-leaf species), *Lolium, Nardus, Phleum,* and *Phragmites.* Eggs overwinter. Eggs hatched in S. Germany (Schmutterer, 1952b) at end of May; adults developed by early July and started ovipositing at end of July. Kozár (1971) found in Hungary ovipositing females during second half of June, and by end of August adult females of a second generation. Two generations were observed in NE USA (Patch, 1905) and Poland; only 1 generation was noted in S Germany, where females laid 382–1209 eggs. Both in USA and Ukraine mass infestations were observed. In Ukraine 636 ovisacs per m^2 were found on *Agropyron*; about 18–20 ovisacs per stem. Occasionally of economic importance, because heavily infested pastures cannot be grazed.

Natural enemies. A number of species given in the literature are apparently hyperparasites, but no attempt was made here to identify them. It is also possible that the records of parasites and hosts were based on misidentifications. Parasites: *Beocharis pascuorum, Cerapterocerus mirabilis, C. multiradiatus, C. pilicornis, Cerchysius subplanus, Choreia maculata, Coccophagus rjabovi, C. scutellaris, Discodes cercopiformis, Marietta picta, Metaphycus parvus eriopelti, Pachyneuron coccorum, P. concolor, Subprionomitus cantabricus, Trichomasthus albimanus, T. cyaneus, T. cyanifrons, T. frontalis.* Predators: *Hyperaspis campestris, Leucopis annulipes, L. szepligeti, L. vorax, Leucopomyia silesiaca, Nephus bipunctata* (Schmutterer, 1955c; Kosztarab, 1959; Trjapitzin, 1978; Kozár and Sugonjaev, 1979).

Associated ant. *Myrmica ruginodis* (Schmutterer, 1952b).

Note. Because of a high degree of morphological variation found in this species in different parts of the Holarctic Region by a number of recent authors, a species complex could be present.

99. **Eriopeltis lichtensteini** Signoret – Lichtenstein's Cottony Scale

Syn. & Lit.: *Eriopeltis lichtensteini* Signoret, 1876:607; Herberg, *1918:1; Řehaček, 1954a:141, 1960:31; Borchsenius, 1957:144; Koteja, 1971b:323; Danzig 1975d:809; Kosztarab & Kozár, 1978:97; Tereznikova, *1981:102.

Ovisac feltlike, bulging on lateral sides, tapering anteriorly, white; normally 7–10, rarely up to 14 mm long, 3.5–3.8 mm wide. Postreproductive female elongate-oval, broadly rounded anteriorly, slightly tapering caudally. Teneral female usually 5–6.5 mm, rarely up to 10 mm long, 2–3 mm wide; anal cleft 1½ times longer than anal plates. *Venter.* Antennae 8-segmented according to Borchsenius (1957), ca 234 μ long, 3rd segment longest; labium 85 μ long, 118 μ wide (Koteja, 1974a); legs short and stout; each anterior spiracular band with 29–60 quinquelocular pores, each posterior band with 22–43 pores; multilocular pores 7–8 μ in diameter with 5–12 loculi, form transverse rows on abdominal sternites, those on anterior half of abdomen with 5–7 loculi, others on posterior half with more loculi; tubular ducts 18–22 μ long, most numerous in a submarginal band. *Dorsum.* Truncate setae few, most on anterior and posterior ends, and some along body margin, on anterior margin they form a partial double row, and the distance between setae equals their length; dorsal marginal setae short, 10–27 μ long; simple disc pores scattered on entire surface; discoidal pores form a wide longitudinal medial band from head to anal plates; tubular ducts ca 4 μ in diameter, with heavily sclerotized rim, numerous, except few or absent in medial longitudinal band of discoidal pores; anal ring with 2 rows of pores and 8 setae, most 217–250 μ long; 2 of these shorter and thicker than others; anal plates elongate-conical. Adult male described by Borchsenius (1957), male test and nymph by Řehaček (1960), all stages described (except adult male) in much detail by Herberg (1918).

Distribution. A moderately frequent boreal species. In CE: Bu, Cz, Fr, Ge, Hu, Po, USSR; also in Denmark, England, Finland, Holland, Ireland, Spain and Sweden.

Biology. On leaves of *Agropyron, Agrostis, Brachypodium, Bromus, Calamagrostis, Elymus, Festuca, Holcus, Luzula(?), Melica, Phalaris, Phragmites, Poa,* and *Stipa.* Eggs overwinter. Herberg (1918) reported 5 developmental stages for females that appear more than usual for the group. In Finland Tiensuu (1951) found eggs hatching during second half of June; ovisacs were formed from end of July, egg laying started in early August; each female layed 741–1502 eggs (Herberg, 1918).

Natural enemies. Parasites: *Beocharis pascuorum, Coccophagus lycimnia, Marietta picta, Metaphycus parvus, Microterys herbaceus, Pachyneuron coccorum, Trichomasthus albimanus, T. cyaneus* [latter destroyed 30% of scales (Danzig, 1959a)], *T. cyanifrons.* Predators: *Leucopis alticeps, L. nigricornis, L. vorax, Leucopomyia silesiaca,* (Fulmek, 1943; Zak-Ogaza, 1961; Kozarzhevskaya and Reitzel, 1975; Trjapitzin, 1978). Herberg (1918) reared 329 chalcids from 297 females, and each parasitized female served as host for 1 to 9 chalcids; he also observed two associated fungi: *Cladosporium aphidis* and *C. coccidarum.*

100. **Eriopeltis stammeri** Schmutterer – Stammer's Cottony Scale

Syn. & Lit.: *Eriopeltis stammeri* Schmutterer, 1952b:555; Řehaček, 1956:47; Borchsenius, 1957:167; Danzig, 1975d:812; Kosztarab & Kozár, 1978:97; Tereznikova, 1981:104. *E. plumeus* Borchsenius 1956a:409. *E. festucae,* Borchsenius, 1957:154 (not Fonscolombe, 1834, in part); Koteja, 1971b:319.

Ovisac fluffy, cottony, with protruding long wax threads, white, ca 6 mm long, 3–3.5 mm wide. Postreproductive female elongate-oval, convex. Teneral female 2.2–4.7 mm long, 1.2–2 mm wide; length of anal cleft about as long as length of anal plates. *Venter.* Antennae 7- or 8-segmented, ca 150 μ long, 3rd segment as long as 4th and 5th combined; legs reduced, 180–200 μ long, base of claws wide; spiracular disc pores predominantly quinquelocular, occasionally with 6 or 7 loculi, 16–29 pores in each band; only up to 10 marginal setae on each lateral margin, without disc pores at their base; multilocular pores with 10 loculi, most located in a band surrounding vulva, others in transverse rows and bands on preceding abdominal sternites; tubular ducts scattered on entire body, fewer in medial area of venter. *Dorsum.* Truncate setae numerous on surface except absent in medial longitudinal band, containing discoidal pores, latter 6–9 μ in diameter; simple disc pores 4–5 μ in diameter, scattered on entire surface; anal ring with 8 setae, most ca 330 μ long, 1 pair shorter than others. N_1 described by Schmutterer (1952b).

Distribution. Danzig (1975d) suggested that it occurs in most of the Palearctic Region. In CE: Cz, Ge, Hu (Kozár's unpublished record), Po, USSR; also in Denmark. Common in steppe and forest habitats.

Biology. On narrow-leaved grasses: *Corynephorus, Deschampsia, Festuca, Nardus.* Overwinters either as N_2 or as young females in S Germany (Schmutterer, 1952b), eggs are laid from early June, began hatching from early July. One generation per year.

Natural enemies. Parasite: *Eunotus cretaceus* (Schmutterer, 1955c). Predator: *Leucopomyia silesiaca* (Koteja, 1971b).

Genus **Eulecanium** Cockerell

Syn. & Lit.: *Eulecanium* Cockerell, 1893a:54, Leonardi, 1920:285; Gomez-Menor, 1937:266; Balachowsky, 1939:155; Schmutterer, 1952b:545, 1954;62; Borchsenius, *1957:386; Kawecki, 1958:135, Řehaček, 1960:57; Morrison & Morrison, 1966:72; Kosztarab & Kozár, 1978:98; Danzig, 1980a:275; Tereznikova, 1981:152. *Lecanium* (in part) Signoret, 1873:395 (225); Newstead, 1903:75; Šulc, 1908a:36, 1932:1; Richards 1958:305; Williams & Kosztarab, 1972:68.

Type-species: *Coccus tiliae* Linnaeus, 1758.

Postreproductive female, heavily sclerotized, oval to broadly circular, strongly convex, rarely almost spherical, usually yellowish or brownish; 2.5–11 mm long. Teneral female oval to circular from light-yellow to greenish-grey, often with darker transverse and longitudinal bands; anal cleft short. *Venter.* Antennae 5–8-segmented; antennae and legs small in relation to body size of mature females; labium 2-segmented, with 10 setae; spiracles and peritreme large, larger than coxae; spiracular furrows with 1 row or a narrow band of normally quinquelocular pores; 2 or 3 stout spiracular spines in each group, these often shorter than marginal setae; marginal setae spinelike conical, few species also with hairlike setae; multilocular pores form transverse bands on abdomen and bands or groups on thorax; tubular ducts numerous in a submarginal band, a few in medial area of venter; rarely with some sievelike pores. *Dorsum.* With scattered disc pores and tubular ducts; setae spinelike or slender and hairlike; derm of anal area often heavily sclerotized, striated or tessellated; anal ring large, normally with 2 rows of pores and 6–10 large setae; anal plates well-developed with apical and subapical setae.

Distribution. According to Borchsenius (1957), there are ca 60 species world-wide with most occurring in the Holarctic Region; 6 in CE; some assignments are questionable.

Biology. Feed on deciduous trees and shrubs, rarely on conifers (e. g. *sericeum*). In CE second instars overwinter, there is 1 generation per year. Some are pests of fruit and ornamental trees. Biology of four species is given by Schmutterer (1952b).

Key to species of *Eulecanium*

1. Marginal setae of two types, hairlike setae on anterior and posterior margins, 80–180 μ long, spinelike setae on lateral margins, 27–45 μ; postreproductive female strongly convex dorsally (figs 8, 85) . . . . *tiliae,* p. 189
   - All marginal setae spinelike, normally 27–48 μ long; postreproductive female not strongly convex dorsally 2
2. Spiracular setae large, stout, longer and stouter than marginal setae; antennae 5–6-segmented; on fir, *Abies* spp. only . . . . *sericeum,* p. 189

Spiracular setae not larger and longer than marginal setae; antennae 7- or 8-segmented; on deciduous trees and shrubs only 3

3. Length of antennae ca 320 μ, of labium 157 μ, diameter of hind spiracle peritreme 125 μ; postreproductive females in dorsal view circular or wider than long, with flattened top from side view; females in transverse position on twigs .... *ciliatum*, p. 186

Length of antennae up to 260 μ, of labium less than 130 μ, diameter of hind spiracle peritreme up to 115 μ; postreproductive females in dorsal view oval, not flattened on dorsum; females normally in longitudinal position on twigs 4

4. With quinquelocular pores along dorsal margin of anal cleft; diameter of hind spiracle peritreme up to 86 μ long; postreproductive females small, 3.5–5 mm long, 2–4 mm wide; on Ericaceae only .... *franconicum*, p. 188

Normally without quinquelocular pores along dorsal margin of anal cleft; diameter of hind spiracle peritreme more than 86 μ long; postreproductive females of medium or large size, 5–9 mm long, 5–6 mm wide; not known from Ericaceae 5

5. Anal ring with 8–10 setae, of these 2–4 shorter and stouter than others; large ventral tubular ducts 30–35 μ long; derm around anal plates strongly sclerotized; postreproductive female yellow or light brown; only on *Caragana* spp. .... *caraganae*, p. 185

Anal ring with 6 stout setae and 1–2 slender short setae; large ventral tubular ducts ca 25 μ long; derm around anal plates about as sclerotized as rest of derm; postreproductive female dark or greenish-brown; not known from *Caragana* spp. .... *douglasi*, p. 187

101. **Eulecanium caraganae** Borchsenius – Pea Tree Scale

Syn. & Lit.: *Eulecanium caraganae* Borchsenius, 1953:285, 1957:408; Danzig, 1959a:448; 1974:71; Matesova 1968a:118; Sheldeshova, 1972:51; Vu Quang Kon, 1974:8; Vu Quang Kon & Sugonjaev, 1975a:1, 1975b:1488; Kosztarab & Kozár, 1978:100; Tereznikova, 1981:153.

Postreproductive female, according to Borchsenius (1957), oval, strongly convex, yellow to light brown, along sides with 2 rows of depressed points; 5.5–7 mm long, 5–6 mm wide, 4–5 mm high. Teneral female oval, live specimen greenish-grey with darker transverse and longitudinal bands. *Venter.* Antennae 7- or 8-segmented, ca 240 μ long; claw long and narrow, with denticle near tip, claw digitules similar; diameter of hind spiracular peritreme 90–115 μ; each spiracular furrow with 16–28 quinquelocular pores; spiracular setae often undifferentiated from marginal setae, if present, 2 in each group, these stout, with broadly rounded tips and shorter (18–24 μ long) than marginal setae; rarely with marginal setae between spiracular setae; marginal setae, 36–40 μ long (rarely shorter), with acute apex, space between setae 1 ½–2 time length of setae; short body setae sparsely scattered on surfaces; multilocular pores form a small cluster near each spiracle and each antennal base, larger groups near bases of front legs, and form transverse bands between hind coxae and on abdomen; number of loculi vary from 6 to

14, rarely 4 or 5; large tubular ducts, 30–35 μ long and 5 μ wide, form a submarginal band, with few small, dark-rimmed pores 3 μ in diameter; smaller tubular ducts near mouth parts. *Dorsum.* With anal area heavily sclerotized; disc pores, 5–6 μ in diameter, and spinelike setae 12–15 μ long, scattered on dorsal surface; anal ring with 8–10 setae, of these 2–4 shorter and stouter than others. First and second instars described by Borchsenius (1957) and Tereznikova (1981).

Distribution. Known only from USSR (Leningrad, Moscow, Donets, and Odessa areas, Altai Region and West Kazakhstan in Central Asia) and Mongolia. It is probably more widely distributed in Europe, but was confused with *Parthenolecanium corni.* A steppe inhabiting species.

Biology. Live on branches of *Caragana* spp. Often becomes a pest in city parks. Males unknown; second instars overwinter in the Leningrad region; adults develop in May and start laying eggs from mid-May, 317–1275 eggs per female; eggs hatch at end of June and in July; most nymphs feed on leaves, some on young twigs (Danzig, 1959a). Sheldeshova (1972) reported nymphal survival at –30°C winter temperatures.

Natural enemies. Parasites: *Blastothrix isomorpha, B. matesovae, Cerapterocerus mirabilis, Cheiloneurus paralia, C. claviger, Coccophagus atterimus* and *Encyrtus infidus.* Some are probably hyperparasites. Predators: *Anthribus nebulosus* (destroyed 30% of the scales, according to Danzig, 1959a), *Brachytarsus scapularis, Calocoris biclavatus* (Danzig, 1959a; Matesova, 1968a); *Chilocorus rubidus* (Hodek, 1973).

102. **Eulecanium ciliatum** (Douglas) – Ciliate Oak Scale

Fig. 86a

Syn. & Lit.: *Lecanium ciliatum* Douglas, 1891a:67; Newstead, *1903:96; Šulc, 1932:92. *L. douglasi,* Šulc 1932:95. *Eulecanium ciliatum,* Schmutterer, 1952b:549, 1972b:406; Borchsenius, 1957:401; Řehaček, *1960:60; Kosztarab & Kozár, 1978:99; Danzig, 1980a:278; Tereznikova, 1981:156.

Postreproductive female broadly rounded, often wider than long; dorsum more flat than in any other species in this genus, with 2 longitudinal rows of large deep punctures; yellow to light brown, but appears greyish from thin layer of powdery wax cover; body margin fringed with numerous short, slender, white wax filaments; diameter 5–8 mm, height 3.5 mm. Usually found in a transverse position on twigs, thus their body often halfway encircles a twig. Teneral female. *Venter.* Antennae 7- or 8-segmented, ca 320 μ long, when 7-segmented, 3rd segment as long as 4, 5 and 6 combined; labium 157 μ long, 142 μ wide (Koteja, 1974a); diameter of hind spiracular peritreme ca 125 μ; spiracular setae with blunt tip, not longer than conical marginal setae, latter 40–48 μ long; space between marginal setae 1½–2½ times their length; spiracular furrows with a row of quinquelocular pores; multilocular pores with 10 loculi, form transverse rows and bands on median areas of metathorax and most abdominal segments, few on other thoracic segments and head near antennae; tubular ducts in a wide submarginal band. *Dorsum.* Covered with numerous simple disc pores and small tubular ducts; setae conical. Adult male, male test, and nymphal stages described by Newstead (1903) and Šulc (1932), characteristic male test and first instar by Řehaček (1960).

Distribution. A Trans-Palearctic mesophilous species. Occurs infrequently. In CE: Au, Bu, Cz, Fr, Ge, Hu, Po, USSR (European parts to the Far East), and Yu; also known from England, Hebrides Islands, Holland, Sweden, and Iraq.

Biology. It was reported from twigs and small branches of *Acer, Alnus, Armeniaca, Betula, Corylus, Crataegus, Juglans, Lonicera, Malus, Populus, Prunus, Quercus, Rhamnus* and *Salix*. Oaks are the most frequent hosts. It has 1 generation per year. Second instars overwinter on twigs; in S Germany, according to Schmutterer (1952b), adults develop from end of April, each female lays ca 430–680 eggs from mid-May, eggs hatch in ca 35 days, nymphs feed on underside of leaves, molt from end of August through mid-September, and second instars move to twigs from mid-October. Associated ants: *Lasius fuliginosus, L. niger niger* (Schmutterer, 1952b).

Natural enemies. Parasites: *Blastothrix britannica* and *B. sericea; Cheiloneurus formosus* (Schmutterer, 1953, 1972b; Trjapitzin, 1978). Predators: *Anthribus nebulosus* (Schmutterer, 1952b); birds destroyed ca 50% of females in the spring (Newstead, 1903).

Note. This species is in need of a detailed morphological description and full illustration.

103. **Eulecanium douglasi** (Šulc) – Currant Soft Scale

Syn. & Lit.: *Lecanium douglasi* Šulc, 1895c:37. *L. zebrium* Green, 1917:203. *L. ciliatum*, Šulc, 1932:95. *Eulecanium douglasi*, Borchsenius, 1957:406; Matesova, 1968a:117; Danzig, 1980:280; Schmutterer, 1972b:407; Kosztarab & Kozár, 1978:100; Tereznikova, 1981:157. *E. longisetum* Borchsenius, 1955c:296. *E. trjapitzini* Danzig, 1967:164. *E. eoum* Danzig, 1967:165. *E. coangustum* Danzig, *1967:166.

Postreproductive female oval, dorsum of anterior half strongly convex, posterior end sloping, lateral sides with many tiny depressed points; brown or greenish-brown, with white wax threads along margin; 5–9 mm long, 4–6 mm wide and 4–5 mm high; young adult females with a longitudinal and several transverse dark tan stripes. Teneral female oval to circular. *Venter.* Antennae 7- or 8-segmented, 190–260 μ long; labium 128 μ long, 133 μ wide (Koteja, 1974a); claw with very small denticle near tip, claw digitules slender, similar; spiracles with very wide peritreme, diameter ca 105 μ; spiracular furrows with a row of quinquelocular pores; 2–3 spiracular setae in each group, setae stout, 18–24 μ long, rarely with marginal setae between them; marginal setae sharp-pointed, rarely slightly bent, 30–40 μ long, space between setae 1.5–4 times length of setae; body setae short and stout; interantennal setae 45–50 μ long; multilocular pores with 10 loculi, few on thorax, numerous on abdomen; large tubular ducts, ca 25 μ long, 5 μ wide, form a wide submarginal band, smaller ducts on thorax and head. *Dorsum.* With many scattered small disc pores, diameter 4–7 μ, tubular ducts smaller than those on venter and sparsely distributed; few thick, spinelike setae, 12–15 μ long; marginal setae acute and long; area around anal plates similar to rest of derm; anal ring with pores in partly double rows, 6 stout setae, and 1–2 slender and shorter setae. Male test described by Borchsenius (1957), test and adult male by Green (1917); 1st instar (*longisetum*) by Borchsenius (1957), 1st and 2nd instars by Šulc (1895c).

Distribution. Trans-Palearctic. A mesophilous forest species adapted to winters of N Karelia in USSR. In CE: Cz, Po, and USSR; also known from England and Mongolia.

Biology. Feed on twigs and branches of *Alnus, Betula, Populus, Ribes* and *Salix*. Seldom pest of currants. One generation per year. Adult male emerged at beginning of June, and 1st instars were observed at end of June in England (Green, 1917), 2nd instars overwintered in Bohemia (Šulc, 1895c).

Natural enemies: Parasites: *Blastothrix trjapitzini, Coccophagus aterrimus, Discodes aeneus,* and *Encyrtus infidus* (Trjapitzin, 1978). Matesova (1968a) reported other parasites predators and visiting ants from *Rosa,* but because this is an unusual host for this species, we deleted those records.

Note. Apparently, there is no complete illustration on the teneral female of this species.

104. **Eulecanium franconicum** (Lindinger) – Heather Soft Scale

Syn. & Lit.: *Lecanium franconicum* Lindinger, 1912:90; Šulc, 1932:95; Kawecki, 1971:255. *L. rubellum* Lindinger, 1907:138 (preoccupied name). *L. slavum,* Kawecki, 1961:66. *Eulecanium franconicum,* Schmutterer, 1952b:550, 1972b:408. Borchsenius, 1957:406; Řehaček, 1960:61; Danzig, 1968:304, 1980a:282; Kosztarab & Kozár, 1978:100; Tereznikova, 1981:158. *E. f.* forma *vaccinicola, E. f.* forma *calluneti,* Danzig, *1961:571, 573. *E. rhododendri* Danzig, 1967:162.

Postreproductive female, oval to almost circular, convex, wider anteriorly, narrowing and steeply sloping posteriorly; brown to dark brown, shiny, with longitudinal carina dorsally, and with many depressed points laterally, 3.5–5 mm long, 2–4 mm wide. Body shape, size, and color of live females may vary on different host genera (Danzig, 1961). Teneral female light yellow, with 1 longitudinal and several transverse brown bands. *Venter.* Antennae 7–8-segmented, ca 240 μ long; labium 109 μ long, 119 μ wide (Koteja, 1974a); legs thickened, claw with small denticle; hind spiracular peritreme 56–86 μ in diameter; spiracular furrows poorly developed, at most with 1 row of quinquelocular pores; spiracular setae similar to marginal setae; marginal setae elongate conical, 30–32 μ long on head, space between setae equals their length, or rarely twice as long; multilocular pores form transverse bands on abdomen, few scattered on thorax near bases of legs and antennae; large tubular ducts, ca 22 μ long, 3 μ wide, form a submarginal band, a group of smaller ducts in median area of abdomen and around mouthparts; disk pores form an irregular row medially from submarginal band of tubular ducts. *Dorsum.* With numerous short spinelike setae and disc pores; some scattered discoidal pores; quinquelocular pores along margin of anal cleft; anal ring with 2 rows of pores and 6 stout setae. Male test described by Šulc (1932), Borchsenius (1957), and Řehaček (1960); latter also described the first instar.

Distribution. A rare boreal Trans-Palearctic species. In CE: Bu, Cz, Fr, Hu, It, Po, and USSR; also known from England.

Biology. Restricted to Ericaceae; live on twigs of *Calluna, Erica, Rhododendron,* and *Vaccinium.* Develops 1 generation per year; 2nd instars overwintered; adults developed by mid-May in S Germany, and layed eggs from early June through end of June; eggs hatched mid-June through late June; first instars molted from mid-August (Schmutterer 1952b). Associated ants: *Lasius niger alienus* (Schmutterer, 1952b).

Natural enemies. Parasites: *Blastothrix truncatipennis* [very effective according to Danzig (1961)], *Cheiloneurus elegans, Coccophagus aterrimus, Discodes aeneus, D. minor,* and *Encyrtus infidus* (Trjapitzin, 1978; Danzig, 1959).

105. **Eulecanium sericeum** (Lindinger) – Fir Twig Scale

Syn. & Lit.: *Lecanium sericeum* Lindinger, 1906b:147; Šulc, 1932:101; Balachowsky, 1934b:71. *Eulecanium sericeum*, Leonardi, 1920:310; Schmutterer, 1952b:550, 1972b:408; Borchsenius, *1957:394; Řehaček, *1960:58; Kawecki, *1961:1; Kosztarab & Kozár, 1978:99; Tereznikova, 1981:160.

Postreproductive female almost spherical, yellowish-brown, shiny, diameter 7–10 mm. Younger adult female covered with white cottonlike waxy secretion laterally and with white powdery wax dorsally. Teneral female strongly sclerotized on dorsum around anal plates and with radial folds forming a reticulate pattern. *Venter.* Antennae 5–6-segmented, 250–280 μ long; labium 104 μ long, 119 μ wide (Koteja, 1974a); legs, claw and claw digitules slender; each spiracular furrow with a row of disc pores with 5 or 6 loculi; 1–3 spiracular setae in each group, thick and apically blunt; marginal setae sharply pointed, slender, 28–32 μ long, shorter than spiracular setae, distance between marginal setae 3–9 times their length; multilocular pores with 10 loculi, form wide transverse bands on abdomen and in median groups on thorax; tubular ducts, 5–6 μ wide, form a submarginal band, and groups in medial area of thorax, and in area of spiracular furrows; body setae few; paired setae on 6th abdominal sternite 120–135 μ long. *Dorsum.* With tubular ducts ca 6 μ wide, interspersed among scattered setae 42–50 μ long; disc pores with indistinct 5 loculi, diameter 4–5 μ; anal ring with partly double row of pores and 8 setae, each ca 160 μ long. First and second instars described by Kawecki (1961), first instar by Leonardi (1920) and Řehaček (1960).

Distribution. A rare montane species. In CE: Cz, Ge, It, Po and USSR; also known from Corsica.

Biology. Live on thin twigs of firs, *Abies* spp. One generation per year; few N_2 overwinter on needles, most on twigs; in S Germany females develop by end of April; each female lays 1,500–3,000, rarely over 8,000 eggs (Kawecki, 1961), from beginning of June; eggs hatch from mid-July, nymphs move to needles to feed, molt at end of August, and most return to twigs during end of October, and early November (Schmutterer, 1952b). Associated ants: *Formica pratensis major, F. rufa, Lasius fuliginosus,* (Schmutterer, 1952b).

Natural enemies. Parasites: *Encyrtus sylvius*. Predators: *Anthribus nebulosus, Loensia fasciata* and undetermined fungi (Kawecki, 1961).

106. **Eulecanium tiliae** (Linnaeus) – Nut Scale
Figs 8, 85, 86b

Syn. & Lit.: *Coccus tiliae* Linnaeus, 1758:456. *C. coryli* Linnaeus, 1758:456 (in part). *C. mali* Schrank, 1781:295, *Lecanium capreae* (L.), Newstead, 1903:105. *L. coryli*, Marchal, 1908:295; *Kawecki, 1961:1; Phillips, 1965:232. *Eulecanium coryli*, Silvestri, 1919b:127, 1920:514; Leonardi, 1920:296; Schmutterer, 1952b:545, 1972b:407; Řehaček, 1960:62. *Lecanium tiliae*, Signoret, 1873:431 (261); Šulc, *1932:87; Richards, 1958:310. *Eulecanium ibericum* Hadzibejli, 1960:316; in Danzig, (1977b:74). *E. gyrcanicum* Hadzibejli, 1967:719 in Danzig, (1977b:74). *E. mali*, Borchsenius, 1957:420. *E. tiliae*, Borchsenius, 1957:423; Kosztarab & Kozár, 1978:102; Tereznikova, 1981:161. Only the most important synonyms are cited here. A more complete listing of ca 130 synonyms cited from the literature is given by Kawecki (1961:2).

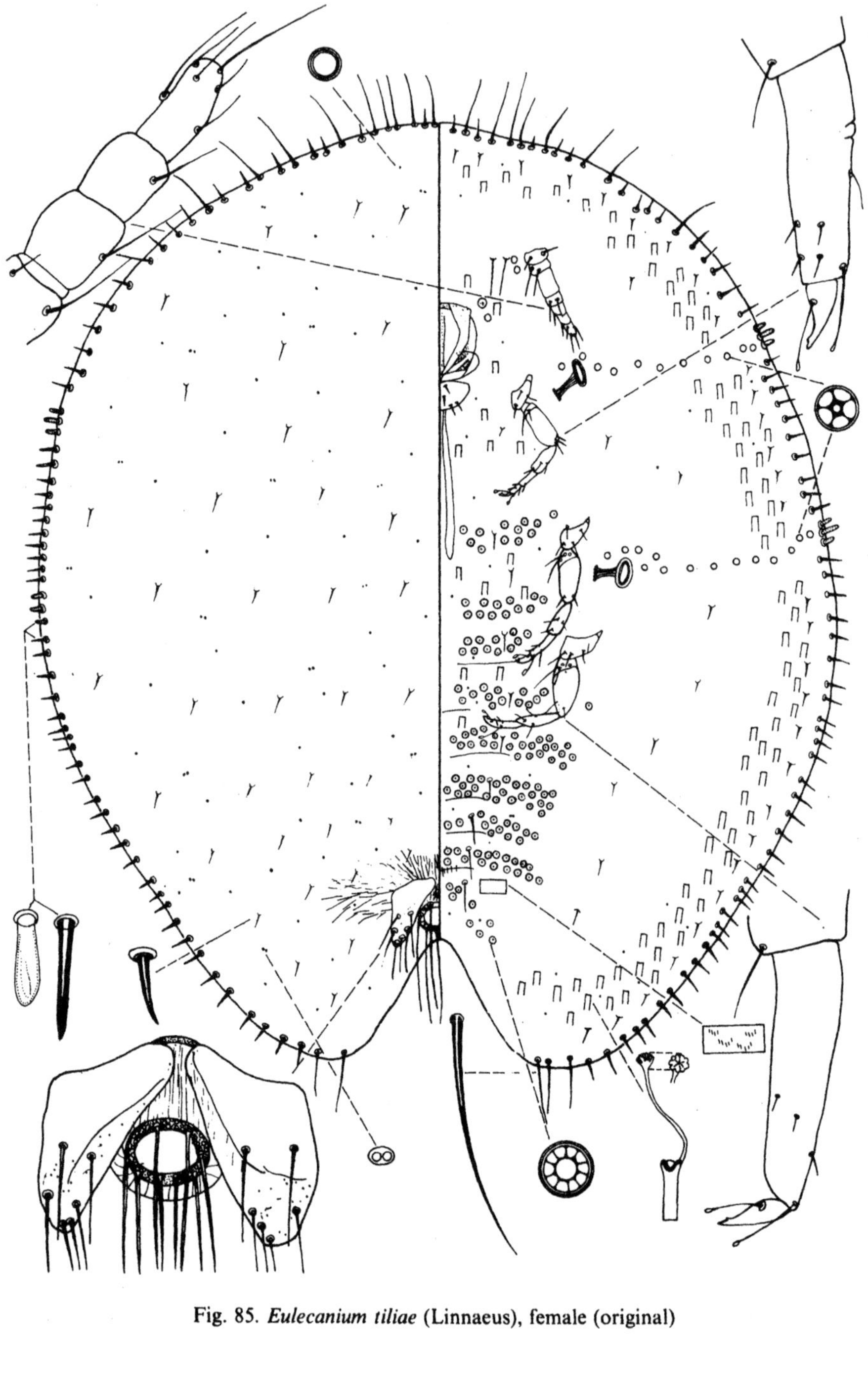

Fig. 85. *Eulecanium tiliae* (Linnaeus), female (original)

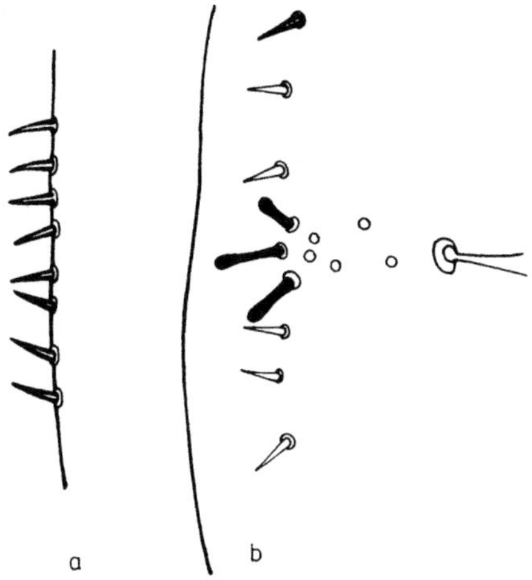

Fig. 86. Spiracular setae at the spiracular furrow of (a) *Eulecanium ciliatum*, (b) *Eulecanium tiliae* (after Kosztarab and Kozár, 1978)

Postreproductive female short-oval, strongly convex, yellowish-brown to dark-brown, 4–6.5 mm long, 3.5–6 mm wide, and 3–5 mm high. Teneral female variable in shape and color, normally broadly oval, light-yellow to cream-colored, with irregular brown or rarely reddish bands. *Venter*. Antennae 6 or 7, rarely 8-segmented, 220–278 μ long; labium 99 μ long, 118 μ wide (Koteja, 1974a); 2 pairs of interantennal setae, longer setae 50–75 μ long; hind femur 135–150 μ, tibia 115–135 μ, tarsus 95–110 μ long; claw thickened at base, with small denticle; spiracular furrows with 1 row of quinquelocular pores; 3 spiracular setae in each group, clublike, 13–27 μ long; marginal setae of 2 types; spinelike setae on lateral margins 27–45 μ long, hairlike setae 80–180 μ long on anterior and posterior margins, often interspersed among spinelike setae; multilocular pores with 6–12 loculi form wide transverse bands in medial area of abdomen and metathorax, in large groups in medial area on rest of thorax; tubular ducts form a submarginal band, and scattered in medial area of abdomen, thorax and head. *Dorsum*. With scattered simple disc pores, and small spinelike setae; derm of anal area heavily sclerotized, striated or tesselated into elongate cells; anal ring with 2 rows of pores and 8 setae. Adult male was described by Silvestri (1920), Šulc (1932), Kawecki (1961) and Giliomee (1967); male test by Šulc (1932), Řehaček (1960) and Kawecki, (1961). Nymphal stages were described by Marchal (1908), Leonardi (1920), Silvestri (1920), Šulc (1932) and Řehaček (1960).

Distribution. A common species, probably present throughout Europe. In CE: Au, Bu, Cz, Fr, Ge, Hu, It, Po, Ro, Sw, USSR and Yu; also known from Denmark, England, Sweden, Turkey, Israel, Iran, Iraq, N Africa, Japan, western N America and Tasmania.

Biology. It is a polyphagous species. Kawecki (1961) listed ca 70 host plant species from various literature sources. Since this species was often misidentified in the past, we are listing here only the most common hosts from more recent records: *Acer*,

Aesculus, Alnus, Betula, Cornus, Corylus, Crataegus, Cydonia, Fraxinus, Juglans, Malus, Pavia, Populus, Pyrus, Quercus, Rosa, Salix, Sorbus, and *Tilia.* One generation per year; according to Kawecki (1961) an obligatory biparental species; second instars overwinter on branches; in S Germany (Schmutterer, 1952b) adults develop by end of April and beginning of May; males are present in large numbers; the mated females lay ca 1356 (638–1833) eggs from beginning of May; eggs hatch in about 40 days from mid-June; N_1 moves to the underside of leaves and feed, molt during August and N_2 migrates to twigs from mid-October. It is common on ornamental trees, and often is a pest. Associated ants: *Lasius bruneus, L. niger niger, Formica fusca fusca,* and *F. rufa pratensis* (Schmutterer, 1952b).

Natural enemies. Parasites: Kawecki (1961) listed 38 parasite species from the literature. Because we are assuming some host misidentifications in the old literature, we are giving here only the records resulting from more recent determinations: *Blastothrix britannica, B. sericea* (latter was found most effective by a number of authors), *Coccophagus lycimnia, C. scutellaris, Encyrtus infidus, E. obscurus, Metaphycus dispar, M. fulvifrons, M. melanostomatus, M. punctipes, Microterys aeneiventris, M. duplicatus, M. lunatus, M. steinbergi,* and *M. sylvius;* also the parasitic fungus *Cordyceps piscilariaeformis.* Predators: *Anthribus fasciatus, A. nebulosus, A. scapularis,* and *Chilocorus bipustulatus* (Silvestri, 1919b; Kawecki, 1961; Schmutterer, 1972b; Canard, 1980).

Genus **Lecanopsis** Targioni-Tozzetti

Syn. & Lit.: *Lecanopsis* Targioni-Tozzetti, 1868:729; Newstead, 1893a:205; Leonardi, 1920:325; Steinweden, 1929:237; Goux, 1937c:95; Borchsenius, *1952:280, 1957:91; Řehaček, 1960:22; Danzig, 1967:828, 1980a:256; Ali, 1971:55; Kosztarab & Kozár, 1978:103; Tereznikova, 1981:110; Boratynski et al., 1982:517. *Rhizobium* Targioni-Tozzetti, 1867:23.

Type-species: *Lecanopsis rhyzophila* Targioni-Tozzetti.

Ovisac thick, cottony, white, encloses egg-laying female. Postreproductive female oval, convex dorsally, concave ventrally, up to 5 mm long, and 3 mm wide, with short anal cleft. Teneral female with membranous derm. *Venter.* Antennae short, 5–8-segmented, 1st and sometimes 2nd segment much thicker than others; legs stout, tibiotarsal articulation free and scleroses present; spiracles large, each located in funnel-shaped depression, and normally with a group of disc pores anteriorly; spiracular furrows with disc pores; spiracular setae absent; a row of slender marginal setae; body setae short, inconspicuous; small simple disc pores, ca 3 μ in diameter, in submarginal area; multilocular pores on abdominal sternites often of different sizes and with varying number of loculi; tubular ducts normally of 2 sizes. *Dorsum.* Seldom with scattered small disc pores; usually numerous discoidal pores of different sizes in medial area; tubular ducts of 2 sizes on entire surface; anal ring wide with circular pores and 6–8 setae; anal plates small, with lateral angle rounded.

Distribution. Genus with 14 species: 12 in the Palearctic, and 2 in the Oriental Region; in CE 4 species.

Biology. Most species live on grasses, all species infest underground parts of the host and are often attended by ants. Females have 3 nymphal instars.

Note. Horváth (1897) recorded *Lecanopsis rhyzophila* from Hungary. This species was known previously from Italy (Florence) only. Because of its uncertain status and because no voucher specimens were found for either records, we deleted this species from this study. This genus is in need of revision.

Key to species of *Lecanopsis*

1. Tarsus strongly curved, less than ½ as long as tibia; adult female 3.8–5.4 mm long; derm with lateral submarginal fold .... *formicarum*, p. 194
   Tarsus not strongly curved, more than ½ as long as tibia; adult female normally less than 3.8 mm long, except in *festucae* up to 4.2 mm; derm without lateral submarginal fold 2
2. Group of disc pores (except 1 or 2) absent in front of anterior spiracles; 3rd antennal segment longest .... *festucae*, p. 193
   A large elongate group of disc pores present in front of anterior spiracles, 1st or last antennal segment longest..3
3. Antennae 6-segmented, last segment longest; discoidal pores on dorsum form a wide longitudinal band, occupying medial one-half or more of body width; discoidal pores large, 9–10 μ in diameter (fig. 87) ...... *porifera*, p. 194
   Antennae 8-segmented, 1st segment longest; discoidal pores on dorsum form a narrow longitudinal band in medial area; discoidal pores small, 5–6 μ in diameter ....*terrestris*, p. 196

107. **Lecanopsis festucae** Borchsenius – Borchsenius' Grass Scale

Syn. & Lit.: *Lecanopsis festucae* Borchsenius, 1952:288, 1957:103; Zak-Ogaza, 1966:81; Danzig, 1967:829; Koteja & Zak-Ogaza, 1969:365; Danzig, 1977a:201; Kosztarab & Kozár, 1978:105; Tereznikova, 1981:116.

Postreproductive female oval, up to 4.2 mm long. Teneral female. *Venter.* Antennae 7- or 8-segmented, 1st segment widest, 3rd longest; legs stout, claws curved with digitules expanded at apex; spiracular furrows with many disc pores, with 5–8 loculi; group of disc pores anterior of spiracles absent, seldom 1 or 2 pores present; a row of slender marginal setae combined with hairlike setae into a submarginal band; body setae short, inconspicuous; 2 interantennal setae 24–30 μ long; multilocular pores with 6, rarely 5 or 7-loculi, on abdominal sternites, more numerous on posterior segments; few large tubular ducts on prosoma, numerous on abdomen, in a submarginal band and randomly distributed. *Dorsum.* Discoidal pores, ca 10 μ in diameter, form a medial longitudinal band from head to anal plate, others scattered on rest of dorsal surface; few large tubular ducts, ca 30 μ long and 6 μ in diameter, few on head and thorax, more numerous on abdomen especially toward posterior end; small tubular ducts, ca 5 μ long and 2 μ wide, scattered on entire surface; anal ring with 6 setae and 2 rows of circular pores. N_2 and N_3 described by Borchsenius (1957); 3 nymphal stages by Tereznikova (1981).

Distribution. An occasionally found steppe-inhabiting xerophilous species. In CE: Bu, Hu, Po(?), Ro and USSR; also in Mongolia.

Biology. On crown roots of *Agrostis, Dactylis, Elytrigia, Festuca* and *Poa*. Often under stones. Males unknown. One generation per year. Adults or N_3 overwinter.

Natural enemy. Parasite: *Hoplopsis minuta* (Trjapitzin, 1978).

108. **Lecanopsis formicarum** Newstead – Newstead's Soft Scale

Syn. & Lit.: *Lecanopsis formicarum* Newstead, 1893a:206, 1893b:138; Leonardi, 1920:326; Green, 1921:193; Steinweden, 1929:237; Gomez-Menor, 1937:291; Goux, 1937c:95; Fulmek, 1943:51; Schmutterer, 1952b:560; Borchsenius, 1957:98; Řehaček, 1960:22; Koteja & Zak-Ogaza, 1969:365; Kozár, 1972:281; Kozarzhevskaya & Reitzel, 1975:24; Kosztarab & Kozár, 1978:103; Boratynski et al., 1982:517. *L. brevicornis* Newstead, 1896:59. *L. butleri* Green, 1917:208.

Ovisac of loose woolly wax threads. Postreproductive female elongate-oval, convex, with 2 dark longitudinal lateral bands along entire body; anal cleft short, about 1/10 length of body. Teneral female elongate-oval, flat; live light reddish-yellow; 3.5–4.5 mm long, 2–3 mm wide. *Venter.* Antennae 400—464 μ long, 5-, 6- or 7-segmented, 1st segment very wide, 3rd segment longest, apical segment with 8–10 setae; legs stout, 0.9–1.1 mm long, tarsus less than half as long as tibia, claws strongly curved; spiracles large each with a band of disc pores with 7–9 loculi, extending to body margin; similar disc pores inside each atrium; marginal setae absent; other body setae small and sparse; few multilocular pores around vulva, in transverse rows on 3–4 preceding segments; tubular ducts scattered on both body surfaces. *Dorsum.* Small setae and simple disc pores scattered on entire surface; discoidal pores form a mesal longitudinal band; anal ring with pores and 8 setae, each 173–185 μ long; anal plates conical, ca. 148 μ long, 99 μ wide, with 3–4 terminal setae. Male test and N_1 described by Řehaček (1960), N_1 and N_3 by Schmutterer (1952b).

Distribution. Western Palearctic. In CE: Cz, Fr, Ge, Hu, It, Po, Sw; also known in Denmark, England, and Spain.

Biology. On rootcrown of grasses: *Agropyron, Dactylis, Elymus, Festuca* and *Poa*. One generation per year. N_3 overwintered in S. Germany (Schmutterer, 1952b); adults developed by mid-April, each female laid 238–418 eggs during May, eggs hatched within 2–3 weeks; N_1 molted in July. Associated ants: *Formica nigra, Lasius flavus, L. niger alienus, L. n. niger,* and *Tetramorium caespitum* (Schmutterer, 1952b).

Natural enemies. Parasites: *Choreia inepta, Coccobius annulicornis, Dusmetia ceballosi* and *Microterys sceptriger, M. tessellatus* (Fulmek, 1943; Trjapitzin, 1978; Kozarzhevskaya and Reitzel, 1975; Schmutterer, 1955c).

109. **Lecanopsis porifera** Borchsenius – Pore-bearing Soft Scale
Fig. 87

Syn. & Lit.: *Lecanopsis porifera* Borchsenius, 1952:282, *1957:94; Rogojanu, 1966b:433; Danzig, 1967:828; Kozár, 1972:281; Kosztarab & Kozár, 1978:104; Tereznikova, *1981:113. *L. formicarum* Kiritchenko, 1931:318 (in part), not *L. formicarum* Newstead.

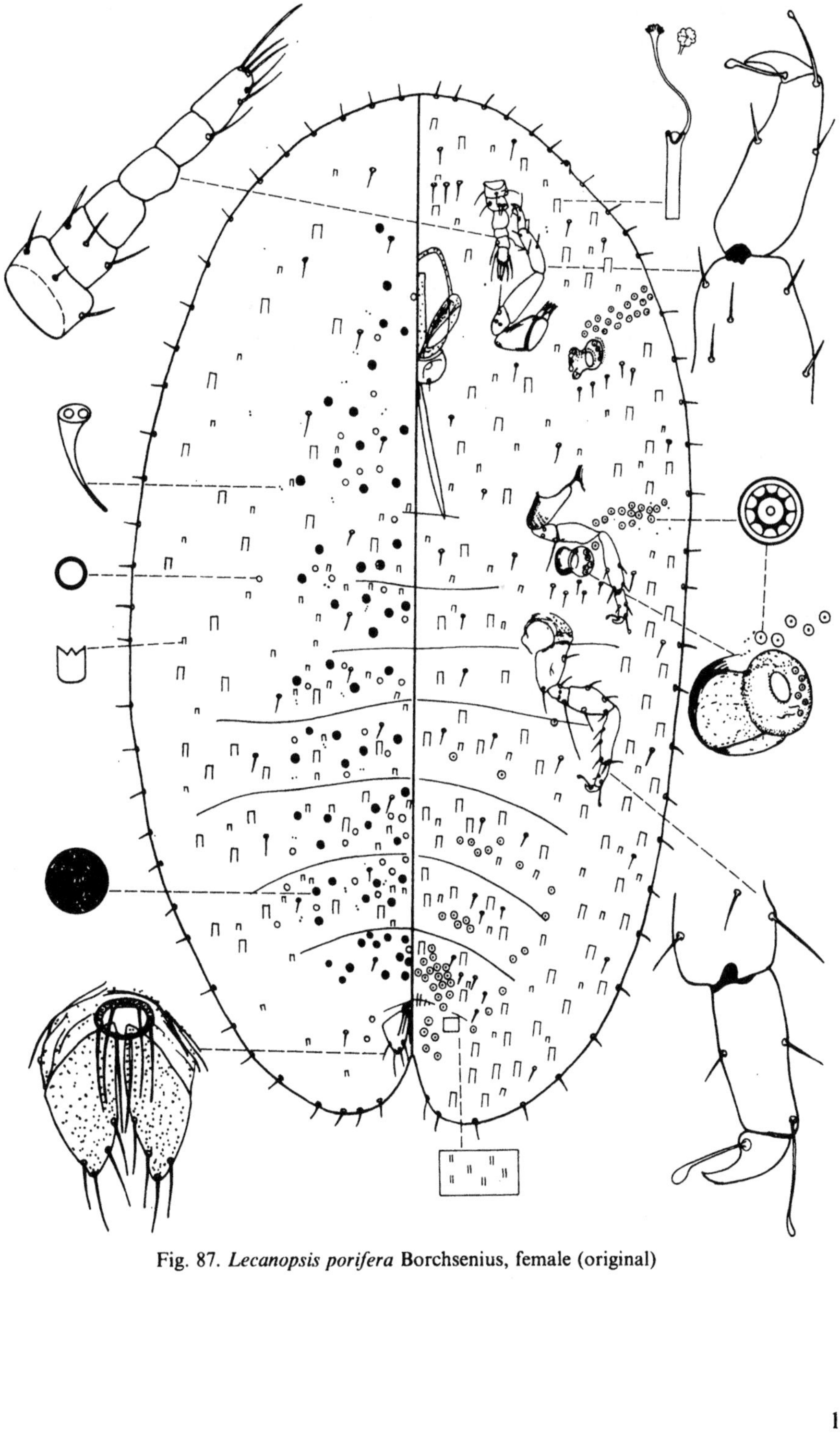

Fig. 87. *Lecanopsis porifera* Borchsenius, female (original)

Postreproductive female elongate-oval, anal cleft parallel-sided, anal lobe angle broadly rounded. Teneral female 2.4–2.6 mm long, ca 1.2 mm wide. *Venter*. Antennae 6-segmented ca 263 μ long, 1st segment much thicker than others, last segment longest; legs relatively large, moderately stout; claw strongly curved, digitules strongly expanded at apex; each hind spiracular furrow with 14–20 multilocular disc pores, these ca 6 μ in diameter; spiracular peritreme with a row of similar pores; marginal setae, 18–24 μ long; submarginal setae 9–12 μ long, in an irregular row; other body setae short and inconspicuous except 3 pairs of interantennal setae, where median pair 20 μ long; multilocular pores, ca 10 μ in diameter, with 7–10 loculi, form transverse rows or bands on most abdominal sternites; large tubular ducts, ca 21 μ long, 7 μ in diameter, found scattered, or in transverse rows; small tubular ducts, ca 7 μ long and 2 μ in diameter, common on entire surface. *Dorsum*. Discoidal pores, 9–10 μ in diameter, form a wide longitudinal band in medial area from antennae to anal plate; large tubular ducts fewer on anterior, more dense on posterior surface; small tubular ducts common; anal ring with 2 rows of pores and 6–8 setae, when with 8 setae, 2 setae shorter and stouter than others. All 3 nymphal stages described by Borchsenius (1957) and Tereznikova (1981).

Distribution. A rare species. In CE: Hu, Ro, USSR and Yu.

Biology. On rootcrowns of grasses: *Agropyron, Bromus, Dactylis, Elytrigia, Festuca, Lolium, Secale*. One generation per year. Adult females found in May; 3 nymphal instars recorded. Males unknown.

110. **Lecanopsis terrestris** Borchsenius – Subterranean Soft Scale

Syn. & Lit.: *Lecanopsis terrestris* Borchsenius, 1952:287, 1957:102; Tereznikova, 1963:49; Danzig, 1967:829; Kosztarab & Kozár, 1978:104.

Postreproductive female oval, segmentation well marked by sclerotized folds at segmental borders. Teneral female ca 3 mm long. *Venter*. Antennae 8-segmented, segments 1 and 2 wider than others, 1st segment longest; legs large and stout; 1 or 2 rows of disc pores in each spiracular peritreme; an elongate group of 37–48 disc pores each with 4–8 loculi, in front of each anterior spiracle, 5–6 disc pores at each posterior spiracle; body setae small, inconspicuous; 2 interantennal setae 25–30 μ long; multilocular pores ca 10 μ in diameter and with 7 loculi (rarely with 6 or 8) in a wide band around vulva and in smaller groups on preceding sternites; large tubular ducts, 33–36 μ long and 6–10 μ in diameter occur singly or in groups; small tubular ducts, ca 3 μ in diameter, numerous on entire surface. *Dorsum*. Discoidal pores ca 5–6 μ in diameter, form a narrow median longitudinal band on metathorax and abdomen; large tubular ducts, 33–36 μ long and 9–10 μ in diameter scattered irregularly on most of surface, but absent from anterior end; small tubular ducts common on entire surface more conspicuous on body margin where these form a wide band. Last nymphal instar described by Borchsenius (1957).

Distribution. A rare steppe-inhabiting species. Known only from Ukraine and Chkalovsk District in USSR.

Biology. On underground parts of *Agropyron, Agrostis* and *Poa*. Males unknown. Apparently 3rd instars and adults overwinter.

Genus **Lichtensia** Signoret

Syn. & Lit.: *Lichtensia* Signoret, 1873:27; Fernald, 1903:144; Leonardi, 1920:347; Gomez-Menor, 1937:289; Morrison & Morrison, 1966:110; Ben-Dov, 1975:113. *Filippia*, Targioni-Tozzetti, 1868:33; 1869:726; Lichtenstein, 1881:114; Lindinger, 1912:336; Steinweden, 1929:236; Borchsenius, 1957:187; Řehaček, 1960:35; Kosztarab & Kozár, 1978:102; Tereznikova, 1981:106,

Type-species: *Lichtensia viburni* Signoret, 1873.

It was described as a monotypic genus. Since the generic assignments of the subsequently included species are uncertain, we consider the genus monotypic here. Therefore, the description of *L. viburni* is the same as for this genus.

111. **Lichtensia viburni** Signoret – Viburnum Cushion Scale
Fig. 88

Syn. & Lit.: *Lichtensia viburni* Signoret, 1873:28(204); Newstead 1895a:165, 1902:33; Leonardi, 1920:348; Gomez-Menor, 1937:289; Ben-Dov *1975:113. *Filippia viburni*. Řehaček, 1954:219, 1960:36; Borchsenius, 1957:189; Koteja, 1974a:108; Kosztarab & Kozár, 1978:103; Tereznikova, 1981:108.

Ovisac fluffy, white, oval, encloses mature female, sac 4–7 mm long, 2.5–4 mm wide, narrowed at anterior end where sometimes body exposed through small opening.

Postreproductive female oval, slightly convex, brown, up to 6.0 mm long and 3.5 mm wide. Teneral female oval, live lemon-yellow or greenish-yellow, 1.8–2.8 mm long, 1.1–2.0 mm wide. *Venter*. Antennae slender, 8-segmented, rarely 7 or 9, 300–540 μ long; labium 95 μ long and wide; stylet loop reaches line between mesocoxae; legs large, tibia ca twice as long as tarsus, tibio-tarsal articulation free and sclerosis present, claw with denticle at its base, and robust knobbed digitules; spiracular furrows each with 29–35 disc pores, with 5, 6 to 8 loculi, in band 2–3 pores wide; 3 spiracular setae in each spiracular indentation, these much stouter than marginal setae and with blunt apex, median seta much longer than laterals, 24–45 μ long; lateral setae about as long as marginal setae, 18–27 μ long; most marginal setae 14–24 μ long, chisel-shaped, notched on tip, slightly constricted at base; submarginal setae hairlike, 15–18 μ long, form a row; dark-rimmed pores form a submarginal band; multilocular pores with 7–12 loculi, sparse on first 2 abdominal sternites, in transverse rows on preceding 3 sternites; body setae of various lengths, longest ones in median area of body; tubular ducts in a submarginal band, few scattered in median area. *Dorsum*. Dorsal setae slender, tapering, ca 8–9 μ long; small bilocular ducts form a narrow band from mouthparts to line between hind coxae; tubular ducts evenly distributed on dorsum; anal ring with 2 rows of pores and 8 setae; anal plates triangular, anterior-lateral margin ca 99 μ, inner margin 105–148 μ, posterior-lateral margin 90–123 μ long; each plate with 3 apical setae, largest ca 25 μ long, and 1 seta on inner margin; each operculum with 3–4 subapical setae and 2 fringe setae, inner seta short; anal ring with 6 setae. Adult male and test described by Newstead (1902), Leonardi (1920), Borchsenius (1957), adult by Giliomee (1967), test and N_1 by Řehaček (1960), N_1 by Leonardi (1920), N_1 and N_2 by Ben-Dov (1975).

Distribution. A rare Mediterranean species. In CE: Cz, Fr, Ge, It, USSR and Yu; also known from England, Wales, Spain, Portugal, Sicily and USA.

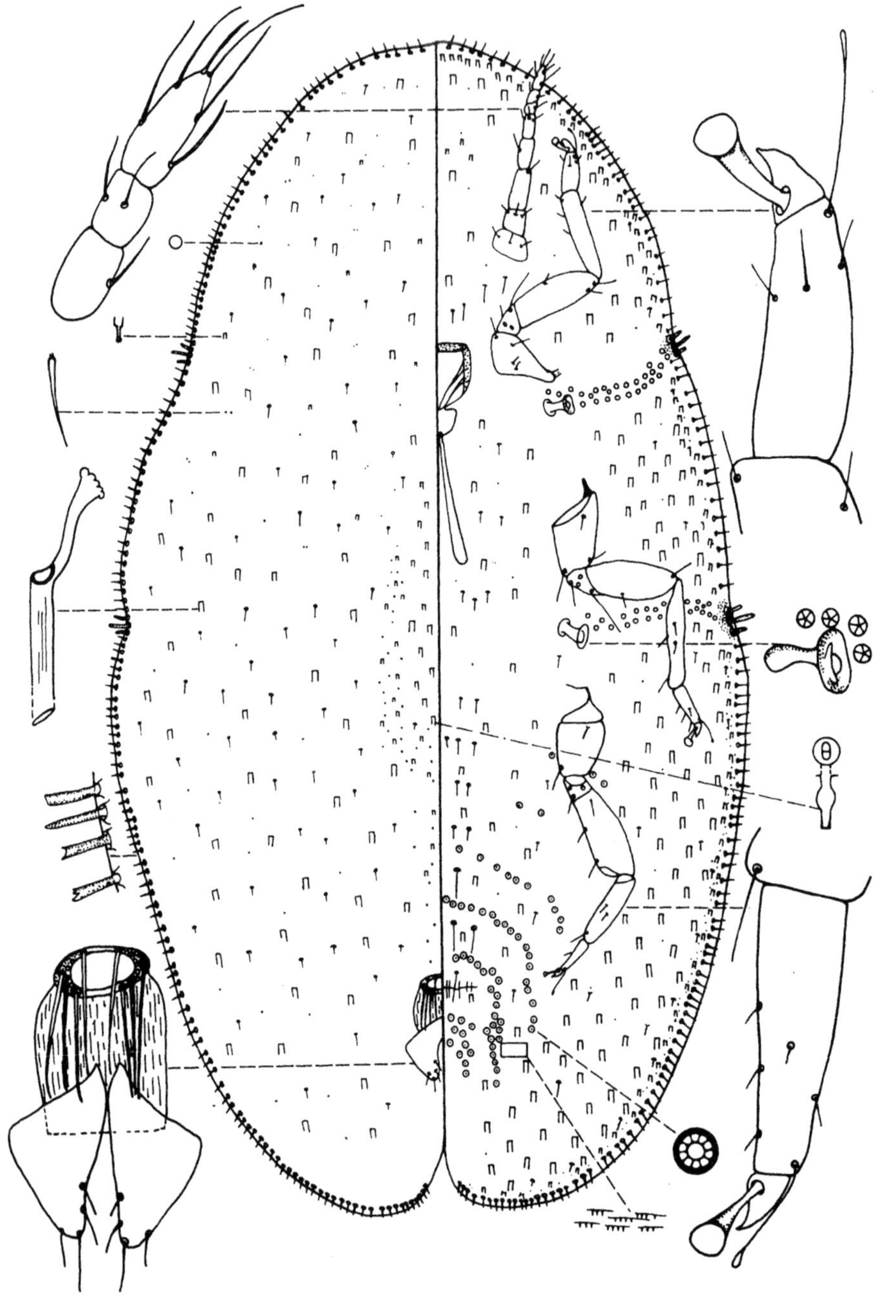

Fig. 88. *Lichtensia viburni* (Signoret), female (original)

Biology. Infests the lower side of leaves, twigs and stems of *Arbutus, Hedera, Ilex, Phillyrea, Prunus, Rhamnus* and *Viburnum*. The host list may include inaccuracies, because of the frequent confusion of this species with *F. oleae* (Costa). One generation per year; N_2 overwinter, ovisac produced from May in Crimea, each female lays up to 587 eggs until end of July (Borchsenius, 1957). Umnov (1940) reported it as a pest on evergreen snowballs.

Natural enemies. Parasites: *Coccophagus palaeolecanii, Microterys masii* (Trjapitzin, 1978).

Genus **Luzulaspis** Cockerell

Syn. & Lit.: *Luzulaspis* Cockerell, 1902b:25; Green, 1928a:24; Steinweden, 1929:230; Balachowsky, 1932c:200; Goux, 1937c:95; Borchsenius, 1952:273, 1957:118; Takahashi, 1955b:25; Řehaček, 1960:23; Danzig, 1967:038, 1980a:247; Ali, 1971:55; Kosztarab & Kozár, 1978:105; Koteja & Howell, 1979:334; Koteja, 1978:317, *1979a:585; Tereznikova, 1981:116. *Signoretia* Targioni-Tozzetti, 1868:727 (preoccupied).

Type-species: *Aspidiotus luzulae* Dufour, 1864.

Ovisac elongate, almost parallel-sided, compact, white, 4–30 mm long; encloses female. Postreproductive female sclerotized, it shrinks and falls out of ovisac, except when parasitized. Teneral female elongate, parallel-sided, rounded or slightly tapered at both ends, dorsum slightly convex, venter almost flat; live usually yellowish, with 2 red dorsal stripes. *Venter.* Antennae slender, 8-segmented; labium cup-shaped, with 5 pairs of setae; stylet loop about as long as width of labium; legs slender, anterior legs always shorter, tibio-tarsal articulatory scleroses present, claw digitules normally large with expanded apical knob; spiracular pore bands mostly of quinquelocular pores; 2 subequal spiracular setae in each group, very rarely 1 or 3; marginal setae spinelike or hairlike, usually intervals between setae about equal to length of setae or greater; body setae of various lengths, interantennal setae up to 150 μ long; micropores normally form a marginal row; 1–2 rarely 3, campaniform pores, 2–3 μ in diameter, at base of each antenna; multilocular pores usually with 8–10 loculi form transverse bands on abdominal sternites 6–8, rarely on anterior segments; tubular ducts of various sizes, fewer than on dorsum; microducts, 1–1.5 μ in diameter, form a submarginal band, also on medial area of head, thorax and in some species on abdomen. *Dorsum.* Body setae of various shapes and sizes, from small hairlike to large conical; minute simple disc pores, ca 2 μ in diameter scattered over entire surface; discoidal pores, with sievelike structure, 3–6 μ in diameter, in a medial longitudinal band on thorax and abdomen; tubular ducts numerous on body surface (100–200 ducts on 6th tergite); anal ring 45–80 μ in diameter with 6 (rarely 7–9) setae, each 110–180 μ long; anal plates triangular, each with 4 apical setae.

Distribution. Holarctic, with 17 species; 8 in CE.

Biology. On leaves of Juncaceae and Cyperaceae, usually on *Carex, Juncus, Luzula*. Ovisacs often produced on non-host plants. One, rarely 2 generations per year; 3 nymphal instars in females; probably in all species eggs overwinter; oviposition from June to September.

Note. Records on the occurrence of *Luzulaspis grandis* Borchsenius in Germany (Schmutterer, 1955b) and Poland (Zak-Ogaza and Koteja, 1964) are probably the result of misidentifications (Koteja, 1979a).

Key to species of *Luzulaspis*

1. All dorsal setae short, spinelike, conical or almost parallel-sided........ 2
   Dorsal setae elongate, hairlike, except in *pieninica* with few elongate setae on median part of head 4
2. Two apical marginal setae on head and apical seta of anal lobe thick, twice as wide at base as other marginal setae; apical anal lobe seta ca 30 μ long; setae on 6th abdominal sternite 40–50 μ long; interantennal setae less than 75 μ long (fig. 90) .... *rajae,* p. 206
   Apical marginal setae on head and anal lobes similar to other marginal setae; apical anal lobe seta less than 27 μ long; setae on 6th abdominal sternite longer than 91 μ; interantennal setae longer than 84 μ 3
3. Teneral female less than 3 times as long as wide; antennae 370–460 μ long; hind legs 710–890 μ long; spiracular setae about twice as long as marginal setae ....*dactylis,* p. 200
   Teneral female more than 3 times as long as wide; antennae 456–648 μ long; hind legs 971–1274 μ long; spiracular setae only slightly longer than largest marginal setae (fig. 91b) .... *scotica,* p. 206
4. Teneral female at least 6.7 mm long; antennae over 500 μ long; most dorsal setae short, conical, but a few, especially on abdomen elongate spinelike, or short hairlike; anal plates ca 134 μ long ....*pieninica,* p. 205
   Teneral female up to 4.7 mm long; antennae less than 500 μ long; dorsal setae elongate hairlike; anal plates up to 110 μ long 5
5. Antennae located halfway between anterior margin of head and clypeolabral shield; on Juncaceae, e.g. *Juncus, Luzula.* 6
   Antennae located closer to clypeolabral shield than to anterior margin of head; on Cyperaceae (e.g. *Carex*) 7
6. Lateral hairlike seta on apical antennal segment up to 1.5 times as long as segment; spiracular setae rarely branched; live females with 2 wide dark red longitudinal bands covering most of dorsum (fig. 91a) ... *luzulae,* p. 204
   Lateral hairlike seta on apical antennal segment 2–3 times as long as segment; spiracular setae often bifurcated; live females with 2 narrow longitudinal light red bands. .... *nemorosa,* p. 204
7(5). Head strongly produced forward; spiracular setae sharp-pointed, slender, bent, some sickle-like; head with dorsal setae 10–14 μ long, marginal setae 9–20 μ long.................................... *frontalis,* p. 201
   Head not produced forward; spiracular setae stout, bent, blunt-pointed; head with dorsal setae 14–26 μ long, marginal setae 20–53 μ long (fig. 89).... .... *kosztarabi,* p. 202

112. **Luzulaspis dactylis** Green – Green's Soft Scale

Syn. & Lit.: *Luzulaspis dactylis* Green, 1928a:24; Borchsenius, 1957:121; Koteja, 1978:323, *1979a:611. *L. montana* Schmutterer, 1955a:163; 1955b:99; Borchsenius, 1957:126; *Danzig, 1959a:393; Řehaček, 1960:26; Zak-Ogaza & Koteja, 1964:427; Danzig, 1964:639; Kosztarab & Kozár, 1978:107.

Ovisac elongate, convex dorsally, posterior end tricarinate, white, 4–8 mm long, 1.5–2 mm wide. Postreproductive female elongate-oval. Teneral female 2.3–3.8 mm long, 0.8–1.5 mm wide. *Venter*. Antennae halfway between anterior margin of body and clypeolabral shield 370–460 μ long, 3rd segment longest; legs well-developed, hind 710–890 μ long; spiracular pore bands of quinquelocular pores; spiracular setae stout, of various shapes, slightly bent, rarely bifurcate, ca twice as long as marginal setae; marginal setae spinelike, conical, rarely hairlike or bent, intervals between setae normally greater than their length, form 1 or 2 rows; body setae large in medial area; micropores form a marginal row, scattered on rest of venter; multilocular pores, with 5–10 loculi, on segments 6–8; microducts on head and medial area of prothorax and form a submarginal band around body; filaments of tubular ducts long, about as wide as ducts, ducts scattered on entire surface. *Dorsum*. Body setae spinelike with blunt or sharp apex; discoidal pores in medial groups or band from mesothorax to anterior of anal plates; microducts scattered over entire surface; tubular ducts numerous, scattered on surface, except absent on antero-medial area of head; diameter of anal ring 43–48 μ, with 6 setae, each 132–156 μ long; apical setae of anal plates shorter than medial setae (Koteja, 1979a).

Distribution. A boreal species. In CE: Cz, Ge, It, Po and USSR (Leningrad Region); also in England and Greece.

Biology. Feed on leaves of *Carex* spp., but ovisacs were often found on other non-host plants, such as *Dactylis, Festuca, Galium, Leontodon, Senecio* and *Sieglingia*. One generation per year; eggs overwinter; adults were present in S Slovakia during July and August.

113. **Luzulaspis frontalis** Green – Long-headed Soft Scale

Syn. & Lit.: *Luzulaspis frontalis* Green, 1928a:25; Schmutterer, 1952b:557; 1955b:100; Borchsenius, 1957:128; Řehaček, 1960:25; Danzig, 1964:639; Kosztarab & Kozár, 1978:108; Koteja, *1966a:17, *1979a:620.

Ovisac very slender, 6–11 mm long, ca 1.2 mm wide. Postreproductive female flat dorso-ventrally, light brown. Teneral female 3.2–4.7 mm long, 0.7–1.2 mm wide. *Venter*. Head strongly produced anteriorly; antennae situated much closer to clypeolabral shield than to anterior margin of head; antennae 340–440 μ long; legs slender, posteriors 690–756 μ long; spiracular pores 24–40 in each band, with 5 or 6 loculi, few pores scattered in front of anterior bands; spiracular setae sharp-pointed, slender, bent; marginal setae hairlike, on head 9–20 μ long, form 1–2 rows, much shorter than spiracular setae, intervals between them 2–3 times greater than their length; body setae of various lengths; micropores form a marginal row; most multilocular pores with 10 loculi, ca 10 pores on 6th, 80 on 7th and 60 on 8th abdominal sternite; tubular ducts form a submarginal band, others scattered on entire surface, filaments of tubular ducts about as wide as ducts. *Dorsum*. Body setae small, hairlike, randomly distributed; micropores scattered on entire surface; discoidal pores form 2 narrow medial longitudinal bands from mesothorax to anterior of anal plates; tubular ducts densely distributed, absent from anteriomedial area of head, except few along exterior margin, duct filaments only 1/3 as wide as ducts; diameter of anal ring ca

48 μ, setae ca 144 μ long; apical seta of anal plates needlelike, other setae hairlike (Koteja, 1979a). All life-stages, including adult male, described by Koteja (1966a).

Distribution. A rare boreal species. In CE: Au, Cz, Ge and Po; also in England. The records on its occurrence in Hungary and records from France (Rhône) are probably based on misidentifications (Koteja, 1979a).

Biology. Prefers shady, moderately damp mixed forests in CE (Koteja, 1966a). Infest leaves of *Carex* spp.; migrating females produce ovisacs and lay eggs on other plants, that probably are non-hosts. One generation per year; first instars appeared during end of April and early May in Poland (Koteja 1966a), molted mid-June; second molt end of June and early July, third molt during second part of July; adult males appeared second part of July; egg-laying observed one month after copulation; Schmutterer (1952b) reported 125–451 eggs, Koteja (1966a) 20–107 eggs per female.

Natural enemies. Parasites: *Baeocharis pascuorum, Coccophagus insidiator, C. lycimnia* (Koteja, 1966a) and *Microterys rogenhoferi* (Schmutterer, 1955c).

114. **Luzulaspis kosztarabi** Koteja & Kozár – Hungarian Sedge Scale
Fig. 89

Syn. & Lit.: *Luzulaspis kosztarabi* Koteja & Kozár 1979:121; Koteja, 1979a:622; Kozár, 1981:320.

Ovisac 4–9 mm long. Postreproductive female strongly convex. Teneral female elongate-oval, derm membranous, segmentation obscure, 3.8–4.7 μ long, 1.3–1.5 μ wide. *Venter.* Antennae 350–370 μ long; situated closer to clypeolabral shield than to anterior margin of head; labium 60 μ long, 96 μ wide; legs relatively large, posteriors 811–881 μ long; pores in spiracular bands normally quinquelocular, rarely with 4–7 loculi, 33–40 in each band; spiracular setae stout, curved, blunt or sharp-pointed; marginal setae hairlike, long, arranged in an irregular row, intervals between them exceed 2–4 times their length; body setae much longer in medial area than on rest of surface; micropores, 1–2 μ in diameter, form a marginal row; multilocular pores with 7–9 loculi, on abdominal sternites 6–8; microducts, 5–7 μ long, scattered on entire surface, more numerous around labium; tubular ducts, of various sizes, their filaments as wide as ducts or wider. *Dorsum.* Body setae hairlike, long, pointed, usually bent; micropores scattered on entire surface; discoidal pores, 3–6 μ in diameter, form a medial longitudinal band on 3–4 abdominal tergites anterior to anal plates; microducts ca 10 μ long scattered on entire surface; tubular ducts numerous, except absent on anterior of head, 14–17 μ long and 3–4 μ in diameter, with inner filament ca 25 μ long and 1.5 μ in diameter; anal ring ca 48 μ in diameter, setae ca 120 μ long; anal plates twice as long as wide, apical seta spinelike, other setae needle or hairlike.

Distribution. A xerophilous rare species; known only from Hungary from karstic localities, at 300–500 m above sea level.

Biology. On *Carex* sp. Adults were present in July and August and layed eggs by August.

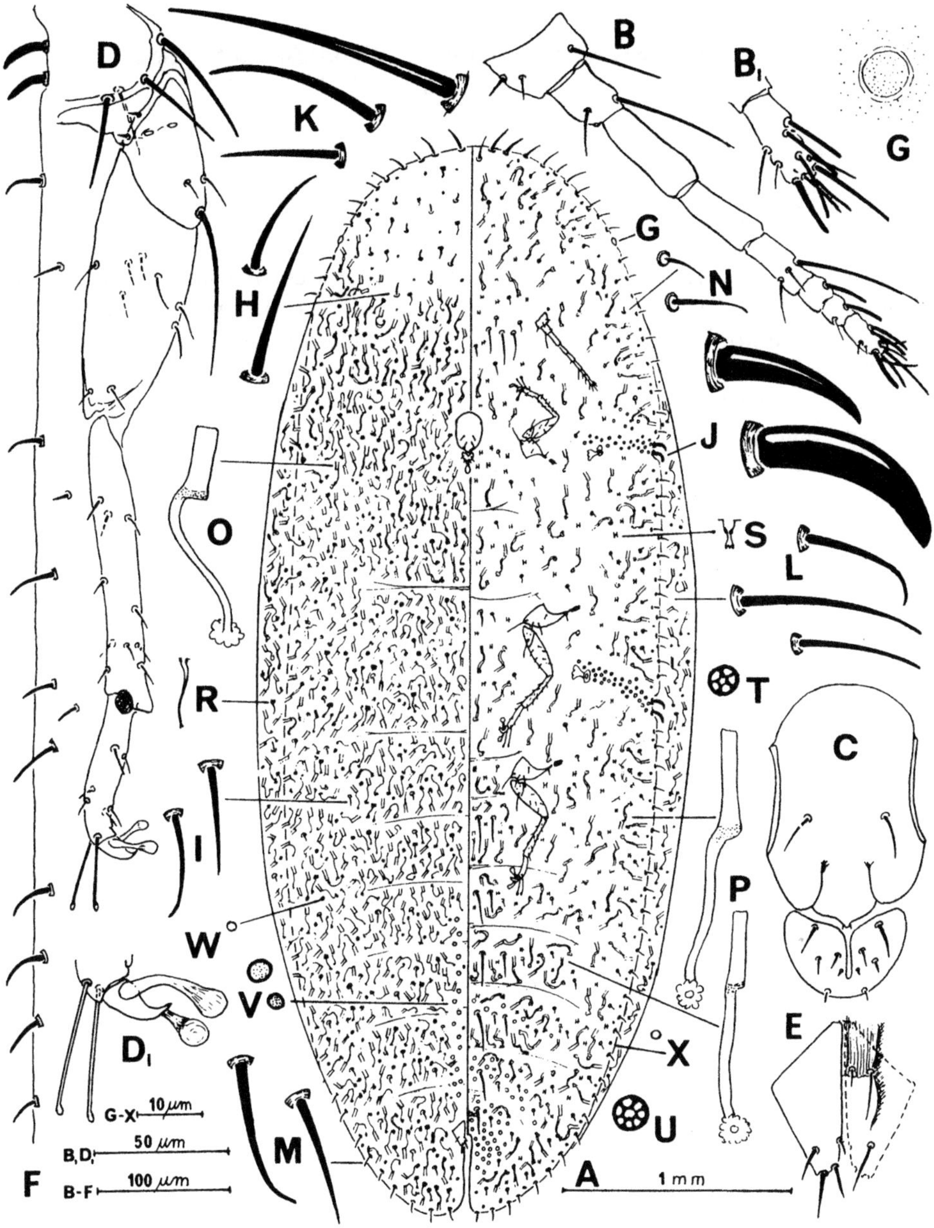

Fig. 89. *Luzulaspis kosztarabi* Koteja & Kozár, female (after Koteja and Kozár, 1979)

Note. The letter designations referring to various structures in figure 89 are given in the article by Koteja and Kozár (1979).

115. **Luzulaspis luzulae** (Dufour) – Woodrush Soft Scale
Fig. 91a

Syn. & Lit.: *Aspidiotus luzulae* Dufour, 1864:207. *Signoretia luzulae*, Newstead, 1903:27. *Luzulaspis luzulae*, Steinweden, 1929:230; Schmutterer, 1952b:557; Borchsenius, 1952:274, 1957:120; Řehaček, 1960:24; Danzig, 1964:638, 1980a:250; Tsalev, 1968:208; Kosztarab & Kozár, 1978:105; Koteja, *1979a:614; Tereznikova, 1981:118.

Ovisac narrows anteriorly, 4–6.5 mm long. Postreproductive female elongate-oval, color dark orange. Teneral female parallel-sided, live specimen with 2 wide dark red longitudinal bands; 2.6–3.6 mm long, 1.1–1.2 mm wide. *Venter*. Antennae 325–425 μ long, located halfway between anterior margin of head and clypeolabral shield; labium 58 μ long, 82 μ wide; hind legs 660–787 μ long; each spiracular band with 14–33 quinquelocular pores, few pores scattered in front of anterior bands; spiracular setae stout, bent; marginal setae short hairlike, in 1–2 rows, intervals between setae 2–3 times greater than their length; body setae long in medial area of venter; micropores in a marginal row; multilocular pores with 6–10 loculi on abdominal sternites 6–8; microducts absent from medial area of abdominal sternites; tubular ducts scattered on entire surface, their filaments slightly narrower than ducts, ca 50 on 6th abdominal sternite. *Dorsum*. Body setae hairlike, sparsely scattered; micropores scattered on entire dorsum; discoidal pores form a narrow longitudinal band from head to anal plates; tubular duct filaments half as wide as ducts, ca 100 ducts on 6th abdominal tergite, absent from anterior part of head, except few along anterior margin; diameter of anal ring ca 40 μ, setae 108 μ long; anal plates ca 91 μ long, 43 μ wide, apical setae short, needle-like (Koteja, 1979a). All developmental stages described by Newstead (1903), by Koteja (1965, 'unpublished), adult male by Giliomee (1967); male test and first instar by Řehaček (1960).

Distribution. A common boreo-montane Trans-Palearctic species, frequenting moist shady forests. In CE: Bu, Cz, Fr, Ge, Po, USSR (Leningrad Region, Ukraine, and Far East); also England. Many other distribution records need confirmation.

Biology. On leaves of *Juncus* and *Luzula*. Other host records based on ovisac associations are questionable, eg. *Agrostis, Bellis, Carex, Nardus, Potentilla, Scirpus* and *Sieglingia*. One, rarely 2 generations per year. Adults of both sexes appeared at end of June, and early July; ovisac was formed from early August; females developed on upper part of leaves, males on underside.

Natural enemies. Parasites: *Coccophagus lycimmia* (Zak-Ogaza, 1958), *Metaphycus zebratus* (Trjapitzin, 1978).

116. **Luzulaspis nemorosa** Koteja – Koteja's Soft Scale

Syn. & Lit.: *Luzulaspis nemorosa* Koteja, *1966b:45, 1969:8, 1979a:618; Zahradnik, 1977:119; Kosztarab & Kozár, 1978:107; Kozár et al., 1979:130.

Ovisac 4.5–6.5 mm long, ca 1.5 mm wide. Postreproductive female elongate-oval, brown. Teneral female almost parallel-sided 2.4–4 mm long, 1.1–1.2 mm wide; dorsum with 2 narrow light red longitudinal bands. *Venter*. Antennae 287–362 μ long, located halfway between anterior margin of head and clypeolabral shield, terminal segment with 9 setae, up to 40 μ long, and one 75–107 μ long; labium 60 μ long, 82 μ wide; legs well-developed, hind legs 648–710 μ long; spiracular pores with 5, rarely 3–6 loculi, 20–41 pores in each band; spiracular setae variable in shape, usually curved, often bifurcate or branched; marginal setae hairlike, form 1 irregular row, intervals between them 2–3 times greater than their length; submarginal setae similar to latter, but shorter; body setae variable, long setae in medial area, short setae form 4 irregular longitudinal rows laterad of spiracles and coxae; micropores in an irregular marginal row; multilocular pores with 8–11 loculi form transverse bands on abdominal sternites 7 and 8, scattered on sternite 6; tubular ducts 7–11 μ long, ca 2.5 μ wide, sparsely scattered on surface. *Dorsum.* Body setae 5–10 on each tergite; micropores, ca 2 μ in diameter, scattered on entire surface; a total of 85–172 discoidal pores form 2 narrow longitudinal bands from pro or mesothorax to anal plates; microtubular ducts over venter; tubular ducts, 7–11 μ long and ca 3.5 μ wide, densely distributed, except absent from anterior of head; diameter of anal ring 48 μ, setae 120 μ long; anal plates 103 μ long, 48 μ wide, apical seta spinelike (Koteja, 1966b). All developmental stages, including adult male, described by Koteja, (1965 unpublished).

Distribution. A frequently collected boreal species, that often forms large colonies. In CE: Bu, Cz, Fr, Ge, Hu, Po, and USSR (Transcarpathian Region). Its occurrence in England has not been confirmed (Koteja, 1979a).

Biology. On leaves of *Luzula nemorosa.* One generation per year; reproduction biparental. Occurs in beech and fir-beech forests.

117. **Luzulaspis pieninica** Koteja & Zak-Ogaza – Sedge Soft Scale

Syn. & Lit.: *Luzulaspis pieninica* Koteja & Zak-Ogaza, *1966:322; Koteja, 1978:324, *1979a:629; Kosztarab & Kozár, 1978:108; *L. grandis*, Zak-Ogaza & Koteja, 1964:427 (misidentification).

Ovisac 10–15 mm long. Postreproductive female convex, heavily sclerotized. Teneral female spindle-shaped, live specimen with narrow reddish-brown longitudinal band, 6.7–9.5 mm long, 1.3–1.7 mm wide. *Venter.* Antennae 550–575 μ long, situated 2/3 distance from anterior margin of head to clypeolabral shield; labium 65 μ long, 96 μ wide; legs slender, hind legs 926–1013 μ long; each spiracular band of 30–40 pores with 5, rarely 4 or 6 loculi, a few pores in front of anterior bands; spiracular setae moderately thick, 17–25 μ long, curved, and pointed, rarely forked, rarely only 1 seta developed; marginal setae hairlike, except thicker on head, form 2 irregular rows, at lateral margins intervals between setae 2–8 times their length; body setae much enlarged in the medial area; interantennal setae often with expanded tip; micropores form a marginal row, size larger than on dorsum; multilocular pores with 7–10 loculi, 5–6 μ in diameter, form transverse bands on abdominal sternites 6–8, few scattered on sternites 2–5; few microducts scattered on surface; filaments of tubular ducts as wide as ducts. *Dorsum.* Body setae in medial area of head conical, on rest of surface spinelike and bent; micropores scattered on entire surface; ca 150 discoidal pores form a medial longitudinal band from metathorax to anal plates; tubular ducts with long narrow filaments, densely scattered on surface, except absent from anterior of head region;

diameter of anal ring 60 μ, setae 168 μ long; anal plates twice as long as wide, apical seta spinelike (Koteja, 1979a).

Distribution. A rare xerophilous species. In CE: Cz, Ge and Po.

Biology. On leaves of *Carex* spp. Adult females present in July.

118. **Luzulaspis rajae** Kozár – Raja's Soft Scale
Fig. 90

Syn. & Lit.: *Luzulaspis rajae* Kozár, 1981:315.

Ovisac and postreproductive female unknown. Teneral female elongate-oval, 2.5–3 mm long, 1–1.2 mm wide. *Venter.* Antennae 360–400 μ long, located nearer to clypeolabral shield than to anterior margin; labium ca 50 μ long, 90 μ wide; stylet loop as long as labium; posterior legs 780 μ long; pores in spiracular bands with 5-loculi, each band with 15–20 pores; spiracular setae thick, with blunt tip, each ca 30 μ long; marginal setae form 1 irregular row, spinelike, ca 12 μ long, 2 on anterior margin of head thicker, conical, ca 25 μ long, on anal lobes 1 seta thicker, ca 30 μ long, intervals between setae 30–50 μ; most body setae short, 8 interantennal setae up to 75 μ long; micropores scattered on surface; multilocular pores with 5–10 loculi, ca 7–9 on each abdominal sternite 6 and 7; microducts scattered on surface; tubular ducts of various sizes numerous on entire surface. *Dorsum.* Body setae conical, ca 10 μ long; micropores scattered on entire surface; discoidal pores form a medial longitudinal band on abdomen anterior to anal plates; microducts distributed on entire surface; tubular ducts numerous, scattered; diameter of anal ring ca 60 μ, setae ca 130 μ long; anal plates 85–110 μ long, ca 60 μ wide.

Distribution. Known only from northern Hungary.

Biology. On leaves of *Carex montana*. Apparently one generation per year. Adult females developed by end of May.

119. **Luzulaspis scotica** Green – Scottish Soft Scale
Fig. 91b

Syn. & Lit.: *Luzulaspis scotica* Green, 1926:179, 1928a:27; Borchsenius, 1957:126; Ossiannilsson, 1959:197; Koteja, 1978:324, *1979a:608. *L. borchsenii* Řehaček, 1959:176, 1960:26; Danzig, 1964:639; Kosztarab & Kozár, 1978:106; Tereznikova, 1981:117.

Ovisac 5–8, rarely 10 mm long, 1–1.5 mm wide. Postreproductive female elongate spindle-shaped. Teneral female almost parallel-sided; 3.7–6.3 mm long, 1.1–1.7 mm wide. *Venter.* Antennae 456–648 μ long, located halfway between anterior margin of head and clypeolabral shield; labium 72–84 μ 115–120 μ wide; legs slender, hind legs 971–1274 μ long; spiracular pore bands with 5–7 loculi, some scattered in front of anterior bands; spiracular setae large, straight or bent, rarely sicklelike, only slightly longer than marginal setae; marginal setae conical, variable in size, intervals between setae on anterior margin less than their length, on lateral margins about equal; body setae of various sizes, longer on head region, longest in midregion of body, metathorax

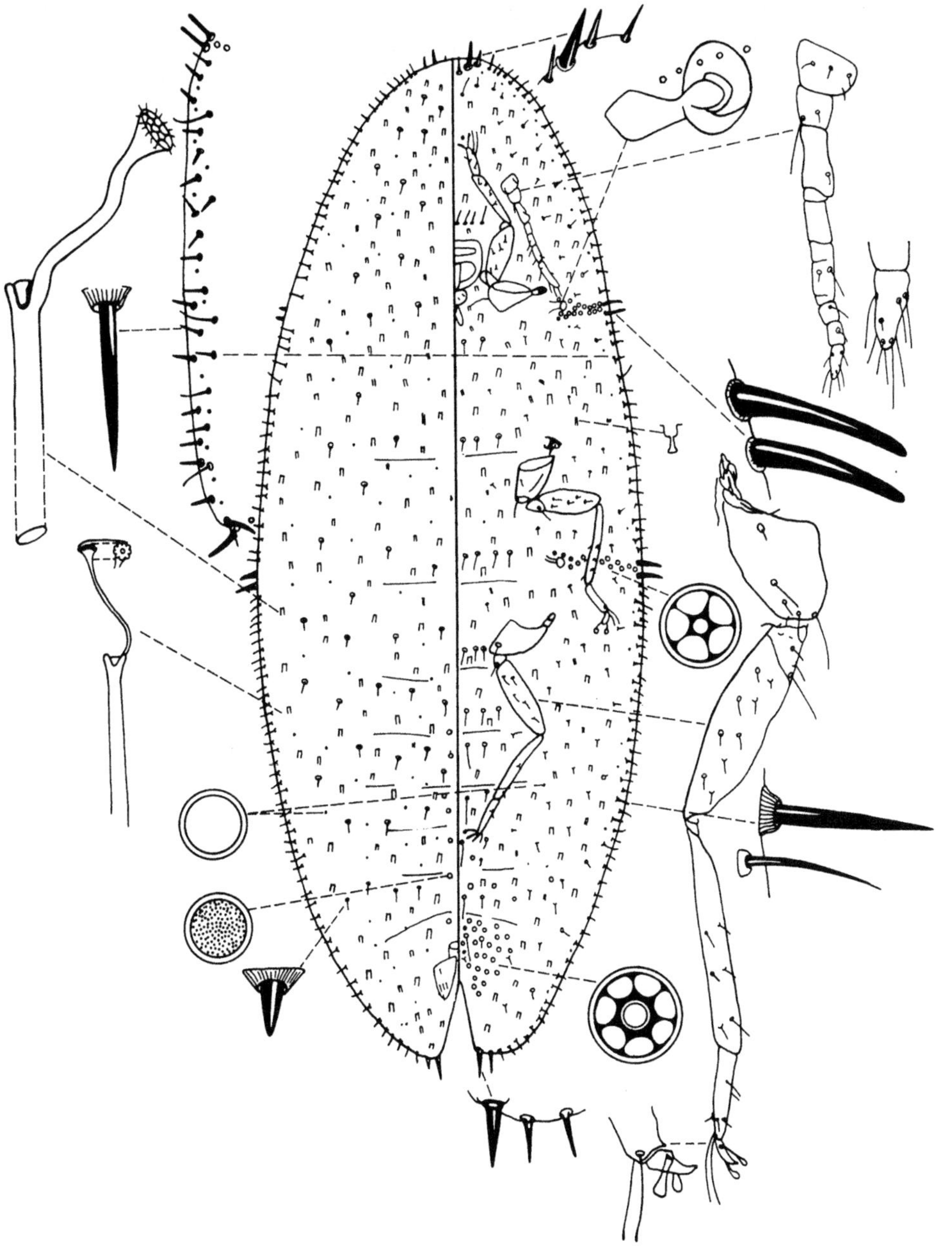

Fig. 90. *Luzulaspis rajae* Kozár, female (after Kozár, 1982)

with 5–18 setae, those 70–140 μ long; micropores scattered on entire surface; multilocular pores with 5–12 loculi; 4–8 μ in diameter; form transverse rows on terminal abdominal sternites; tubular ducts scattered, their filaments slightly thinner than ducts, ca 60 ducts on 6th abdominal sternite. *Dorsum*. Body setae spinelike, most conical, rarely on head and thorax with blunt apices; micropores scattered on entire surface; discoidal pores, 3–6 μ in diameter in clusters forming a medial longitudinal

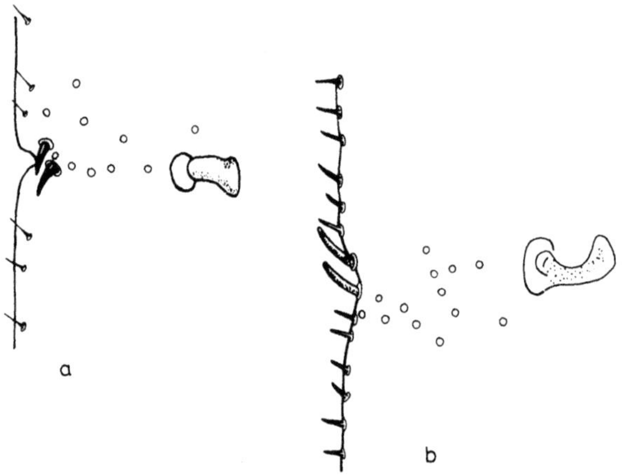

Fig. 91. Spiracular and marginal setae of (a) *Luzulaspis luzulae* (Dufour) female, (b) *Luzulaspis scotica* Green female (after Kosztarab and Kozár, 1978)

band from metathorax to anal plates; tubular ducts numerous except absent from anterior-medial part of head, filaments ca 1/3 as wide as ducts; diameter of anal ring ca 60 μ, its setae 170–180 μ long; anal plates twice as long as wide, apical and exterior subapical setae spinelike (Koteja, 1979a). Male test and first instar described by Řehaček (1960).

Distribution. A boreal species, found in swampy stands. In CE: Cz, Fr, Hu, Po, Ro and USSR (Transcarpathian Region); also England, Scotland, Wales and Sweden (Koteja, 1979a).

Biology. On leaves of *Carex* spp. and *Eriophorum vaginatum*. One generation per year; adults present during July–August, eggs were laid in Leningrad Region during August (Danzig, 1959a).

Natural enemy. Parasite: *Trichomasthus danzigae* (Trjapitzin, 1978).

Genus **Nemolecanium** Borchsenius

Syn. & Lit.: *Nemolecanium* Borchsenius, 1955c:289, 1957:378; Řehaček, 1960:57; Kosztarab & Kozár, 1978:108; Tereznikova, 1981:163.

Type-species: (*Nemolecanium abietis* Borchsenius, 1955) = *Physokermes graniformis* Wünn, 1921.

Postreproductive females similar to *Parthenolecanium* females, but with lateral margins strongly bent underneath, elongate convex, 3–8 mm long, 1.7–5 mm wide. Teneral females. *Venter*. Antennae 7- or 8-segmented; marginal setae short, interval between setae far exceeds their length; spiracular furrow lined with a row of disc pores, normally quinqueloculars; usually 3 stout spiracular setae in each group; marginal setae slender; multilocular pores form transverse bands on most sternites posterior of front legs; tubular ducts form a submarginal band. *Dorsum*. Derm around anal area either more sclerotized or similar to rest of dorsal derm. Sievelike pores and slender setae scattered on entire surface; anal ring normally with 2 rows of pores and 6 setae.

Distribution. Borchsenius (1957) placed 4 species in the genus; Mediterranean and Western Palearctic distribution; 1 species in CE.

Biology. On firs, *Abies* spp. Feed on needles, rarely on new shoots.

Note. This genus is in need of a revision. We did not include *N. adventicium* Borchsenius here. Although described from specimens collected in the Moscow Botanical Garden, Borchsenius assumed that the species was introduced from Germany.

120. **Nemolecanium graniformis** (Wünn) – Fir Needle Scale

Syn. & Lit.: *Physokermes graniformis* Wünn, 1921:29. *Eulecanium graniforme*, Schmutterer, 1952a:14. *Lecanium graniforme*, Kawecki, 1957:195. *Nemolecanium graniformis* Borchsenius, 1957:379; Řehaček, 1960:57; Kosztarab & Kozár, 1978:108. *N. graniforme* Schmutterer, 1972b:408; *N. abietis* Borchsenius, 1955c:290; Tereznikova, *1981:163.

Postreproductive female (as given under *N. abietis* by Borchsenius, 1955c, 1957) irregularly oval, grain seed-like, body narrowed and sloping toward end of abdomen, convex with 3 keels converging toward anal plates, dorsum with lateral sides strongly bent underneath, greyish-brown, ca 6 mm long, 4 mm wide and 3.5 mm high; anterior end always toward base of needle. Teneral female, yellowish-grey with a striking lighter longitudinal band. *Venter*. Antennae 8-segmented, ca 350 μ long; labium 95 μ long, 109 μ wide with 10 setae (Koteja, 1974a): legs and claws slender; diameter of spiracular peritremes 85–100 μ; spiracular furrows narrow, broader at two ends, each furrow with 26–30 pores, mostly in single row of quinquelocular pores, rarely with few trilocular pores; 3 spiracular setae in each group, median seta slightly stouter and longer (15–18 μ) than laterals (12–15 μ); marginal setae short and slender, at anterior end of body 10–12 μ long, at posterior end 15–18 μ long, intervals between setae 60 to 200 μ; submarginal setae hairlike, 8–12 μ long, on anal lobes 15 μ long; other body setae stout, of various lengths; multilocular pores with 8–10 loculi form transverse bands in median area of abdomen, and in clusters at bases of legs and antennae; dark-rimmed pores distributed in submarginal bands on thorax and head; tubular ducts ca 22 μ long, 5 μ wide in a submarginal band on entire venter, in a small group around mouthparts. *Dorsum*. Strongly sclerotized, around anal plates with radial striations, but without reticulated pattern; dorsal surface with scattered sievelike pores, ca 3 μ in diameter and slender setae, 10–12 μ long; anal plates elongate, with outer angles broadly rounded; anal ring wide with 2 rows of pores and 6 setae ca 300 μ long. Adult male (*abietis*) and male test described by Borchsenius (1957).

Distribution. A rare species. In CE: Cz, Fr, Ge, Po, Sw and USSR.

Biology. At base and underside of needles of firs, *Abies* spp., rarely on shoots; second instars overwinter in Slovakia (Řehaček, 1960); adults develop by May, and start egglaying in mid-June; eggs hatch in July. Borchsenius (1957) reported mass flight of males at end of May in Crimea.

Natural enemies. Parasites: *Aphycoides merceti, Blastothrix sericea,* and *Coccophagus lycimnia.* (Schmutterer 1972b).

Note. The original description of *N. graniformis* by Wünn included only general characters of postreproductive females. Detailed description of teneral females is unavailable for this species, except under *N. abietis,* which was synonymized with *N. graniformis* by Schmutterer (1972b). The only doubt we have concerning the synonymy is the diverse geographical locations of the two; *N. graniformis* is considered a montane species in CE, normally occurring at 500 to 1400 m elevations, while *N. abietis* was found on the southern shores of Crimea.

Genus **Palaeolecanium** Šulc

Syn. & Lit.: *Palaeolecanium* Šulc, 1908:36, 1932:82; Borchsenius, 1957:345; Řehaček, 1960:51; Kosztarab & Kozár, 1978:108; Tereznikova, 1981:165.

Type-species: *Lecanium bituberculatum* Targioni-Tozzetti, 1868.

A monotypic genus. For generic description see *P. bituberculatum.*

Postreproductive females can be separated from the related *Parthenolecanium* by the knoblike tubercles on dorsum in *Palaeolecanium*; the tubercles are absent from members of *Parthenolecanium.* The teneral females in *Palaeolecanium* have smaller spiracular setae than in *Parthenolecanium* and a narrow and long longitudinal band of sievelike discoidal pores extending from prothorax to anal plates, these pores form a smaller elongate group in *Parthenolecanium.*

Distribution. As given for *P. bituberculatum.*

Biology. Eggs overwinter. In all related groups the second instars overwinter.

121. **Palaeolecanium bituberculatum** (Targioni-Tozzetti) – Bituberculate Scale
Figs 92, 93

Syn. & Lit.: *Lecanium bituberculatum* Targioni-Tozzetti, 1868:731; Signoret, 1873:414(245); Newstead, 1903:101; Šulc, 1932:82; Kawecki, 1935:1. *Eulecanium bituberculatum,* Leonardi, 1920:303; Schmutterer, 1952b:548; Săvescu, 1961:39. *Palaeolecanium bituberculatum,* Borchsenius, 1957:344; Smolyannikov, 1959:80; Řehaček, 1960:51; Goidanich, 1962:269; Rogojanu, 1963:284; Vashinskaya, 1969:742; Dziedzicka, & Marchewczyk, *1970:28; Kosztarab & Kozár, 1978:109. Tereznikova, 1981:166. *Eulecanium kosswigi* Bodenheimer, 1953c:100.

Postreproductive female short-oval, only moderately convex, margins sloping, with 2 pairs of knoblike tubercles on dorsum, anterior tubercles more prominent, larger than posterior ones, latter sometimes weakly developed, with a longitudinal carina on mid-dorsum, brown, ca 5–8 mm long, 4–6 mm wide, and up to 2.5 mm high. Teneral female. *Venter.* With 7-, rarely 6-segmented antennae, ca 260 μ long; labium 85 μ long,

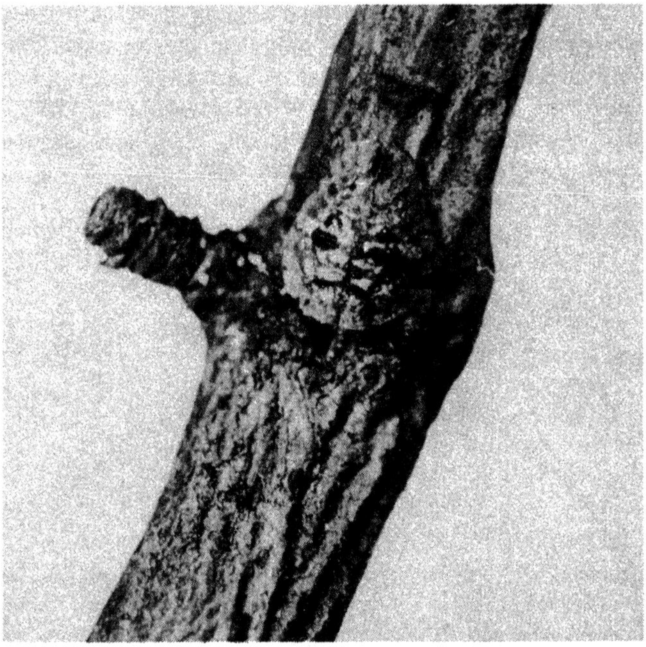

Fig. 92. *Palaeolecanium bituberculatum* (Targioni-Tozzetti), female on *Crataegus* twig (Photo F. Kozár)

85 μ wide, with 10 setae (Koteja, 1974a); claw digitules dissimilar; diameter of peritreme of posterior spiracles ca 45 μ; each spiracular furrow with 15–23 disc pores normally forming a single row, pores with 5 or 6 loculi; spiracular setae stout, with blunt tip; median seta ca 27–30 μ long, lateral setae 21–27 long; marginal setae slender, sharp-pointed, ca 27–45 μ long, space between setae 2 to 5 times their length; other body setae short and few, 4 between antennae, 2 of these ca 35 μ long; dark-rimmed pores interspersed in a submarginal band with tubular ducts; few tubular ducts around mouthparts; multilocular pores with 10–12 loculi in small groups or rows near mouthparts, at bases of legs and in transverse rows or bands on each abdominal sternite. *Dorsum.* With many scattered small tubular ducts and few small spinelike setae, except absent from narrow longitudinal median band, sievelike discoidal pores extending from prothorax to anal plates (Borchsenius, 1957). Adult male described by Newstead (1903), Giliomee (1967); male test (characteristically formed) and first instar by Šulc (1932) and Řehaček (1960); first and second instars by Leonardi (1920); and all stages by Dziedzicka and Marchewczyk (1970).

Distribution. A common Western Palearctic species. In CE: Bu, Cz, Fr, Ge, Hu, It, Po, Ro, Sw, USSR and Yu; also found in Denmark, England, Holland, Sweden, Afghanistan, Iraq, Iran, Turkey and N Africa. Record from USA was based on specimens found on imported plant (Dietz and Morrison, 1916).

Biology. Nymphs on leaves and adults on thin branches are often concealed at the base of a thorn or bud. Hosts are all Rosaceae (except *Juglans* and *Corylus*): *Armeniaca, Crataegus, Cydonia, Malus, Mespilus, Prunus* and *Pyrus*. One generation per year. Eggs overwinter; hatch around the blooming of apple trees and N_1 move to leaves; females migrate back to branches at end of June and mature, start egg laying in early August in Hungary or mid-September in Germany (Schmutterer, 1952b), and complete egg laying by end of October in N Italy (Goidanich, 1962). According to Schmutterer (1952b) each female lays 283–1,003 eggs. Only small populations found in Hungary; apparently without economic importance.

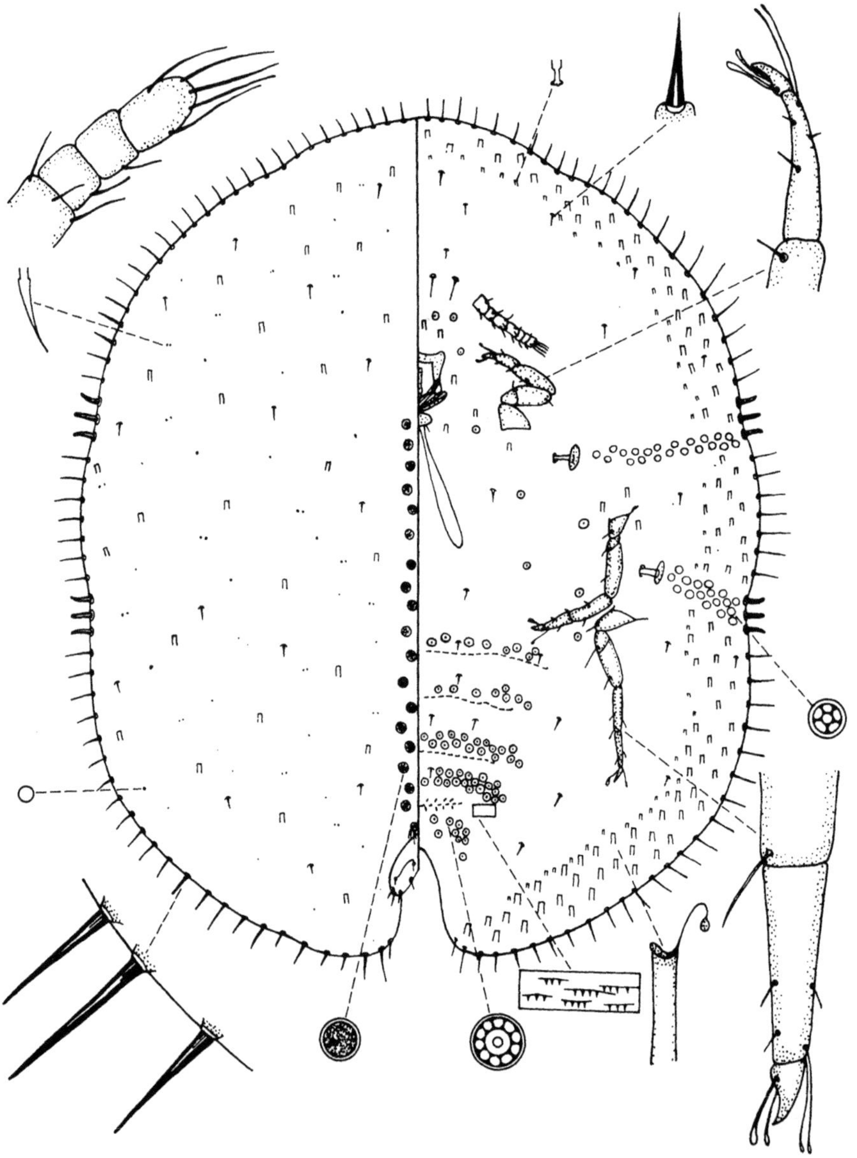

Fig. 93. *Palaeolecanium bituberculatum* (Targioni-Tozzetti), female (original)

Natural enemies. It is often effectively controlled by its parasites. Rate of parasitization reached 90 to 100 % in Hungary, during 1970. Parasites: *Coccophagus differens, C. lycimnia, C. insidiator, C. palaeolecanii, C. scutellaris, Microterys duplicatus, Trichomasthus albimanus* (Borchsenius, 1957; Dziedzicka and Marchewczyk, 1970; Trjapitzin, 1978; Kozár and Sugonjaev, 1979).

Genus **Parafairmairia** Cockerell

Syn. & Lit.: *Parafairmairia* Cockerell, 1899:332; Green, 1922a:19; Steinweden, 1929:229; Schmutterer, 1952b:558; Borchsenius, 1952:271, 1957:133; Řehaček, 1960:27; Morrison & Morrison, 1966:144; Danzig, 1967:826, 1980a:254; Koteja & Rosciszewska, *1970:234; Kosztarab & Kozár, 1978:110; Tereznikova, 1981:120.

Type-species: *Fairmairia bipartita* Signoret, 1876.

Test of female of translucent wax sheaths, with rooflike, coneshaped extensions centrally; ovisac poorly developed, compressed between female body and test; latter protects the eggs. Postreproductive female elongate-oval, strongly convex above, concave ventrally, 2.5–4.5 mm long. Teneral female elongate, almost parallel-sided, somewhat pointed at both extremities, with short anal cleft. *Venter.* Antennae 8-segmented; legs stout and long, claw digitules stout, strongly expanded at apex; tibio-tarsal sclerosis present; spiracular furrows lined with disk pores with 3 to 7 loculi; 1 or 2 spiracular setae, stouter than marginal setae and usually curved; marginal setae pointed, usually curved; submarginal setae hairlike; dark-rimmed pores in a wide marginal band and in groups on head; multilocular pores form transverse rows or bands on terminal abdominal sternites; large tubular ducts common on both body surfaces. *Dorsum.* Body setae absent. Minute simple disc pores scattered on entire surface; discoidal pores form transverse bands on abdomen; besides large tubular ducts, minute ducts form rows; anal ring with 6 setae and a row of pores; anal cleft short; anal plates conical, thick and wide.

Distribution. In the Holarctic Region with 4 species, 2 in CE.

Biology. On leaves of sedges and grasses; especially common on *Carex* and *Eriophorum* (Cyperaceae). Reproduction biparental, overwinter as eggs; one generation per year.

Key to species of *Parafairmairia*

Discoidal pores form 2 pairs of transverse bands, on dorsum of abdomen only; 2 spiracular setae in each posterior group; second central waxy cone of test higher than third one; young female yellow with red patches on dorsum (figs 94, 95) .... *gracilis*, p. 214

Discoidal pores form 6 pairs of transverse bands on entire dorsum; 1 spiracular seta in each posterior group; third central waxy cone of test higher than second one; young female yellow with brown patches on dorsum .... *bipartita*, p. 214

122. **Parafairmairia bipartita** (Signoret) – Two-humped Soft Scale

Syn. & Lit.: *Fairmairia bipartita* Signoret, 1872c:34, 1874:98. *Parafairmairia bipartita*, Schmutterer, 1955b:100; Borchsenius, 1957:134; Gomez-Menor, 1958:8, 14; Zak-Ogaza & Koteja, 1964:427; Koteja & Rosciszewska, *1970:252; Kosztarab & Kozár, 1978:111; Kozár et al., 1979:130; Danzig, 1980a:254; *P. delicata* Borchsenius, 1952:272; Tereznikova, *1981:120.

Test of female of glassy wax plates, composed of 4 cones, and 12 subrectangular lateral plates; anterior cone smallest, second larger and third cone largest, fourth cone small; test 3–4.2 mm long, 1.4–1.8 mm wide. Teneral female elongate, narrow, flattened laterally, according to Koteja and Rosciszewska (1970) yellow before oviposition with some brown patches; 2–3.8 mm long, 0.6–1.2 mm wide. *Venter.* Antennae 8-segmented, 243–322 μ long; labium short, globular, 81 μ long, 85 μ wide (Koteja, 1974a); legs 555–855 μ long, claw stout, digitules 25–35 μ long; spiracular pore bands with 5–7 locular pores, 14–28 pores in each anterior band, 18–51 in each posterior; spiracular setae stout, curved, 17–40 μ long, 2 in each anterior group and 1 at each posterior group; marginal setae stout often with hairlike tips, 15–27 μ long; submarginal setae smaller, hairlike; small hairlike setae scattered on most of venter; minute disc pores scattered on entire body; dark-rimmed pores in a wide submarginal band, and in transverse groups posterior of labium; multilocular pores with 10–12 loculi in transverse bands on sternites, form a large group around vulva; large tubular ducts, 18–20 μ long, 3 μ in diameter, common on both body surfaces. *Dorsum.* Discoidal pores, 3–4 μ in diameter form 2 pairs of transverse bands on abdomen; anterior bands with 12–70, posteriors with 23–79 pores on each side; minute tubular ducts, ca 5 μ long, 1 μ wide, form rows aligned with border of wax plates; anal ring ca 60 μ in diameter with pores and 6 setae, each 130–180 μ long; anal plates conical, 85–100 μ long, ca 50 μ wide, each with 4 apical setae. Adult male described by Koteja and Rosciszewska (1970).

Distribution. A Trans-Palearctic, steppe-inhabiting, mesophilous species. In CE: Bu, Fr, Ge, Hu, Po and USSR (Ukraine to Far East).

Biology. On leaves of *Agropyron, Agrostis, Brachypodium, Carex, Equisetum* (on stem), *Festuca, Juncus, Mesembryanthemum* and *Stipa.* Collected at up to 800 mm elevation in Poland. Two nymphal stages observed in Poland; oviposition in July and August; eggs overwintered (Koteja and Rosciewska, 1970).

123. **Parafairmairia gracilis** Green – Elongate Soft Scale
Figs 94, 95

Syn. & Lit.: *Parafairmairia gracilis* Green, 1916:24, 1922a:19; Schmutterer, 1952b:558; Řehaček, 1956:48, 1960:28; Borchsenius, 1957:133; Ossiannillson, 1959:197; Danzig, 1967:829; Koteja & Zak-Ogaza, 1969:366; Koteja & Rosciszewska, *1970:240; Kozarzhewskaya & Reitzel, 1975:25; Kozár et al., 1977:72; Kosztarab & Kozár, 1978:112; Tereznikova, 1981:122.

Test of female of glassy wax plates, formed of 4 wax cones, 2nd central cone taller than 3rd, and 7–8 pairs of lateral plates; test 3.9–6.5 mm long, 0.8–1.5 mm high. Teneral female, according to Koteja and Rosciszewska (1970), elongate, narrow, flattened laterally, before oviposition yellow with 4 red patches on cones and 7 pairs

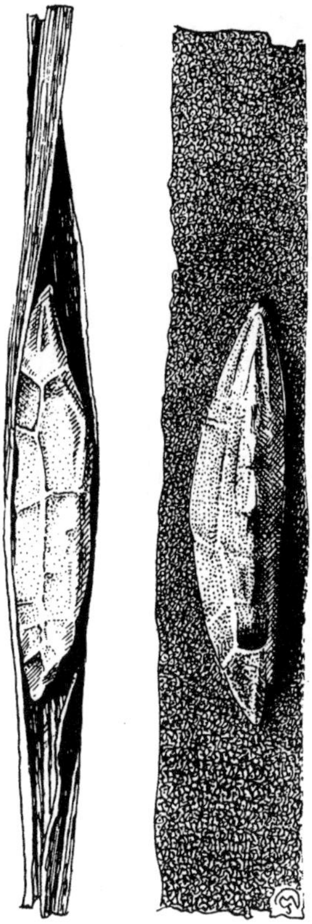

Fig. 94. *Parafairmairia gracilis* Green, ovisacs, dorsal and lateral view (after Kosztarab and Kozár, 1978)

laterally, female 2.7–4.5 mm long, 0.8–1.2 mm wide. *Venter*. Antennae 8-segmented, 250–430 μ long; labium short, globular, 95 μ long, 109 μ wide (Koteja, 1974a); legs well-developed 530–1,060 μ long, claw digitules 20–35 μ long; disc pores in spiracular pore bands with 3–7 loculi, each anterior band with 17–26, each posterior with 15–34 pores; 2 spiracular setae in each group, stout, curved, 20–35 μ long; marginal setae, spinelike, often with a hairlike tip, 20–30 μ long; submarginal setae smaller, slender, often hairlike, 5–10 μ long; 2 pairs of interantennal setae, up to 150 μ long, 1–2 similar setae on median area of each abdominal sternites 6–8; minute disc pores on both body surfaces, larger ones form a narrow marginal band; dark-rimmed pores in a wide submarginal band and in groups on head, and thorax; multilocular pores with 11 loculi on abdomen, rare on anterior segments, form wide bands on posterior segments; large tubular ducts, 18–20 μ long, ca 3 μ in diameter, in a wide submarginal band, others scattered on surface. *Dorsum*. Discoidal pores, 3.5–4 μ in diameter, form 6 pairs of transverse bands, on dorsum; minute disc pores scattered on entire surface; minute tubular ducts, ca 5 μ long and 1 μ in diameter with long interior extension aligned with borderlines of wax plates; large tubular ducts, similar to those on venter, scattered on most of dorsum; anal ring, ca 60 μ in diameter, with pores and 6 setae, each ca 150 μ

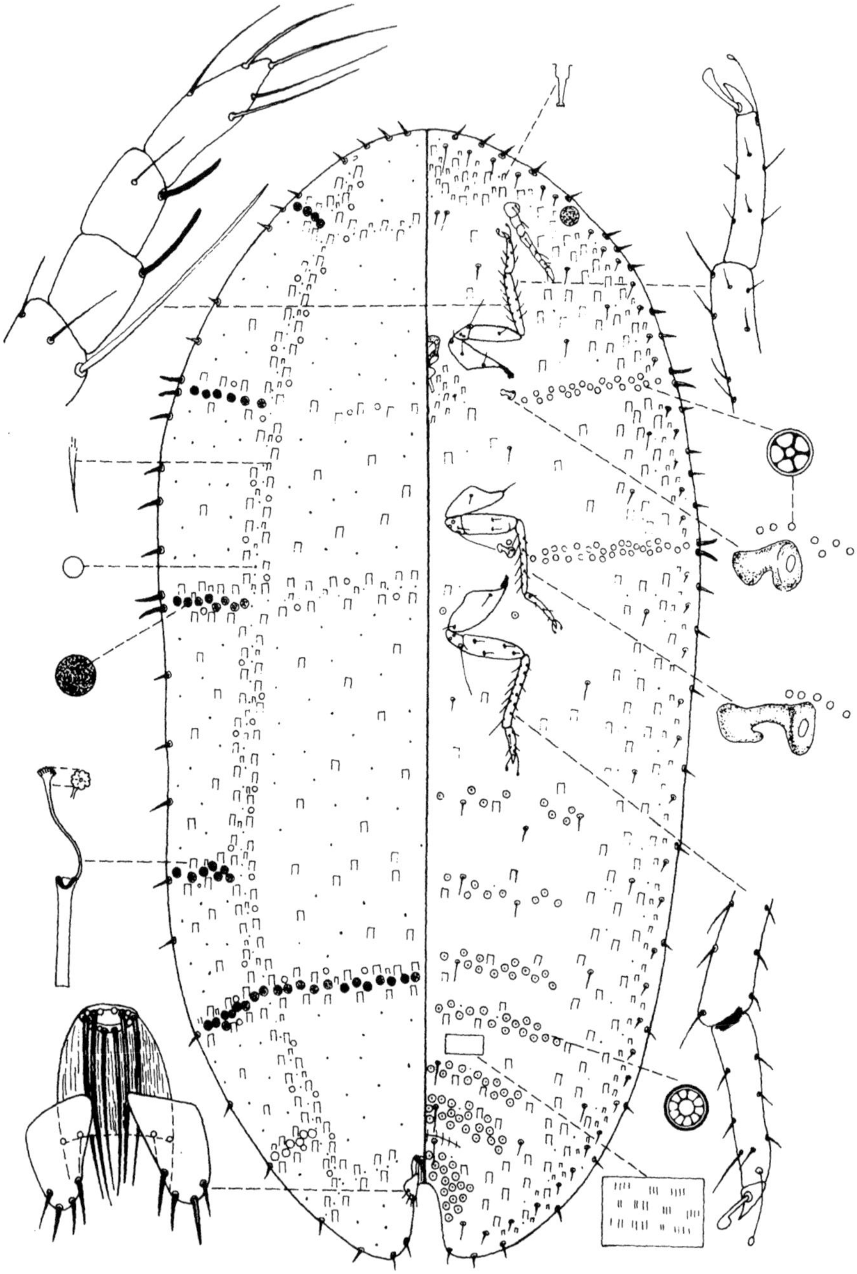

Fig. 95. *Parafairmairia gracilis* Green, female (original)

long; anal plates conical, 80–100 μ long, ca 50 μ wide, each with 4 apical setae. Adult male and nymphal instars described by Koteja and Rosciszewska (1970), male test and nymphs by Řehaček (1960), N_1 and female test by Schmutterer (1952b).

Distribution. A rare western-Palearctic, boreal, hygrophilous (xerophilous in Poland) species. In CE: Cz, Fr, Ge, Hu, Po, USSR and Yu; also in Denmark, England and Sweden. Recently found in Austria (Kosztarab, unpublished).

Biology. On *Agropyron, Carex, Eriophorum, Juncus,* and *Typha*; ovipositing females found on a variety of plants, and non-plant materials. One generation per year, but 2 reported from Ukraine by Tereznikova (1963); eggs overwintered and hatched in S Germany at end of April through early May; N_1 molted in last third of May; adults developed by end of June through early July; each female laid 16–64 eggs (Schmutterer, 1952b).

Genus **Parthenolecanium** Šulc

Syn. & Lit.: *Parthenolecanium* Šulc, 1908:64; Borchsenius, 1957:347; Řehaček, 1960:52; Danzig, 1967:151, 1980:268; Schmutterer, 1972b:408; Kosztarab & Kozár, 1978:112; Tereznikova, 1981:167; Danzig & Kerzhner, 1981:147. *Lecanium* (in part), Šulc, 1932:1; Dziedzicka, 1968:125; Williams & Kosztarab, 1972:68. *Eulecanium* (in part) Leonardi, 1920:285; Gomez-Menor Ortega, 1937:266; Schmutterer 1952b:547; 1954:62; Paik, 1978:241.

Type-species: *Lecanium corni* Bouché, 1844.

Postreproductive females elongate-oval, convex, dorsum strongly sclerotized, usually brownish, venter thin, membranous, body 3–10 mm long; eggs deposited under abdomen. Teneral female. *Venter.* Antennae 6 to 9-segmented (usually 7 or 8); labium 2-segmented, with 10 setae; legs slender, tibia markedly longer than tarsus, claw usually with a small denticle; spiracular furrow normally with quinquelocular pores (rarely with 6–8 loculi); 3 spiracular setae in each spiracular depression, median usually larger than laterals; marginal setae normally form a row, seldom a double or a partial double row; multilocular disc pores form transverse bands in medial area of abdomen; dark-rimmed pores often interspersed among tubular ducts that form a wide submarginal band. *Dorsum.* With or without submarginal tubercles; setae distributed on entire derm; most spinelike conical and small, but large stout conical setae normally form a longitudinal row in medial area; simple and bilocular pores normally scattered on entire dorsum; an elongate group of discoidal pores on abdomen anterior to anal plates; 2 anal plates usually triangular, each with 4 apical setae; anal ring large and wide, with 6 or 8 setae and 2 irregular rows of pores.

Distribution. About 20 species in the world, most in the Holarctic Region, 8 in the Palearctic Region, and 6 in CE. The systematic status of a number of species need clarification.

Biology. Most feed on woody plants; some are polyphagous, others monophagous in food habit. Normally have one generation per year; females with 3 sometimes 4 stages, males with 5 stages; 2nd instars overwinter on twigs and branches in CE, become adults by end of spring or early summer; eggs are laid under females' body; the 1st instars normally settle on underside of leaves for feeding and molting. Reproduction is biparental and parthenogenetic. A number of species are considered pests of cultivated fruit trees and ornamentals. The females are visited by ants for their honeydew. Some species, e. g. *P. corni* and *persicae*, are often effectively controlled by natural enemies.

Note. Most of the species can be determined through their male tests. More attention should be paid to the utilization of male tests in the future.

Key to species of *Parthenolecanium*

1. With 24–42 dorsal submarginal tubercles; marginal setae hairlike, 39–62 μ long; tibio-tarsal sclerosis present (fig. 98). . . . . *persicae*, p. 223
   With none or less than 24 dorsal submarginal tubercles; marginal setae spinelike, relatively small, with blunt or rounded tips, 13–25 μ long; tibio-tarsal sclerosis absent 2
2. Marginal setae equal in length (ca 13 μ long), or slightly shorter than lateral spiracular setae (ca 15 μ long); normally with less than 10 discoidal pores anterior of anal plates; live on Cupressaceae, rarely on Pinaceae . . . . *fletcheri*, p. 222
   Marginal setae 13–25 μ long, definitely shorter than lateral spiracular setae (27–42 μ long); with more than 10 discoidal pores anterior to anal plates; not restricted to Cupressaceae and Pinaceae 3
3. Marginal setae relatively few, ca 60 around body, form one definite row only, set far apart from each other; live on *Taxus* spp. only. . . . . *pomeranicum*, p. 225
   Marginal setae numerous, ca 150 or more around body, form an irregular double or partly double row; live on various hosts 4
4. Marginal setae stout, conical; anterior end of body more narrowed than posterior end; old females with highest point of dorsal carina at 2/3 of body length; primary host oaks . . . . *rufulum*, p. 225
   Marginal setae slender; anterior and posterior ends similar; old females with highest point of dorsal carina at 1/3 of body length; polyphagous or on *Thymus* only 5
5. Medial spiracular setae 41–45 μ long; anal ring setae 115–172 μ long; widely distributed; polyphagous (fig. 97) . . . . *corni*, p. 218
   Medial spiracular setae 55–60 μ long; anal ring setae 80–90 μ long; known from S Poland only; live on *Thymus* sp . . . . *smreczynskii*, p. 227

124. **Parthenolecanium corni** (Bouché) – Brown Scale, European Fruit Lecanium
Figs 96, 97

Syn. & Lit.: *Lecanium corni* Bouché, 1844:298; Marchal, 1908:264; Green, 1930:12; Kawecki, 1962a:20 (lists most important synonyms); Williams & Kosztarab, 1972:76. *L. vini* Bouché, 1851:112. *L. wistariae* Signoret, 1873:433 (263); Danzig, 1980b:596. *L. robiniarum* Douglas, 1890:318. *L. coryli* (L.), Sulč, 1932:64. *L. tiliae*, Richards, 1958:310. *Palaeolecanium xylostei* (Schrank), Lindinger, 1934:76 (lists 130 synonyms). *Eulecanium corni* var. *robiniarum*, Jablonowski, 1916:56. *E. corni*, Leonardi, 1920:287; Balachowsky & Mesnil, 1935:716; Gomez-Menor, 1937:274; Schmutterer, *1954:63; Canard, *1958:185; Paik, 1978:245. *E.*

corni corni, Schmutterer, 1952b:546. *Parthenolecanium corni*, Řehaček, 1960:54; Borchsenius, 1957:356; Saakyan-Baranova et al., 1971:1; Blahutiak, 1977:9; Zahradnik, 1977:119; Kosztarab & Kozár, 1978:116; Danzig, 1980a:270; Kawai 1980:163; Tereznikova, 1981:168. *P. wistariae*, Komeili-Birjandi, 1981:47.

Only the synonyms pertinent to Central Europe are listed here.

Postreproductive females extremely variable in shape, size and coloration, depending on the host plant species and their age; old specimens heavily sclerotized, strongly convex, brown, ca 6 mm long, 4 mm wide. Teneral females oval, those from *Robinia pseudoacacia* tan, mottled with radiating lines of black blotches, 3–6.5 mm long, 2–4 mm wide. *Venter*. Antennae 7-, rarely 6- or 8-segmented, 248–332 μ long; labium 95 μ long, 104 μ wide (Koteja, 1974a); length of legs: pro- 442–540 μ, meso- 489–598 μ, metathoracic 493–586 μ; claw with small denticle; spiracular pore bands 2–3 pores wide, each band with 30–47 quinquelocular pores; spiracles 82–117 μ long, diameter of peritreme 50–75 μ; 3 spiracular setae, median slender, curved, blunt-tipped, 41–45 μ long, normally 1.5 times longer than lateral setae; marginal setae 150–200, straight or slightly curved, blunt-tipped, 17–23 μ long, form an irregular, partial double row; body setae of 2 types, those in median region very slender, others scattered on entire venter, short, stout; multilocular pores (most with 10-loculi) form transverse bands on abdomen, few in median area of thorax; tubular microducts, ca 4.4 μ long, form a submarginal band, few scattered over most of venter; large tubular ducts, 17–24 μ long in a submarginal band, sparsely scattered in median region. *Dorsum*. Normally with a submarginal row of tubercles, numbering up to 18; dorsal setae of 2 types: those irregularly scattered on entire surface, 10–12 μ long, stout, straight and blunt, and those forming an irregular double row in mid-dorsal area ca 19

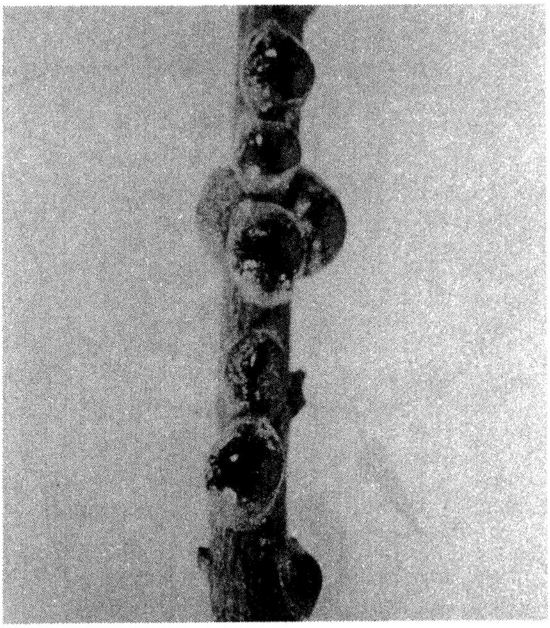

Fig. 96. *Parthenolecanium corni* (Bouché), adult females (Photo A. B. Hamon)

μ long, stout, tapering to a blunt tip; simple disc pores sparsely scattered over dorsum; discoidal pores in a group of 15–30 anterior to anal plates; bilocular pores usually in irregular radiating lines extending from mid-dorsum to margin; tubular ducts sparsely scattered over surface; anal plates triangular, ca 141–168 μ long, 78–113 μ wide, each with 4 apical and 2 subapical setae; anal fold with 2 pairs of fringe setae; anal ring ca 86 μ long, 74 μ wide, with 8 setae ca 115–172 μ long, and 2 irregular rows of translucent

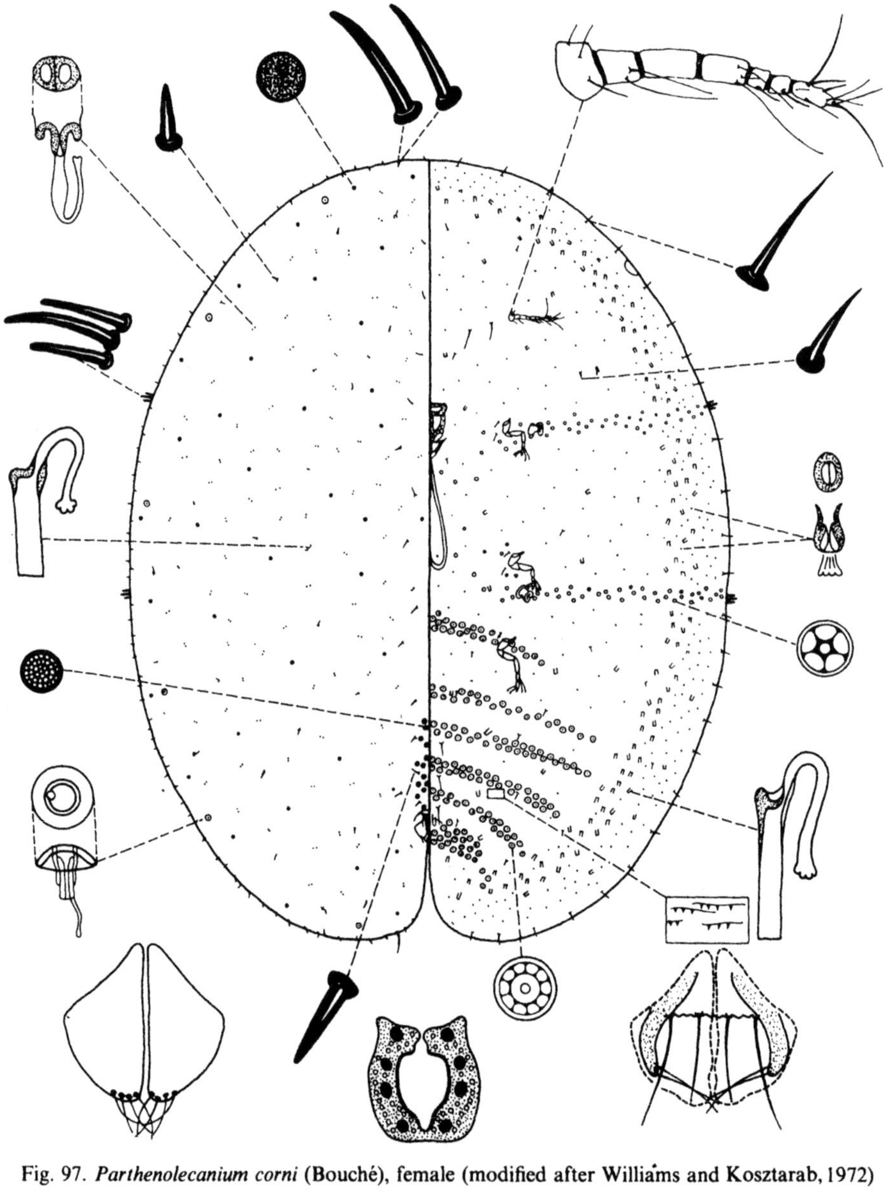

Fig. 97. *Parthenolecanium corni* (Bouché), female (modified after Williams and Kosztarab, 1972)

pores. Adult male, male test and various immature stages described by Šulc (1932), Habib (1956a, 1957), Borchsenius (1957), Canard (1958), Kawecki (1962a), Saakyan-Baranova et al. (1971), 2nd instar female by Dziedzicka and Sermak (1967), and adult male by Giliomee (1967).

Distribution. Almost world-wide in temperate regions. A very common species in all countries in CE.

Biology. A polyphagous species. Kawecki (1962a) listed from literature sources ca 350 host plants, of these 130 were found by him. Reproduction biparental, but also parthenogenetic. There is 1 generation per year in northern Hungary and 2 south of this area. Farther south 3 generations are reported from Caucasus and Central Asia (Borchsenius, 1957). Besides the climate, the number of generations is also affected by the host. For example, in the USSR (Krasnodar region) there is 1 generation on plum, 2 on peach and 3 on *Robinia pseudoacacia* (Borchsenius, 1957). Second instars overwinter; adults in CE normally develop by end of May, each female lays 31 (on *Crataegus*), to 5,124 eggs (on *Gleditsia*); eggs hatch from second half of June; N_1 migrate to leaves, where they settle and feed mainly on the underside, molt, and N_2 return to the woody parts for overwintering in the fall. Mass outbreaks have been reported in 10–11 year intervals since 1880 in CE, in the last outbreak of 1955 its most effective parasite (*Blastotrix confusa*) was discovered. According to Sugonjaev (personal correspondence 1980), this species is synonymous with *B. longipennis*, and was introduced from North America. This parasite apparently is responsible for the cessation of mass outbreaks of the brown scale in CE after 1955, although a population increase was noted in 1975, but in a much smaller scale than previously. European fruit lecanium is an important pest in orchards and on ornamental and forest trees, rarely on herbaceous plants.

Natural enemies. According to Saakyan-Baranova et al. (1971) there are many questionable literature records on the over 30 parasites reported for the brown scale, and only the records marked with an asterisk were recently verified. These are as follows: **Blastotrix confusa, Cerapterocerus mirabilis, Cheiloneurus claviger, C. paralia, *Coccophagus lycimnia, Eupelmus microzonus, E. urozonus, Eusemion cornigerum, Marietta picta, *Metaphycus insidiosus, *Microterys sylvius, Pachyneuron solitarium, Prospaltella lutea,* and **Trichomasthus albimanus.* Predators: *Adalia bipunctata, Anthribus nebulosus, Brachytarsus fasciatus, Chilocorus bipustulatus, C. renipustulatus, Chrysopa perla, Coccinella 10-punctata, C. septempunctata, Exochomus quadripustulatus, Hyperaspis campestris, Leucopis annulipes, L. nigricornis, L. silesiaca, Plastophora rufa, Sympherobius elegans* and the mite *Allothrombium fuliginosum.* Entomophagous fungi, *Cordiceps clavulata* (conidial form *Issaria lecaniicola, C. piscillariaeformis.* (Schmutterer, 1952b; Borchsenius, 1957; Canard, 1958; Kawecki, 1962a; Blahutiak, 1977).

Associated ants. *Dolichoderus quadripunctatus, Formica cinerea, F. fusca fusca, F. rufa pratensis, Lasius brunneus L. fuliginosus, L. emarginatus, L. niger niger, Myrmica laevinodis* (Schmutterer, 1952b).

125. **Parthenolecanium fletcheri** (Cockerell) – Arborvitae Soft Scale, Fletcher Scale
Fig. 9

Syn. & Lit.: *Lecanium fletcheri* Cockerell, 1893c:221; Kawecki, 1954:9; Dziedzicka, 1968:125; Williams & Kosztarab, 1972:82; Johnson & Lyon, 1976:80. *L. arion* Lindinger, 1911b:379. *Eulecanium arion*, Schmutterer, 1952b:549; 1952d:1; 1954:75. *P. fletcheri*, Borchsenius, 1957:370; Řehaček, 1960:55; Rogojanu, 1962:283; Tsalev, 1965:25; Dziedzicka, *1968:125; Schmutterer, 1972b:412; Zahradnik, 1977:119; Kosztarab & Kozár, 1978:116; Tereznikova, *1981:174.

Postreproductive female oval, somewhat extended at both extremities, strongly convex, almost semi-spherical, yellowish-brown to dark brown, 2–3. 5 mm long, 1.5–3 mm wide and 1.5–2.5 mm high; dorsal derm strongly sclerotized. Teneral female oval light brown, with a white longitudinal stripe on dorsum; 1.5–3.5 mm long, 1–3 mm wide. *Venter*. Antennae 6–8-segmented, 222–258 μ long; labium 95 μ long and wide (Koteja, 1974a); legs small in relation to body size; claw stout, with a small denticle; claw digitules longer than claw; spiracles larger than in related species, peritreme 42–50 μ in diameter; each spiracular furrow with a band of 23–36 mostly quinquelocular pores; 3 spiracular setae in each depression rarely 2, median seta 28–36 μ long, lateral setae ca 15 μ long; when only 2 setae, anterior longer; 60–80 marginal setae, ca 13 μ long, larger and thicker than submarginal setae of dorsum; numerous multilocular pores, (most with 9- or 10-loculi, rarely with 3-, 6-, 7-loculi), irregularly distributed in medial area of abdominal venter, few between legs on thorax; tubular ducts form a wide submarginal band. *Dorsum*. Without submarginal bicylindrical duct tubercles; 2 rows of submedial and 1 row of submarginal setae; a group of up to 10 discoidal pores anterior of anal plates; simple disc pores irregularly scattered on entire dorsum; dorsal ducts absent; each anal plate with 4 apical setae, 2 subapical setae; anal fold with 2 fringe setae; anal ring with 6 or 8 setae. First and second instars described by Schmutterer (1954), Řehaček (1960) and Dziedzicka (1968).

Distribution. In CE: Au, Bu, Cz, Fr, Ge, Hu, Po, Ro, Sw, USSR; also known from Denmark, England, Holland, Sweden, Canada and USA.

Biology. Live on shoots and underside of leaves of *Biota, Cupressus, Juniperus, Thuja* and *Tsuga*. Males unknown. One generation per year; N_2 overwinter; adult females developed by May in Slovakia (Řehaček, 1960). Females laid 43–1,346 eggs during second half of May and in June; eggs hatched in ca 40 days, last ones by mid-July in S Germany. Nymphs molted during end of August, through first half of September (Schmutterer, 1954).

Natural enemies. Parasites: *Asaphes vulgaris, Aphycus pulvinariae, Blasthotrix confusa, B. hedquisti, B. longipennis, B. sericea, Cheiloneurus albicornis, Coccophagus fletcheri, C. lycimnia, Encyrtus bicolor* and *Microterys flavus*. (Borchsenius, 1957; Dziedzicka, 1968; Williams and Kosztarab, 1972; Trjapitzin, 1978).

Note. A pest of *Thuja* spp. and *Taxus* spp. in North America. It is closely related to *P. pomeranicum*, which is common on *Taxus* in CE.

126. **Parthenolecanium persicae** (Fabricius) – Peach Scale, European Peach Scale
Figs 10, 98

Syn. & Lit.: *Chermes persicae* Fabricius, 1776:304. *Coccus persicorum* Sulzer, 1776:112. *C. costatus* Schrank, 1781:296. *C. clematis* Gmelin, 1789:2220. *Lecanium cymbiformis* Targioni-Tozzetti, 1868:730. *L. persicochilense* Targioni-Tozzetti, 1868:731. *L. berberidis* Signoret, 1873:403; *L. elongatum* Signoret, 1873:404. *L. genistae* Signoret, 1873:405. *L. mori* Signoret, 1873:407. *L. rosarum* Signoret, 1873:427. *L. sarothami* Douglas, 1891:65. *Eulecanium magnoliarum* var. *hortensiae* Cockerell 1903a:19. *E. cecconi* Leonardi, 1908:178. *E. persicae*, Leonardi, 1920:306. *L. persicae goidanichi* Kawecki, 1962:15. *Parthenolecanium thymi* Danzig, 1967:152. *L. persicae* Marchal, 1908:285; Green, 1922:16; Šulc, 1932:75; Ezzat & Hussein, 1967:408; Kawecki, 1971:258; Williams & Kosztarab, *1972:91. *Parthenolecanium persicae*, Rogojanu, 1963:286; Borchsenius, *1957:350; Boratynski, *1970:63; Kosztarab & Kozár, 1978:113; Kawai, 1980:162; Danzig, 1980a:272; Tereznikova, 1981:175.

Postreproductive females of various shapes and colors, most often elongate oval, dorsum with a longitudinal median dorsal keel, reddish brown. Teneral female oval, constricted at spiracular setae, yellow, streaked with light brown, 5–10 mm long, ca 3 mm wide. *Venter.* Antennae 8- or 9-segmented, 411–504 μ long; length of legs: pro- 606–758, meso- 676–832, and metathoracic 669–830 μ; claw with denticle near apex, claw digitules knobbed, wider than tarsal digitules; with tibio-tarsal sclerosis; anterior spiracular furrow with 57–80 quinquelocular pores in band 2–4 pores wide, posterior furrow with 66–76 pores; median spiracular seta (50–82 μ long) wider and usually longer than lateral setae (35–55 μ long); marginal setae hairlike, 39–62 μ long, slightly curved; submarginal setae in double row, setae in outer row 37–59 μ long, in inner row 15–17 μ long; body setae small, hairlike, scattered over derm; multilocular pores (10–12 locular) in transverse bands or rows on abdomen, in small group mesally of pro- and mesothoracic legs; four types of tubular ducts: (1) with heavily sclerotized rim and a slender filament anterior of antennae, (2) with broad inner filament, in a wide submarginal band on thorax and abdomen, (3) with filament expanded at tip, interspersed among second type in submarginal band, and scattered in median area, (4) a small heavily sclerotized duct with short slender filament scattered over entire derm. *Dorsum.* A row of 24–42 submarginal tubercles around body; ca 60 stout large conical setae in a longitudinal band in mid-dorsal area; smaller setae scattered over derm; bilocular pores and simple small disc pores distributed over entire derm; ca 24 large discoidal pores in median area, anterior of anal plates; anal plates triangular, 160–188 μ long, 78–137 μ wide, each with 4 apical and 3–4 subapical setae; anal fold with 4 fringe setae; anal ring horseshoe-shaped, with 6 large setae ca 248 μ long, 2 smaller setae and 2 irregular rows of translucent pores. Adult male and 2nd instar described by Borchsenius (1957), male test by Šulc (1932); female nymphs by Boratynski (1970).

Distribution. Almost world-wide. In CE: Au, Bu, Fr, Hu, It, Ro, Sw, USSR and Yu. Distribution in Cz uncertain (Zahradnik, 1977).

Biology. A polyphagous pest species, known from: *Albizzia, Amygdalus, Armeniaca, Berberis, Citrus, Clematis, Cydonia, Daphne, Diospyros, Elaeagnus, Ficus, Fraxinus, Gleditsia, Juglans, Maclura, Mallotus, Malus, Mespilus, Morus, Parthenocissus, Persica, Prunus, Punica, Pyrus, Ribes, Robinia, Rosa, Sophora, Tamarix, Vitis* and *Wistaria. Berberis, Clematis,* and *Vitis* are most commonly infested in CE. Nymphs feed on underside of leaves; overwinter on branches in N_2 stage; adults develop by end of May and each female lays 1000–2600 eggs. One generation per year in CE, but

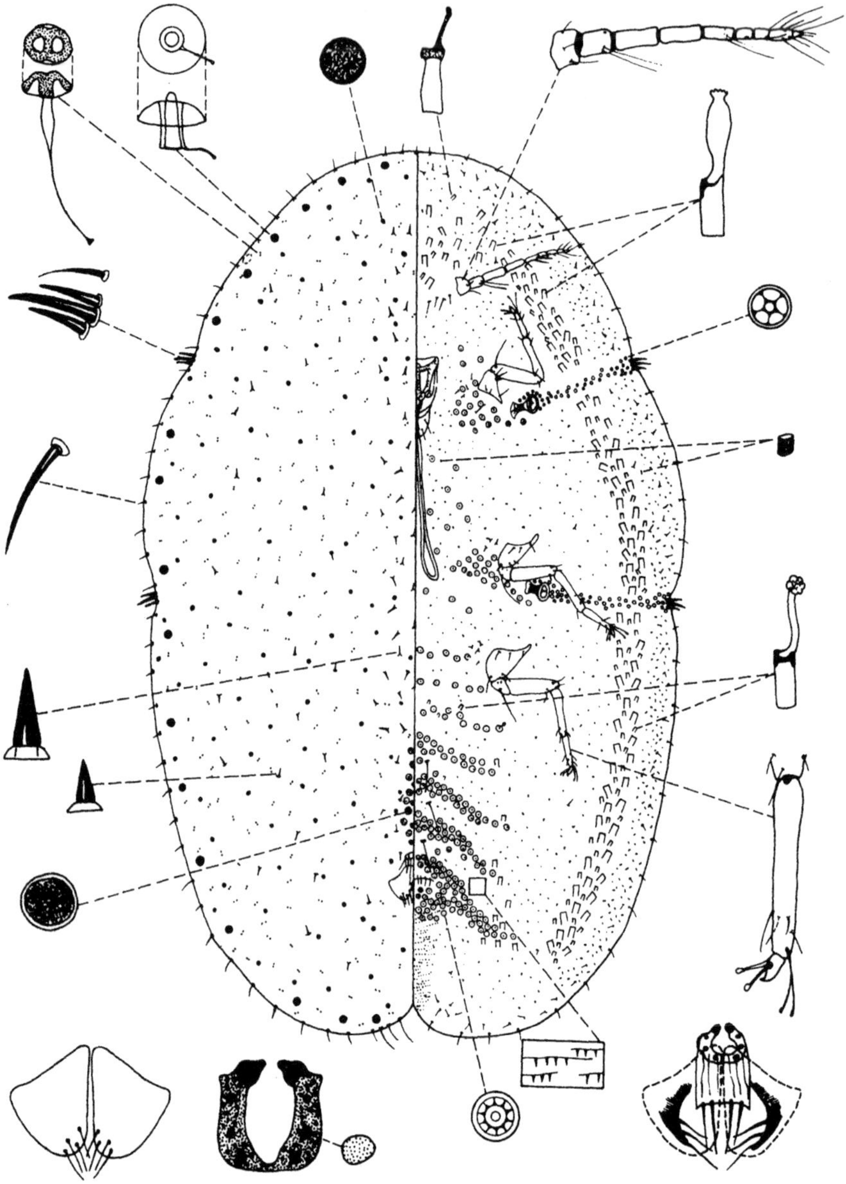

Fig. 98. *Parthenolecanium persicae* (Fabricius), female (modified after Williams and Kosztarab, 1972)

Borchsenius (1957) reported two generations from Central Asia. Reproduction is parthenogenetic and/or biparental (Williams and Kosztarab, 1972).

Natural enemies. Parasites: *Aphycus timberlakei, Blastothrix hungarica, Cheiloneurus formosus, Coccophagus lycimnia, C. pulchellus, C. scutellaris, Eunotus cretaceus, E. obscurus, Homalotylus quaylei, Metaphycus dispar, M. parvus, M. punctipes,*

Microterys sylvius, M. temporaris, M. xanthopsis, Muscideopsis lecanii, Phaenodiscus aeneus, Trichomasthus albimanus (Trjapitzin, 1978). Predators: *Chilocorus bipustulatus, Leucopis alticeps* (Babaev, 1970); *Holcocera iceryaella*, (Borchsenius, 1957).

127. **Parthenolecanium pomeranicum** (Kawecki) – Yew Scale

Syn. & Lit.: *Lecanium pomeranicum* Kawecki, 1954:9; Dziedzicka, *1968:152. *Eulecanium arion*, Goux, 1945:37. *E. crudum*, Schmutterer, 1954:78. *E. taxi* Habib, 1956b:453. *Parthenolecanium pomeranicum*, Borchsenius, 1957:372; Řehaček, 1960:56; Boratynski & Williams, 1964:107; Giliomee, 1967:113; Kosztarab & Kozár, 1978:116; Tereznikova, 1981:178. *P. crudum*, Schmutterer, 1972b:411.

Postreproductive female broadly oval, slightly tapering toward both ends, convex, similar to *P. corni*, but dorsal keel not obvious, shiny dark brown, 3.8–5 mm long. Teneral female elongate-oval, yellowish with 3 anterior reddish transverse bands well-developed, body somewhat transparent. *Venter.* Antennae 7- or 8-segmented, ca 275 μ long; labium ca 100 μ long and wide (Koteja, 1974a); legs and claws slender, claws without denticle, claw digitules of different thickness; peritreme of hind spiracles ca 55 μ in diameter; each spiracular furrow with 30–35 quinquelocular disc pores; 3 spiracular setae with blunt tip, median ca 50 μ, laterals ca 27 μ long; marginal setae in 1 row, sparse, ca 60, far apart, 17–20 μ long; multilocular pores with 10 loculi, fewer than in *P. corni*, form transverse bands in median area of abdomen; large tubular ducts in a submarginal band. *Dorsum.* Submarginal bicylindrical duct tubercles absent; a longitudinal double row of large, conical setae in median area, other long setae sparsely scattered on entire dorsum; up to 15 discoidal pores in a medial longitudinal group anterior to anal plates; anal plates elongate triangular, each with a group of apical setae. Adult male described by Theron (1958); Giliomee (1967); first and second instars by Kawecki (1954) and Dziedzicka (1968).

Distribution. In CE: Au, Bu, Cz, Fr, Ge, Hu, Po, Ro, Sw, USSR; also in Denmark, England, Holland, Sweden and Egypt.

Biology. Lives on *Taxus* spp. only. One generation per year; N_2 overwinter; adults developed by May; each female lays up to 3000 eggs from end of May through July; nymphs settle on leaves in July, molt into second instars during end of August and in September, then migrate to branches (Schmutterer, 1954).

Natural enemies. Parasites: *Aphycus insidiosus, A. parvus*, and *Coccophagus lycimnia* (Dziedzicka, 1968).

128. **Parthenolecanium rufulum** (Cockerell) – Oak Soft Scale
Fig. 11

Syn. & Lit.: *Eulecanium alni* var. *rufulum* Cockerell, 1903a:21. *Lecanium pulchrum* King, Reh 1903:410; Marchal 1908:304; Schimitschek 1935:217. *Eulecanium pulchrum*, Schmutterer 1952b:548, *1954:72. *Lecanium rufulum*, Dziedzicka, 1968:157; Kawecki, 1971:257. *Parthenolecanium rufulum*, Borchsenius, 1957:373; Řehaček, 1960:55; Boratynski & Williams, *1964:105; Schmutterer, 1972b:412; Kosztarab & Kozár, 1978:116; Danzig, 1980b:594; Tereznikova, 1981:179. *P. wistariae* (Signoret), Ben-Dov, 1975a:115.

Postreproductive female oval, strongly sclerotized, dorsum with an incomplete median carina; highest point of carina at 2/3 of dorsal length; dull chestnut-brown to reddish-brown; up to 6 mm long, 4 mm wide. At oviposition yellowish-brown to chestnut-brown. Teneral female. *Venter*. Antennae 7-segmented, ca 300 μ long; labium globular, 85 μ long, 95 μ wide (Koteja 1974a); stylet reaches halfway between front of median legs; legs slender, 380–400 μ long, excluding coxa; claw with minute denticle, 2 claw digitules of equal length; spiracles large, anterior smaller than posterior ones; each anterior spiracular furrow with 18–43 quinquelocular pores (rarely with 6–8-loculi), posterior furrow with 30–50 pores; 3 spiracular setae, blunt-tipped, median slightly curved, 45–57 μ long, laterals 30–42 μ long; marginal setae stout, acute 14–21 μ long, form a partial double row; 2 pairs of interantennal setae 35–60 μ long; 3 pairs of prevulvar setae, each 70–80 μ long; other body setae scattered, 6–10 μ long; multilocular pores, with 8–11 loculi form transverse bands on abdominal sternites, also in small groups at base of each coxa; dark-rimmed pores interspersed among large tubular ducts, ca 15 μ long, that form a wide submarginal band on most of venter, except on two ends of body. *Dorsum*. Submarginal tubercles absent; dorsal setae of 2 types: thick, 10–14 μ long setae form a longitudinal double row in median area; smaller, 5–6 μ long setae, form 3–4 irregular longitudinal rows on rest of dorsum; simple disc pores and small bilocular pores interspersed among numerous small tubular ducts, ca 10 μ long, on entire dorsum; large discoidal pores, diameter 5–10 μ, arranged in an elongate cluster anterior of anal plates; anal plates triangular, each with 4 apical setae; anal ring with 6 setae, each ca 206 μ long, often with 1 or 2 smaller additional setae, ca 84 μ long (Borchsenius 1957; Boratynski and Williams, 1964). First and second instars described by Schmutterer (1954) and Dziedzicka (1968); first instar also illustrated by Řehaček (1960).

Distribution. A common species. In CE: Au, Bu, Cz, Fr, Ge, Hu, It, Po, Ro, USSR and Yu; also in England and Turkey.

Biology. Live primarily on *Quercus*, but also reported from *Betula, Carpinus, Castanea, Corylus, Diospyros, Fagus, Morus, Robinia, Rosa, Rubus, Sarothamus, Taxus, Thea, Thuja, Ulex, Ulmus* and *Vaccinium*. Some of the host records probably resulted from misidentifications of this insect. Parthenogenetic species with 1 generation per year; overwinter as N_2 and according to Schmutterer (1952b), each female lays 58–1208 eggs on oaks from end of May through end of June; eggs hatch in about 40 days, starting in mid-July; N_1 moves to underside of leaves for feeding, molts from end of August through mid-September; N_2 moves from leaves to woody parts of the tree from mid-October through end of November in S Germany.

Natural enemies. Parasites: *Blasthotrix confusa, Coccophagus lycimnia, Metaphycus insidiosus, Microterys sylvius* (Trjapitzin, 1978). Predators: *Anthribus nebulosus, Chrysopa aspersa* (Schmutterer, 1972b). Fungi, *Cephalosporium lecanii* (Borchsenius, 1957).

Associated ants. *Camponotus ligniperda, Formica cinerea F. fusca gagates, F. rufa pratensis, F. rufa rufo pratensis major, Lasius fuliginosus, L. niger niger*, and *Myrmica laevinodis* (Schmutterer, 1952b).

129. **Parthenolecanium smreczynskii** Kawecki – Thyme Soft Scale

Syn. & Lit.: *Lecanium smreczynskii* Kawecki, 1967:687, 1971:258.

Postreproductive female, according to Kawecki, (1967) oval, somewhat pointed at both ends, light brown, darker brown dorsally, ca 3 mm long, 1.8 mm wide, and 1.3 mm high. Parasitized females somewhat longer, up to 4 mm and more lustrous. Teneral female oval. *Venter.* Antennae 7-segmented, third and fourth segments each ca 50 μ long; anterior and posterior spiracular setae well-developed, median longest, 55–60 μ long; marginal setae on head ca 25 μ long, spaced at intervals of 45–50 μ; submarginal setae small, 15–18 μ long; 6 interantennal setae, pair nearest antennae shortest, ca 7 μ, next pair ca 36–48 μ, mesal pair 60–75 μ long; numerous ventral tubular ducts on submarginal area, 15–25 μ long; genital region with numerous multilocular pores, most with 10 loculi, each ca 8 μ in diameter. *Dorsum.* A row of 7 pairs of submarginal tubercles, each ca 25 μ in diameter; dorsal setae 12–16 μ long; anal ring with 8 setae each 80–90 μ long; anal plates triangular, each with a small group of apical setae. First instar partly described and distinguished from *P. corni* by Kawecki (1967). Since there were few distinguishing characters given for the adult female, we are summarizing here the morphological characteristics given for the first instars by Kawecki (1967) that could more readily separate this species from the closely related *P. corni* (see Table).

Distinguishing characters of First Instars	*P. corni*	*P. smreczynskii*
Length of body	320–350 μ	ca 500 μ
Disc pores in each spiracular furrow	3	usually 4, rarely 3
Length of frontal marginal setae	10–12μ	16–20 μ
Dorsal bilocular pores	present	absent
Trilocular pore size	very small, visible with oil immersion only	fairly large, diam. ca 2.4 μ
Stylet loop reaches anal ring	yes	no

Distribution. Known only from Zakopane area in S Poland.
Biology. Feed on *Thymus,* probably *T. pulegioides.* Biology and males unknown.
Note. The species is in need of a detailed redescription and illustration to clarify its status.

Genus **Phyllostroma** Šulc

Syn. & Lit.: *Phyllostroma* Šulc, 1942:5; Schmutterer, 1952b:554; Kawecki, 1957:197; Borchsenius, 1957:280; Řehaček, 1960:41; Morrison & Morrison, 1966:155; Danzig, 1967:831; Kosztarab & Kozár, 1978:116.

Type-species: (*Pulvinaria ericae* Löw, 1883) = *Lecanium myrtilli* Kaltenbach, 1874. A monotypic genus; for generic description see *P. myrtilli.*

Postreproductive and teneral females as in *P. myrtilli.* It is closely related to *Rhizopulvinaria.*

Distribution and biology. As in *P. myrtilli.*

130. **Phyllostroma myrtilli** (Kaltenbach) – Heather Cushion Scale
Fig. 99

Syn. & Lit.: *Lecanium myrtilli* Kaltenbach, 1874:420. *Pulvinaria ericae* Löw, 1883c:115. *Phyllostroma myrtilli,* Schmutterer, 1952b:554, 1955:160; 1972:413; Kawecki, 1957:197; Řehaček, 1954a:138, 1960:42; Borchsenius, 1957:281; Danzig, 1959a:399; Koteja, 1974a:109; Kosztarab & Kozár, 1978:117; Tereznikova, *1981:136.

Ovisac slightly convex, oval, bluish-white, ca 4.5 mm long, 2.5 mm wide. Postreproductive female short-oval, moderately convex, yellow to greenish or brownish-yellow; with dark longitudinal bands before oviposition. Teneral female oval, up to 3.2 mm long. *Venter.* Antennae 7–8, rarely 6 or 9-segmented, ca 344 μ long; stylet loop reaches line between mid-coxae; legs thick, tibio-tarsal articulation free and sclerosis present, claw stout and curved, claw digitules thick, strongly widened at tip; each spiracular furrow with a row or band of 32 to 41 6-locular disc pores; 3 spiracular setae in each group, thick with blunt tip, sometimes slightly curved, median seta ca 1 ½–2 times longer than lateral setae; marginal setae shorter than lateral spiracular setae, acute, distance between marginal setae ca 2 times their length; body setae stout, of various lengths; 10 interantennal setae, 4 much longer than others, 40–50 μ long; prevulvar setae 3, 60–95 μ long; multilocular pores with 6, 7, or 8 loculi, form transverse rows on abdominal sternites 2 to 7; large tubular ducts, ca 20 μ long, 5–6 μ wide, form a wide submarginal band, and interspersed with multilocular pores, and scattered in median area of thorax. *Dorsum.* Body setae small, stout, conical, 5–7 μ long. Anal ring with 10 setae, 4 shorter and more slender than others; large tubular ducts in a submarginal band and scattered on rest of dorsum; small tubular ducts, ca 7 μ long, 2.5 μ wide, in 2 elongate medial groups anterior of anal plates, few also interspersed with large ducts in submarginal band. Adult male described by Giliomee (1967); male test by Borchsenius (1957) and Řehaček (1960); N_1 by Řehaček, (1960).

Distribution: A rare boreal western Palearctic species. In CE: Au, Cz, Ge, Hu, It, Po, Ro and USSR (Ukraine to Leningrad Region).

Biology. Oligophagous, known from leaves and stems of Ericaceae: *Arctostaphylos, Calluna, Erica,* and *Vaccinium.* According to Schmutterer (1952b) in S Germany N_2 overwintered; adults developed by mid-May, but in Czechoslovakia Řehaček (1960) found adults in April; in S Germany each female laid 127–697 eggs in ovisac on underside of leaves from early June which hatched within 40–45 days.

Natural enemies. Parasites: *Aphycus punctipennis,* and *Trichomasthus albimanus.* Predator: *Hyperaspis campestris* var. *concolor* (Borchsenius, 1957; Zak-Ogaza, 1961; Trjapitzin, 1978).

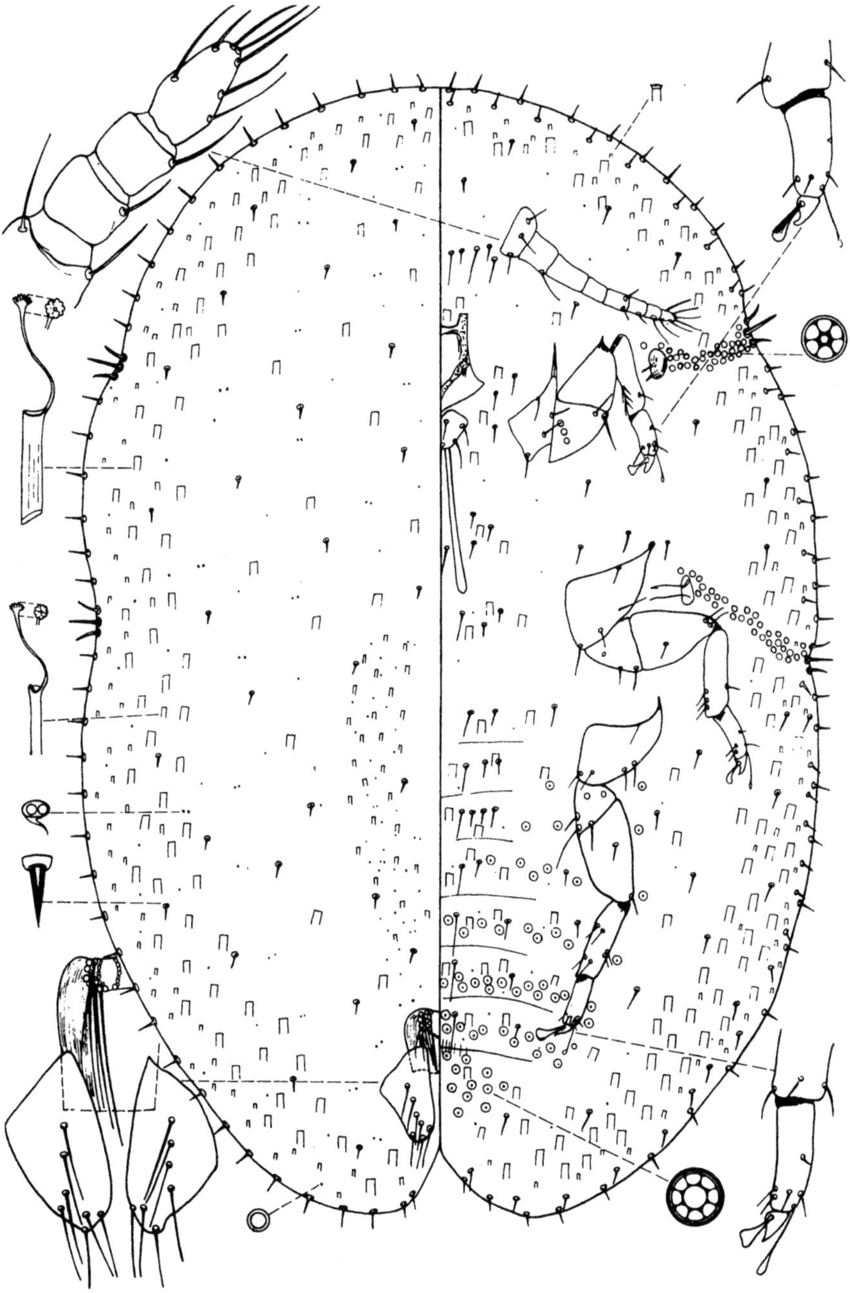

Fig. 99. *Phylostroma myrtilli* (Kaltenbach), female (original)

Genus **Physokermes** Targioni-Tozzetti

Syn. & Lit.: *Physokermes* Targioni-Tozzetti, 1868:41 (separate, 1869:734); Signoret, 1874:87(279); Leonardi, 1920:316; Steinweden, 1929:240; Silvestri 1939:741; Schmutterer, *1956c:454, 1965:300; Borchsenius, 1957:439; Řehaček, 1960:64; Morrison & Morrison, 1966:156; Danzig 1967:169, 1980a:290; Williams & Kosztarab, 1972:105; Kosztarab & Kozár, 1978:117; Tereznikova, 1981:182.

Type-species: *Coccus hemicryphus* Dalman, 1925.

Postreproductive females nearly spherical, budlike, or kidney-shaped, often higher than wide, shiny brown, heavily sclerotized, diameter 3.5–6.5 mm. Teneral female. *Venter.* Antennae reduced, rarely divided, often tubercle-like; legs reduced, often to 2–3-segmented stubs, or absent; spiracular setae absent; disc pores with 5 or more loculi, form groups and transverse bands; rarely some discoidal pores also present; tubular ducts normally in submarginal area. *Dorsum.* Dark-rimmed discoidal pores and tubular ducts scattered on most of dorsum; anal ring small, narrow, without pores, with 2–6 very short setae; anal plates absent, a unique feature of this genus; small body setae distributed on both body surfaces.

Distribution. Holarctic with 9 species; 3 in CE.

Biology. Only on conifers: *Picea* spp., rarely on *Abies* spp. One generation per year; second instars overwinter. Females with 3, males with 5 developmental stages. Produce honeydew and tended by ants.

Note. The genus is in need of a revision.

Key to species of *Physokermes*

1. Dorsum of each anal lobe in teneral females with a wide longitudinal band of long setae; antennae 2-segmented; first instars with 2 stout spiracular setae in each group (fig. 101) .... *inopinatus*, p. 232
   Dorsum of anal lobe in teneral females with a small group of long setae; antennae unsegmented; first instars without spiracular setae 2
2. Teneral females with 41–58 disc pores in each anterior and 31–43 in each posterior spiracular pore band; antennae 40–50 μ, legs ca 70 μ long. Diameter of postreproductive females up to 8 mm. Second instar female with 70 or more thick rimmed 6-locular disc pores on both surfaces of body margin; 10–20 long tubular ducts on vertex and posterior of eyes....... .... *piceae*, p. 234
   Teneral females with 9–25 disc pores in each anterior and 2–10 in each posterior spiracular pore band; antennae 29–36 μ, legs ca 50 μ long. Diameter of postreproductive females up to 4.5 mm. Second instar female with 0–15 thick-rimmed 6-locular disc pores on both surfaces of body margin; 2–6 long tubular ducts on vertex and posterior of eyes. .... *hemicryphus*, p. 231

131. **Physokermes hemicryphus** (Dalman) – Small Spruce-bud Scale
Fig. 100

Syn. & Lit.: *Coccus hemicryphus* Dalman, 1826:369. *Chermes abietis rotundus* Geoffroy, 1762:507. *Coccus racemosus* Ratzeburg, 1843:204. *Physokermes abietis*, Leonardi, 1920:316. *Physokermes piceae*, Borchsenius, 1957:440; Schmutterer, 1952b:553. *P. hemicryphus*, Schmutterer, *1956c:445, 1965:300, 1972b:413; Řehaček, 1960:65; Williams & Kosztarab, 1972:105; Halmágyi, 1974:454; Kosztarab & Kozár, 1978:120; Tereznikova, 1981:182.

Postreproductive female budlike, shiny, dark brown, up to 4.5 mm long, 4 mm wide. Teneral female (according to Schmutterer 1956c), after last molt ca 1 mm long, 0.8 mm wide. *Venter*. Antennae 29–36 μ, legs ca 50 μ long, both reduced, unsegmented, with many setae; labium as in *P. piceae*; spiracular pore bands with multilocular pores with 10 loculi each; 9–25 in each anterior, 2–10 in each posterior band; few body setae; multilocular pores in median area of thorax and abdomen less numerous than in *P. piceae*; distribution of discoidal pores and of tapering cylindrical ducts similar to *P. piceae*, but discoidal pores larger and more numerous here; microducts generally absent from head, but rarely a few present in the submarginal area. *Dorsum*. Thick-rimmed 6-locular disc pores always absent; pore and duct types and their distribution

Fig. 100. *Physokermes hemicryphus* (Dalman), postreproductive female on twig of *Picea* sp. (Photo M. Kosztarab)

as in *P. piceae,* but fewer; anal ring, anal tube, and dorsal setal pattern as in *P. piceae*; anal lobes smaller than in latter. Adult male, first and second instars described by Schmutterer (1956c); first instar (under *abietis*) by Leonardi (1920); first and second instars (under *piceae*) by Borchsenius (1957).

Distribution. Probably generally distributed in Europe. Records were recently verified in CE for: Au, Bu, Cz, Ge, Hu, Po, Ro, USSR and Yu. Also known from USA.

Biology. On *Picea* spp., rarely on *Abies alba* and *A. pectinata.* One generation per year. Females develop under bud scales; males develop at base and underside of needles. Second instars overwinter, in S Germany (Schmutterer, 1956c) adults developed by second half of May; females laid (38–335) eggs from early June through early July; eggs hatched in 5–6 weeks, from second half of July through first half of August; most N_1 molted during second half of September. Ants listed under *P. piceae* may also apply to this species.

Natural enemies. Parasites: *Aphycoides clavellatus, A. merceti, Coccophagus insidiator, Euaphycus* sp., *Metaphycus picearum, Microterys cyanocephalus* var. *obscuricornis, M. fuscipennis, M. lunatus,* and the most effective species, *Pseudorhopus testaceus,* (Schmutterer, 1965, 1972b; Trjapitzin, 1978). Predators: Same as at *P. piceae* (Schmutterer, 1972b).

Note. Often becomes a pest, especially on physiologically weakened ornamental trees. Schmutterer (1972b) points out their beneficial role with the mass production of honeydew, especially during May and June, which the bees collect and process into honey.

132. **Physokermes inopinatus** Danzig & Kozár – Hungarian Spruce Scale
Figs 12, 101

Syn. & Lit.: *Physokermes inopinatus* Danzig & Kozár, 1973:832; Kosztarab & Kozár, 1978:119; Tereznikova, 1981:184.

Postreproductive female kidney-shaped, dorsal carina indistinct; shiny brown, diameter 5–8 mm, under crowded conditions only 3 mm. Teneral female pinkish-yellow. *Venter.* Antennae 2-segmented; legs reduced, rarely tibia and tarsus separated; claw digitules much longer than claw; spiracular pore band wide with multilocular pores, most with 10 loculi; few quinquelocular pores near spiracles; numerous multilocular pores on medial region of abdomen and area between legs and spiracles; numerous dark-rimmed discoidal pores and large tubular ducts on entire body surface; body setae scattered on both body surface; long apical setae on anal lobes. *Dorsum.* Dark-rimmed discoidal pores numerous on anal lobe, also scattered on other parts of dorsum; small tubular ducts scattered on prothorax and head; area around anal ring sclerotized; in teneral females each anal lobe with a wide longitudinal band of long setae along cleft. First instar with 2 spiracular setae in each group; other morphological characters given by Danzig and Kozár (1973) and Tereznikova (1981).

Distribution. Known only from Hungary.

Biology. Only on *Picea abies* and *P. glauca.* Often produced high populations and damage. One generation per year; reproduction biparental; adults develop by April in Hungary; males develop on needles, females at base of twig forkings and base of needles; lay eggs, often several thousand, from end of May; first instars appear from end of June, beginning July, and crawl to needles; at high population levels, more males are produced than females.

Natural enemies. Parasites: *Aphycoides clavellatus*, *Coccophagus lycimnia*, *C. scutellaris*, *Microterys lunatus* (Kozár and Sugonjaev, 1979). Predators: *Anthribus nebulosus*, *Chilocorus bipustulatus*, *Forficula auricularia*, *Scymnus* sp. Parasites were only moderately effective in natural control, while among the predators, *Anthribus nebulosus* provided over 50% control. *F. auricularia* destroyed many females and occurred at high population levels on the scale infested trees (Kozár, unpublished). Kosztarab introduced *A. nebulosus* from Hungary into USA for the control of other Coccidae species in 1975, and field-released it in 1978.

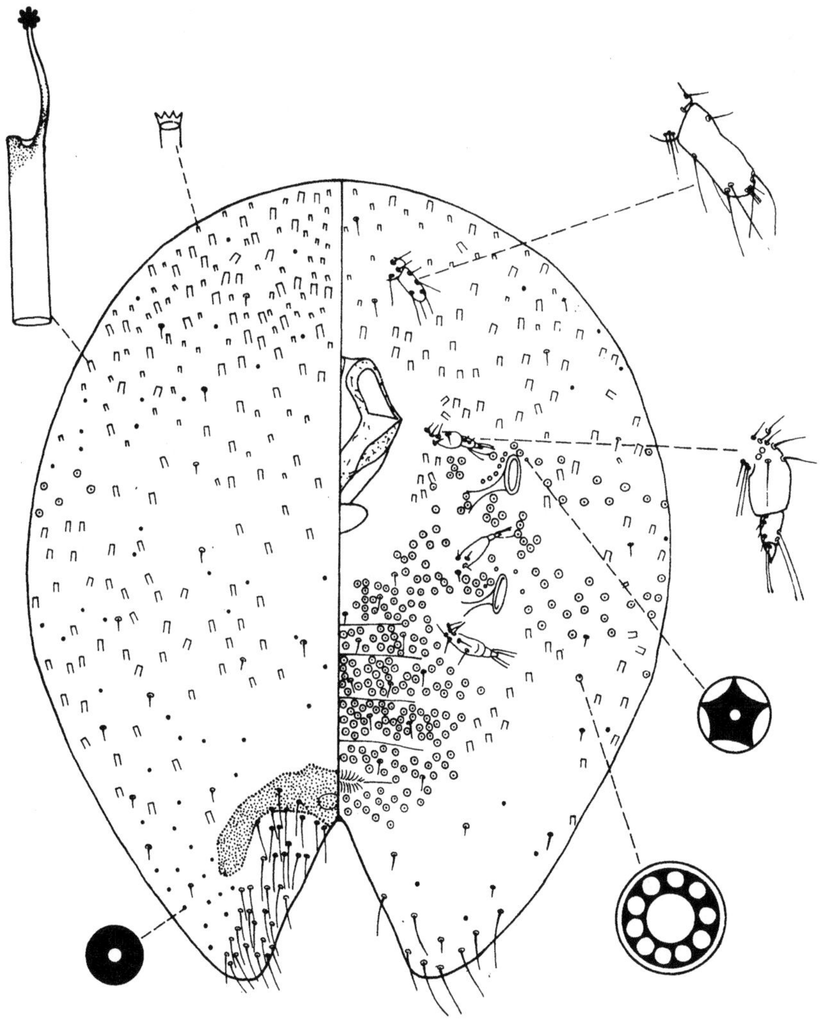

Fig. 101. *Physokermes inopinatus* Danzig and Kozár, adult female, (modified after Danzig and Kozár, 1973)

133. **Physokermes piceae** (Schrank) – Spruce-Bud Scale

Syn. & Lit.: *Coccus piceae* Schrank, 1801:146. *C. racemosus* Ratzeburg, 1843:204. *Physokermes hemicryphus*, Signoret, 1874:88(280). *Physokermes piceae*, Schmutterer, *1956c:445, 1965:300; 1972b:416; Danzig, 1959a:401; Řehaček, 1960:65; Săvescu, 1961:38; Rogojanu, 1962:285; Halmágyi, 1974:454; Kosztarab & Kozár, 1978:119; Tereznikova, 1981:180. *P. latipes* Borchsenius, 1957:445.

Postreproductive female budlike or kidney-shaped, yellowish-brown, up to 8 mm long and 7.5 mm wide. Teneral female (according to Schmutterer, 1956c) more or less oval, reddish, ca 1.2 mm long, 1 mm wide. *Venter*. Antennae 40–50 μ, legs ca 70 μ long, both reduced, unsegmented, deformed; antennae with many long setae; labium 2-segmented, 95 μ long, 109 μ wide, with 5 pairs of setae (Koteja, 1974a); disc pores in spiracular pore bands with 5–7 loculi, 41–58 pores each in anterior and 31–43 in each posterior band; only few body setae, medially from each antenna 1 longer and 1 shorter interantennal seta; numerous multilocular pores, normally with 10-loculi, in median area of venter from labium to vulva; discoidal pores in a submarginal band; numerous microducts on head and thorax forward from anterior spiracular furrows; large tapering cylindrical ducts on anterior 2/3 of body, absent from posterior third. *Dorsum*. Thick-rimmed 6-locular disc pores often present in submarginal area of anterior end of both surfaces; few discoidal pores in submarginal area of posterior half of body; numerous invaginated, wide tubular ducts on dorsum, except fewer on posterior abdominal segments; anal ring at apex of short anal tube; with 2 small setae (6 in N American specimens); anal area heavily sclerotized; anal lobes padlike, each with many long terminal setae. Adult male described by Jancke (1955) and Giliomee (1967); male test by Řehaček 1960; first and second instars by Schmutterer (1956c).

Distribution. A common species. In CE: Au, Bu, Cz, Fr, Ge, Hu, It, Po, Ro, Sw, USSR and Yu. Also known from W, N, and S Europe and N America.

Biology. Occurs on *Picea* spp. only. Females develop under bud scales and on twigs, males on underside of needles. One generation per year. Reproduction biparental. Second instars overwinter; in S Germany (Schmutterer, 1956c) adults developed during April, latest during first half of May; each female laid 672–1987 eggs from mid-April through first third of June; eggs hatched within 5–6 weeks, from end of June through first third of July; N_1 molted from mid-August through mid-September. An occasional pest on ornamental spruce.

Natural enemies. Parasites: *Aphycoides clavellatus, A. merceti, Cheiloneurus formosus, Coccophagus insidiator, C. lycimnia, Euaphycus* sp., *Eusemion cornigerum, Metaphycus picearum, M. stagnarum, Microterys fuscipennis, M. lunatus.* Latter species was the most common in Germany and USSR (Schmutterer, 1965; 1972b; Trjapitzin, 1978). Predators: *Anthribus nebulosus, Driobates major, Exochomus quadripustulatus, Sciurus vulgaris,* and *Scymnus abietis* (Schmutterer, 1965).

Associated ants. Because of the large quantities of honeydew produced, ants are common visitors: *Camponotus ligniperda, Formica exsecta* var. *pressilabris, F. fusca gagates, F. f. fusca, F. rufa pratensis, F. r. rufa, F. r. rufopratensis major, F. r. r. minor, F. sanguinea, Lasius brunneus, L. emarginatus, L. fuliginosus, L. niger alienus, L. n. niger, Myrmica laevinodis, M. ruginodis* (Schmutterer, 1952b, 1965).

Note. Old records on natural enemies and ants could apply to *P. hemicryphus* as well, since these two species were only recently clearly separated by Schmutterer (1956c).

Genus **Poaspis** Koteja

Syn. & Lit.: *Poaspis* Koteja, 1978:320, *1979:451.

Type-species: *Luzulaspis jahandiezi* Balachowsky, 1932.

Postreproductive female elongate-oval, slightly convex. Teneral female flattened dorso-ventrally, derm slightly sclerotized. *Venter.* Antennae 8-segmented; legs well-developed, tibia and tarsus freely articulated and with sclerosis, tarsal digitules slender; spiracular setae 0–4 in each group, when present thick, slightly curved; spiracular pore bands normally with quinquelocular pores; marginal setae spinelike with sharp apex or hairlike, distance between setae shorter than their length; ventral body setae long and numerous; multilocular pores, with 8–10 loculi, in transverse bands on posterior abdominal sternites; tubular ducts up to 20 μ long and 6 μ in diameter, fewer than dorsum. *Dorsum.* Setae spinelike, sharp-pointed, numerous on entire surface; discoidal pores in median longitudinal band on thorax to anal plates; various types of tubular ducts scattered on entire dorsum.

Distribution. Trans-Palearctic, with 7 species; 1 in CE.

Biology. On various grasses, also on *Luzula* (?). One generation per year; eggs overwinter.

Note. Related to *Exaeretopus* and *Luzulaspis*.

134. **Poaspis jahandiezi** (Balachowsky) – Balachowsky's Soft Scale

Syn. & Lit.: *Luzulaspis jahandiezi* Balachowsky, 1932c:197; Borchsenius, 1957:112; Řehaček, 1957:16; 1960:26; Tsalev, 1968:208; Koteja, 1974a:110; Danzig, 1975c:137; Kosztarab & Kozár, 1978:107. *Poaspis jahandiezi*, Koteja, 1978:321, 1979:463. *Exaeretopus hellenicus* "Green," Bodenheimer 1928:192 (nomen nudum).

Ovisac felted, white, entirely covers female. Postreproductive female elongate-oval, 8–9 mm long, 3–4 mm wide; live female red, of pomegranate color (Balachowsky, 1932b). Teneral female, according to Koteja (1979), 4–5.7 mm long, 1.6–2.5 mm wide. *Venter.* Antennae 8-segmented, 460–560 μ long, 3rd segment 120–150 μ long; labium 70–85 μ long, ca 110 μ wide; legs well-developed, 1.2–1.5 mm long, claw curved without denticle, claw digitules thick, robust, with large apical knob; disc pores in spiracular pore bands with 5 or 6 loculi, each anterior with 19–45, posterior with 15–49 pores; 2 spiracular setae in each group, stout and often bent at tip, shorter than marginal setae; marginal setae spinelike, sharply pointed 20–90 μ long, arranged in 2 irregular rows, distance between setae less than their length; body setae of various lengths, 35–70 interantennal setae up to 198 μ long, others 24–100 μ; multilocular pores with 6–10 loculi, on abdominal sternites 5–8; tubular ducts of variable size. *Dorsum.* Setae straight or bent, with sharp or blunt apices, numerous, 15–45 μ long; discoidal pores 3–4.5 μ in diameter in a medial longitudinal band from prothorax to anal plate; tubular ducts numerous, of variable size and shape; anal plates triangular, with pointed lateral angle; anal ring ca 70 μ in diameter, with 8 setae, each ca 170 μ long;

Distribution. Western Palearctic; in CE: Bu, Cz, Fr, Hu; also in Greece and Cyprus. A rare thermophilous species.

Biology. On grasses: *Agropyron, Agrostis, Andropogon, Calamagrostis, Phragmites, Piptaterum* and other undetermined grass species. Řehaček (1960) in Slovakia found one generation per year; eggs overwintered; adults developed by July. Biparental reproduction.

Genus **Psilococcus** Borchsenius

Syn. & Lit.: *Psilococcus* Borchsenius, 1952:269; 1957:113. Danzig, 1959a:398, 1967:826, 1980a:255; Koteja, 1969a:33, 1978:313; Kozár & Danzig, 1976:67; Kosztarab & Kozár, 1978:120; Tereznikova, 1981:124.

Type-species: *Psilococcus ruber* Borchsenius, 1952.

A monotypic genus, for generic description see *P. ruber*.

Note. Borchsenius reported that based on the adult females, *Psilococcus* is related to *Lecanopsis* from which it can be separated by the broad submarginal band of quinquelocular pores on venter which is absent in *Lecanopsis*. Koteja (1969) found this genus to be related to the *Eriopeltis* group based on the adult males.

135. **Psilococcus ruber** Borchsenius – Red Sedge Scale
Fig. 102

Syn. & Lit.: *Psilococcus ruber* Borchsenius, 1952:270; Danzig, 1980a:255. *P. parvus* Borchsenius, 1957:115; Danzig, 1959:398; Koteja, *1969a:24, 1974a:111; Kosztarab & Kozár, 1978:121; Tereznikova, 1981:124.

Ovisac small, covers eggs only. Postreproductive female elongate, flattened dorsoventrally; after oviposition light brown. *Teneral female.* According to Koteja (1969a) body narrowly-elongate; live specimen yellowish-red; 3.5–6.4 mm long, 1–1.4 mm wide. *Venter.* Antennae 6-, rarely 5- or 4-segmented, 85–124 μ long, base of antennae halfway between anterior end of body and anterior coxae; labium semi-globular, cup-shaped, 76 μ long and 95 μ wide; stylet loop 155–186 μ long; legs short, often reduced, fore legs 252–285 μ, median 260–282, hind 255–302 μ long; claw digitules dissimilar, one stout, other slender; anterior spiracular peritreme 29–37 μ, posterior 44–46 μ in diameter; quinquelocular pores form a group at anterior part of each depression of spiracles; spiracular pore bands wide but short, pores with 5, 3, 4 or 6 loculi; spiracular setae absent; marginal setae needle-like, distance between setae often much more than their length; body setae of various lengths; minute simple disc pores on head, in submarginal band, and on abdominal sternites in transverse groups or bands; dark-rimmed disc pores scattered on entire venter, but more common anterior of antennae; multilocular pores with 7–11 loculi on posterior abdominal sternites; other disc pores with 5-, rarely 3, 4 or 6 loculi in a narrow submarginal band; tubular ducts on posterior abdominal segments numerous, extending in submarginal band to anterior spiracle; microducts numerous on head, in clusters by mouthparts, anterior spiracles, thoracic legs and few scattered on thorax and abdomen. *Dorsum.* Body setae stout, rarely spinelike, apex sharp; discoidal pores in a wide medial group anterior of anal plates; minute simple disc pores on body margin and in transverse

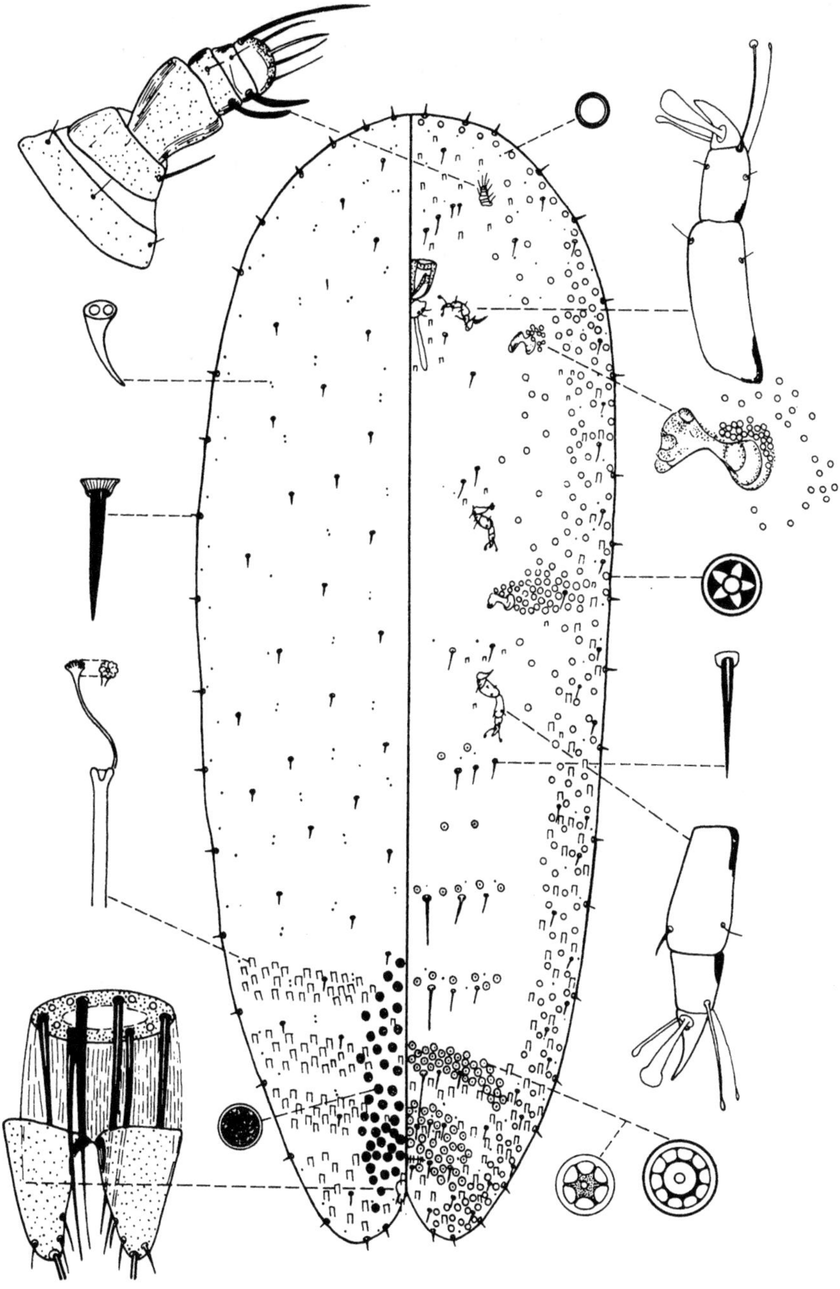

Fig. 102. *Psilococcus ruber* Borchsenius, female (original)

bands on last few abdominal tergites; oval bilocular pores in transverse bands and groups on entire dorsum; tubular ducts on posterior abdominal segments; anal ring sclerotized, circular, diameter ca 30 μ, with ca 30 pores and 6 setae, each seta 75–87 μ long; and plates conical, 50–60 μ long, 30–50 μ wide at base, each with 3 apical setae. Adult male, test, and all nymphal stages described by Koteja (1969a).

Distribution. A rare boreal species. Known from Hu, Po, Korea, and USSR (European part to Far East).

Biology. In rootcrown and under leaf sheaths of *Carex* spp. One generation per year; overwinter as eggs; females develop through 3 nymphal stages; adults develop by second half of summer (Koteja, 1969a).

Natural enemy. *Trichomasthus albimanus* (Kozár and Sugonjaev, 1979).

Genus **Pulvinaria** Targioni-Tozzetti

Syn. & Lit.: *Pulvinaria* Targioni-Tozzetti, 1866:146, 1867:13; Newstead, 1903:50; Leonardi, 1920:329; Steinweden, 1929:226, 1946:3; Šulc, 1942:5; Borchsenius, 1952:296, 1953:287, 1957:226; Schmutterer, 1952b:544; Bodenheimer, 1953:94; Takahashi, 1955:148; Řehaček, 1960:40; Hodgson, 1967:198, 1968:141; Ali, 1971:49; Williams & Kosztarab, 1972:118; Kosztarab & Kozár, 1978:121; Danzig, 1967:830, 1980a:263; Tereznikova, 1981:137.

Types-species: *Coccus vitis* Linnaeus, 1758.

Ovisac large, strongly convex, white. Postreproductive female, oval or circular, moderately convex, wrinkled; body tilted upward, to almost vertical position after ovisac fully formed, remains on plant fastened to ovisac. Teneral female oval, slightly convex, with marginal fold well-developed, 3–9.5 mm long, anal cleft of moderate length. *Venter.* Antennae slender, 8-segmented, rarely 7 or 9; tibio-tarsal articulation free and scleroses present, claw often with small denticle; claw digitules wide, strongly expanded at apex; spiracles large; spiracular pore bands narrow, of quinquelocular pores; spiracular setae 3, rarely 4 in each group, median seta 2 or 3 times longer than laterals; marginal setae pointed, form 1 or 2 rows; body setae of various lengths; multilocular pores on abdominal sternites; large tubular ducts numerous especially in submarginal band. *Dorsum.* Submarginal tubercles present or absent; body setae few, small, spinelike, scattered on entire surface; discoidal pores usually developed, form a medial group or band anterior of anal plates; small tubular ducts, when present, normally located in circular or oval weakly sclerotized cells; anal ring with 8, rarely 6 setae and with oval or circular pores; anal plates triangular, with apical setae.

Distribution. The genus, according to Borchsenius' (1957) concept, includes 26 Palearctic and 12 Nearctic species; another 14 species are reported from the Ethiopian Region (Hodgson, 1968). The taxonomic status of a number of species from other zoogeographical regions remains uncertain. There are 4 species in CE.

Biology. Most species are polyphagous and feed principally on woody plants, normally on the above ground parts. A few are important pests on fruit and ornamental trees and shrubs.

Note. *Pulvinaria sericea* (Fourcroy), a poorly understood oak-infesting Mediterranean species, was not included in this work.

Key to species of *Pulvinaria*

1. Each spiracular band with more than 77 quinquelocular pores 2
   Each spiracular band with less than 77 quinquelocular pores. 3
2. Legs more than 800 μ long; labium 133 μ long, 142 μ wide; 105–370 (normally more than 200) disc pores in each spiracular band; only on poplar, *Populus* spp. . . . . *populi,* p. 242
   Legs less than 600 μ long; labium 114 μ long, 133 μ wide; 78–115 disc pores in each spiracular band; polyphagous, feeds on a variety of trees (fig. 104) . . . . *betulae,* p. 239
3(1). Median spiracular setae 60–84 μ long; hind leg ca 850 μ long; labium 95 μ long, 109 μ wide; only on currants, *Ribes* spp. . . . . *ribesiae,* p. 243
   Median spiracular setae 90–95 μ long; hind leg ca 750 μ long; labium 109 μ long, 142 μ wide; polyphagous, feeds on a variety of trees and shrubs . . . . *vitis,* p. 243

136. **Pulvinaria betulae** (Linneaus) – Cottony Birch Scale
Figs 103, 104

Syn. & Lit.: *Coccus betulae* Linnaeus, 1758:455. *Pulvinaria betulae* Signoret, 1873:31(207); Lindinger, 1912:84; Jablonowski, 1916:42; Leonardi, 1920:329 (in part); Schmutterer, 1952b:554, 1972b:417 (in part); Borchsenius, 1957:229; Řehaček, 1960:40; Săvescu, 1961:45; Danzig, 1967:830, 1980:263; Koteja, 1974a:111; Dziedzicka, 1977:76; Kosztarab & Kozár, 1978:112; Tereznikova, 1981:137. *Coccus carpini* Linnaeus, 1758:455. *Pulvinaria betulae* var. *alni* Douglas, 1891b:100. Lindinger (1912:376) lists 17 additional synonym names that need verification.

Ovisac strongly convex, smooth, white, ca 8 mm long. Postreproductive female short, heart-shaped, almost circular, wrinkled, dark brown. Teneral female short oval, dark yellow to brown with small grayish tubercles up to 5–7 mm long. *Venter.* Antennae 8-, rarely 7- or 9-segmented, ca 506 μ long; labium 114 μ long, 133 μ wide (Koteja, 1974a); legs slender, their length: fore 510, middle 585, hind 575 μ (Dziedzicka, 1977); according to Borchsenius (1957), each spiracular pore band with 78–115 quinquelocular pores, each 5–6 μ in diameter, bands 3–4 pores wide; 3 spiracular setae in each group, stout, medial 90–95 μ, twice as long as laterals (42–51 μ); marginal setae slender, acute, form 1 or 2 rows, distance between setae equal or less than their length; body setae of various lengths, 9 interantennal setae, 2 longest 155–165 μ long, 3 prevulvar setae 115–270 μ long, other dermal setae 15–30 μ long; multilocular pores, 7–8 μ in diameter with 10 or more loculi, in transverse rows on abdominal sternites, few at bases of middle and hind coxae; large tubular ducts 21–27 μ long, 4 μ wide, numerous in a submarginal band, and medial area few on head. *Dorsum.* Body setae spinelike, 6–15 μ long, scattered over entire surface; discoidal pores form a 6–10 pores wide medial longitudinal band on abdomen, anterior of anal plates; short tubular ducts numerous, scattered, each in small circular cell-like area, space between cells 2–3 times their diameter; anal ring with 8 setae of similar size. Adult male described by Giliomee (1967), male test by Borchsenius (1957) and Řehaček (1960); N_1 by Řehaček (1960).

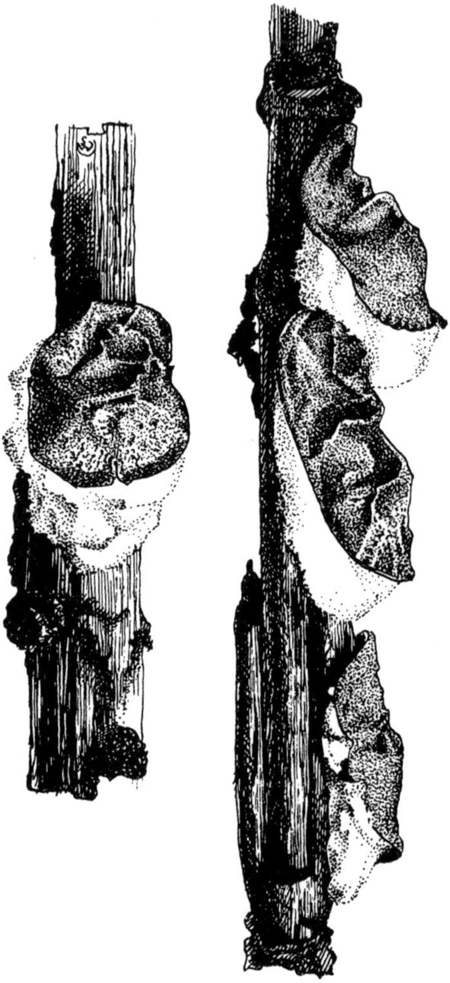

Fig. 103. *Pulvinaria betulae* (Linnaeus), female with ovisacs (after Kosztarab and Kozár, 1978)

Distribution. A common wide-spread Holarctic species, that was often confused with *P. vitis*; therefore early distribution records need verification. In CE: Au, Bu, Cz, Fr, Ge, Hu, It, Po, Ro, Sw, USSR (European parts to Far East) and Yu; in other parts of Europe in Denmark, England, Finland, Holland, Norway, Portugal, Spain, Sweden; in Asia, Iran, Japan, Mongolia, Turkey.

Biology. Polyphagous. Because of the frequent misidentifications based on postreproductive females, the host list presented here may not be complete or some records may not be correct. Host genera from recent records: *Acer, Aesculus, Alnus, Betula, Carpinus, Corylus, Crataegus, Cydonia, Euonymus, Fagus, Fraxinus, Juglans, Malus, Mespilus, Populus, Prunus, Pyrus, Ribes, Rosa, Salix, Sorbus, Tilia,* and *Vitis*. Often a pest on grapes. Adult females overwinter and lay eggs in S Germany from

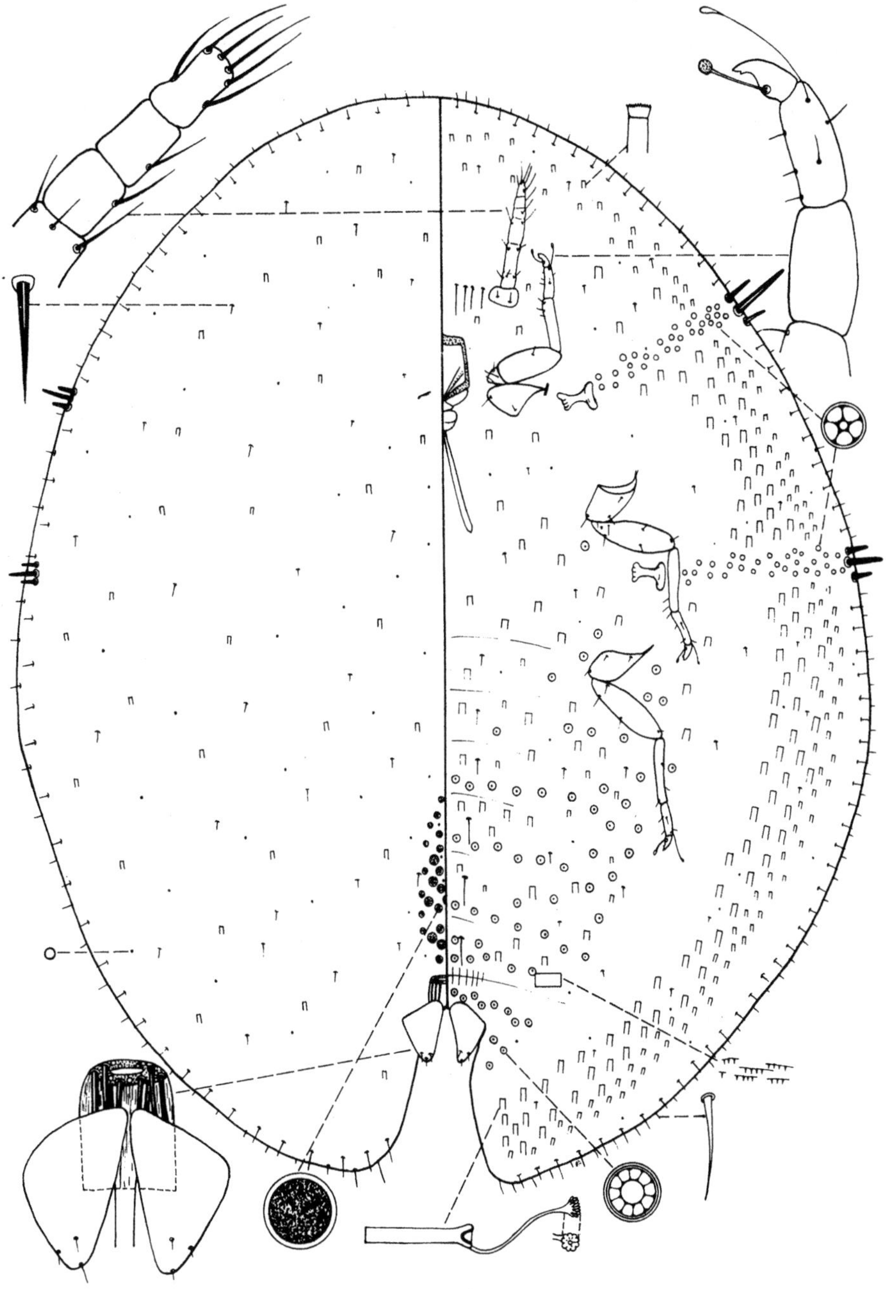

Fig. 104. *Pulvinaria betulae* (Linnaeus), female (original)

second half of April through early June. Nymphs hatched from both fertilized and unfertilized eggs; according to Jablonowski (1916) and Schmutterer (1952b) an uniparental race and a biparental race were present. Females on *Crataegus* deposited an average of 3486 eggs (range 1238–5108); and eggs hatched within 35–40 days. First instars molted during second half of July, and N_2 molted at end of August. Winged males appeared during first half of September. During second half of October adult females moved from leaves to young twigs for overwintering. Associated ants: *Formica rufa pratensis, Lasius niger alienus, L. n. niger* (Schmutterer, 1952b).

Natural enemies. Parasites: *Cheiloneurus formosus. Coccophagus insidiator, C. lycimnia, C. scutellaris, Encyrtus albitarsis, E. merceti, E. obscurus, E. swederi, Microteris cneus, M. duplicatus, Pachyneuron concolor*. Predators: *Anthribus fasciatus, A. nebulosus, Leucopis anulipes, L. membe, L. nigricornis, L. silesiaca* (Aczél, 1936; Kosztarab, 1959; Schmutterer, 1972b; Trjapitzin, 1978; Kozár and Sugonjaev, 1979; Ceianu, 1980). These and other records on natural enemies might be based on misidentification of host; it was often confused with *P. vitis*.

137. **Pulvinaria populi** Signoret – Cottony Poplar Scale

Syn. & Lit.: *Pulvinaria populi* Signoret, 1873:42(218); Borchsenius, 1955d:863, 1957:234; Danzig, 1967:830; Tsalev, 1968:209; Koteja, 1974a:111; Dziedzicka, 1977:76; Kosztarab & Kozár, 1978:121; Tereznikova, 1981:141.

Ovisac strongly convex, white, longitudinal carinae absent, 7–9 mm long. Postreproductive female brown, with numerous transverse wrinkles. Teneral female short oval, nearly circular, 7–9 mm long, 6–7 mm wide. *Venter.* Antennae 7- or 8-segmented, ca 542 μ long, 3rd segment much longer than others; labium 133 μ long, 142 μ wide (Koteja, 1974a); legs slender, their lengths: fore 835, middle 990, hind 890 μ (Dziedzicka, 1977). According to Borchsenius (1957) spiracular pore bands mostly 4–7 pores wide, in places 2–3 pores wide, 105–370 (normally over 200) quinquelocular pores in each band; 3 slightly bent, stout, spiracular setae in each group, pointed at tip, medial (102–120 μ) 2 or 3 times longer than lateral setae (36–45 μ); marginal setae hairlike and 48–60 μ long in one row, in a second row setae shorter, 36–48 μ long, irregularly placed, intervals between marginal setae 1 ½–2 times their length; hairs on posterior end 60–75 μ long. *Dorsum.* Discoidal pores, ca 4 μ in diameter, form an elongate group anterior of anal plates. In other morphological characters adult females similar to *P. betulae*. Adult male, male test, and N_1 described by Borchsenius (1957).

Distribution. A rare Western-Palearctic species. In CE: Bu, Fr and USSR (only European parts).

Biology. On twigs of *Populus* spp. One generation per year. Probably mated females overwinter. Eggs laid in May, hatched in June, and adults developed by end of August in Krasnodar area of USSR (Borchsenius, 1957).

Note. Only inadequate descriptions are available for this species at present.

138. **Pulvinaria ribesiae** Signoret – Cottony Currant Scale, Woolly Currant Scale

Syn. & Lit.: *Pulvinaria ribesiae* Signoret, 1873:43(219); Borchsenius, 1957:243; Drozdovskiy, 1966:402; Boratynski & Williams, 1964:108; Koteja, 1964:182, 1974a:112; Danzig, 1967:830, 1972c:205; Matesova, 1968a:121; Tsalev, 1968:209; Dziedzicka, 1977:76; Kosztarab & Kozár, 1978:123. *Pulvinaria vitis* var. *ribesiae* Newstead, 1903:55. *Pulvinaria betulae* Lindinger, 1912:376.

Ovisac large, strongly convex, white. Postreproductive female nearly circular, dark brown. Teneral female short oval, reddish-brown to dark brown, with raised light yellow longitudinal band in median area of dorsum, laterally with transverse wrinkles and stippling, 3.5–4 mm long, 2.8–3 mm wide. *Venter.* Antennae normally 8-segmented, rarely 6 or 7. Labium 95 μ long, 109 μ wide (Koteja, 1974a); legs slender, their length, fore 840, middle 775, hind 850 μ, (Dziedzicka, 1977); according to Borchsenius (1957), spiracular pore bands each with 55–76 quinquelocular pores; 3 stout spiracular setae in each group, median 60–84 μ long, laterals 18–27 μ long; marginal setae slender, acute in a staggered alignment, space between setae, except those at caudal end, at least 1 ½ times length of setae; setae along margin of head and thorax 15–30 μ long, at caudal end 27–36 μ long. In other morphological characters similar to *P. betulae*.

Distribution. A rare boreal species. Only few confirmed records are available on their distribution. In CE: Bu, Fr, Ge, Hu, Po and USSR. Also known from England, Sweden and Mongolia.

Biology. On twigs and leaves of currants, *Ribes* spp. Adult females overwintered in England and laid eggs at end of April or early May. N_1 hatched and crawled to leaves and young shoots for feeding from early June to early July and molted in mid-July; N_2 molted in August. Winged males appeared in September after N_3 females molted a third time and adult females emerged (Newstead, 1903). In the Moscow area no males were found. The species overwintered primarily as N_2, seldom as adult females. Eggs laid at end of June through early July, hatched at end of July, through early August (Danzig, 1972e). One generation per year was observed at both locations.

Note. According to Drozdovskiy (1966) chromosome number 18, in *P. betulae* 16.

139. **Pulvinaria vitis** (Linnaeus) – Cottony Grape Scale

Syn. & Lit.: *Coccus vitis* Linnaeus, 1758:456. *Pulvinaria vitis,* Signoret, 1873:45(221); Newstead, 1903:51; Leonardi, 1920:329; Green, 1922:17; Balachowsky & Mesnil, 1936:727; Gomez-Menor, 1937:279; Steinweden, 1946:13; Schmutterer, 1952b:554; Borchsenius, 1957:232; Phillips, *1962:497, *1963:372; Rogojanu, 1963:279; Danzig, 1967:830; Tereznikova, 1981:142. *Pulvinaria betulae,* Bodenheimer, 1944a:88; Kosztarab & Kozár, 1978:122.

Ovisac wide, convex, cottony, white, longer than adult female. Postreproductive female oval, dark brown, dorsum wrinkled and strongly sclerotized. Teneral female 2.5–6.5 mm long, 1.5–6.5 mm wide, dorsal derm with circular or oval areas. *Venter.* Antennae 8-segmented; labium 109 μ long, 142 μ wide (Koteja, 1974a); legs well-developed, length of legs: fore 760, middle 860, hind 750 μ; medial spiracular seta 2 or 3 times longer (90–95 μ long) than lateral setae; spiracular pore bands 2–4 pores wide, each with 40–75 quinquelocular pores; marginal setae slender, acutely pointed, 35–50 μ long, with 1 or 2 irregular rows in staggered order, intervals between setae equal or exceed their length by 1½ to rarely 3 times, multilocular pores with 10 loculi, 8–9.5 μ in diameter, clustered in anal region, and form transverse bands or rows on abdominal sternites, few near each coxa; tubular ducts of 3 types, in a wide submarginal band, others scattered on rest of venter, but very few on submargin of head. *Dorsum*. Small submarginal tubercles 0–8, normally 5–6 around body; setae slender, 8–16 μ long, with acute tip; 45–75 discoidal pores medially anterior of anal plates; minute bilocular pores interspersed with small simple disc pores scattered over entire surface; anal ring, almost rectangular, 66 μ long, 63 μ wide, with 8 setae and 2 irregular rows of pores; anal plates triangular, each with 4 apical and 3 subapical setae. Nymphal stages (3) described by Phillips (1962).

Distribution. A common Holarctic species. In CE: Cz, Fr, Ge, Hu, It, Ro, USSR and Yu. Also known from countries in the Mediterranean area, Denmark, England, Sweden, Middle East and N America.

Biology. Polyphagous; a pest on grapes, *Vitis* spp., in Europe where it infests roots, stems and leaves, and a pest of peaches in Ontario, Canada. Some host records reported in the early literature may be based on misidentification of *P. betulae*. One generation per year. Four instars in females. Mated females overwinter in France on grape twigs and lay eggs in May; eggs hatch in June; nymphs feed on leaves; adults mate in October. Reproduction is parthenogenetic in Ontario, Canada, where each female lays an average of 4,000 eggs.

Natural enemies. Parasites: *Aphycus maculipes,* and *Coccophagus lycimmia.* Predators: *Chrysopa carnea, Hyperaspis binotata, H. proba proba* (Phillips, 1963).

Genus **Rhizopulvinaria** Borchsenius

Syn. & Lit.: *Rhizopulvinaria* Borchsenius, 1952:301, 1957:257; Danzig, 1967:831; Canard, 1967:159, 1968:90; Kosztarab & Kozár, 1978:123; Tereznikova, 1981:142.

Type-species: *Rhizopulvinaria virgulata* Borchsenius, 1952.

Ovisac strongly convex, oval or circular, straight or curved, normally white. Postreproductive female, somewhat shriveled, hardened, remains fastened to ovisac. Teneral female short, oval or almost circular, strongly convex, dorsum moderately sclerotized, venter membranous, marginal fold well-developed; anal cleft short. *Venter*. Antennae 8- (rarely 6–7- or 9-) segmented, like legs, small in comparison to body size; tibio-tarsal articulation free and scleroses present; spiracular pore band usually wide, rarely narrow; spiracular spines normally present, usually similar in size, 2–3, rarely 1 or 4 in each group; marginal setae in 1 or 2 rows, with blunt or pointed tip; interantennal setae and prevulvar setae on abdominal sternites less than 60 μ long; small body setae scattered on entire surface; multilocular pores on abdominal sternites; large tubular ducts numerous especially in a submarginal band. *Dorsum*. Small, elongate conical setae scattered on entire surface; disc pores usually of different diameters; small tubular ducts in moderate numbers on most of surface; anal ring with 6, rarely 8 setae, with oval or circular pores; anal plates conical or triangular, normally with 4 apical setae.

Distribution. In the Palearctic region, especially common in xerophilous habitats of the Mediterranean area; with ca 23 species; 2 in CE.

Biology. On roots, rootcrowns, rarely on stems of herbaceous and semi-woody plants. Biparental reproduction, one generation per year. Eggs laid in May or June in the studied species. Presumably adult females overwinter.

Key to species of *Rhizopulvinaria*

Spiracular setae shorter than marginal setae; each spiracular pore band with 40–64 quinquelocular pores; 2 spiracular setae in each posterior group .... *artemisiae*, p. 245

Spiracular setae longer than marginal setae; each spiracular pore band with 89–115 quinquelocular pores; 3 spiracular setae in each posterior group (fig. 105)...... .... *spinifera*, p. 246

140. **Rhizopulvinaria artemisiae** (Signoret) – Cottony Wormwood Scale

Syn. & Lit.: *Pulvinaria artemisiae* Signoret, 1873:(207)31; Lindinger, 1912:77; Schmutterer, 1955b:100; Kosztarab, 1959:408. *Rhizopulvinaria artemisiae*, Borchsenius 1952:302, 1957:279; Canard, 1967:165, *1968:91; Koteja, 1974a:81; Zahradnik, 1977:119; Dziedzicka, 1977:77; Kosztarab & Kozár, 1978:123.

Ovisac oval, or circular, strongly convex, 2.4 mm long, 2.3 mm wide (Canard, 1968). Postreproductive female dorsally wrinkled, circular, dark grey, up to 2 mm long (Signoret, 1873). Teneral female oval, 1.4–3.2 long, 0.9–2.2 mm long up to 1.6 mm high. *Venter.* Antennae 8-segmented, ca 268 μ long, 3rd segment longest; labium 90 μ long, 104 μ wide (Koteja, 1974a), stylet loop extends beyond middle of abdomen. Legs slender, length of legs: fore 390 μ, middle 450 μ, hind 444 μ (Dziedzicka, 1977); surface of tarsus undulate, tarsus shorter than tibia; claw with small denticle, its digitules similar with expanded tip; each spiracular pore band with 40–55 quiquelocular pores in anterior, and 50–64 pores in posterior bands, diameter of pores ca 5 μ; 3 spiracular setae in each group, median shortest, laterals ca 15 μ long. Marginal setae spinelike, 17–26 μ long, longer than spiracular setae, on anal lobes setae 24–34 μ long; multilocular pores on abdominal sternites, ca 7 μ in diameter; tubular ducts as given for the genus. *Dorsum.* Simple disc pores 2 μ in diameter; anal plates conical, ca 140 μ long, each with 4 apical setae; other morphological structures same as in generic description. Male test and N_1 described by Canard (1968).

Distribution. A rare xerophilous Mediterranean species. In CE: Cz, Fr, Hu; also found in Iraq.

Biology. On roots of *Artemisia* and *Dianthus,* presumably with one generation per year. Adult females overwinter. Associated ant: *Crematogaster auberti* (Canard, 1968).

141. **Rhizopulvinaria spinifera** Borchsenius – Cottony Carnation Scale
Fig. 105

Syn. & Lit.: *Rhizopulvinaria spinifera* Borchsenius, 1952:305; 1957:261; Matesova, 1968a:121; Kosztarab & Kozár, 1978:124; Tereznikova, *1981:144.

Ovisac strongly convex, curved, white. Teneral female oval; according to Borchsenius (1957), live young female greenish and before reaching maximum size ca 1.9 mm long, 1.1 mm wide. *Venter*. Antennae 8- or 7-segmented, ca 240 μ long; when 8-segmented, 2nd and 3rd segments longest and about equal in size; legs moderately thick, claw with small denticle; spiracular pore bands broad, 3 to 7 pores wide, each band with 89–115 quinquelocular pores 5–6 μ in diameter; spiracular setae 3 in anterior and 2 in posterior group, distinctly stouter and longer than marginal setae, bluntly tipped, 30–42 μ long; most marginal setae 21–27 μ long, at posterior end of body 27–33 μ long; in 2 rows at anterior and posterior margins of body; 4 interantennal setae 10–18 μ long; 3 prevulvar setae 42–45 μ long; multilocular pores 7–8 μ in diameter, with 7 loculi, form sparse transverse rows on abdominal sternites; tubular ducts ca 4 μ in diameter form a wide submarginal band and scattered on rest of venter. *Dorsum*. Small conical setae, 8–11 μ long scattered on entire surface; discoidal pores, 5–6 μ in diameter, in a short longitudinal band anterior of anal plates; small tubular ducts, ca 2 μ in diameter scattered on entire surface; anal ring with 6 setae and oval pores; anal plates each with 4 apical setae up to 48 μ long.

Distribution. A rare, xerophilous Mediterranean species. In CE: Hu (Kozár's unpublished record), in USSR from Ukraine to Gruzia and Kazakhstan.

Biology. On roots and root crowns of *Achillea, Dianthus, Galium, Gypsophila, Helichrysum* and *Silene*. In Kazakhstan each female laid 290 eggs by the end of May which hatched before the end of June (Matesova, 1968a).

Genus **Rhodococcus** Borchsenius

Syn. & Lit.: *Rhodococcus* Borchsenius, 1953:283, 1957:424; Řehaček, 1960:62; Kosztarab & Kozár, 1978:124; Danzig, 1980a:289.

Type-species: *Rhodococcus rosaelutea* Borchsenius, 1953.

Postreproductive female nearly spherical with margins bent underneath body with various grooves and rows of indented spots on dorsum, often with colorful bands. Teneral female. *Venter*. With 6-, rarely 5-, 7- or 8-segmented antennae; 1 or 2 spiracular setae in each group, often not different from marginal setae; marginal setae hairlike, longer on head and posterior end of body or thick, spinelike with blunt tip on lateral margins. *Dorsum*. Derm often sclerotized, especially in anal area; anal ring small, narrow, without pores, often setae absent or poorly developed; anal plates triangular with rounded lateral angles.

Distribution. Palearctic, with 6 species; 2 in CE.

Biology. Except for *R. turanicus,* all species found on Rosaceae only.

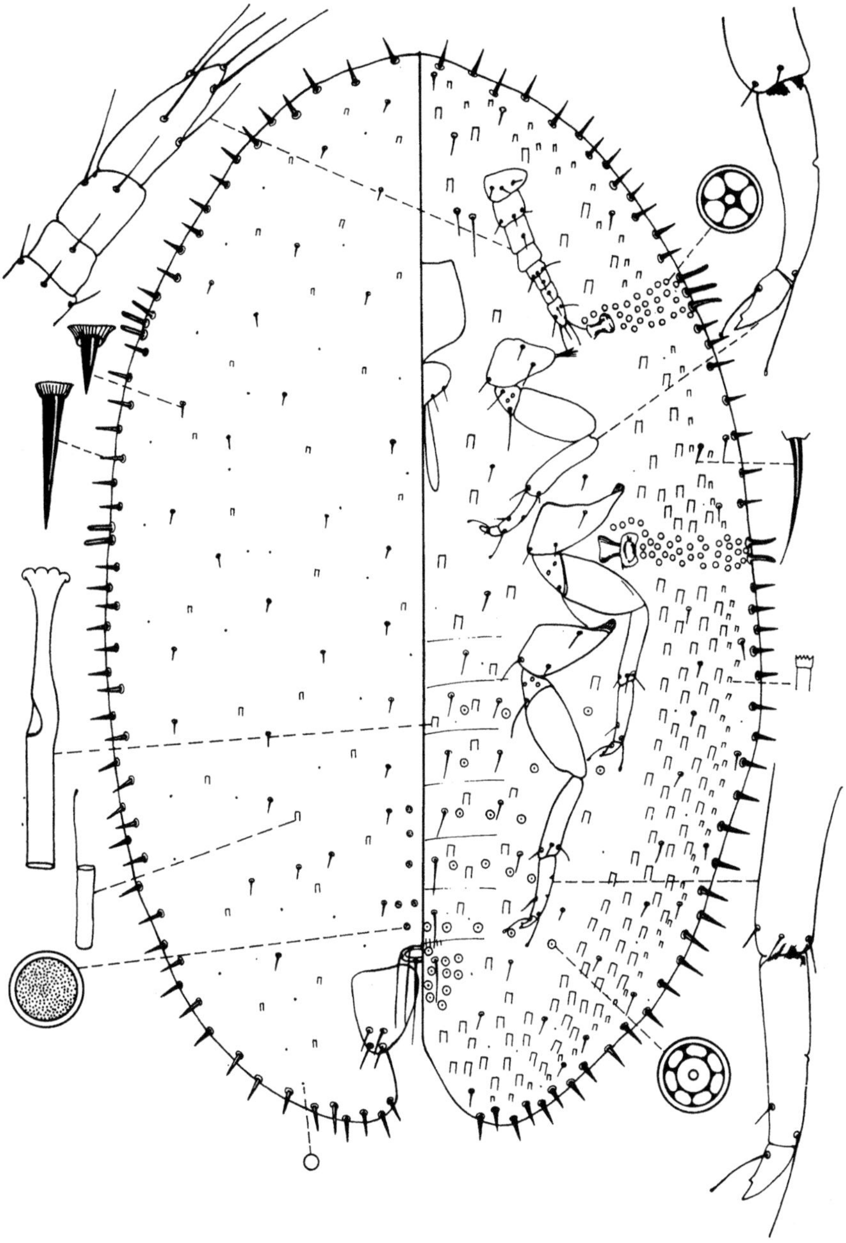

Fig. 105. *Rhizopulvinaria spinifera* Borchsenius, female (original)

Key to species of *Rhodococcus*

142. **Rhodococcus perornatus** (Cockerell & Parrott) – Rose Soft Scale
Figs 13, 106

Syn. & Lit.: *Lecanium perornatum* Cockerell & Parrott, 1899:236. *Eulecanium bulgariense* Wünn 1939:703. **Syn. n.** *Rhodococcus bulgariensis*, Tsalev *1966:3; Kosztarab & Kozár, 1978:125; Danzig, 1980a:289. *Rhodococcus rosophilus* Borchsenius, 1957:436.

Postreproductive female almost spherical, often higher than wide and its shape often deformed by spines of rose, margins bent underneath body, shiny, yellowish-brown with 3 alternate transverse bands of yellow and dark brown, and a longitudinal median dorsal brown band, also with 2 longitudinal rows of indented spots, converging to anal plates; 4–4.5 mm long, 3.5–4.5 mm wide and 3.5–5 mm high. Teneral female. *Venter*. With 7-, rarely 6-segmented antennae, 216–251 μ long; stylet loop ca 309 μ long, reaches line between mid-coxae; legs thick, claw bent, slender, with a denticle; peritreme of posterior spiracles 62–85 μ in diameter; spiracular furrows indistinct, each with 32–46 disc pores with 5–8 loculi; 2 thick blunt-pointed spiracular setae in each group, 24–30 μ long, often far apart; 1 row of hairlike marginal setae, 60–150 μ long, longer on head, distance between setae 1.5–2.5 times their length; body setae short; multilocular pores with 9–12 loculi, form transverse rows, bands, or groups on entire venter; tubular ducts in a wide submarginal band. *Dorsum*. Anal area heavily sclerotized, striations around anal plates replaced outwards by reticulations; small disc pores and setae scattered on entire dorsum; anal ring oval, with rudimentary setae. Adult male described by Wünn (1939), male with male test by Tsalev (1966); first instar (*R. rosophilus*) by Borchsenius (1957), first and second instars by Tsalev (1966).

Distribution. A rare Trans-Palearctic steppe-inhabiting mesophilous species. In CE: Au, Bu, Cz, Hu, It and USSR.

Biology. On branches of *Rosa* spp. only. Occasionally mass infestations are found. It is considered a pest of cultivated roses in Bulgaria. One generation per year. Overwinters as second instars. Adults are present from mid-April in Bulgaria and by end of May in Hungary. Each female lays 436–2287 eggs during mid-May in Bulgaria, and mid-June in Hungary. Eggs hatch during the second half of June in Bulgaria, and in early July in Hungary (Tsalev, 1966; Kosztarab and Kozár, 1978).

Natural enemies. Parasites: *Coccophagus lycimnia, Discodes rubtzovi, Metaphycus bulgariensis, M. insidiosus, M. kozari, M. parvus, Microterys bellae,* and *M. sylvius*. (Tsalev, 1966; Nikolova, 1969; Trjapitzin, 1978; Kozár and Sugonjaev, 1979). Predators: *Brachytarsus fasciatus, B. nebulosus, B. scabrosus, Coccinella septempunctata, Coreus marginatus* and *Oratocelis communimaculata* (Wünn, 1939; Tsalev, 1966; Nikolova, 1969).

Note. We studied the type specimens of this species in the USNM Collection and concluded that *R. bulgariensis* was a junior synonym.

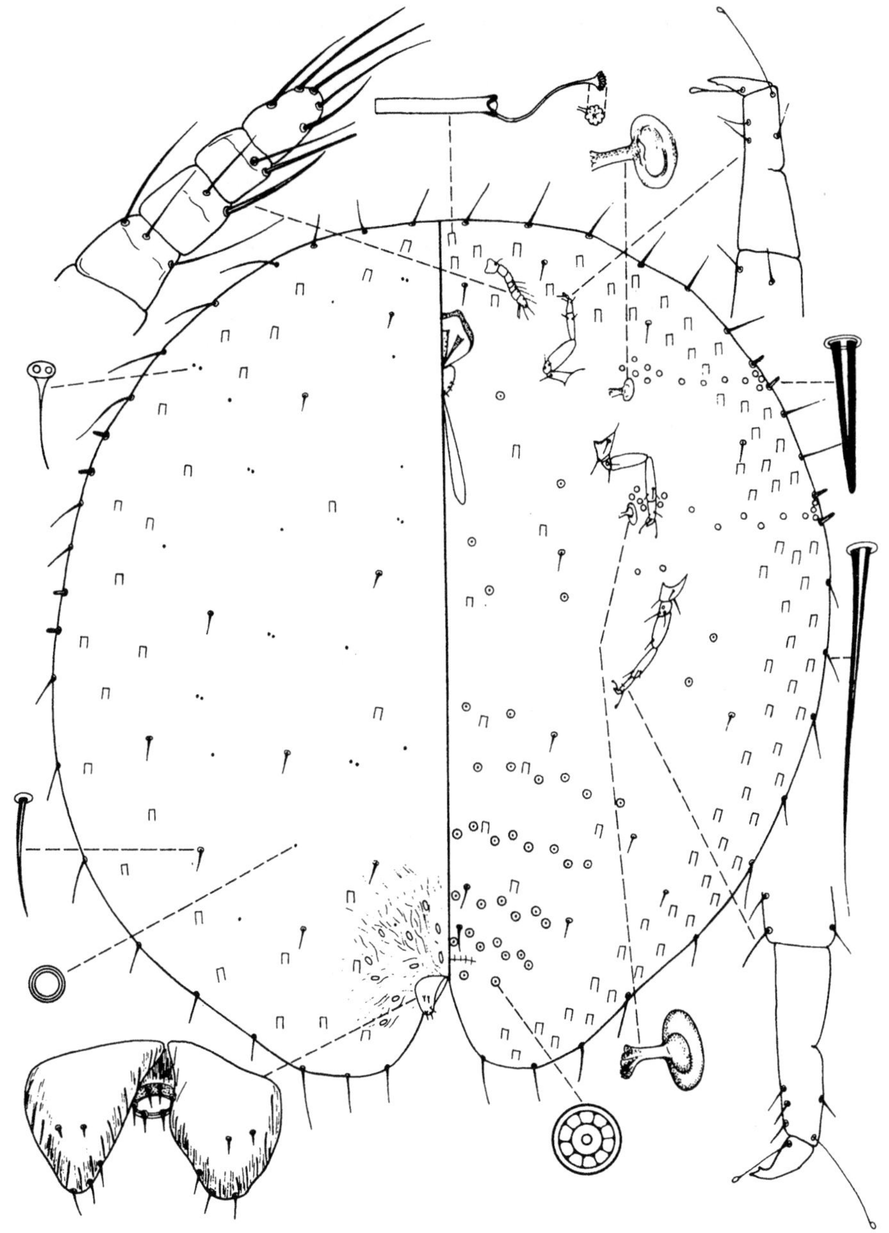

Fig. 106. *Rhodococcus perornatus* (Cockerell & Parrott), female (original)

143. **Rhodococcus spiraeae** (Borchsenius) – Spirea Soft Scale

Syn. & Lit.: *Eulecanium spiraeae* Borchsenius, 1949:340. *Rhodococcus spiraeae* Borchsenius, 1957:428; Řehaček, 1960:62; Matesova, 1968a:120; Kosztarab & Kozár, 1978:125; Danzig, 1980a:290; Tereznikova, *1981:186.

Postreproductive female nearly spherical, with margins bent underneath body on dorsum, a longitudinal median groove, also 6 longitudinal rows of indented spots, light-brown to brown, rarely with a median darker band on mid-dorsum; ca 2.5–3 mm long, 2.5 mm high, but on thick twigs and at base of stems near roots usually larger, reaching 4.5 mm in diameter and 4 mm in height. Teneral female. *Venter*. With small, thick 6-segmented antennae, 148–179 μ long; labium 2-segmented, 85 μ long, 90 μ wide, with 10 setae (Koteja, 1974a); legs relatively small and thick, claw with a denticle near tip; diameter of peritreme of posterior spiracles 57–75 μ; spiracular furrows narrow, each with 9–15 disc pores with 5 or 6-loculi, rows of disc pores often interrupted in fully developed females; 1–2 spiracular setae in each group, moderately stout, bent, ca 25–40 μ long, slightly different from marginal setae; marginal setae spinelike and hairlike, 75–135 μ long on anterior and posterior ends of body, much shorter on rest of body; multilocular pores with 9–14 loculi in large groups near bases of legs and in wide transverse bands on abdominal segments; tubular ducts ca 27 μ long, form a submarginal band. *Dorsum*. Heavily sclerotized, with scattered tubular ducts mostly in submarginal area; discoidal pores and short setae scattered on entire dorsum; anal ring narrow, small, without pores, with 6 short slender setae. Adult male described by Borchsenius (1957), Giliomee (1967) and Tereznikova (1981); male test by Řehaček (1960); first instar by Borchsenius (1957), Řehaček (1960) and Tereznikova (1981).

Distribution. A rare steppe-inhabiting Trans-Palearctic species. In CE: Cz and USSR; also known from Mongolia.

Biology. Monophagous. On twigs and stems of *Spiraea* spp. One generation par year. Second instars overwinter. Adults develop by May (Řehaček, 1960) and females lay eggs in July which hatch from August.

Natural enemies. Parasites: *Blastothrix isomorpha, B. nikolskajae, B. sericea, Discodes rubtzovi, Lounsburyia spiraeae, Marietta picta, Microterys* sp. Predators: *Brachytarsus nebulosus, B. scapularis.* (Borchsenius, 1957; Matesova, 1968a; Trjapitzin, 1978).

Genus **Scythia** Kiritchenko

Syn. & Lit.: *Scythia* Kiritchenko, 1938:229; Šulc, 1941:16; Borchsenius, 1957:178; Řehaček, 1960:34; Danzig, 1967:826; Kosztarab & Kozár, 1978:125; Tereznikova, 1981:126. *Mohelnia* Šulc, 1941:1.

Type-species: *Scythia cranium-equinum* Kiritchenko, 1938.

Ovisac oblong, dense, feltlike, up to 8 mm long, includes nymphal exuviae divided in 2 parts; female entirely enclosed in sac, but in *S. festuceti,* after oviposition, female drops from ovisac. Postreproductive female elongate-oval, convex, up to 5.5 mm long. Teneral female with membranous derm and short anal cleft. *Venter*. Antennae reduced to 1–4-segmented stubs; legs reduced, stublike; spiracles large; spiracular pore band

and spiracular setae absent; marginal setae hairlike, far apart; body setae short, inconspicuous; multilocular pores with 8–10 loculi form transverse rows and bands on abdominal sternites; tubular ducts common on both surfaces. *Dorsum*. Body setae short, scattered on entire surface; discoidal pores in a medial group or in a longitudinal band anterior of anal plates; numerous tubular ducts on entire surface; anal ring with rows of pores and 6–10 setae; anal plates triangular, well-developed.

Distribution. Trans-Palearctic, with 2 species, both in CE.

Biology. On grasses. Reproduction biparental. Eggs overwinter.

Key to species of *Scythia*

Antennae 3–4 segmented; anal ring with 8–10 setae; dorsal conical setae on head and on anal lobes sharp-pointed (fig. 107) .... *craniumequinum*, p. 251

Antennae 1 or 2-segmented; anal ring with 6 setae; dorsal conical setae on head and on anal lobes with truncate tip .... *festuceti*, p. 253

144. **Scythia craniumequinum** Kiritchenko – Horsehead-shaped Scale
Figs 14, 107

Syn. & Lit.: *Scythia cranium-equinum* Kiritchenko, 1938:229; Šulc 1941:6; Borchsenius, 1957:179; Danzig, 1964:639; Kosztarab & Kozár, 1978:126; Tereznikova, 1981:126.

Ovisac 5–8 mm long, brownish-gray, resembles an upturned horse head, with remnants of glassy nymphal exuviae on lateral sides. Postreproductive female elongate-oval, flat ventrally, convex dorsally. Teneral female reddish-brown when live, 2.5–5 mm long, 1.2–2.5 mm wide and high. *Venter*. Antennae stout, 3–4-segmented, apical segment with several setae; labium 95 μ long and wide (Koteja, 1974a); legs stublike, 2–3-segmented, claw digitules shorter than claw; a semicircular cluster of disc pores with 5 or 6 loculi by spiracular peritreme, also 2–5 such pores anterior of each spiracle; body setae scattered on venter, longer setae along anal cleft; multilocular pores form a row between hind coxae, and narrow transverse bands on each of first 3 abdominal sternites, and wide transverse bands on other abdominal sternites; tubular ducts numerous on both body surfaces. *Dorsum*. Body setae conical, apex sharp, few, ca 35 on head, and 13–27 longer setae on apex of each anal lobe; translucent pores, 3–4 μ in diameter, scattered on entire surface; discoidal pores, ca 6 μ in diameter, form a wide medial longitudinal band anterior of anal plates; anal ring with 8 or 10 setae, 2 or 4 setae shorter and stouter than others; each anal plate with 4 long slender apical and 4 subapical setae. Adult male and male test described by Borchsenius (1957).

Distribution. A rare steppe-inhabiting Palearctic species. In CE: Hu (Kozár, unpublished record) and USSR (from Ukraine to Central Asia); also in Mongolia (Danzig, 1974).

Biology. On stems and leaves of *Festuca* and *Stipa*; often in large colonies. Adults of both sexes found in Hungary during September; eggs overwinter, apparently one generation per year.

Natural enemy. Parasite: *Marietta picta* (Kozár, unpublished record).

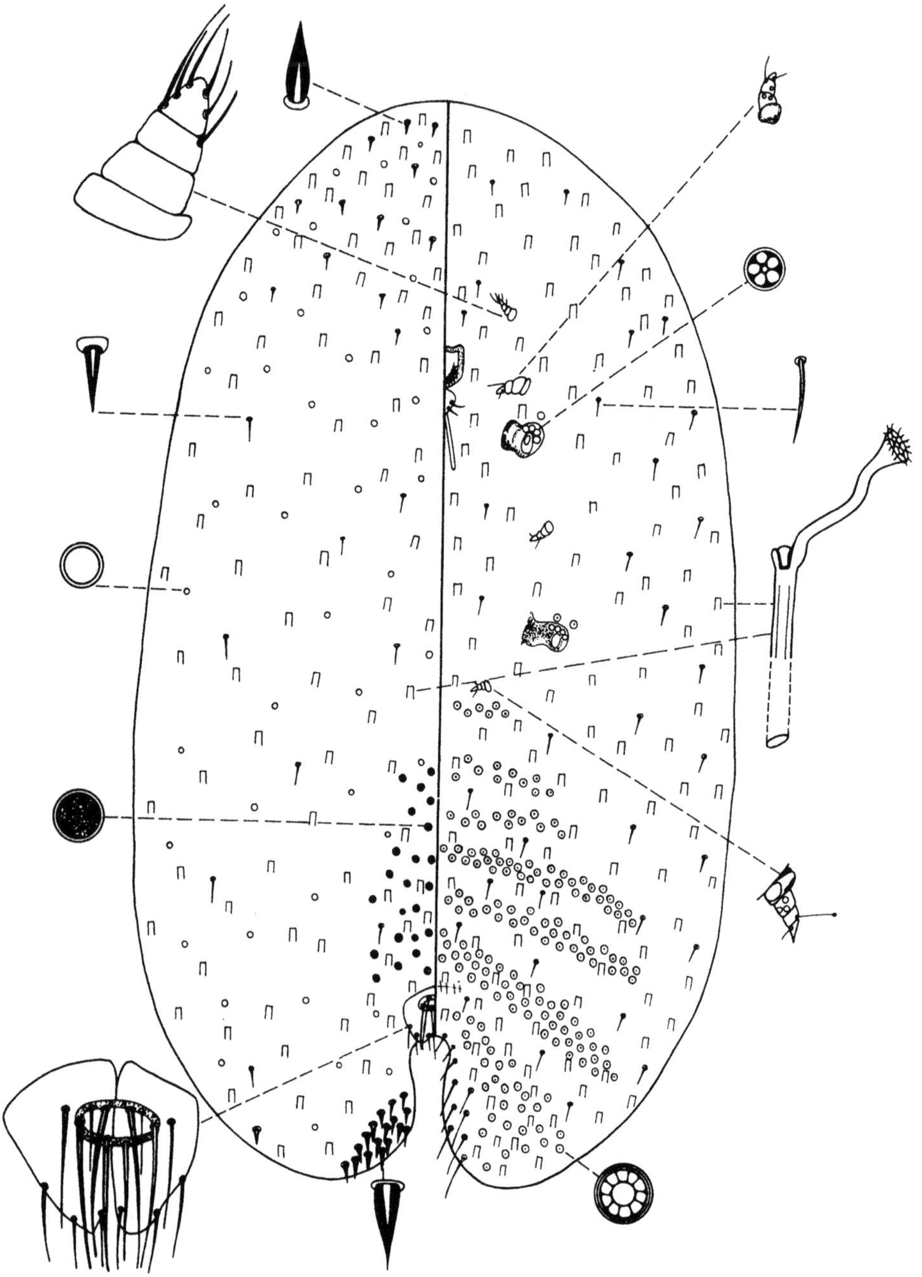

Fig. 107. *Scythia craniumequinum* Kiritchenko, female (original)

145. **Scythia festuceti** (Šulc) – Fescue Soft Scale
Fig. 108

Syn. & Lit.: *Mohelnia festuceti* Šulc, *1941:2. *Scythia festuceti*, Borchsenius, 1957:182; Řehaček, *1960:34; Danzig, 1964:639; Kosztarab & Kozár, 1978:127; Tereznikova, 1981:128.

Ovisac elongate, spindle-shaped, dense, feltlike, up to 7 mm long, 1.6 mm wide, yellowish-gray, with remnants of dull glassy nymphal exuviae bulging on lateral sides. Postreproductive female elongate-oval, somewhat pointed at both extremities. Teneral female 5–5.5 mm long, 1.5 mm wide, with short anal cleft. *Venter.* Antennae small, stublike, 1–2-segmented, apex with a tuft of setae; labium 81 μ long and wide (Koteja, 1974a); legs much reduced, stublike with claw and few short hairs; spiracles large with sclerotized frame; each with a few associated quinquelocular pores; body setae short and few; multilocular pores form transverse rows and bands on abdominal sternites; tubular ducts numerous on both surfaces. *Dorsum.* Conical short setae with truncate tip form 2 groups on head and 1 group on each anal lobe; discoidal pores, 8–9 μ in diameter, form a cluster anterior of anal plates; anal ring wide with 2 rows of pores and 6 short setae, each ca 120 μ long; anal plates conical, each with 4 stout setae. Adult male, test, N_1 and N_2 described by Šulc (1941); N_1 and male test by Řehaček (1960).

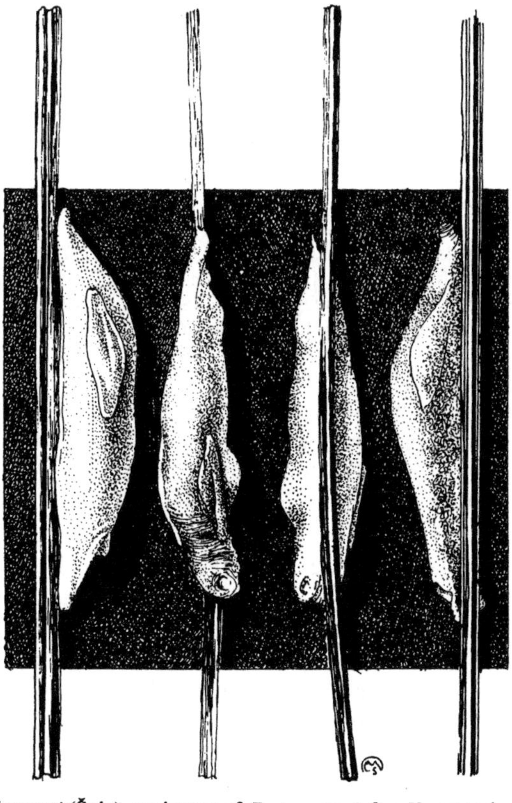

Fig. 108 *Scythia festuceti* (Šulc), on leaves of *Festuca* sp. (after Kosztarab and Kozár, 1978)

Distribution. A rare steppe-inhabiting Western-Palearctic species. In CE: Cz, Hu and USSR (Ukraine to Northern Caucasus).

Biology. On stems of *Festuca*. One generation per year. In Czechoslovakia eggs overwinter and hatch in May, and the adults develop by end of July (Řehaček, 1960).

Natural enemy. Parasite: *Coccophagus rjabovi* (Trjapitzin, 1978).

Genus **Sphaerolecanium** Šulc

Syn. & Lit.: *Sphaerolecanium* Šulc, 1908a:36, 1932:78 Silvestri, 1919:70, 1920:501; Borchsenius, 1957:322; Řehaček, 1960:46; Kosztarab & Kozár, 1978:127. *Sphaerolecanium* Leonardi, 1908:180 (a homonym) considered a junior synonym of *Eulecanium* Cockerell by Silvestri, (in Leonardi, 1920:312).

Type-species: *Coccus prunastri* Fonscolombe 1834. A monotypic genus, for generic description see *S. prunastri*.

The almost circular body outline, strongly spherical body shape, dark brown to almost black body color and small size in postreproductive females, the lack of tubular ducts, and presence of a wide medial longitudinal band of sievelike discoidal pores and long hairlike setae on the dorsum of teneral females, will distinguish this taxon from other genera.

Distribution and biology. As given at *S. prunastri*.

146. **Sphaerolecanium prunastri** (Fonscolombe) – Plum Lecanium, Globose Scale
Figs 15, 109

Syn. & Lit.: *Coccus prunastri* Fonscolombe, 1834:211. *Lecanium blanchardii* Targioni-Tozzetti, 1868:731. *L. rotundum* Signoret, 1873:428. *L. prunastri*, Signoret, 1873:423; Šulc 1932:78. *Eulecanium piligerum* Leonardi, 1920:552. *Sphaerolecanium prunastri;* Silvestri, 1919a:70, 1920:501; Balachowsky & Mesnil, 1935:395; Schmutterer, 1952b:552, 1972b:418; Borchsenius, 1957:323; Řehaček, 1960:47; Ben-Dov, 1968:615; Kawecki, 1968:689, 1972:857; Tranfaglia, 1973:145; Argyriou & Paloukis, 1976:230; Kosztarab & Kozár, 1978:128; Kozár & Viktorin, 1978:401; Tereznikova, *1981:192.

Postreproductive female short oval, almost circular, strongly convex, lateral sides nearly vertical, slightly widened toward lower surface, dark brown or almost black, shiny, very finely punctate, ca 2–3.5 μ wide. Teneral female. *Venter*. With 8-, rarely 7- or 6-segmented antennae, ca 270–309 μ long; labium 100 μ long, 118 μ wide, with 10 setae (Koteja, 1974a); legs short and thick, claw with small denticle, claw digitules unequal, one stout, with expanded tip, other slender; spiracular furrows 3–5 pores wide, each with 50–91 quinquelocular pores (rarely some pores with 3 to 6 loculi) in band 3–5 pores wide; 3 spiracular setae in each group, rarely 2 or 1, setae similar in length, with blunt tip, stouter and normally shorter than most marginal setae; marginal setae slender, spinelike or hairlike, 15–45 μ long, form an irregular band 2–4 setae wide, space between setae normally less than length of longest seta; other body setae, of various lengths, 8 long interantennal setae, medial pair longest; medial pairs of setae on sternites 5–7, 110–120 μ long; oval dark-rimmed pores, ca 3 μ long, form a submarginal band on abdomen, scattered in medial area of thorax and abdominal sternites 1–3; multilocular pores form transverse rows and bands in medial area of abdomen;

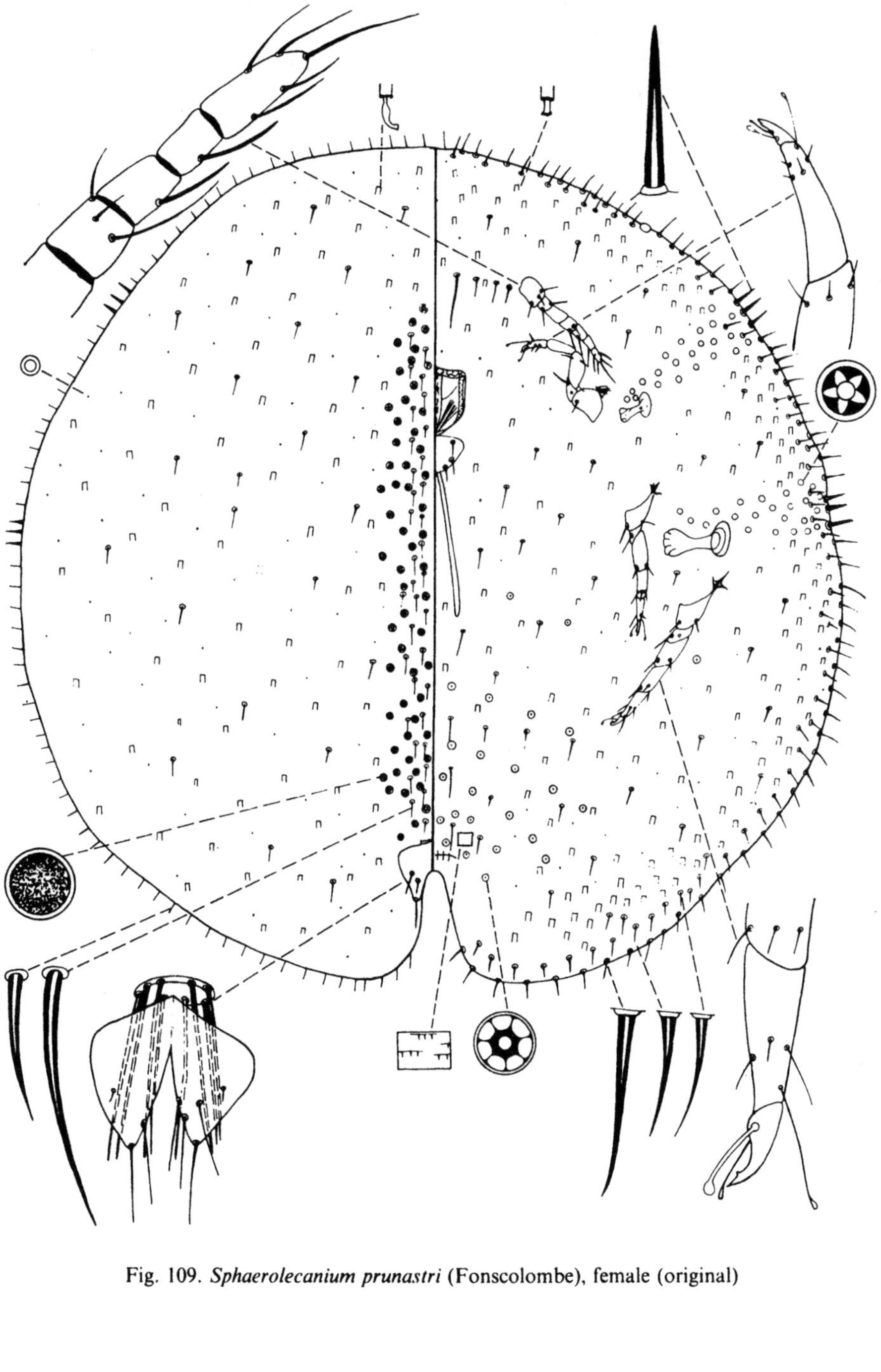

Fig. 109. *Sphaerolecanium prunastri* (Fonscolombe), female (original)

apparently tubular ducts absent. *Dorsum*. With scattered simple disc pores, setae slender, hairlike or spinelike, ca 12–23 μ long; a wide longitudinal medial band of sievelike discoidal pores and 50–120 μ long hairlike setae on head to anal plates; no tubular ducts visible in our specimen on dorsum; anal ring with 8 stout long setae and 2–4 slender and shorter setae; anal plates triangular with rounded lateral angles; ca 186–210 μ long, and 4 apical setae each (Borchsenius, 1957). Adult males and tests described by Silvestri (1919a), Leonardi (1920), Šulc (1932) and Borchsenius (1957); male by Giliomee (1967). Male test characteristic for species. N_1 and N_2 described by Silvestri (1919a), Leonardi (1920), Šulc (1932), and Borchsenius (1957); N_3 by Ben-Dov (1968).

Distribution. A common species in Palearctic Region. In CE: Au, Bu, Cz, Fr, Ge, Hu, It, Po, Ro, Sw, USSR and Yu. Also known from Greece, Spain, Turkey, Israel, Lebanon, Iran, China, Japan, and N America.

Biology. Infests the branches and trunks, and prefers rosaceous fruit trees, especially stone fruits. Hosts include: *Amygdalus, Armeniaca, Cerasus, Cydonia, Malus, Persica, Prunus, Pyrus*, and *Vitis*. One generation per year. Females develop through 4, and males through 5 stages. Second instars overwinter. Adults develop by end of April through early May in Hungary, mate, and females lay up to 3,000 eggs (Borchsenius, 1957). According to Schmutterer (1952b) eggs hatch in a few hours after oviposition, from end of June; N_1 settle on branches, not on leaves as in related genera; and molt into N_2 from September 5 through early October (Schmutterer, 1952b). Often a pest on stone fruits, especially on plums and peaches, as in CE (Kawecki, 1972; Kozár and Viktorin, 1978). Its common native host in CE is *Prunus spinosa*. *P. cerasifera* var. *Pissartii*, which also serves as a host, is a common ornamental tree in the area. Both are often reservoirs for reinfestation of fruit trees. The infestations on these 2 hosts are not controlled.

Natural enemies. Parasites. It is heavily parasitized in CE by a large number of chalcid wasp species whose nomenclatural position is often uncertain. The reliability of a large number of early parasite determinations is questionable. Some are hyperparasites. Parasites: *Cerapterocerus mirabilis, Cheiloneurus formosus, Coccophagus differens, C. excelsus, C. lycimnia, C. proximus, C. pulchellus, C. scutellaris, Discodes coccophagus, Encyrtus lunatus, E. masii, E. sylvius, Eucomys lecaniorum, Habrolepis zetterstedti, Marietta picta, M. zebrata, Metaphycus silvestrii, Microterys hortulanus, Pachyneuron coccorum, P. concolor, Perissopterus zebra* and *Phaenodiscus aeneus*. Predators: *Chilocorus bipustulatus, Chrysopa* sp., *Coccidiphila gerasimovi, Exochomus quadripustulatus, Hyperaspis campestris., Oratocelis communimacula* (this noctuid was very effective in Crimea), *Pullus subvillosus*, and parasitic fungus *Cephalosporium lecanii*. (Borchsenius, 1957; Argyriou and Paloukis, 1976; Podsiadlo, 1981a).

Associated ants: *Formica fusca fusca, F. rufa pratensis, Lasius fuliginosus, L. niger niger* (Schmutterer, 1952b).

Genus **Vittacoccus** Borchsenius

Syn. & Lit.: *Vittacoccus* Borchsenius, 1952:271, 1957:137; Koteja, *1969:8, 1970:223; Danzig, 1975c:138, 1980a:251; Kosztarab & Kozár, 1978:128.

Type-species: *Lecanopsis longicornis* Green, 1916.

Ovisac and postreproductive female as in *V. longicornis*. Teneral female. *Venter.* Antennae 8-segmented; legs and spiracular pore bands well-developed; disc pores of simple, dark-rimmed type and with 3 to 9 loculi; 2 types of tubular ducts; a row or band of marginal setae. *Dorsum.* Disc pores with 3–6 loculi in submarginal band or in groups; tubular ducts on entire surface; anal ring with few pores and 6–8 setaa: anal plates triangular with 3 apical setae.

Distribution. Palearctic, with 2 species, 1 in CE.

Biology. On leaves of *Carex* and grasses.

147. **Vittacoccus longicornis** (Green) – Long-horned Scale
Fig. 110

Syn. & Lit.: *Lecanopsis longicornis* Green, 1916:26. *Vittacoccus longicornis*, Borchsenius, 1952:271, 1957:137; Koteja, *1969:8, 1970:223, 1974a:111; Danzig, 1975a:138, 1980a:251; Zahradnik, 1977:119; Kosztarab & Kozár, 1978:128. *V. ordinatus*, Danzig, 1971a:1416.

Ovisac elongate, stout, cylindrical, white. Postreproductive female elongate, moderately convex, often covered with loose white woolly secretion. Teneral female elongate, oval, almost parallel-sided, narrowing to posterior apex; before oviposition, yellow with two red, broad longitudinal bands on dorsum (Koteja, 1969), 3.5–4.8 mm long, 1.3–2 mm wide. *Venter.* Antennae 8-segmented, 400–482 μ long; labium 75 μ long, 82 μ wide (Koteja, 1974a); legs elongate, tibio-tarsal articulation free and sclerosis present; 690–860 μ long; claw digitules stout, 27–35 μ long; spiracles small, spiracular pore band with partly double rows of pores, with 5 (rarely 3–6) loculi, 16–32 pores per band; spiracular setae absent; marginal setae hairlike, 7–15 μ long; submarginal setae similar in shape and size to marginal setae; interantennal and abdominal median setae up to 100 μ long; simple disc pores along body margin and scattered on both surfaces; dark-rimmed pores in preantennal area and in a submarginal band on abdomen; quinquelocular pores, rarely with 3–6 loculi, form a marginal band; multilocular pores with 7–9 loculi form a broad horseshoe-shaped band around genital opening and a wide transverse band on preceding sternite; large tubular ducts, ca 15 μ long and 4 μ in diameter, numerous on most of venter, absent from anal lobes and medial area; narrower tubular ducts anterior to vulva, ca 18 μ long and 2.5 μ in diameter. *Dorsum.* Body setae ca 5 μ long, stout, spinelike, not conical; disc pores with 3–6 loculi in a marginal band, extending from venter; discoidal pores in a longitudinal median band from head to anal plates; large tubular ducts scattered on entire surface; anal ring with a few pores and 6–8 setae, 130–170 μ long, 2 setae shorter and thinner than others; anal plates triangular, ca 100 μ long; 60 μ wide, each with 3 apical setae. Adult male described in detail by Koteja (1970).

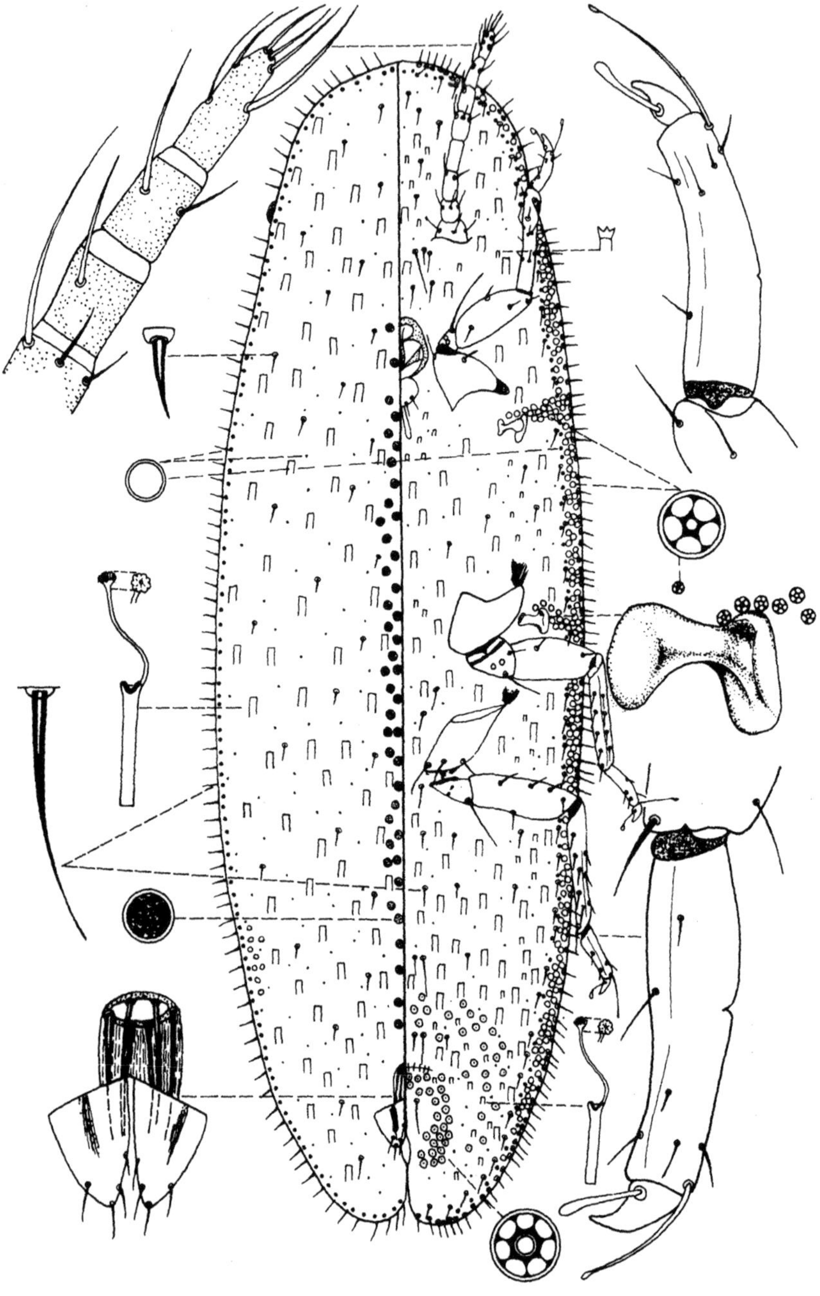

Fig. 110. *Vittacoccus longicornis* (Green), female (original)

Distribution. A rare boreal Trans-Palearctic species originally described from England. In CE: Cz, Po and USSR (Far East). Kozár (unpublished) also found it in the European part of USSR.

Biology. At base of *Carex* leaves. In Poland it hibernates in egg stage; larvae hatch in May; first, second and third molt during May, June, July; egg laying in August, sometimes in September (Koteja, 1985, personal comm.).

FAMILY **ACLERDIDAE** – FLAT GRASS SCALES

Syn. & Lit.: Aclerdini Cockerell, 1905:197; Balachowsky, 1948:255. Aclerdinae (in Coccidae) Koteja, 1974a:55, 113. Aclerdidae, Teague, 1925:432; Ferris, 1937:6, 1951:2; Balachowsky, 1942:43; McConnell, *1943:104, 1953:20; Borchsenius, 1950:156, 1960:248; Koteja & Liniowska, 1976:671; Kosztarab & Kozár, 1978:129.

Female body exposed or covered with thin glassy transparent test or with wax powder. Adult female oval, flat, light yellow to brown, heavily sclerotized at maturity, especially at two body extremities, 1.5–15 mm long; antennae reduced to tubercles with setae, rarely with ca 4 irregularly-shaped segments; labium 1-segmented, short, semicircular, wider than long; legs absent or represented by sclerotized spots with groups of setae; spiracles large, spiracular atrium expanded, with quinquelocular pores; anal complex includes a telescoping anal tube evaginated from base of anal cleft, forming single anal plate that covers anal ring and bases of anal ring setae; sclerotized posterior margin of abdomen includes ridges and furrows and with the anal complex gives the characteristic appearance to members of this family; simple pores normally scattered on dorsum; trilocular to multilocular pores on venter only, normally in a wide submarginal band and in clusters near spiracles; tubular ducts of 2 forms, macrotubular ducts form bands on dorsal and ventral submargins, rarely absent or distributed over both surfaces, microtubular ducts, 2–4 μ in diameter, 1–3 times as long as wide, form bands on ventral submargin, occasionally in bands on dorsal submargin or scattered over dorsum; marginal setae vary from dome-shaped to elongate and stout, arranged in a single row or in bands on body margin; body setae of various shapes and sizes on both surfaces. Adult male and nymphs described by McConnell (1953) and Borchsenius (1960). Metamorphosis of *Aclerda berlesei* studied in detail by La Face (1916).

Distribution. World-wide with ca 50 species in 3 genera; in CE 1 genus with 1 species.

Biology. Poorly known. Feed primarily on Gramineae; except 2 species feed on Orchidaceae, 1 on Cyperaceae, and 2 on Combretaceae. These insects are usually on plant stem under leaf sheaths, rarely on roots.

Genus **Aclerda** Signoret

Syn. & Lit.: *Aclerda* Signoret, 1874:96; Leonardi, 1920:367; Teague, 1925:432; Kuwana, 1932:57; Balachowsky, 1939:55; McConnell, *1954:23; Borchsenius, 1950:156, 1960:259; Ferris, 1955:3; Kosztarab & Kozár, 1978: 129; *Pseudolecanium* Cockerell, 1896b:19.

Type-species: *Aclerda subterranea* Signoret, 1874.
Morphological description of females as given at the family Aclerdidae.
Distribution. World-wide, with ca 47 species; 1 in CE.
Biology. Feed on Gramineae, except 2 species on Orchidaceae, and 1 on Cyperaceae.

148. **Aclerda subterranea** Signoret – Flat Root Scale
Fig. 111

Syn. & Lit.: *Aclerda subterranea* Signoret, 1874:97; Lindinger, 1912:60, 172, 216; Teague, 1925:439; Balachowsky, 1939:55. McConnell, *1954:89; Borchsenius, 1950:157, 1960:261; Kosztarab & Kozár, 1978:131. *A. signoreti* Lindinger, 1912:170, 1937:178; Balachowsky, 1939:56.

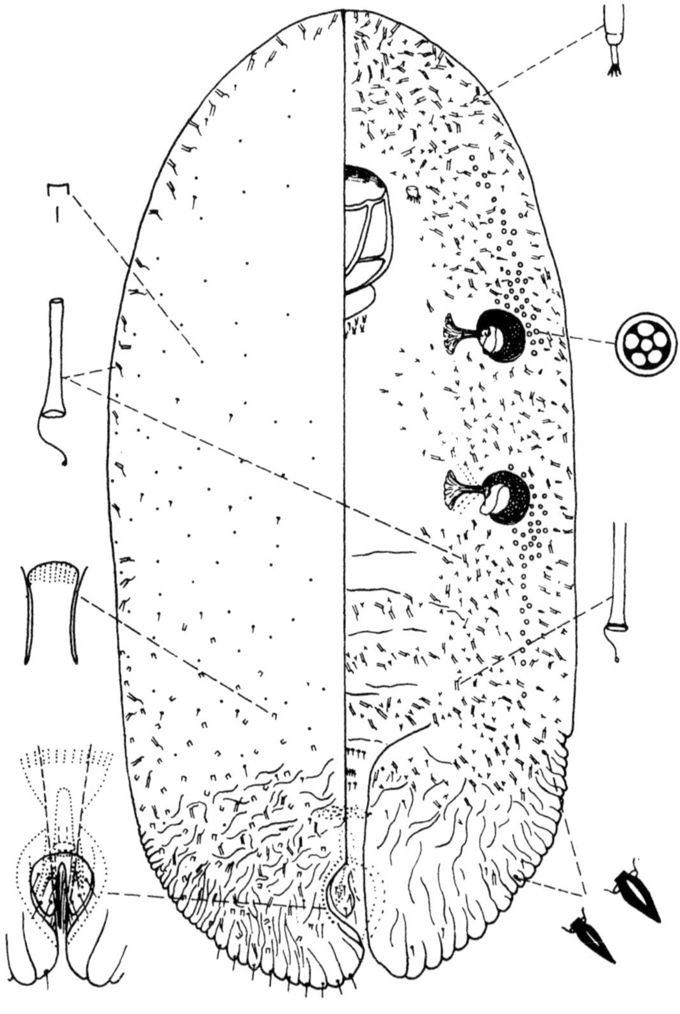

Fig. 111. *Aclerda subterranea* Signoret, female (modified after McConnell, 1953).

Female with white powdery wax along margin, and with a transparent glassy wax covering entire body (Borchsenius, 1960). Adult female oval or elongate oval, flat, 2–6 mm long, old specimens up to 15 mm long; antennae as unsegmented tubercles with several apical setae; labium about 3 times as wide as long; legs absent; spiracles very large, with slender vase-shaped bar; anal ring with ca 20 setae; anal plate split almost to its base, as wide as long; anal cleft ca 2.5–3 times as long as anal plate; simple disc pores numerous over entire dorsum; quinquelocular pores in large elongate clusters of 25–125 each near spiracles; macrotubular ducts form wide band along submargin of both surfaces, others in 4 transverse bands in median area of abdomen; microtubular ducts scattered on venter mixed with macrotubular ducts, but more numerous; marginal setae pointed, far apart, in a single row; ca 50–60 dorsal invaginated setae on each side of abdomen; body setae variable in size and shape, more on venter. First instar described by McConnell (1954) and Borchsenius (1960).

Distribution. Known from Au, Fr, and It. Rare.

Biology. Lives on roots and stems of *Agropyron* and *Milium* (Teague, 1925). Apparently a xerophilous Mediterranean species. Biology unknown.

Natural enemies. Parasites: *Aschitus subterraneus* (Trjapitzin, 1978); *Anabrolepis zetterstedti; Cheiloneurus elegans, Microterys ferrugineus, M. matritensis* and *Paraphaenodiscus subterraneus*. Parker (1960) noted 68% parasitization, principally by the latter species. Predator: *Eublemma jucunda*.

FAMILY **KERMESIDAE** – GALL-LIKE SCALES

Syn. & Lit.: Kermesidae Signoret, in Lobdell, 1929:762; Schmutterer, 1972c:418; Koteja, 1974b:299, 1974a:49; Kosztarab & Kozár, 1978:85; Danzig, 1980a:235. Hemicoccinae (Coccidae), Leonardi, 1920:266. Kermesini (Lecanoidea; Kerminae) Balachowsky, 1948:254. Kermococcidae Balachowsky, in Borchsenius, 1950:127, 1960:9. Dactylopiidae (in part), Ferris, 1955:69. Eriococcidae (in part), Ferris, 1955:192, 1957:87.

Postreproductive females of *Kermes* spp., as given at that genus. Females of *Olliffiella* spp., because they live in oak leaf galls, their derm is not heavily sclerotized. Teneral females described at genus *Kermes*. For accurate species determination it is essential to collect and mount on slides young females that just completed their last molt, thus their body is still soft and flat. This stage was seldom described in the past, therefore determinations for most species are still based on postreproductive females that are often hard to distinguish from each other.

Distribution. Holarctic; with ca 70 species in 6 genera. Only 1 genus with 5 species is represented in CE.

Biology. All members of this family feed on Fagaceae, but only on oaks, *Quercus* spp., in Europe. Biology of only a few species was studied. Seldom become pests on oaks. Because they produce large quantities of honeydew, ants and other hymenoptera are common visitors. Numerous species of parasites and predators normally keep the populations at low density.

Note. Kermesidae is one of the least studied families among the scale insects. A monograph on the Nearctic fauna was prepared in 1985 by Bullington, Baer and Kosztarab. Teneral females for most of the Palearctic species need to be collected, slide-mounted, described and illustrated.

Genus **Kermes** Boitard

Syn. & Lit.: *Kermes* Boitard, 1828:171; Signoret, 1875:547; King, 1900:78; Lindinger, 1912:284; Kuwana, 1931:15; Balachowsky, 1950a:342, 1950b:739, 1953:181; Schmutterer, 1952b:419; Ferris, 1955:192; Hoy, 1963:136; Koteja, 1974b:299, 1974a:49; Kosztarab & Kozár, 1978:85. *Chermes* Geoffroy, 1762:498. *Talla* von Heyden, 1860:90; Lindinger, 1933:143. *Kermococcus* Silvestri, 1911:148; Leonardi, 1920:266. Borchsenius, 1950:128, 1960:9.

Type-species: *Coccus variegatus* Gmelin 1789, a junior synonym of *Chermes roboris* Fourcroy, 1785.

Postreproductive females heavily sclerotized, spherical, heart, kidney, bud or gall-shaped; from light brown to black; 2–10 mm long; often with furrows, pits and bumps, also with variously colored bands and spots or rarely with white wax powder on derm. Teneral females subcircular or oval; antennae 2–9-segmented; legs short, normally 5-segmented; labium long, slender, 3-segmented, with 12–14 setae; spiracles large, especially posterior ones, with associated multilocular pores in their peritreme; no spiracular furrows or spiracular setae; anal ring flat, normally with setae, but without pores; anal ring surrounded by a large band of 3- to 9-locular pores extending to anal lobes; multilocular pores normally in transverse bands of pore clusters on abdominal venter, also in bands or groups in the area of spiracles; tubular ducts of at least two types, micro- and macroducts; microducts normally scattered on entire dorsum; macroducts form a wide band along dorsal margin; slender setae on both derms; setae often surrounded by a group of disc pores; large group of setae often form a horseshoe-shaped group near anal ring; often with sclerotized irregular spots on both sides of abdomen. First instars described and used in classification by Kuwana (1931), and Balachowsky (1953d). All immature stages and adult male of *K. kingi* Cockerell described by Hamon et al., 1976; adult male of *K. quercus* by Koteja and Zak-Ogaza (1972).

Distribution. Holarctic, between the Northern latitudes of 25–60 (Balachowsky, 1950a) with ca 67 species; ca 30 in the Palearctic Region, with 5 species in CE.

Biology. Infest bark and bark crevices of Fagaceae, seldom leaves (in North America). One yearly generation in species studied; lay up to 6,000 eggs per female; first instars overwinter; males known in most species.

Key to species of *Kermes*
(Based primarily on postreproductive unmounted females)

1. Dorsal surface smooth or with small pits, never with deep furrows; at most with 1 median transverse or longitudinal wide and shallow depression 2
   Dorsal surface not smooth, but with deep longitudinal, and often with transverse furrows 3
2. Body spherical; venter almost flat; dorsum normally with 3–4 wide black transverse bands, with 6–10 black oval spots between bands; legs ca. 400–420 μ long; widely distributed in Europe (fig. 18).... *roboris,* p. 267
   Body not spherical, somewhat horseshoe-shaped; venter bulging; dorsum with more than 5 irregular black transverse bands and oval spots, bands often

supplemented by rows of oval spots, occasionally only 2 spots in a row; legs ca. 300 μ long; known from USSR and possibly Triest area in Italy . . . . *corticalis*, p. 264

3(1). Only with longitudinal furrow on derm, rarely even this developed only on posterior half of derm; body heart or kidney-shaped, venter often wedge-shaped, bulging; in bark crevices of trunk and older branches; common in CE (fig, 17, 113).. *quercus*, p. 265

Derm with both longitudinal and transverse furrows; body globose; venter flat, not bulging; on twigs and branches of southern oaks; rare in CE......4

4. Derm one-colored, brownish-black or black, shiny; dorsum with longitudinal and 2–3 transverse shallow furrows, few pits and bumps; 5.5–6 mm long; a circum-Mediterranean species; besides *Quercus cerris*, infests a number of S European oaks (fig. 112)........................ *bacciformis*, p. 263

Derm bicolored, dark reddish brown; dorsum with few circular black spots, with black longitudinal and 4–6 black, deep and wide furrows; 3–5 mm long; known from Austria and Hungary only; on *Quercus cerris* .... *gibbosus*, p. 265

Key to first instars of three species of *Kermes*

1. Dorsal marginal setae long spinelike, sharp-pointed; only 2 longitudinal rows of marginal setae developed; tarsus of midleg more than twice as long as tibia; anal lobes well developed .... *roboris*, p. 267

Dorsal marginal setae short, bullet-shaped, blunt-tipped; 4 or 6 longitudinal rows of dorsal setae developed; tarsus of midleg less than twice as long as tibia; anal lobes poorly developed.............................. 2

2. Abdominal venter with 2 submedian longitudinal rows of disc pores; dorsal submedian setae present on head; dorsal thick setae more than 1 ½ times as long as wide; third antennal segment as long as sixth or longer .... *bacciformis*, p. 263

Abdominal venter without longitudinal rows of disc pores; dorsal submedian setae absent from head; dorsal thick setae less than 1 ½ times as long as wide; third antennal segment shorter than sixth .... *quercus*, p. 265

149. **Kermes bacciformis** Leonardi – Pearl-like Kermes
Fig. 112

Syn. & Lit.: *Kermes bacciformis* Leonardi, 1908:156; Lindinger, 1912:284; Balachowsky, 1950b:7, 1953d: 182; Hoy, 1963:140; Kozár, 1974:535; Kosztarab & Kozár, 1978:88. *Kermococcus bacciformis* Leonardi, 1920:273; Gomez Menor, 1937:246; Borchsenius, 1960:72.

Postreproductive female spherical, somewhat flattened on venter; dorsum divided into two halves by a longitudinal furrow, and further divided by 2–3 transverse furrows; normally furrows less obvious, but body, according to Leonardi (1908), often blackberry-like; some small pits and humps also present; at anal region in these pits a large "gland" opening surrounded by derm of polygonal structure with scattered small disc pores; derm brilliant tarlike black, shiny; body 5.5–6 (rarely 7) mm long, ca 5 mm wide and high. Teneral female apparently never described due to lack of suitable specimens for slide-mounting. Leonardi (1908) illustrated antennae with 6 segments, 3rd segment longest, terminal segment with ca 7 setae; leg short, thick, 5-segmented, tarsus much longer than tibia; dorsum with heavily sclerotized areas surrounding multilocular pores in their center. First instar described in detail by Leonardi (1908, 1920) and Balachowsky (1950b).

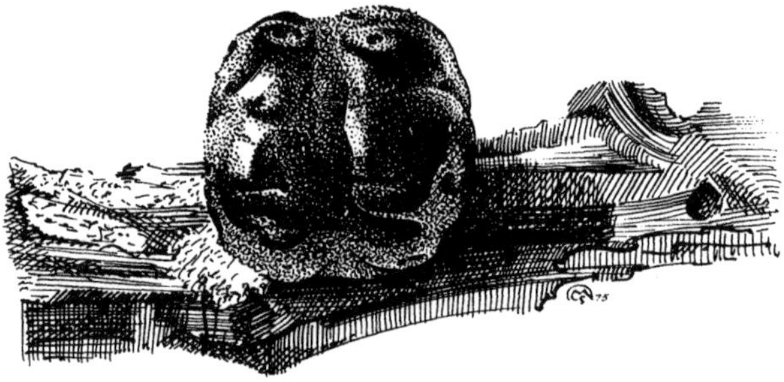

Fig. 112. *Kermes bacciformis* Leonardi, female on a twig of *Quercus* sp. (after Kosztarab and Kozár, 1978)

Distribution. A rare circum-Mediterranean species, known in CE from It, Fr, Hu; also from Spain and Morocco.

Biology. It infests branches and twigs of *Quercus cerris, Q. ilex, Q. suber* and *Quercus* sp. Biology unknown, except that this is the only Mediterranean *Kermes* species that has succeeded in adapting to oaks with deciduous leaves (Balachowsky, 1950a). No records on males.

Natural enemies. Parasites: *Aenasioidea hispanica, Blastothrix erythrostetha, Encyrtus chalcostomus, Metaphycus hirtipennis, Psilophrys longicornis* (Borchsenius, 1960).

150. **Kermes corticalis** (Nassonov) – Southern Kermes

Syn. & Lit.: *Kermes variegatus corticalis* Nassonov, 1908b:490. *K. roboris*, Lindinger, 1912:369; Bodenheimer, 1931:243. *Talla roboris*, Lindinger, 1957:549. *Kermococcus corticalis*, Borchsenius, 1950:130, 1960:46. *K. corticalis*, Kosztarab & Kozár, 1978:87.

Postreproductive female irregular, somewhat horseshoe-shaped; derm smooth, shiny, light or dark brown, with several black irregular transverse bands and 3 rows of oval spots, occasionally only 2 spots in a row; venter bulging, covered with white powdery wax; body shape varies according to bark surface; body 4–6 mm long, 6–7 mm wide, 6–8 mm high. Teneral female, according to Borchsenius (1960), with 5- or 6-segmented antenna, third segment much longer than others, terminal segments with fleshy sensory setae; labium 3-segmented; legs thick, 300 μ long, short as antennae in proportion to body size, claw with 2 thin digitules; spiracles large, diameter of atrium 135–150 μ, each spiracle surrounded by numerous disc pores; anal ring open on two ends, with 2–3 small bumps, each with 1 short spinelike seta; disc pores form 7 bands on venter; body margin with a row of 44–50 groups of pores, each group of 3 to 11 disc pores surrounding a marginal seta; tubular ducts in large number, form a horseshoe-shaped band on venter; numerous microtubular ducts on dorsum; 2 bands of setae, each 36–70 μ long, near anal ring. Some records on first instar given by Borchsenius (1960).

Distribution. A rare species. Known from USSR (Ukraine, West Georgia, to N Caucasus), also from Triest area of Italy if we accept Borchsenius' (1960) suggestion that *K. cordiformis* is a junior synonym of *K. corticalis*.

Biology. It infests bark crevices on trunk and branches of *Quercus* spp. Often forms larger colonies and occasionally becomes a pest. Males unknown; its biology poorly known.

Natural enemies. Parasites: *Blastothrix erythrostetha*, and *Psilophrys longicornis* (Borchsenius, 1960; Trjapitzin, 1978).

Note. According to Nassonov (1908b), the coloration of postreproductive females and the morphology of first instars is very similar to *K. roboris* (called *K. variegatus* Gmelin by him).

151. **Kermes gibbosus** Signoret – Humped Kermes
Fig. 16

Syn. & Lit.: *Kermes gibbosus* Signoret 1875:552; Lindinger 1912:285; Bodenheimer, 1931:244; Hoy 1963:145; Kosztarab & Kozár, 1978:89. *Kermococcus gibbosus*, Borchsenius, 1960:73; Tsalev, 1964:23.

Postreproductive female spherical, without body segmentation, dorsum divided by 1 longitudinal and 4–6 black, deep and wide transverse furrows; derm bicolored, dark reddish-brown with a few circular black spots, often covered with white wax powder; shape and size varies, especially at high population density when often deformed; 3–5 mm long, 5–6 mm wide, 3–4 mm high (Lindinger, 1912). Teneral females unknown, apparently never collected.

Distribution. A rare species; known from Au, Bu, Ge and Hu in CE.

Biology. No records available on its biology or males. Collected from *Quercus cerris* only.

152. **Kermes quercus** (Linnaeus) – Striped Kermes
Figs 17, 113

Syn. & Lit.: *Coccus quercus* Linnaeus, 1758:455. *C. cambii*, Ratzeburg, 1844:194. *K. reniformis* Signoret, 1875:553. *K. lindingeri* King, 1914:34. *K. cordiformis* Lindinger, 1912:286. *K. quercus*, Lindinger, 1912:285; Balachowsky, 1950a:343; Schmutterer, 1952b:419, 1972c:419; Hoy, 1963:154; Pfugfelder, 1970:397; Koteja & Zak-Ogaza, 1972:199; Saakyan-Baranova & Muzafarov, 1972:498, 697; Kozár, 1974:534; Kosztarab & Kozár, 1978:88. *Kermococcus quercus*, Borchsenius, 1960:65.

Postreproductive female heart or kidney-shaped, venter wedge-shaped, bulging; derm only with transverse furrows base color dark-brown; with 4–8 black transverse bands on dorsum; body 3–4 mm long, 3.5–5.0 mm wide, 4–6 mm high. Teneral female with 4-segmented antennae; labium 3-segmented, with 12 setae, labial stylets long; legs

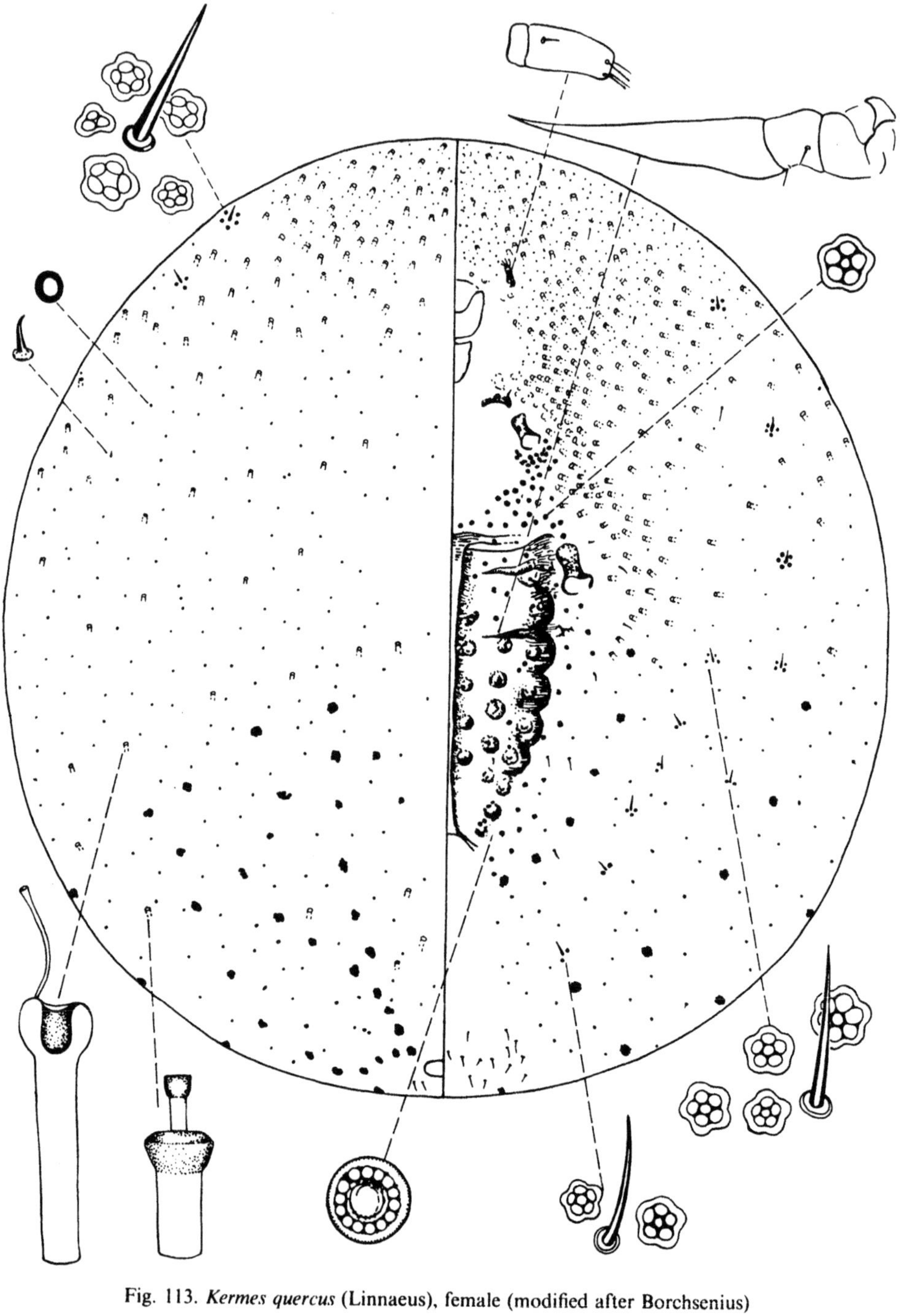

Fig. 113. *Kermes quercus* (Linnaeus), female (modified after Borchsenius)

small, segmentation not clear, appears 3-segmented; claw digitules longer than claws; spiracles large, with associated multilocular pores, these with 6–8 loculi; anal ring circular with 2 short setae on posterior margin; multilocular pores form longitudinal rows or bands on abdominal venter; numerous microtubular ducts distributed on dorsum; macrotubular ducts form a horseshoe-shaped band on anterior part of venter, a few similar ducts present on dorsum also; 18–22 groups of disc pores, with 5–6–7-loculi, each surrounding a spinelike seta, form row on body venter; heavily sclerotized areas form rows on venter and dorsum; a horseshoe-shaped group of setae (each seta 20–70 μ long) near anal ring. Adult male described by Borchsenius (1960), also by Koteja and Zak-Ogaza (1970). Nymphs described by Balachowsky (1950b), Borchsenius (1960), Saakyan-Baranova and Muzafarov (1972).

Distribution. A common species in Europe. In CE: Au, Bu, Cz, Fr, Ge, Hu, Po, Ro, USSR (European part), and Yu. Also known from the British Isles, Denmark, Holland, Finland, Portugal and Sweden.

Biology. It infests different *Quercus* species. Primarily lives in bark crevices of trunk and on thick branches, but at high density is found on thinner branches also. It has 1 yearly generation; second instars overwinter, molt to adults at end of April or first third of May in S Germany, mate and start laying eggs, average of 1,500, max. 2,500, from end of May (Schmutterer, 1972c); eggs hatch in 2–3 weeks.

Occasionally, at high population density damage trees, e. g. at end of 19th century in Hungary (Ratkovsky, 1899), in Russia (Nassonov, 1910), and recently in Hungary (Kozár, 1974) and in Poland (Koteja, 1984a). Ants and bees are their common visitors for the large quantities of honeydew produced.

Natural enemies. Parasites: *Encyrtus chalcostomus, Microterys ferrugineus,* and *Tetrastichus pachyneurus.* Predators: *Exochomus quadripustulatus* and larvae of *Sesia vespiformis* (Kosztarab, 1959; Schmutterer 1972c; Trjapitzin, 1978).

Associated ants. *Formica cinerea, Lasius fuliginosus, L. niger, Myrmica laevinodis, M. ruginodis, Leptothorax acervorum* (Schmutterer, 1952b).

153. **Kermes roboris** (Fourcroy) – English Oak Kermes
Fig. 18

Syn. & Lit.: *Chermes roboris* Fourcroy, 1785:229. *Lecanium quercus* Bouché, 1851:112. *Kermes pallidus* (Reaumur), *K. variegatus* (Gmelin, 1789), (Balachowsky, 1950b). *K. roboris* var. *simplex* Lindinger 1949:213. *Kermococcus roboris,* Leonardi 1920:269; Borchsenius, 1950:130, *1960:41; Kosztarab, 1959:404; Tsalev, 1964:23. *Talla roboris,* Kawecki, 1936:322. *K. roboris,* Lindinger, 1912:287; Balachowsky, 1950a:343, 345, 1950b:740, 743; Hoy 1963:157; Schmutterer, 1972c:419; Kosztarab & Kozár, 1978:87.

Postreproductive female nearly spherical, almost flat on venter; derm smooth, shiny, normally light brown, rarely dark brown, with 3–4 transverse irregular black bands made of spots and 2–10 oval spots between bands; 5.0–7.5 mm long, 6.8–8.5 mm wide, 5.5–8.0 mm high, rarely larger. If female crowded or in a twig forking, often deformed, not spherical. Teneral females oval, quite flat in side view, resembling Coccidae; antennae 6-, rarely 5-segmented, ca 220—260 μ long, third segment much longer than others; labium 3-segmented; legs thick, short, 400–420 μ long, claw with 2 thin claw digitules; spiracles surrounded by numerous disc pores; anal ring slightly sclerotized normally without spinelike setae, or the most 2–2 present; multilocular disc pores on abdominal venter form 7 transverse bands composed of pore clusters, other multilocular pores with 6–7 loculi form small clusters around setae that form a row (ca 44–50) on body margin; tubular ducts numerous on dorsum, form a wide horseshoe-shaped band on venter; near anal ring 2 rows of setae, each seta 14–45 μ long, setal rows united posteriorly (Borchsenius, 1960). First instars described by Leonardi (1920), Balachowsky (1950b) and Borchsenius (1960).

Distribution. A common species in Europe south of 60° North latitude. In CE: Au, Bu, Cz, Fr, Ge, Hu, It, Po, Ro, USSR, and Yu. Also known from England, Portugal, Spain, Turkey and Morocco.

Biology. Females found singly on thin branches of *Quercus* spp. According to Schmutterer (1972c) it has one yearly generation; lay eggs in May in CE; commonly visited by ants; no records on males.

Natural enemies. Parasites: *Blastothrix erythrostetha, Cheiloneurus formosus, Microterys ferrugineus, Phaenodiscus aeneus, Psilophrys longicornis, Tetrastichus pachyneurus* (Borchsenius, 1960).

FAMILY **CRYPTOCOCCIDAE** – BARK-CREVICE SCALES

Syn. & Lit.: Cryptococcidae Kosztarab 1968:12; Koteja, 1974b:50; Howell & Williams, 1976:187; Kosztarab & Kozár, 1978:82. Dactylopiidae (in part), Ferris, 1955:69. Eriococcidae (in part), Hoy 1963:53; Danzig, 1980a:234.

Adult females in bark crevices of trees, covered with fine, white, waxy secretion or feltlike ovisac. Female oval, derm sclerotized at full maturity; antennae 1–6 segmented; clypeolabral shield (Koteja and Liniowska, 1976) pentagonal, clypeal setae absent, labral ridge very short, marginal ridges strong; labium short and broad, 3-segmented, with 8 pairs of setae; legs present, or entirely absent or reduced to unsegmented stubs; spiracles surrounded by a sclerotized frame, with a group of associated quinquelocular pores; normally a cluster pore plate just below each posterior spiracle; anal ring heavily sclerotized, with 4–6 short setae or hairs, with or without pores, surrounded by large setae; no cruciform pores; simple disc pores in dorsal submargin; normally trilocular pores on dorsum; quinquelocular pores on both derms; seldom quadrilocular pores scattered near spiracles; tubular ducts on both derms, often of two types. Adult males present in *Pseudochermes fraxini* only.

Distribution. Holarctic, with 1 exception, *C. nudatus* Brittin, in New Zealand. Total of 6 species in 2 genera; in CE 3 species in 2 genera.

Biology. Feed in bark crevices of deciduous trees. Apparently with one yearly generation.

Key to genera of Cryptococcidae

Legs developed and hind coxae with translucent pores; no cluster pore plates posterior of hind spiracles; antennae 6-segmented; normally on *Fraxinus* (fig. 115) .... *Pseudochermes*, p. 272

Legs absent or reduced to unsegmented stubs; with cluster pore plate posterior of each hind spiracle; antennae 1–5-segmented; on other trees than *Fraxinus* (fig. 114) .. .. *Cryptococcus*, p. 269

Genus **Cryptococcus** Douglas

Syn. & Lit.: *Cryptococcus* Douglas 1890a:155; Borchsenius, 1949:370; Ferris, 1955:83, 1957:85; Hoy 1962:25, 1963:53; Kosztarab & Hale, 1968:7; Kosztarab, 1968:12; Kosztarab & Kozár, 1978:84; Danzig, 1980a:234.

Type-species. *Coccus fagi* Baerensprung, 1849.

Adult females covered with fine, white, waxy secretion. Female oval, 0.5–1.0 mm long; antenna normally 1–2, rarely 4–5-segmented; labium 3-segmented; legs absent, rarely reduced to unsegmented stubs; spiracles with sclerotized frame and a few associated quinquelocular pores; with a cluster pore plate below each hind spiracle; anal ring with 4–5 setae, rarely with pores; anal tube well developed, sclerotized; with simple, trilocular and quinquelocular pores, seldom with quadrilocular pores; without cruciform pores; tubular ducts, normally two types on both derms.

Distribution. World-wide with 5 species; 2 in CE.

Biology. Live under bark flakes or in crevices of bark. One yearly generation in CE.

Key to species of *Cryptococcus*

Antennae 4-segmented; anal ring with 6 thin setae; 10 setae surround anal ring; each spiracle with 6–9 associated quinquelocular pores; on *Acer* spp., seldom on *Pyrus* or *Tilia* .... *aceris*, p. 269

Antennae 3-segmented; anal ring with 4 thick setae; 6 setae surround anal ring; each spiracle with 3–6 associated quinquelocular pores; on *Fagus* spp. (fig. 114) .... *fagisuga*, p. 270

154. **Cryptococcus aceris** Borchsenius – Maple Bark Scale

Syn. & Lit.: *Cryptococcus aceris* Borchsenius, 1937:62, 1939a:45; 1949:372; Schmutterer, 1955a:160; Tereznikova, 1959:448; Kosztarab & Hale, 1968:7; Danzig, 1971:1415; Zahradnik, 1972:405.

Adult female covered by white waxy secretion. Female, according to Borchsenius (1949), hemispherical, yellow, up to 0.9 mm long; antenna 4-segmented, first segment much wider than others; labium broad, stylet loop extends to abdominal sternites 7 or 8; legs absent; each spiracle with 6–9 associated quinquelocular pores; anal ring almost circular, with 5–7 pores and 6 thin, 12 μ long, setae; anal ring surrounded by 10 setae; quinquelocular pores few, located singly along dorsal margin, also near labium and spiracles; tubular ducts form transverse rows on dorsum; body setae few.

Distribution. A rare species. In CE: Ge, Hu, USSR (including the Transcarpathian Region).

Biology. Feed on trunk of trees in the genera: *Acer*, *Pyrus* and *Tilia*. Biology poorly known. In S Germany it infested the trunk of *Acer pseudoplatanus* (Schmutterer, 1955a).

155. **Cryptococcus fagisuga** Lindinger – Beech Scale
Figs 19, 114

Syn. & Lit.: *Coccus fagi* Baerensprung (preoccupied) 1849:174; *Cryptococcus fagi*, Newstead 1903:215; Schmutterer, 1952b:417; Ferris, 1955:83, 1957:85; Kosztarab & Hale, 1968:7. *C. fagisuga*, Borchsenius, 1949:371; Kosztarab, 1959:402; Hoy, 1963:54; Zahradnik, 1972:403; Kosztarab & Kozár, 1978:85.

Female covered with white cottony wax secretion. Adult female circular or oval, light, yellow, 0.5–1.0 mm long; antenna stublike, 3-segmented, segmentation weakly developed; labium broad, stylet loop extends to middle of abdomen; fore and mid-legs absent, each hind leg reduced to an oval pore plate; spiracles heavily sclerotized, each with 3–6 associated quinquelocular pores; anal ring wide, platelike, with 5–6 pores and 4 thick setae, each 6–8 μ long; ring surrounded by 6 large thick setae; anal tube well developed; cruciform pores absent; a group of quinquelocular pores in area of anal ring; other quinquelocular pores mostly in marginal area of venter; micro- and macrotubular ducts distributed on dorsum and along ventral margin, macroducts of two sizes; conical setae, 3–4 μ long, distributed on dorsum; those on venter hairlike, 3 μ long. Morphological structures of first and second instars described by Bareth & Liefferinge (1976).

Distribution. A common species in beech forests. In CE: Au, Bu, Cz, Fr, Ge, Hu, It, Po, Ro, Sw, USSR and Yu. Also known from W and N Europe, Middle East, and Eastern N America.

Biology. Lives on trunk, branches and tree roots of *Fagus* spp., at up to 1200 m elevantion in CE. In one location found on *Pinus sylvestris* in N Hungary.* Occasionally becomes a pest, especially with the transmission of beech bark disease, caused by *Nectria coccinea* var. *faginata* and *N. galligena* (Schmutterer et al., 1957; Shigo, 1976). It has one yearly generation; normally first instars overwinter; adults develop by May and start laying yellow eggs (6–60 per female) end of June and continue through first part of August; eggs start hatching after 5–6 weeks (Zahradnik, 1972); crawlers settle in cracks of bark, produce a filamentous waxy secretion, covering entirely their body; develop through 3 stages. Although Reh (1903) reported males, no one else ever found these. His observations could have been based on males of another species.

Natural enemies. Predators: *Chilocorus bipustulatus, C. stigma, C. renipustulatus, Chrysopa ciliata, C. ventralis, Enicmus minutus, Exochomus quadripustulatus, Fumea casta, Histiogaster entomophagus, Microphysa pselaphiformis.* (Schmutterer, 1952b, Schmutterer et al., 1957).

* This unusual record is from crevices of *Pinus sylvestris* bark at Lászlótanya, Nagymilic Mtn., North Hungary, 13 Aug. 1980 (1 female), coll. M. Kosztarab and F. Kozár (Slide No. 1278); also 2 females on 2 slides, from same host and locality, by same collectors on 7 Aug. 1975; deposited in the collections of the Hungarian Research Institute for Plant Protection (HRIPP) (Slide No. 534a), Budapest, and the Virginia

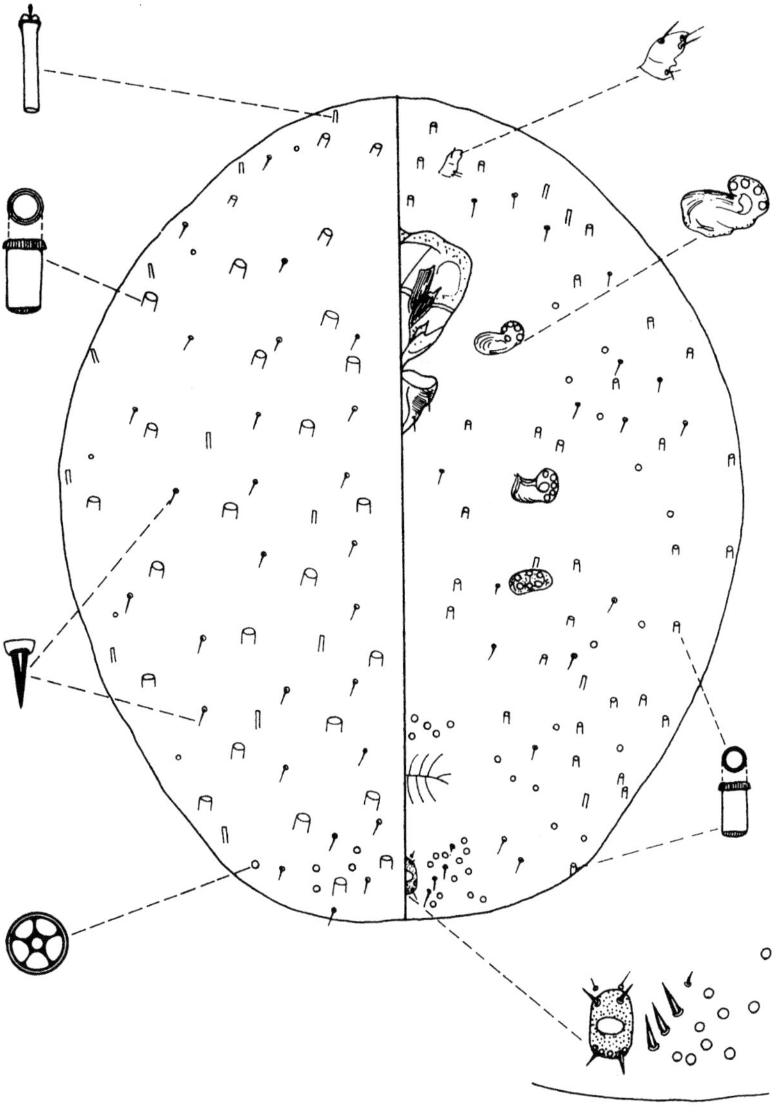

Fig. 114. *Cryptococcus fagisuga* Lindinger, female (modified after Kosztarab and Kozár, 1978)

Polytechnic Institute and State University, Blacksburg (Slide No. 534b). The adults somewhat resemble those of *C. aceris*. When an adequate sample is available for study this population (if not an accidental infestation from the nearby *Fagus*-inhabiting typical *fagi* population) might turn out to represent a new species or subspecies.

Genus **Pseudochermes** Nitsche

Syn. & Lit.: *Pseudochermes* Nitsche (in Judeich & Nitsche, 1895:1249). *Apterococcus* Newstead 1898:97. *Fonscolombia* Cockerell 1899:278; Schmutterer, 1952b:418. *Pseudochermes*, Borchsenius, 1949:365; Ferris, 1955:178; Hoy, 1963:187; Morrison & Morrison, 1966:164; Danzig, 1980a:231; Afifi, 1968:193; Kosztarab & Kozár, 1978:81.

Type-species: *Chermes* (?) *fraxini* Kaltenbach, 1860.

Ovisac feltlike, oval, cream colored, covers female completely. Adult female ovoid, with body segmentation well visible; antennae 6-segmented; labium 3-segmented with 8 setae; eyes present; hind coxa enlarged and with a group of translucent pores; claw without denticle; without cruciform pores; only 1–3 disc pores associated with each spiracle; disc pores, predominantly quinqueloculars, found on both derms; tubular ducts of two types, micro- and macroducts; anal ring with 6 stout setae, with few pores, no pores or setae anteriorly, anal lobes not distinct, each with 1 apical seta and a few short setae; conical setae slender, not enlarged, they form transverse rows on dorsal segments; hairlike setae on venter only.

Distribution. In the Palearctic; with 1 species only.

Biology. See under *P. fraxini*.

156. **Pseudochermes fraxini** (Kaltenbach) – Ash Bark Scale
Fig. 115

Syn. & Lit.: *Chermes fraxini* Kaltenbach, 1860:259. *Eriococcus fraxini* Newstead, 1891:165. *Ripersia fraxini* Newstead, 1896:57. *Apterococcus fraxini* Newstead, 1903:210. *Fonscolombia fraxini* Lindinger, 1912:159. *Fonscolombia fraxini* Šulc, 1912:33. *Pseudochermes fraxini*, Borchsenius, 1949:365; Schmutterer, 1952b:418; Kosztarab, 1959:402; Hoy 1963:188; Danzig, 1967:823, *1980a:233; Afifi, 1968:193; Zahradnik, 1972:403; Kosztarab & Kozár, 1978:82.

Ovisac feltlike, cream colored, ca 1.0 mm long, covers entire body. Adult female ovoid, orange-red, up to 0.75 mm long, 0.5 mm wide; antenna 6-segmented, ca 102–106 μ long; labium broad, 3-segmented, stylet loop extends to near apex of abdomen; leg segments thick, hind coxa with a group of translucent pores, femur 54–60-, tibia 32–42-, tarsus 35–42 μ long; quinquelocular disc pores form rows on both derms, except on prosomal venter, where these scattered; macrotubular ducts dispersed on ventral and dorsal derm, form a band along body margin; microtubular ducts on dorsum only; setae slender, larger ones 13–16 μ long, form transverse rows on terminal abdominal tergites, smaller ones 6 μ long, found on all tergites; ventral hairlike setae short and few; anal ring almost circular, with 6 thick setae, each 22 μ long, positioned in a semicircle, with pores in a single row; anal lobes hardly detectable, with apical seta ca 47–60 μ long, and 2 short setae on anterior of anal lobe dorsum. Adult male described by Afifi (1968).

Distribution. A common Trans-Palearctic species. In CE: Au, Bu, Cz, Fr, Ge, Hu, Po, Ro, Sw, USSR (Ukraine to Far East), and Yu; also known from S, W and N Europe and Middle East.

Biology. Primarily in crevices of trunk and branches of ash, *Fraxinus* spp., but also reported from *Sorbus* and *Syringa*. Often becomes a pest in towns. According to

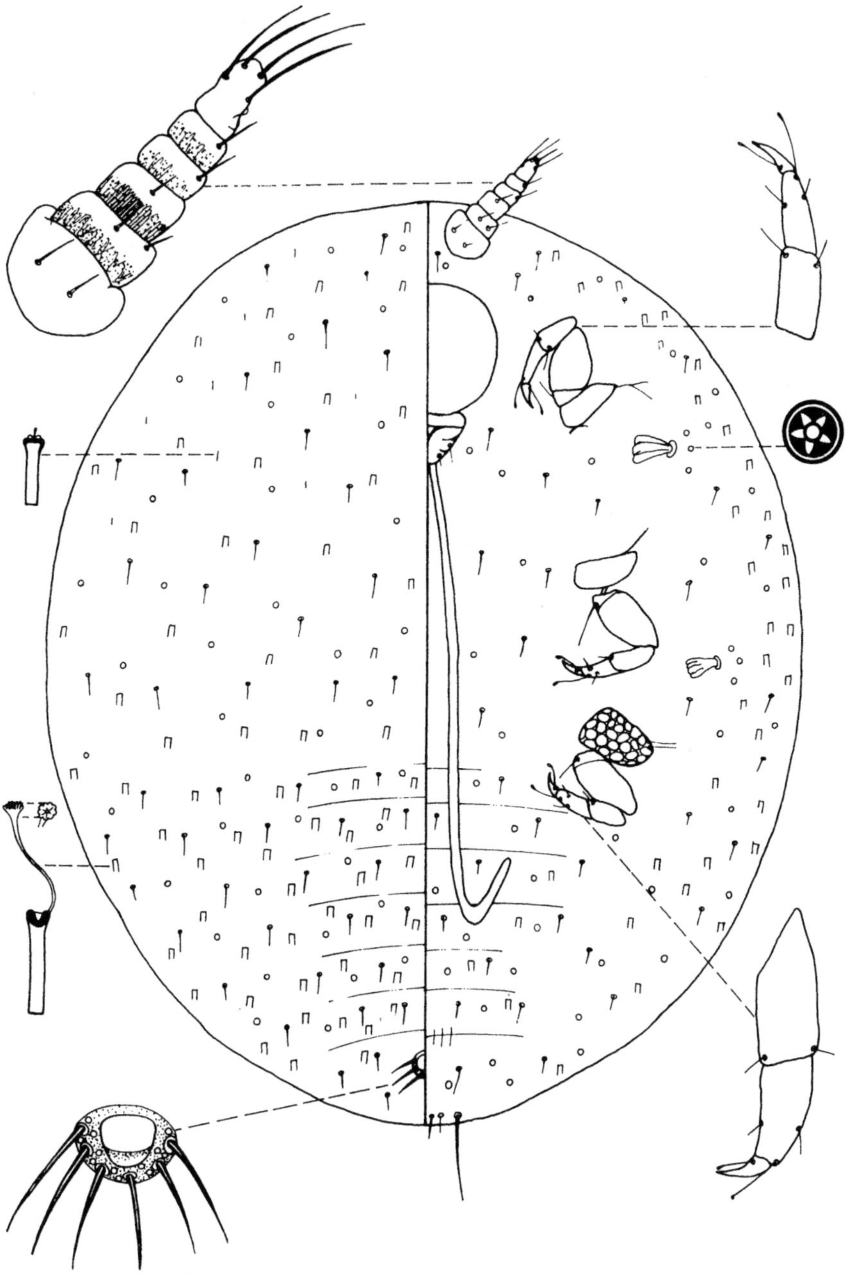

Fig. 115. *Pseudochermes fraxini* (Kaltenbach), female (original)

Schmutterer (1952b) it has one yearly generation, overwinters in N_2 stage; adults appear in March and April, lay 6–93 eggs per female in May, eggs hatch in June.

Natural enemies. Predators (Schmutterer, 1952b): *Anthocoris minki, Chilocorus bipustulatus, Exochomus quadripustulatus,* and *Platybunus pinetorum.*

FAMILY **ERIOCOCCIDAE** – FELT SCALES, ERIOCOCCID SCALES

Syn. & Lit.: Acanthococcites Signoret, 1875:314. Acanthococcidae (in part) Maskell, 1887:88. Dactylopites Signoret, 1875:339. Dactylopiidae, (in part) Fernald, 1903:39, Ferris, 1955:69. Eriococcini Cockerell, 1899b:389. Eriococcidae, Hoy, 1963:1. More below.

The status of this diverse family, its definition and borderlines, constitute much controversy among coccidologists today. The confusion with the selection of the proper family name, and type-genus for it, was described in the following major works Hoy (1962:11, 1963:5); Morrison and Morrison (1966:1, 69); Miller and McKenzie (1967:480); Afifi (1968:167); Miller (1969:47); Williams (1969:318, 324, 325); Koteja (1974b:295); Miller and Williams (1976:118); Koteja and Zak-Ogaza (1981:501). Some authors (e.g. Hoy, 1963) treat the genera *Acanthococcus* and *Rhizococcus* as junior synonyms of *Eriococcus*. See for more information Notes at the genus *Acanthococcus*.

The body of the adult female and her eggs are often entirely (*Eriococcus* and others) or seldom partially (*Gossyparia*) enclosed in a feltlike sac, thus the common name "felt scales".

Adult female. The morphology in eriococcids is very diverse, therefore we have used a combination of characters to define most genera. Female usually elongate-oval, seldom circular. Derm usually sclerotized. Most species with 6- or 7-segmented antennae; labium normally 3-segmented with 16–20 setae; normally with 5-segmented legs, these sometimes reduced, leg setae reduced in number, claws rarely without denticle; translucent pores normally on hind coxa and femur, rarely on tibia; derm seldom with bilocular, trilocular or simple pores; if trilocular pores present, those different from pseudococcid type pores; with quinquelocular, 7 or 9-locular disc pores; often with oval disc (cruciform) pores; derm normally with distinctive microtubular ducts; invaginated macrotubular ducts, often of 2 sizes; with enlarged conspicuous conical setae, and large protruding anal lobes; anal ring normally sclerotized and with pores and 6 or 8 setae, rarely reduced; dorsal ostioles, ventral circuli, and anal plates absent.

Adult males. Described by Newstead (1903), Jancke (1955), Afifi (1968) and some others. They can be winged or wingless; body stout, robust, in wingless species usually spindle-shaped; in wingless forms simple eyes absent and ocelli reduced to small spots or tubercles; arms of midcranial ridge always joined together to form a cruciform structure; antennae usually 10-segmented (9-segmented in *E. buxi*), often with numerous short sensory setae; post-occipital ridge large and conspicuous; scutum usually sclerotized; sometimes with 2 small clear areas near scutellum; hind legs longest, claws with broad base and sharp tip; penial sheath usually short, but complete; usually with 1 pair of pore clusters.

Immature stages. Neglected in the past. These were described in a few, e. g.

Gossyparia spuria by Herbert (1924), *Eriococcus orariensis* Hoy by Hoy (1961), and *Eriococcus* spp. by Schmutterer (1952b) and Miller (1969). First instars and second instar females similar to adult females, but smaller and with fewer structures, including only few microtubular ducts, usually without macrotubular ducts. Second instar males usually with macrotubular ducts and 7-segmented antennae.

Distribution. World-wide about 350 species in about 52 genera. In CE we recognize 24 species in 7 genera. Only few species are of economic importance, e. g. *Acanthococcus aceris, Eriococcus buxi* and *Gossyparia spuria*. The following 3 species are known in CE from greenhouses and homes only: *Eriococcus cactearum* Leonardi, *E. coccineus* (Cockerell), and *Rhizococcus araucariae* (Maskell).

Biology. Females develop through three stages, males through five. They feed on a variety of woody and herbaceous plants. Some are monophagous, others oligophagous or polyphagous. Those feeding on trees are usually found on the bark of trunk, branches or twigs, often in crevices. Species of *Greenisca* in CE are found on grasses only; feed on leaves, on stems, or under the leaf sheaths. A few infest the roots. Usually have one yearly generation. Most species in CE overwinter as nymphs or eggs.

Key to genera of Eriococcidae

1. Disc pores on both body surfaces 2
   Disc pores on venter only 3
2. Antennae at some distance from head margin; anal ring narrow, oval, twice as large as in *Kaweckia*; cruciform pores on dorsum absent (fig. 121 b, c) .. .. *Greenisca*, p. 291
   Antennae near margin of head; anal ring broad, circular; cruciform pores on dorsum present (fig. 122) .... *Kaweckia*, p. 294
3. Transverse rows or bands of enlarged conical setae absent from all or most dorsal body segments 4
   Transverse rows or bands of enlarged conical setae on most dorsal segments 5
4. Enlarged conical setae only on dorsum of anal lobes; antero-lateral seta of anal lobe dorsum small, spinelike, about as long as wide at base .... *Anophococcus*, p. 286
   Enlarged conical setae normally found along entire body margin; antero-lateral seta of anal lobe dorsum long, never small spine-like (fig. 123) ... . *Rhizococcus*, p. 298
5. With unusual macrotubular ducts on head near eyes; dorsal enlarged conical setae unusually sharp-pointed; normally 3 hind leg segments with translucent pores; on *Buxus* (fig. 119) .... *Eriococcus*, p. 287
   Without unusual macrotubular ducts on head near eyes; dorsal conical setae not unusually sharp-pointed; 0–2 hind leg segments with translucent pores; on hosts other than *Buxus* in CE 6
6. Large macrotubular ducts form transverse rows or bands on dorsum; dorsal enlarged setae normally of more than 1 size; entire female body covered by ovisac (fig. 117)........................... *Acanthococcus*, p. 276

Large macrotubular ducts absent from dorsomedial area of abdomen, they form a band along body margin of dorsum and venter; 1 size of enlarged seta with blunt tip over entire dorsum; only marginal part of female dorsum covered by ovisac; common on elms in CE (fig. 120). . *Gossyparia*, p. 289

Genus **Acanthococcus** Signoret

Syn. & Lit.: *Acanthococcus* Signoret, 1875:34; Borchsenius, 1949:331; Boratynski, 1962:55; Danzig, 1964:632, 1975a:62, 1980a:205; Koteja, 1974a:47; Dziedzicka, 1977:5; Kosztarab & Kozár, 1978:67; Tereznikova, 1981:14; Koteja & Zak-Ogaza, 1981:501. *Eriococcus* (in part) Newstead, 1903:195; Leonardi, 1920:426; Green, 1922b:347; Gomez-Menor, 1937:345; Ferris, 1955:94; Hoy, 1963:64.

Type-species: *Acanthococcus aceris* Signoret, 1875.

Ovisac ovoid, feltlike, white or gray, completely encloses female body. Adult female elongate-oval, narrowed posteriorly, with anal lobes conical and normally heavily sclerotized; antennae 6-, 7-, rarely 8-segmented; labium 3-segmented, with 16 setae (of these 10 on apical segment); stylets short or very long, forming a double loop; legs well-developed, coxa often with translucent sensory pores, inner side of hind tibia with only 2 setae, claw usually with a denticle; spiracles often with a few associated disc pores; disc pores on venter only, most usually quinquelocular, but the number of loculi vary between 3 to 9; oval disc pores (or cruciform pores) absent from dorsum, but often on prosomal venter in a marginal band; tubular ducts of 2 types: micro- and macrotubular ducts; microtubular ducts ca 7 μ long and 2 μ wide, scattered or form transverse rows or bands on dorsum, often associated with dorsal conical setae; macrotubular ducts often of 2 sizes, usually form transverse rows or bands on body surfaces; enlarged conical setae normally present at least in dorsal margin, but often on entire dorsum where they form transverse bands or rows; hairlike setae on venter only; anal ring well-developed, sclerotized with partly double row of pores and 8, rarely 6, anal ring setae, latter often as long as apical seta on anal lobes; each anal lobe with a long apical seta and usually with 3 short dorsal conical setae, seldom with more, but at least with 2, ventral hairlike setae also present.

Distribution. World-wide with about 245 species; in the Palearctic Region, about 75 species, in CE we recognize 8 species.

Biology. Feed on woody and herbaceous plants, often on grasses; they infest leaves, stems and roots; but one species is usually specific to one part of the plant, e. g. *A. aceris* and *A. roboris* are found at the bases of twig forkings, while *A. greeni* is a grass infesting species found on the upper surface of leaves. Most species in CE have 1 yearly generation; usually overwinter in egg or second nymphal stage; males normally are present.

Note. We accepted Borchsenius' (1948a, 1949) conclusion on the recognition and separation of the genera *Acanthococcus* and *Eriococcus*, which was followed by Boratynski (1962), Dziedzicka and Koteja (1971), Danzig (1975a), Tereznikova (1981), Koteja and Zak-Ogaza (1981) and others in Europe, as was proposed in the petition to the International Commission on Zoological Nomenclature by Miller and Williams (1976). Genus *Acanthococcus* represents a heterogeneous assemblage that requires further study.

Reviewer's remark. *Acanthococcus aceris*, *A. roboris*, etc. with narrow, almost cylindrical spines, and trees being their host plants, represent one unit; *A. munroi*, *A. erinaceus*, etc. form another group that is associated with dicotyledonous shrubs and herbaceous plants; *A. greeni* and presumably some others that feed on grasses are closely related to species that are currently placed in *Rhizococcus*.

Key to species of *Acanthococcus*

1. Dorsal enlarged setae in fully grown specimens dense, distance between them usually less than their length, only rarely same as or more; known from *Achillea millefolium* . . . . *erinaceus*, p. 281
   Dorsal enlarged setae not dense, distance between them more than their length, only on body margin closer to each other 2
2. Small conical setae on ventral margin form a single row, only occasionally doubling, or absent; all dorsal conical setae with pointed apex; feed on grasses only . . . . *greeni*, p. 281
   Conical setae on ventral margin of prosoma and occasionally on abdomen form bands composed of two or more rows; not all dorsal conical setae with pointed apex; rarely on grasses, more often on other herbaceous or woody plants 3
3. Dorsal enlarged conical setae of two kinds, large truncate and small blunt; live female purple. . . . . *devoniensis*, p. 280
   Dorsal enlarged conical setae large, of one kind, either pointed (at least on anterior end of body), truncate or blunt; live females not purple 4
4. Dorsal enlarged setae, at least on anterior end of body, with pointed tip . 5
   Dorsal enlarged setae with truncate or blunt tip. 6
5. Margin of each abdominal tergite 1–7 with 1–2 large and 2–3 smaller conical setae on each side; apical seta 230–300 μ long; live female orange-yellow (fig. 117) . . . . *munroi*, p. 282
   Margin of each abdominal tergite 1–7 with only 2 large conical setae on each side; apical seta ca 190 μ long; live female dark red . . . . *uvaeursi*, p. 285

6(4). Dorsal enlarged setae with truncate tip; anal ring setae more than 130 μ long; only on Fagaceae, such as *Quercus*, rarely on *Castanea* . . *roboris*, p. 284
   Enlarged dorsal setae with blunt tip; anal ring setae less than 130 μ long; on herbaceous plants or trees 7

7. Hind tibia 42–60 μ wide, 114–156 μ long; without strong seta on mid-tibia; apical setae less than 200 μ long; live adult female chestnut colored; on trees . . . . *aceris*, p. 278
   Hind tibia 62–74 μ wide, 109 μ long; with a strong seta on mid-tibia; apical setae over 230 μ long; live females light pink; on herbaceous plants. . . . . *desertus*, p. 279

157. **Acanthococcus aceris** Signoret – Maple Felt Scale
Figs 20, 116, 118a

Syn. & Lit.: *Acanthococcus aceris* Signoret 1875:35; Borchsenius, *1949:347; Kosztarab, 1959:402; Rogojanu, 1966a:324; Tsalev, 1968:207; Dziedzicka, 1977:5; Kosztarab & Kozár, 1978:72; Koteja & Zak-Ogaza, 1979:674; Tereznikova, *1981:15. *Eriococcus aceris*, Fernald, 1903:70; Marchal, 1908:251; Lindinger, 1912:54; Goux, 1931a:63; Schmutterer, 1952b:406; Hoy, 1963:66; Zahradnik, 1972:401.

Ovisac compact, grayish, up to 4 mm long, 2.2 mm wide. Adult female oval, according to Borchsenius (1949), chestnut colored, covered with white wax powder, ca 2.2 mm long and 1.5–1.7 mm wide; antenna 6-, rarely 7-segmented, ca 249–284 μ long; stylet loop extends to base of mesothoracic legs; coxa without translucent pores; hind leg 490–560 μ long; claw with denticle; disc pores scattered on prosoma, oval disc pores in a marginal band; quinquelocular pores form transverse bands on most abdominal sternites; microtubular ducts numerous, form transverse bands on dorsum of prosoma, rows on abdomen; macrotubular ducts of two sizes in a marginal band on venter, and in transverse rows on abdominal sternites 4 to 7; all dorsal setae conical and with rounded apex, large and medium-sized setae form a row along body margin,

Fig. 116. Sooty mold blackening leaves of *Acer campestre*; it develops on honeydew produced by *Acanthococcus aceris* Signoret (Photo M. Kosztarab)

which on abdominal segments on each side form a group of 2 large and 2–3 medium-sized setae, other dorsal setae form transverse segmental bands; conical ventral setae with pointed apex form a submarginal row, hairlike setae in submedian area of prosoma and abdomen; anal ring circular, sclerotized, with pores and 8 anal ring setae, each 100–124 μ long; each anal lobe sclerotized, serrate on inner margin, with 4 dorsal conical setae, each ca 50–60 μ long, apical setae 155–185 μ long.

Distribution. A common species. In CE: Au, Bu, Cz, Fr, Ge, Hu, Po, Sw, and European USSR, also Yu (Kozár's unpublished record), in addition known from Holland and Middle East.

Biology. Feed on trees; ovisacs usually found at base of twig forks, or between the corky bark strips and crevices of trunk on *Acer* spp., also *Aesculus, Carpinus, Elaeagnus, Fagus, Platanus, Quercus, Salix* and *Ulmus* (Hoy, 1963). Bisexual; second instars overwinter; according to Schmutterer (1952b), adults develop by first half of April and complete egg laying by end of May; lay about 82–378 red eggs per female; eggs hatch in about 30–35 days; first instars feed on leaves, from where they return to bark for overwintering in September or October. In Hungary, all eggs hatched by mid-June 1981 in Budapest. Rarely a pest in urban situations. The species is found in mountains at up to 1,500 m elevation (Zahradnik, 1972). Schmutterer (1952b) listed 5 species of honeydew feeding ants from *A. aceris* colonies.

Natural enemies. Parasites: *Arrhenophagus chionaspidis, Coccophagus lycimnia, Microterys lineola* (Kosztarab, 1959; Zahradnik, 1972); *M. trjapitzini, Zaomma eriococci* (Trjapitzin, 1978); *Anagyrus schoenherri, Pachyneuron concolor* (Kozár and Sugonjaev, 1979). Predators: *Exochomus quadripustulatus, Allothrombium fuliginosum* (Schmutterer, 1952b).

158. **Acanthococcus desertus** Matesova – Desert Felt Scale
Fig. 118c

Syn. & Lit.: *Acanthococcus desertus* Matesova, 1957:168; 1968a:113; Kosztarab & Kozár, 1978:76. *Eriococcus desertus*, Hoy, 1963:84.

Ovisac smooth, felted, cream-colored. Adult female oval, constricted toward posterior end, light pink, 1.4–2.3 mm long, 0.8–1.2 mm wide; antennae 7-segmented, 235–237 μ long, third segment longest; eyes near bases of antennae; labium wide, stylet loop 186 μ long; hind coxae with group of translucent pores, femur distinctly thick, 62–74 μ wide, 109 μ long, claw with a small denticle, with a large seta on mid-tibia; trilocular disc pores scattered on prosoma; quinquelocular and 7-locular pores form wide transverse bands on abdominal sternites, they increase in number toward posterior end; microducts scattered on dorsum; macroducts form segmental transverse bands on dorsum and rows on venter; dorsal conical setae with blunt tip, except a few on head acute, enlarged and medium size setae form bands along body margin, with 5–8 setae, 2 of these large, on margin of each abdominal tergite 1–7; they also form transverse rows or bands and a longitudinal band along median line of dorsum; on venter smaller ones form a submarginal band; ventral hairlike setae of various thicknesses and lengths; anal ring with pores and 8 setae, each 99–124 μ long; anal lobes conical, each with apical seta 249–310 μ long, and 1 hairlike seta and 3 conical setae.

Distribution. Known only from Hungary and USSR (Kazakhstan and Turkmenia); Danzig (1977) reported it from Mongolia.

Biology. A steppe inhabiting xerophilous species. It infests bases of leaves and rootcrown of plants in the genera: *Artemisia, Camphorosma, Kochia, Minuartia, Scabiosa* and *Vicia*. Often pest in Kazakhstan. Bisexual; adults appear in May or June, egg production (12–86) probably depends on host plant (Matesova, 1968a). Associated ant: *Tetramorium semilaevi* (Matesova, 1968a).

Natural enemy. Parasite: *Eucoccidophagus semiluniger* (Kozár and Sugonjaev, 1979).

159. **Acanthococcus devoniensis** (Green) – Heather Felt Scale

Syn. & Lit.: *Rhizococcus* (?) *devoniensis* Green, 1896:260. *Eriococcus devoniensis*, Newstead, 1903:201; Green 1930:12; Goux, 1931:72; Schmutterer, *1952b:413; Hoy, 1963:85. *A. devoniensis*, Borchsenius *1949:337; Ossiannilsson 1959:195; Kosztarab & Kozár, 1978:71; Koteja & Zak-Ogaza, 1979:674. See notes on possible synonymy.

Ovisac strongly convex, creamy-white; up to 2.5 mm long, 1.8 mm wide. Adult female oval, tapering posteriorly, live purple, ca 1.5–2.5 mm long, 1–1.5 mm wide; antenna 7-segmented, ca 190–220 μ long, third and fourth segments longest; labium 111 μ long, 99 μ wide, stylet loop 186 μ long; legs stout, hind coxa with large group of translucent pores, hind coxa 68 μ long, tarsus longer than tibia; oval disc pores on venter; apparently few quinquelocular disc pores scattered on venter of prosoma, they also form wide transverse bands on sternites of abdomen; microtubular ducts scattered along dorsal margin; macrotubular ducts of 2 sizes form transverse rows and bands on dorsal segments, and a wide band along margin of venter; smaller macroducts scattered or form transverse rows on other parts of venter; dorsal enlarged conical setae of two kinds on head, large truncate and small blunt, those with truncate tip form a wide band along body margin, also 5 transverse bands on prosoma, and transverse rows on tergites 1–7 of abdomen; ventral conical setae with blunt tip form wide band along prosomal margin and a row along margin of abdomen; ventral hairlike setae stout, preanal setae ca 35 μ long; anal ring with pores and 8 (rarely 7) setae (only 6 noted by Green, 1896), each 99–117 μ long; apex of anal lobes heavily sclerotized, with apical seta 185–210 μ long and 3 dorsal conical setae on each lobe. Adult male described by Jancke (1955). Eggs oval, pink or pinkish yellow, dusted with white wax powder. First instar described by Schmutterer (1952b).

Distribution. A common mesophilous species. In CE: Au (Boratynski's unpublished record from Lower Austria), Fr, Ge, Hu, It, Po, and Sw; in addition known from Denmark, England, Holland, Spain and Sweden, also from Sardinia, Malta and N Africa.

Biology. Infests stems and branches of plants in the genera: *Artemisia, Calluna, Dianthus, Erica, Ilex, Lavandula, Salicornia, Santolina, Thymelaea, Thymus, Ulex* and *Vaccinium*. Bisexual. In S Germany (Schmutterer, 1952b) eggs overwinter; hatch during first third of May, adults appear by end of June or beginning July; lay 28–82 eggs per female from second half of July. Infested young shoots of *Erica* are often distorted.

Natural enemy. Parasite: *Microterys aeneiventris* (Trjapitzin, 1978).

Notes. Although Green (1930) distinguished the two species, our comparison of *A. devoniensis* specimens from *Erica* in England collected and verified by E. E. Green and D. J. Williams, with specimens of *A. ericae* (Signoret, 1875) from *Erica* in Germany and Corsica (Jaap Coll. Nos 135 and 182) revealed no sufficient differences between the two species. We had no specimens of *A. thymi* (Schrank, 1801) available for comparison, but based on redescriptions by others, this species is probably identical with the earlier two. Type specimens, and/or topotypes could assist someone in the clarification of the suspected synonymy of the above three species. Lindinger (1957) already suggested the synonymy of these three species, and included a fourth one, *E. reynei* Schmutterer (1952b). When the type material from Goux collection will be available for the new *Eriococcus* species he described, we probably will find even more synonyms.

160. **Acanthococcus erinaceus** (Kiritchenko) – Woolly Felt Scale

Syn. & Lit.: *Eriococcus erinaceus* Kiritchenko, 1940b:128; Hoy, 1963:88. *A. erinaceus*, Borchsenius, 1949:335; Tereznikova, 1975:29, 73, 1981:27; Kosztarab & Kozár, 1978:70.

Ovisac pyriform, strongly convex, cream-colored, up to 2 mm long, 1 mm wide. Adult female oval, red or rusty-red, up to 1.3 mm long, 0.8 mm wide; antennae 7-segmented, 230 μ long, third segment longest; stylet loop reaches middle line between first and second coxae; hind coxae with few translucent pores; quinquelocular pores on entire venter of prosoma, 1 or 2 transverse rows on each abdominal sternite; dorsal conical setae with blunt apex, largest 36–45 μ long, smaller few, 24–27 μ long; conical setae form 5 transverse bands on prosoma, a narrower transverse band on abdominal tergites 1–6, a row on tergite 7; setae dense, distance between them normally less than or equal to length of setae; marginal band of large setae and 3 longitudinal bands in middle line of dorsum not distinct; ventral hairlike setae of various lengths, preanal setae ca 45 μ long; anal ring with 8 setae, each ca 125 μ long; anal lobes large, strongly sclerotized, apical seta ca 350 μ long; each anal lobe with 3 dorsal conical blunt-tipped setae.

Distribution. Known only from Ukraine in USSR.

Biology. Collected from leaves of *Achillea millefolium*. Apparently a very rare steppe inhabiting species. Males not known; its biology needs study.

161. **Acanthococcus greeni** (Newstead) – Newstead's Felt Scale

Syn. & Lit.: *Eriococcus greeni* Newstead, 1898:96, 1903:200; Hoy, 1963:93. *A. greeni*, Borchsenius, *1949:340; Kosztarab, 1956b:395 (misidentification); Ossiannilsson, 1959:195; Koteja, 1974a:152; Danzig, 1975a:46, 1980a:212; Dziedzicka, 1977:5; Kosztarab & Kozár, 1978:72; Tereznikova, *1981:27.

Ovisac elongate-oval, of tough, coarse, felted wax, creamy white, 3.5–5 mm long, 1–1.7 mm wide. Adult female elongate-oval, white or yellowish, ca 2.5–3.4 mm long,

1.2–1.7 mm wide; antenna 6- or 7-segmented, 220–250 μ long; stylet loop slightly longer than labium; hind leg 680–740 μ long; hind coxae with a large group of translucent pores; few oval disc pores on head and on submargin of venter; numerous quinquelocular pores form transverse bands on venter, also found scattered on middle of prosoma; microtubular ducts scattered on dorsum; macrotubular ducts of two sizes, form transverse rows on entire body; dorsal conical setae sharp-pointed, of large and medium size, larger 40 μ long, smaller 24 μ long; form striking bands along body margin, with 3 large and 2–3 medium-sized setae per abdominal segmental margin; large setae form a longitudinal band in median line of dorsum, also 2 interrupted rows, these also form transverse rows on tergites; a row of small conical setae, 12–20 μ long, located on margin of venter; ventral hairlike setae short and few, form transverse rows on each abdominal sternite; pre-anal seta 69 μ long, anal ring with pores and 8 setae, each ca 120 μ long; anal lobes sclerotized with apical setae ca 350 μ long, also each lobe with 3 dorsal conical setae (4 setae noted by Danzig, 1975a).

Distribution. A common Trans-Palearctic boreal species. In CE: Fr, Hu, Po and USSR, also known from England, Spain and Sweden.

Biology. Feed on leaves of grasses in genera: *Agropyron, Brachypodium, Calamagrostis, Carex, Corynephorus, Dactylis, Festuca, Koeleria, Lolium, Melica* and *Phleum* (Hoy, 1963; Koteja, 1971b; Podsiadlo and Komosinska, 1976; Danzig, 1980a). Lay eggs in July and August. Males unknown; its biology needs study.

Natural enemies. Parasites: *Eucoccidophagus breviventris, Euonotus acutus, Metaphyus alami, M. nitens.* The record of Kosztarab (1956b) on *Trichomasthus albimanus* applies to *Rhizococcus insignis.*

162. **Acanthococcus munroi** Boratynski – Munro's Felt Scale
Fig. 117

Syn. & Lit.: *Acanthococcus munroi* Boratynski, *1962:56; Komosinska & Podsiadlo, 1967:684; Danzig, *1975a:71, 1980a:218; Kozár & Danzig, 1976:67; Podsiadlo & Komosinska, 1976:88; Kosztarab & Kozár, 1978:72, Tereznikova, 1981:32.

Ovisac, according to Boratynski (1962), elongate-oval, closely felted with threads projecting on dorsum and margin, encloses female completely, white to cream, 1.8–2.5 mm long, 1–1.5 mm wide. Adult female elongate-oval, tapering toward posterior end, orange-yellow, according to Boratynski (1962), 1.5–2.1 mm long, 0.9–1.3 mm wide; antennae 7-segmented, rarely appear 6-segmented, 255–270 μ long; stylet loop slightly longer than labium; legs of moderate size, hind coxae 130 μ long, with 23–30 large, translucent pores, claw with a minute denticle near apex; a few oval disc pores along ventral submargin; most disc pores quinquelocular, some appear 3-, 6- or 7-locular; macrotubular ducts of two sizes on both derms; normally microducts associated with dorsal conical setae; dorsal conical setae of various sizes, arranged in a marginal band and in a series of transverse bands and rows, those on anterior part more sharply pointed than the somewhat blunt setae on posterior part of body, each abdominal segmental margin with 2 large and 2 smaller setae on each side; small conical ventral setae form an irregular longitudinal submarginal row; hairlike setae on venter 26–74 μ long, arranged in segmental transverse rows or groups; anal ring oval, sclerotized,

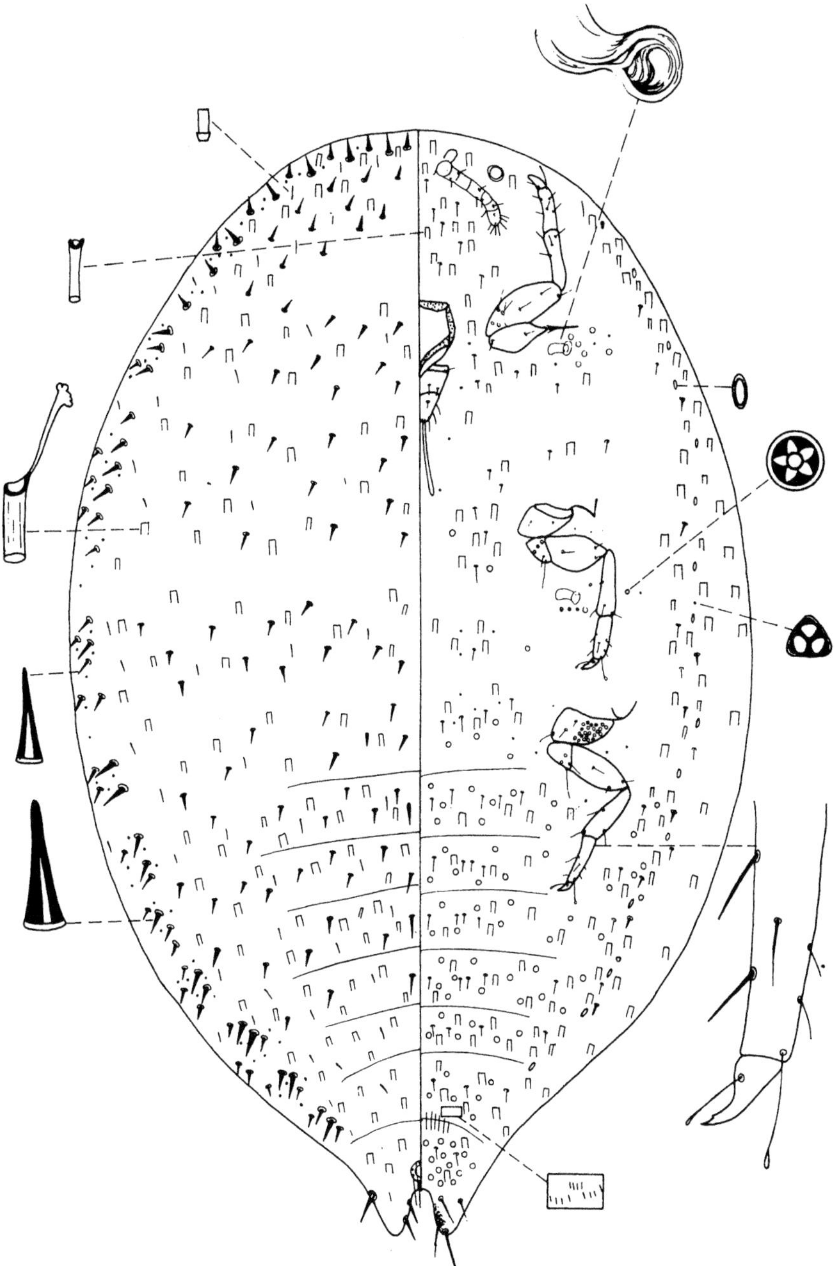

Fig. 117. *Acanthococcus munroi* Boratynski, female (original)

with pores and 8 setae, each seta 90–124 μ long; anal lobes conical, sclerotized along inner margin and at apex, apical seta 230–300 μ long, each lobe with 3 dorsal conical setae of various sizes.

Distribution. A Trans-Palearctic species. In CE: Hu, It, Po, USSR, also known from England.

Biology. On stem, leaves and at root crown area of plants of genera: *Achillea, Euphrasia, Festuca, Fragaria, Hieracium, Hypochoeris, Potentilla, Sassurea, Thymus, Valeriana, Veronica,* and *Vicia.* According to Boratynski (1962) is biparental with 2 yearly generations in England; eggs overwinter, hatch in April, adults appear in June, lay eggs at end of June, eggs hatch in mid-July, adults of second generation appear in September, and females lay the overwintering eggs.

Natural enemy. Parasite: *Anagyrus orbitalis* (Kozár and Sugonjaev, 1979).

163. **Acanthococcus roboris** (Goux) – Oak Felt Scale
Fig. 118b

Syn. & Lit.: *Eriococcus roboris* Goux, 1931a:58; Hoy, 1963:112; Zahradnik, 1972:402. *A. roboris,* Borchsenius, *1949:346; Tereznikova, 1959a:447, 1981:36; Kosztarab, 1959:402; Rogojanu, 1966a:323; Kosztarab & Kozár, 1978:73.

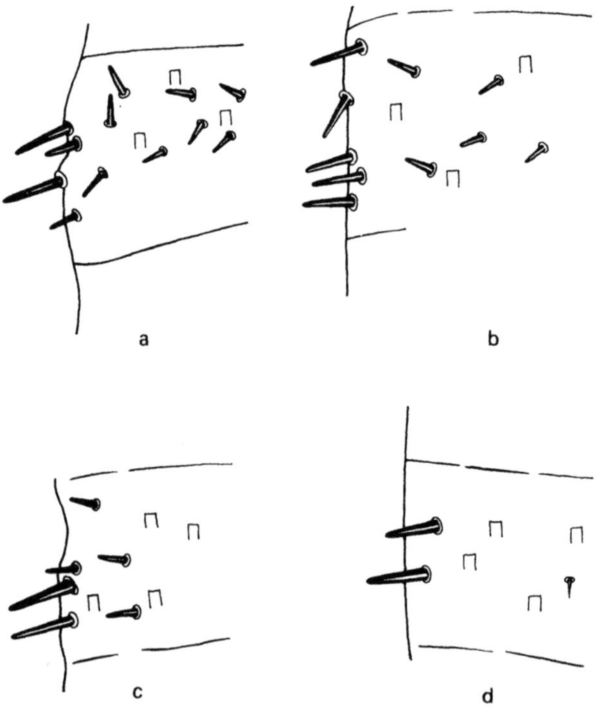

Fig. 118. Margin of 6th abdominal tergite of females; (a) *Acanthococcus aceris* Signoret; (b) *Acanthococcus roboris* (Goux); (c) *Acanthococcus desertus* Matesova; (d) *Rhizococcus agropyri* Borchsenius (modified after Kosztarab and Kozár, 1978)

Ovisac felted, white, up to 9 mm long, 6 mm wide. Adult female oval, according to Borchsenius (1949) of red raspberry color, up to 5 mm long, 4 mm wide; antenna 7-segmented, 285–309 μ long; stylet loop twice as long as labium; legs small, claw with a large denticle; oval disc pores form a submarginal band on venter; quinquelocular pores few, more frequent on terminal abdominal segments; microtubular ducts scattered on dorsum; macrotubular ducts of 2 sizes, larger type distributed on both body surfaces, smaller type on venter only; dorsal conical setae only slightly broadened at base, all with truncate tip; enlarged setae form a row on body margin including 4–5 similar sized setae on each side of each abdominal segment; setae on dorsum form thin transverse bands and rows; ventral setae hairlike, form a double row along body margin; other ventral hairlike setae of various lengths, form transverse segmental rows on abdomen; anal ring with pores and 8 setae, each 136–165 μ long; anal lobes almost cylindrical, heavily sclerotized and with serration on inner margin, terminal setae 235–278 μ long, each with 3, 40–68 μ long, dorsal conical setae. First instar described by Goux (1931a).

Distribution. A rare mesophilous species. Known only from: Fr, Hu, Ro, and USSR.

Biology. Predominantly on *Quercus* spp., but was also reported from *Castanea* by Borchsenius (1949). Usually found at base of buds or twig forkings. Lay eggs in June, these red, ca 370 μ long, hatched at end of July in Hungary; males not known as yet.

Natural enemies. Parasites: *Microterys chalcostomus, Trichomasthus niveicrus* (Kosztarab, 1959); *Zaomma eriococci* (Trjapitzin, 1978).

164. **Acanthococcus uvaeursi** (Linnaeus) – Linnaeus' Felt Scale

Syn. & Lit.: *Coccus uvaeursi* Linnaeus, 1761:266; Fernald, 1903:327. *Eriococcus bezzii* Leonardi, 1920:439. *E. uvaeursi*, Lindinger, 1912:74; Goux, 1936c:352, 1936b:299; Ossiannilsson, 1951:3; Hoy, 1963:123. *Acanthococcus uvaeursi*, Borchsenius, 1949:349; Danzig, 1975a:43, 1980a:215; Kosztarab & Kozár, 1978:76; Tereznikova, 1981:40.

Ovisac oval, cocoon-like, felted, grayish, covers body completely, up to 3 mm long and 2 mm wide. Adult female oval, dark red, about 2 mm long, 1.3 mm wide; antennae 7- or 6-segmented, ca 230 μ long; 4th segment longest; labium wide, stylet loop extends close to median coxae; hind coxae with large group of translucent pores; claw with small denticle; oval disc pores in a band along ventral margin; quinquelocular pores scattered on venter of prosoma and form transverse rows on sternites of abdomen; microtubular ducts scattered on dorsum; macrotubular ducts of two sizes on both body surfaces, fewer on venter; they are scattered on prosoma, form a band along body margin and transverse rows on abdominal sternites; all dorsal conical setae with pointed tip, they form a row along body margin, with 2 large setae on margin of each abdominal segment; shorter setae form transverse rows and bands on dorsum, last few abdominal tergites with only 1 row on each; ventral hairlike setae few; anal ring with pores and 8 setae, each ca 105 μ long; anal lobes elongate, almost cylindrical, heavily sclerotized, apical seta ca 190 μ long, each lobe with 3 dorsal conical setae. First instar described by Leonardi (1920).

Distribution. A rare montane species. Known only from Ge, It and USSR (from European to Far East Provinces), also Sweden.

Biology. Often found on soil-covered stem parts of host plants belonging to genera: *Arbutus, Arctostaphylos, Azalea, Erica, Eubotryoides, Ledum, Rhododendron* and *Vaccinium*. Schmutterer (1955a) reported it from 2,600 m elevation. Males unknown. Little known about its biology.

Genus **Anophococcus** Balachowsky

Syn. & Lit.: *Anophococcus* Balachowsky, 1854:61; Ferris, 1955:94, 1957:85; Hoy, 1962:201, 1963:133; Koteja & Zak-Ogaza, 1981:511. *Rhizococcus*, Danzig, 1962b:532.

Type-species: *Eriococcus inermis* Green, 1915.

There is only the type species assigned at present to this genus (Koteja & Zak-Ogaza, 1981). Therefore, the general description of *inermis* should substitute for a generic description. This genus needs study, and it can be separated only with some difficulty from the related *Rhizococcus* and *Greenisca* by the following combination of morphological characters: in *Anophococcus* the enlarged conical setae restricted to dorsum of anal lobes; anterolateral seta of anal lobe dorsum small, spinelike, about as long as wide at its base; disc pores restricted to venter.

165. **Anophococcus inermis** (Green) – Harmless Felt Scale

Syn. & Lit.: *Eriococcus inermis* Green, 1915b:176, 1922a:20; Balachowsky, 1934c:130. *Greenisca inermis*, Borchsenius, 1949:367. Danzig (1962b) noted that this description applies to another species; Tereznikova, 1959c:683; Hoy, 1963:133. *R. inermis*, Danzig, *1962b:532; Dziedzicka & Koteja, 1971:576; Koteja, 1974a:104; Dziedzicka, 1977:5. *Acanthococcus inermis*, Danzig, 1975a:62, 1980a:226; Kosztarab & Kozár, 1978:67. *Anophococcus inermis*, Koteja & Zak-Ogaza, 1981:511.

Ovisac elongate-oval, white. Adult female elongate-oval, tapering toward both ends; ca 2.6–3 mm long, 1.3 mm wide; antenna 7-segmented, 170–186 μ long; labium ca 100 μ long, 70 μ wide; hind leg ca 540 μ long, hind coxa 99 μ long, with translucent pores, hind femur 104–125; tibia 86–112, tarsus 104–125 μ long; tarsal claw without denticle; disc pores, apparently none on dorsum, numerous, with 5 or 6 loculi, on entire venter; microducts scattered on dorsum and along margin of venter; macrotubular ducts of 2 sizes, larger type form transverse bands on dorsum of prosoma and first abdominal segment, also rows on abdominal sternites 2–7; smaller macroducts scattered on venter; enlarged conical setae with blunt tip, found on dorsum of anal lobes only, 2 larger (16–22 μ long) and 2 small spinelike (6 μ long) on each lobe; body spines minute (3 μ), found along body margin, and form transverse rows on tergites; preanal setae 30 μ long; anal ring with partially double dense row of pores and 8 setae, each 78–110 μ long; anal lobes sclerotized, each with a subapical seta 39 μ, and apical seta, ca 173 μ long.

Distribution. A rare boreal species. In CE: Fr, Ge, Po, USSR (from Transcarpathian and Leningrad Regions to Far East); also known from England.

Biology. Feed on leaves of grasses in the genera *Agrostis, Festuca* and *Koeleria.* Females collected from July to September. Biology unknown; no records on males.

Note. This species was originally the type of the genus *Greenisca* Borchsenius, 1948. Later, Borchsenius and Danzig (1966) replaced the misidentified type species with *Anophococcus gouxi*.

Genus **Eriococcus** Targioni-Tozzetti

Syn. & Lit.: *Eriococcus* Targioni-Tozzetti, 1868:726; Leonardi, 1920:426; Gomez-Menor, 1937:345; Goux, 1948:66; Borchsenius, 1948a:501, 1949:322; Schmutterer, 1952b:405; Ferris, 1955:94; 1957:81; Hoy, 1963:62; Morrison & Morrison, 1966:69; Miller & McKenzie, 1967:481; Afifi, 1968:167; Miller, 1969:48; Koteja, 1974b:297; Tereznikova, 1981:48. *Thekes* Maskell, 1892:28.

Type-species: *Coccus buxi* Fonscolombe, 1834 (as restricted by Signoret, 1872:429). Fernald (1903) designated as type of the genus, *Coccus crispus* Fonscolombe.

Ovisac feltlike, encloses female entirely. Adult female elongate-oval, tapering posteriorly, with well-developed and heavily sclerotized anal lobes; antennae 6- or 7-segmented; hind coxa, femur and tibia often with translucent pores, claw with denticle; disc pores on venter only; no oval disc pores; microtubular ducts scattered on dorsum; macrotubular ducts on both dorsum and venter; some unusually shaped ducts on dorsal margin and on midline of dorsum and on head near eyes; enlarged conical setae with unusually sharp-pointed apex on dorsum; anal ring with pores and 8 setae [according to Gomez-Menor (1937) with 6 setae]; anal lobes long, almost cylindrical each with a long apical seta. See for a more detailed morphological description and records on distribution and biology at *E. buxi*.

Note. Hoy (1963) listed 252 species in this genus, making it the largest and most diverse genus of the Eriococcidae. Because this genus is still in a state of chaos taxonomically, we accepted Borchsenius' (1948a) concept to treat *Eriococcus* as a monotypic genus. Most of the species included in this genus by other authors are treated here in the genera *Acanthococcus* and *Rhizococcus*.

166. **Eriococcus buxi** (Fonscolombe) – Box Scale, Boxwood Felt Scale
Fig. 119

Syn. & Lit.: *Coccus buxi* Fonscolombe, 1834:218. *Eriococcus buxi*, Marchal, 1908:253; Gomez-Menor, 1937:346, 1965:113; Borchsenius, 1949:324; Hoy, 1963:76; Tsalev, 1968:207; Afifi, 1968:175; Dziedzicka, 1977:5; Tereznikova, 1981:48.

Ovisac grayish, 2.5 mm long, 1.5 mm wide. Adult female as noted by Borchsenius (1949), ovoid, dark red, 1.4–1.7 mm long, 0.9–1.2 mm wide; antenna 160–185 μ long, 6- or 7-segmented (3rd segment rarely divided in 2); labium broad, 74 μ long, 99 μ wide, stylet loop extends to mid-coxae; hind coxa, femur and tibia with a group of translucent pores, hind leg ca 425 μ long, hind femur 125–145, tibia 75–95, tarsus 85–100 μ long, claw with a small denticle; no oval disc pores; disc pores with 5 or 7 loculi predominantly scattered on venter of prosoma and form transverse rows on sternites of abdomen; microtubular ducts scattered on dorsal margin and in median line of dorsum; macrotubular ducts of 2 sizes numerous on dorsum, form a band along ventral margin, and transverse rows on abdominal tergites, scattered on venter of prosoma, with unusually shaped long ducts on dorsum especially near eyes, these ca 37 μ long and 6–7 μ wide at sclerotized cap; conical setae with very sharp tip and widened at base form transverse rows and bands on tergites; ventral hairlike setae short; preanal setae 50–70 μ long; anal ring partly with double row of pores, 8 setae, each 120 μ long; anal lobes sclerotized, conical, each with apical seta 178–215 μ long, and 3 setae 41–49 μ long. Adult male described by Afifi (1968).

Distribution. A rare Mediterranean species. In CE: Bu, Cz, Fr, Ge, Sw, USSR (from Crimea to Uzbekistan), and Yu. Also known from Spain.

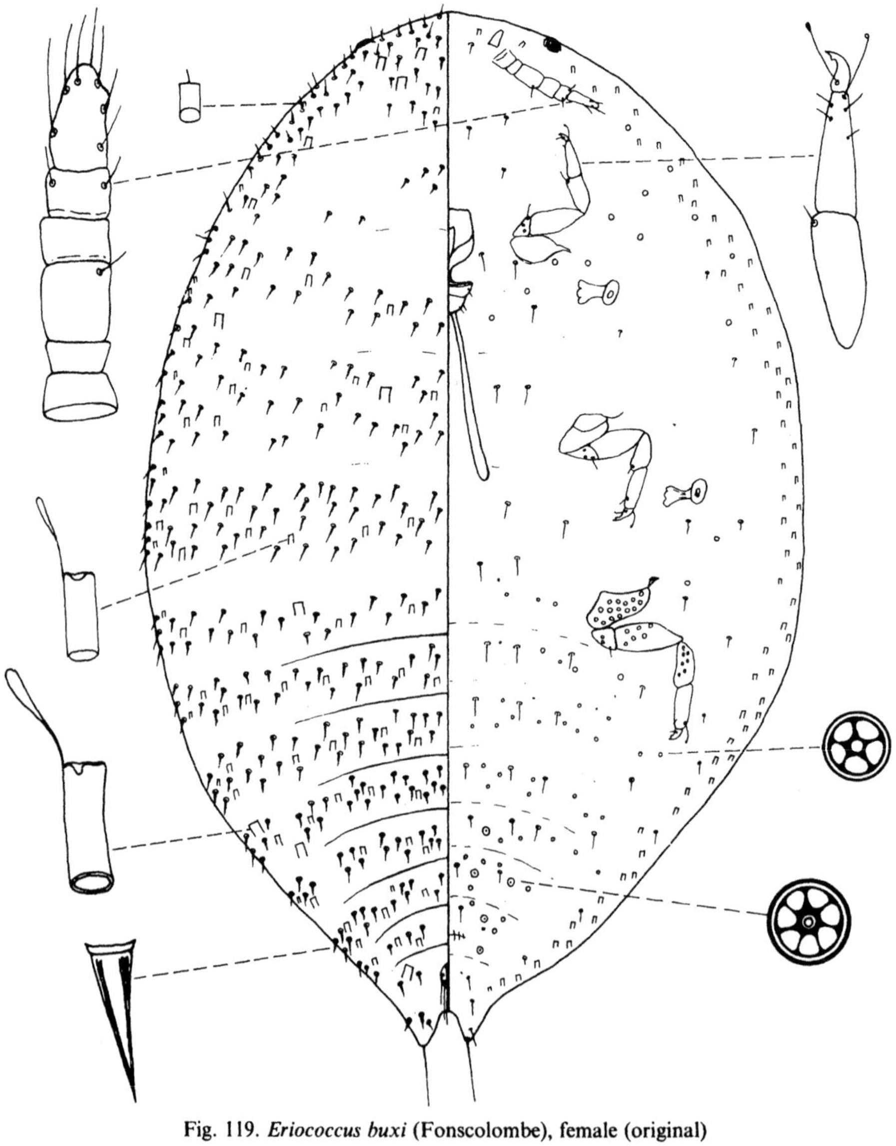

Fig. 119. *Eriococcus buxi* (Fonscolombe), female (original)

Biology. A monophagous species, feed on twigs and leaves of *Buxus* spp. only. According to Umnov (1940) it often becomes a pest and has 2 generations in Crimea, overwinters as second instar; first generation females lay eggs in the first part of June, and second generation females lay in mid-August. Reproduction biparental.

Natural enemy. Parasite: *Metaphycus brachypterus* collected in Hungary (Trjapitzin, 1978).

Genus **Gossyparia** Signoret

Syn. & Lit.: *Gossyparia* Signoret, 1875:20; Leonardi, 1920:464; Borchsenius, 1949:327; Hoy, 1963:127; Morrison & Morrison, 1966:85; Kosztarab & Kozár, 1978:76; Tereznikova, 1981:50. *Eriococcus* (in part), Ferris, 1955:94.

Type-species: *Coccus spurius* Modeer, 1778.

Ovisac feltlike, incomplete; ruptured in middle of dorsum, leaving dorsum of female uncovered. Adult female oval, narrowed posteriorly with large anal lobes; eyes circular, laterad from base of antennae; antennae 6- or 7-segmented; legs well-developed, claw with denticle; oval disc pores on ventral submargin; quinquelocular disc pores on venter only; microtubular ducts normally on both body surfaces; macrotubular ducts on dorsum and venter form a band along body margin; dorsum uniformly covered with thick setae; venter with hairlike setae only; anal ring with pores and 8 setae; anal lobes sclerotized with small teeth on inner margin, each with long apical seta and usually with 3–4 thick setae on dorsal surface.

Distribution. World-wide, 4 species; 2 of these Australian, 1 Palearctic and 1 Holarctic.

Biology. Feed on woody plants; 1 yearly generation; biology of only *G. spuria* was studied.

167. **Gossyparia spuria** (Modeer) – Elm Scale, Elm Bark Louse
Figs 21, 120

Syn. & Lit.: *Coccus ulmi* Linnaeus, 1761:265 (a preoccupied name). *C. spuria* Modeer, 1778:43. *C. lanigera* Gmelin, 1789:2221. *C. gramuntii* Planchon 1864:24. *G. ulmi,* Marchal, 1908:257. *G. spuria,* Leonardi, 1920:465; *Herbert, 1924:1; Green, 1928b:9; Borchsenius, 1949:330; Schmutterer, 1952b:416; Atanasov, 1959:426; Dziedzicka, *1961:203, 1977:5; Hoy, 1963:128; Danzig, 1964:632; Afifi, 1968:182; Zahradnik, 1972:402; Koteja, 1974a:152; Kosztarab & Kozár, 1978:77; Tereznikova, 1981:50. *Eriococcus spurius,* Ferris, 1955:164. The species was also included by earlier authors in the genus *Nidularia* Targ.–Tozz. There are over 400 literature sources on this species in the U. S. Department of Agriculture files. We listed here only the more important and recent ones, especially from Central Europe.

Ovisac ovoid, compact, panlike, grayish-white, covers only venter and body margins, exposing dorsum, up to 3.5 mm long, 2 mm wide. Adult female oval, somewhat tapering posteriorly, live dark red or dark brown, ca 2.5 mm long, 1.5 mm wide; antennae 7, rarely 6-segmented, 221–260 μ long; eyes slightly sclerotized, dome-shaped; labium wide, stylet loop extends to 5th sternite of abdomen; hind femur 136–140, tibia 105–116, tarsus 136–140 μ long; claw thick, with small denticle; oval disc pores on ventral submargin; few trilocular pores in median area of thoracic venter; quinquelocular disc pores scattered on prosomal venter and form transverse rows on abdominal sternites; macrotubular ducts form a wide band on lateral margins of both surfaces; microtubular ducts scattered on dorsum; enlarged setae with blunt apex, form transverse rows or bands, on entire dorsum, 3 or 4 on each abdominal tergal margin; ventral setae thinner, form a row along body margin; ventral hairlike setae in transverse rows on abdominal sternites, some scattered on prosoma; anal ring with

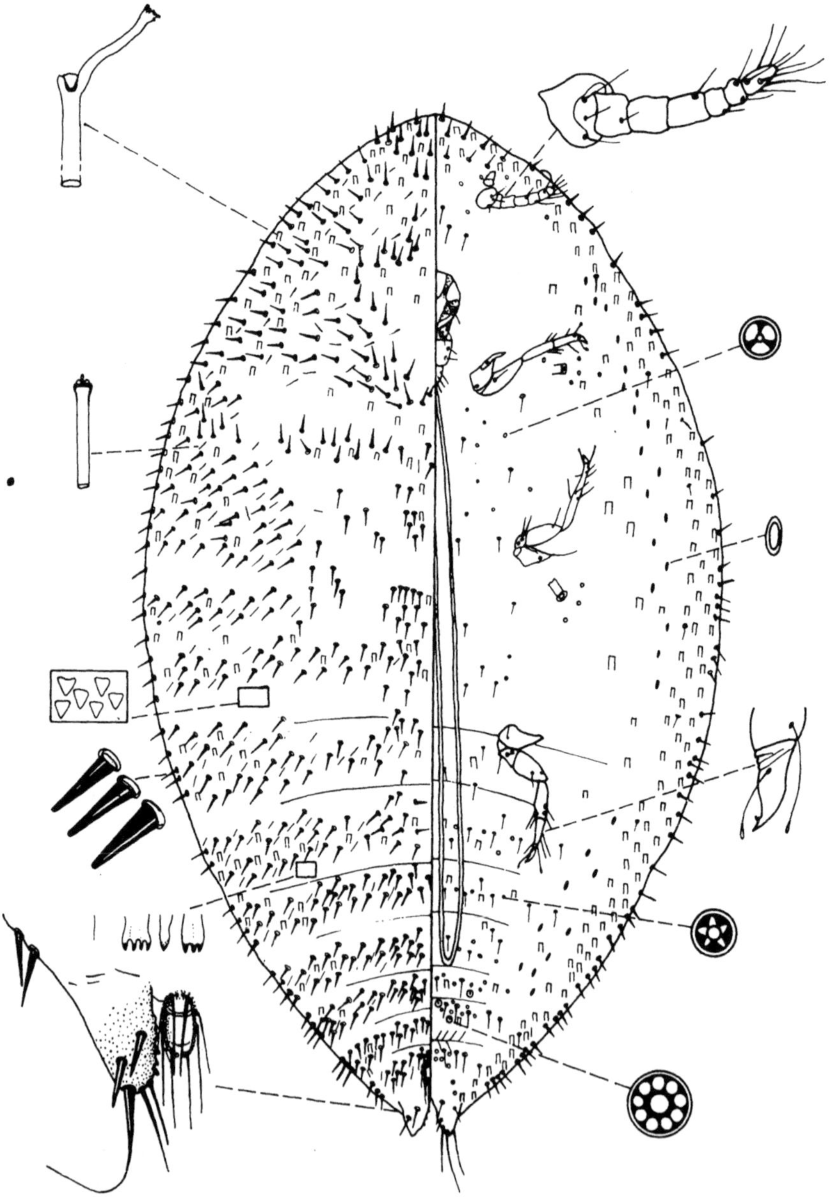

Fig. 120. *Gossyparia spuria* (Modeer), female (after Kosztarab and Kozár, 1978)

partly double row of pores and 8 setae, each 125–136 μ long; anal lobes distinct, heavily sclerotized, with small teeth on inner margin, each with apical seta 230–247 μ long, subapical seta 148 μ long and with 3 enlarged setae. Two winged forms of adult males described in detail by Afifi (1968). Immature stages described by Herbert (1924) and Dziedzicka (1961).

Distribution. A common Holarctic species, in CE in: Au, Bu, Cz, Fr, Ge, Hu, It, Po, Ro, Sw, USSR (European Parts), and Yu; also known from S W and N Europe, N Africa, Middle East, Japan, Canada and USA.

Biology. Adults most common on trunk and branches, frequently in twig forkings of elms, *Ulmus* spp., but also reported from: *Alnus, Corylus, Fraxinus* and *Viscum* (Zahradnik, 1972). It has bisexual or rarely parthenogenetic reproduction; 1 yearly generation; second instars overwinter; adults appear end of April or first part of May; lay 117–416 eggs per female starting in June; eggs hatched from end of July (Schmutterer, 1952b). Herbert (1924) reported from W USA that eggs hatched in less than one hour after laying; first instars migrated to both sides of leaves; often settled near the main veins; molted in about six weeks into second instars; nymphs moved in the fall to crevices of the bark for overwintering; male nymphs started forming cocoons from late January, while female nymphs molted into adults in March and April; egg laying started end of May and lasted for several months, until about mid-August.

Often becomes a pest on ornamental elms. The large quantity of honeydew produced coats the leaves, and on the honeydew sooty mold, *Coniothecium effusum* (Dziedzicka, 1961) forms and covers the foliage. Honeydew feeding ants are common visitors of elm scale colonies; Schmutterer (1952b) listed 5 such species.

Natural enemies. Parasites: *Coccophagus excelsus, C. gossypariae, C. insidiator, Encyrtus feudatorius, Pachyneuron coccorum, Trichomasthus coeruleus, T. cyanifrons, Zaomma eriococci* (Schmutterer et al., 1957; Zahradnik, 1972; Trjapitzin, 1978). Predators: *Chilocorus bipustulatus, Exochomus quadripustulatus, Anthocoris nemorum, Pilophorus cinnamopterus, P. perplexus, Allothrombium fuliginosum.* (Schmutterer, 1952b.) A number of other lady bird beetles and a green lacewing are its predators in North America (Herbert, 1924).

Genus **Greenisca** Borchsenius

Syn. & Lit.: *Greenisca* Borchsenius, 1948a:502, 1949:367; Hoy, 1963:132; Morrison & Morrison, 1966:87; Borchsenius & Danzig, 1966:41; Kosztarab & Kozár, 1978:77; Danzig, 1964:634, 1980a:226; Koteja & Zak-Ogaza, 1981:508. *Anophococcus* (in part) Balachowsky, 1954a:61. *Eriococcus*, (In part) Ferris, 1957:86.

Type-species. *Eriococcus inermis* Green, 1915 (misidentified) = *Anophococcus gouxi* Balachowsky, 1954.

.Ovisac elongate-oval, feltlike, white, or yellowish. Adult female elongate-oval, almost parallel-sided with convex dorsum; antennae 6- or 7-segmented, base some distance from body margin; legs slender, with translucent pores on coxa, hind tibia with 5 setae, 3 on inner margin, claws usually with denticle; stylet loop slightly longer than labium; various types of disc pores on both derms, but cruciform pores on ventrolateral area only; micro- and macrotubular ducts present; enlarged conical setae very few, these with truncate or sharp-pointed apex, and present only on anal lobes, rarely on head; hairlike setae on both surfaces; anal ring oval, narrow, with 6 or 8 setae and pores; anal lobes prominent, occasionally heavily sclerotized. Living females yellowish.

Distribution. Trans-Palearctic, with 3 species, all 3 in CE.

Biology. All species known from leaves of grasses; eggs overwinter, males unknown, biologies need to be studied.

Key to species of *Greenisca*

1. Head with a few enlarged conical setae; apparently with only few disc pores on dorsum .... *placida*, p. 294
   Head without enlarged conical setae; with many disc pores on dorsum .. 2
2. Dorsal trilocular and quinquelocular disc pores less numerous than macrotubular ducts, these scattered, not in groups .... *gouxi*, p. 293
   Dorsal trilocular and quinquelocular disc pores more numerous than macrotubular ducts, form groups arranged in transverse bands on abdominal tergites .... *brachypodii*, p. 292

168. **Greenisca brachypodii** Borchsenius & Danzig – Falsebrome Felt Scale
Fig. 121b

Syn. & Lit.: *Greenisca brachypodii* Borchsenius & Danzig, 1966:43; Komosinska & Podsiadlo, 1967:684; Danzig, 1964:634; Koteja & Zak-Ogaza, 1969:362; 1981:510; Koteja, 1974a:104; Podsiadlo & Komosinska, 1976:89; Dziedzicka, 1977:5; Kosztarab & Kozár, 1978:80.

Ovisac ovoid, white or yellowish white, loosely felted, fluffy, 4.5 mm long, 1.8 mm wide. Adult female elongate-oval, 3.1–3.9 mm long, 1–2.2 mm wide; antennae 7-(rarely 6-) segmented, ca 278 μ long, fourth segment longest; labium 93 μ long, 80 μ

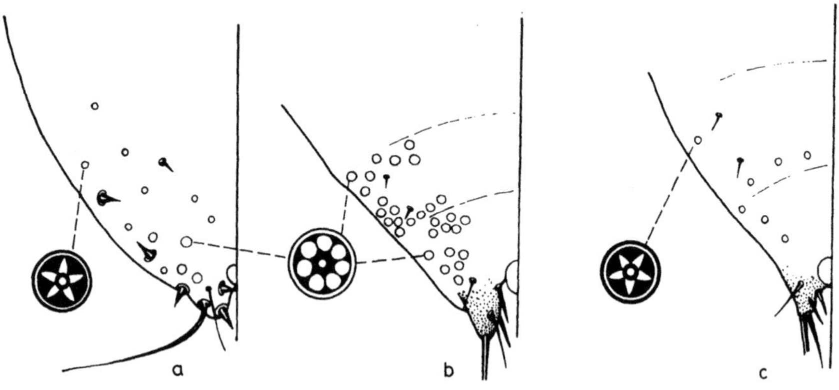

Fig. 121. Dorsal half of the last tergites of (a) *Kaweckia rubra* (Matesova), female, (b) *Greenisca brachypodii* Borchsenius & Danzig, female, (c) *Greenisca gouxi* Balachowsky, female (modified after Kosztarab and Kozár, 1978)

wide, with 18 setae; stylet loop ca 142 μ long; hind leg ca 710 μ long, hind coxa with a group of translucent pores, claw without denticle, its digitules about as long as claw; numerous quinquelocular and trilocular disc pores on dorsum, these form transverse bands or groups; venter with a submarginal band of oval disc pores and quinquelocular pores, and with some multilocular pores with 10-loculi in transverse bands on abdominal sternites 5–8; microtubular ducts on dorsum of prosoma; macrotubular ducts on both derms; hairlike setae short on dorsum, these much longer on venter; no conical setae on dorsum, except on anal lobes; anal ring oval, with 1 row of pores and 8 setae, each ca 173 μ long; anal lobes sclerotized, each lobe with 2 dorsal conical and 1 hairlike setae, 2 ca 50 μ long, 1 only 15 μ long, one subapical seta half as long as anal ring setae, and an apical seta 309–325 μ long.

Distribution. A fairly common steppe inhabiting species. Known in CE from: Cz, Hu, Po and USSR. Koteja and Zak-Ogaza (1969) found it on warm rocky hills.

Biology. Found on leaves of grasses of the genera: *Brachypodium, Bromus, Lolium* and *Stipa*. Adult females were collected during July in Kazakhstan, and in July and September in Poland.

Natural enemy. Parasite: *Eunotus acutus* (Podsiadlo, 1981b).

169. **Greenisca gouxi** (Balachowsky) – Balachowsky's Felt Scale
Fig. 121c

Syn. & Lit.: *Anophococcus gouxi* Balachowsky, *1954a:61; *G. gouxi*, Hoy, 1963:133; Borchsenius & Danzig, 1966:43; Koteja & Zak-Ogaza, 1969:363; Koteja, 1974a:153; Dziedzicka, 1977:5; Kosztarab & Kozár, 1978:80; Danzig, 1980a:228; Koteja & Zak-Ogaza, *1981:510.

Ovisac elongate-oval, white or light cream-colored. Adult female very elongate-oval, tapering posteriorly, yellow, 2.3–4 mm long, 1–1.6 mm wide; antennae 6-, rarely 7-segmented, 278–309 μ long, when 6-segmented third segment longer than 1st and 2nd combined; labium 93 μ long and wide; stylet loop 124 μ long; legs long and slender, hind leg ca 760 μ long, hind coxa and hind femur with a small group of translucent pores, no denticle on claws; disc pores on both derms; oval disc pores with cruciform loculi form a band along ventral margin; dorsum with quinquelocular and trilocular pores scattered on entire surface; venter with marginal and submarginal band of quinquelocular pores and transverse bands of 7 or 10-locular pores on terminal abdominal sternites; microducts along dorsal margin of prosoma; macrotubular ducts in large numbers on dorsum, few on venter; dorsum with very small thin scattered setae; ventral hairlike setae variable in size; anal ring with a row of pores and 8 setae, each 124 μ long; anal lobes heavily sclerotized with teeth on inner margin, 2 conical and 1 hairlike setae on dorsal surface, one preapical seta half as long as anal ring setae, and apical seta 285–325 μ long.

Distribution. A rare Trans-Palearctic species; known in CE: Bu, Fr, Hu, Po and USSR (Georgia to Far East).

Biology. On leaves of grasses in the genera *Brachypodium, Calamagrostis, Carex,* and *Molinia*. Adult females collected from July to September.

170. **Greenisca placida** (Green) – Smooth Felt Scale

Syn. & Lit.: *Ericoccus placidus* Green, 1921:148, 1923:212. *G. placida,* Zahradnik, 1959b:539, 1977:121; Danzig, 1959a:446, 1964:634; Hoy, 1963:134; Boratynski & Williams, 1964:108; Borchsenius & Danzig, 1966:43; Kosztarab & Kozár, 1978:79; Koteja & Zak-Ogaza, 1981:510.

Ovisac strongly convex, grayish-yellow, 3.5–4.5 mm long, 2–2.2 mm wide. Adult female elongate-oval, according to Green (1921), 2.5–3 mm long, 1–1.2 mm wide; antennae 7-segmented, 4th segment longest, and usually much longer than 3rd (Green, 1923); eyes distinct; legs relatively large, tarsus as long or slightly longer than tibia, claw with a small denticle near apex, hind coxa with translucent pores; only few disc pores on dorsum; ventral derm with numerous quinquelocular pores, few triloculars, oval disc pores also numerous; tubular ducts few; conical setae sharp-pointed, very few, confined to head and anal lobes, 3 on each anal lobe, 2 of these at inner margin of anal lobes much longer and stouter than third; 3 or 4 large slender and pointed conical setae on head; most other setae on derm hairlike and of various sizes; anal ring with pores and 8 setae; anal lobes relatively small and only weakly sclerotized; apical setae long and stout.

Distribution. A rare species; known only from Cz (?) and England. It was erroneously reported from the USSR (Danzig, 1964).

Biology. Found on upper surface of grass leaves in the genera: *Agropyron, Brachypodium,* and *Festuca.* Biology and males not known.

Natural enemy. Parasite: *Eunotus acutus* (Trjapitzin, 1978).

Note. This species is known only through the scanty original description. A full redescription, based on type material, is needed.

Genus **Kaweckia** Koteja & Zak-Ogaza

Syn. & Lit.: *Kaweckia* Koteja & Zak-Ogaza, 1981:502.

Type-species: *Eriococcus glyceriae* Green, 1921.

Ovisac oval, or flattened if within leaf sheaths, feltlike. Adult female elongate oval, almost parallel-sided; antennae 6- or 7-segmented, base at body margin; legs well developed, hind tibia with 4 setae, 2 on inner side; stylet loop extends to line between mid-coxae, disc pores on both surfaces, cruciform pores on dorsum and ventral margin; micro- and macrotubular ducts on both surfaces; enlarged conical setae often with truncate apex on 3–5 posterior abdominal segments and on anal lobes; dorsal setae small needle-like, ventral setae in medial area enlarged, hairlike; anal ring circular, broad, twice smaller than in *Greenisca*; anal lobes prominent. Living females red.

Distribution. Trans-Palearctic, with 5 species, 3 in CE.

Biology. On stem, root collar, and in leaf sheaths of various grasses, often below soil surface; eggs overwinter, males unknown.

Key to species of *Kaweckia*

1. Without any quinquelocular or multilocular pores on dorsum .... *laeticoris*, p. 297
   With at least a few such disc pores in transverse rows on dorsum 2
2. All conical setae on dorsal margin of abdominal segments 4–7 and on anal lobes elongate and with truncate apex (fig. 122) ....*glyceriae*, p. 295
   Conical setae on dorsal margin of abdominal segments 4–7 and on anal lobes short, less than twice as long as wide at base and not truncate (fig. 121a) .... *rubra*, p. 297

171. **Kaweckia glyceriae** (Green) – Grass Felt Scale
Fig. 122

Syn. & Lit.: *Eriococcus glyceriae* Green, 1921:146, 1922a:20. *G. glyceriae*, Goux, 1937c:93; Borchsenius, 1949:368, 1956:679; Schmutterer, *1952b:410; Zahradnik, 1959:527; Tereznikova, 1959b:796, 1981:52; Danzig, 1959a:398, 1964:634, 1980a:228; Hoy, 1963:133; Borchsenius & Danzig, 1966:43; Koteja & Zak-Ogaza, 1969:3639 Koteja, 1974a:153; Dziedzicka, 1977:71; Kosztarab & Kozár, 1978:79. *Kaweckia glyceriae*, Koteja & Zak-Ogaza, *1981:506.

Ovisac compact, white to yellowish, up to 4 mm long, 1.6 mm wide; covers female completely. Adult female elongate, twice as long as wide, ca 3 mm long, 1–1.7 mm wide, bright pink, covered with powdery wax; antennae 7-segmented, 217–248 μ long; eyes distinct, dome-shaped; stylet loop reaches line between mid-coxae; hind legs ca 440 μ long; hind coxa with a group of translucent pores; hind femur 112–140, tibia 87–110, tarsus 110–117 μ long; claw with a small denticle; oval disc pores on dorsum and ventral margin; quinquelocular, 7 and 9-locular pores scattered on venter, latter more numerous than first; microtubular ducts scattered on dorsum and margin of venter; macrotubular ducts numerous, form rows and bands on both derms; 8–11 enlarged cylindrical setae with truncate tip and broadened at base form a row on each dorsal margin of abdominal segments 4 to 7; many thin setae on both derms, those on dorsum shorter than those on venter, preanal setae 40–78 μ long; anal ring small, partly with double row of pores and with 8, rarely 6 setae, each 65–95 μ long; anal lobes broad, subapical setae 88–99 μ long, apical setae 195–235 μ long, 3 thick cylindrical setae with truncate tip, each 20–29 μ long, on dorsum of each lobe. First instar described by Schmutterer (1952b) and Tereznikova (1981).

Distribution. A common Trans-Palearctic species. In CE: Cz, Fr, Ge, Hu, Po, USSR (from Ukraine to Far East), and Yu, also in England.

Biology. Lives on leaves, in leaf sheaths and on root crown of grasses under soil surface in the genera: *Agropyron, Agrostis, Anthoxanthum, Arrhenatherum, Calamagrostis, Carex, Corynephorus, Elymus, Elytrigia, Festuca, Glyceria, Hierochloe, Koeleria, Phleum, Piptochaetium, Poa, Secale* and *Triticum*. According to Schmutterer (1952b) it has 1 yearly generation; overwinters as eggs (bright pink), hatch at end of April or last part of May; adults develop by end of June, or beginning of July, and each female lays 15–127 eggs in July and August. Males unknown. Matesova (1968a) considered it a pest in Kazakhstan.

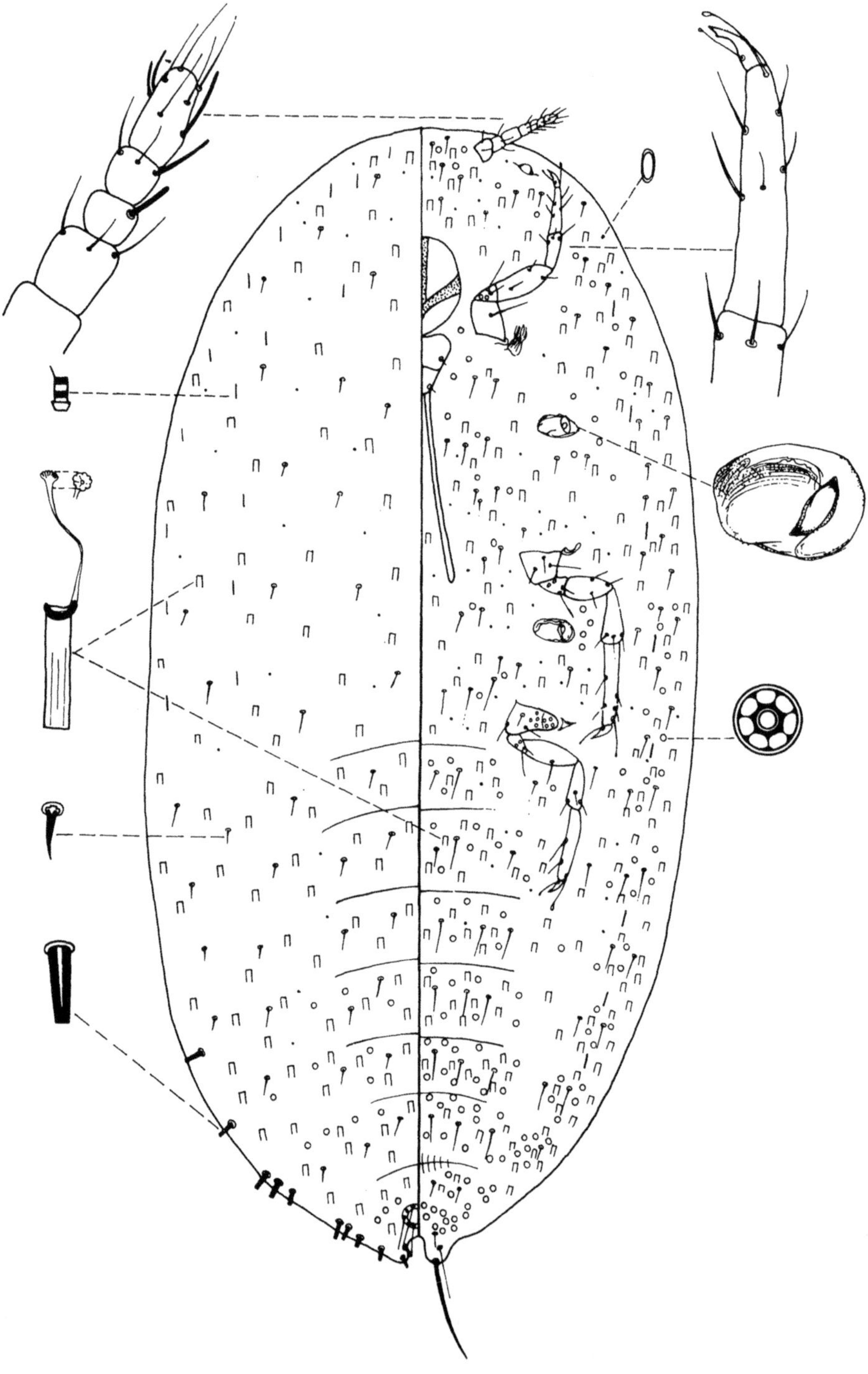

Fig. 122. *Kaweckia glyceriae* (Green), female (original)

172. **Kaweckia laeticoris** (Tereznikova) – Tereznikova's Felt Scale

Syn. & Lit.: *Greenisca laeticoris* Tereznikova, 1965:975, Tereznikova, *1981:55. *G. rubra*, Tereznikova, 1956:956; Koteja, 1974a:77; Dziedzicka, 1977:5. *Kaweckia laeticoris*, Koteja & Zak-Ogaza, 1981:507.

Ovisacs in leaf sheaths, cocoon-like, whitish, flattened, thin. Adult female elongate-oval, reddish, 2.5–4 mm long, 1–1.5 mm wide; antennae 6 or 7-segmented; eyes distinct; legs small, hind coxa with a large group of translucent pores, claws with small denticle; stylet loop reaches line between middle legs; spiracles small, with sclerotized oval rim; a large group of disc pores (5–7-loculars) anterior of each spiracle; oval, thick-rimmed disc pores scattered on entire dorsum; 5–7-locular pores frequent on entire venter, but none on dorsum; microtubular ducts scattered along margin of abdomen on both surfaces; macrotubular ducts of one size also numerous on both surfaces; 2–3 thick conical setae, these ca twice as long as wide at base on margin of abdominal segments 4–7; also on each anal lobe; anal ring circular, with partly double row of pores and 8 setae; anal lobes short and broad with long apical setae.

Distribution. A rare steppe-inhabiting xerophilous species. In CE: Cz, Hu, Po, and USSR (Ukraine).

Biology. In the leaf sheaths of *Agrostis, Elymus, Festuca.* Often found together with *K. glyceriae.* The adult females were collected in July and August.

173. **Kaweckia rubra** (Matesova) – Matesova's Felt Scale
Fig. 121a

Syn. & Lit.: *Greenisca rubra* Matesova, 1960:209; Hoy, 1963:134; Tereznikova, 1956:958; Borchsenius & Danzig, 1966:42; Danzig, 1974:70; *1980a:231; Koteja, 1974a:77; Dziedzicka, 1977:5; Kosztarab & Kozár, 1978:79. *Kaweckia rubra*, Koteja & Zak-Ogaza, 1981:508.

Ovisac cocoon-like, with pink eggs.

Adult female elongate-oval, pink, 2.3–3 mm long, 0.9–1.8 mm wide; antennae 7-segmented ca 280 μ long; eyes distinct; legs relatively small, hind legs 380–400 μ long; hind coxa with translucent pores, claw without denticle; spiracles with sclerotized oval atrial rim; a large group of disc pores anterior to spiracles; oval disc pores scattered on entire dorsum, and along margin of prosomal venter; quinquelocular pores often mixed with 7-locular pores; latter more frequent on terminal abdominal tergites, and in transverse rows or bands on sternites and in a wide submarginal band on venter; microtubular ducts scattered on dorsum; macrotubular ducts of two sizes on both derms, those on venter smaller; few conical setae, less than twice as long as wide at base, on dorsal margin of abdominal segments 5 to 7; anal ring circular, with pores and 8 setae, each seta ca 70 μ long; anal lobes short and broad, each with 3 short conical setae, subapical setae 112 μ long, apical setae ca 210–230 μ long.

Last nymphal instar described by Matesova (1960).

Distribution. A rare steppe inhabiting species. In USSR (from Middle to Far East), also in Mongolia.

Biology. Live in axils of leaves of grasses in the genera: *Agrostis, Elymus, Festuca* and *Stipa.* Infested stems are discolored and die.

Note. Earlier records of this species in CE are erroneous, according to Koteja and Zak-Ogaza (1981).

Genus **Rhizococcus** Signoret

Syn. & Lit.: *Rhizococcus* Signoret, 1875:36; Borchsenius, 1948a:501, 1949:351; Kosztarab, 1959:402; Ossiannilsson, 1959:194; Danzig, *1962b:520, 1964:632; Dziedzicka & Koteja, *1971:557; Koteja, 1974a:47; Dziedzicka, 1977:5; Koteja & Zak-Ogaza, 1981:515. *Eriococcus*, Ferris, 1955:97, 1957:85; Hoy, 1963:65. *Nidularia*, Lindinger, 1937:190. *Acanthococcus*, Kosztarab & Kozár, 1978:67; Danzig, 1980a:205; Tereznikova, 1981:14.

Type-species: *Rhizococcus gnidii* Signoret, 1975.

Ovisac elongate-oval, encloses female completely, white to yellowish, ca 1.5–5 μ long, 0.8–2.2 μ wide. Adult female elongate-oval, tapering posteriorly; antennae 7-, rarely 6-segmented; eyes usually on ventral margin, close to bases of antennae; labium conical, 3-segmented, with 18 setae; legs well-developed, hind coxa, at least on dorsal surface, with translucent pores (Dziedzicka, 1977); claws normally with denticle, all 4 digitules dilated and longer than claws; disc pores on venter only, usually quinquelocular, but sometimes with 3, 4, 6, or 7-loculi; oval disc pores form a band along ventral margin; tubular ducts of 3 types: large macrotubular ducts form transverse bands on dorsum and marginal band on venter, small macrotubular ducts few, found scattered on venter, except margin, microtubular ducts form transverse bands on tergites and a longitudinal marginal band on sternites; dorsal enlarged conical setae of varying sizes: normally large ones form a marginal row, median ones form a submarginal row and small ones form transverse rows or bands on tergites; ventral setae vary: medium-size spinelike conical setae form a row along body margin, small ones in an irregular submarginal band, hairlike setae scattered on venter, but form transverse rows or groups in the median area; anal ring with pores and 8, rarely 6 setae; anal lobes distinct, each with an apical seta, 3 dorsal conical setae and ventral hairlike setae.

Distribution. World-wide about 34 species; probably absent from Ethiopian and Neotropical Regions. Only 7 species in CE.

Biology. Most feed on grasses.

Note. Distinguishing species in this genus is a challenge, since we often have to rely solely on the number, size, shape and distribution pattern of enlarged conical setae, as given by Danzig (1962b) for the species in the USSR, and in an excellent comparative chart by Dziedzicka and Koteja (1971) for the 7 species known from Poland. *R. confusus* Danzig has been known only from the Polish Baltic Coast and the Leningrad Region. It was recently collected in Central Poland, but it was too late to include it in this study. *Rhizococcus* is related to *Acanthococcus* and *Greenisca.*

Key to species of *Rhizococcus*

1. Enlarged setae of abdominal margin slender, not very thick, conical, with truncate or blunt apex; only 2 slender setae on each margin of abdominal segments 1–7 . . . . *agropyri*, p. 299

Enlarged setae of abdominal margin thick, conical; usually with more than 2 setae (except in *palustris* and *cingulatus*) on each margin of abdominal segments 1–7 2

2. Marginal setae on abdomen short and thick, less than 2.5 times as long as wide at base; normally with only 2 enlarged conical setae on each margin of abdominal segments .... *palustris*, p. 304

Marginal setae on abdomen slender, more than 2.5 times as long as wide at base; with 3 to 5 conical setae on margin of each abdominal segment (except in *cingulatus* with 2 setae on some segments) 3

3. With a submarginal row of thick setae on venter of prosomal margin; some abdominal segmental margins with only 2 setae on each side; feed on dicotyledonous plants (except *Koeleria* grass) .... *cingulatus*, p. 300

Without a submarginal row of thick setae on venter of prosomal margin; all abdominal segmental margins with 3 or more setae on each side; feed usually on monocotyledons (grasses) 4

4. Of 4 marginal setae on each side of abdominal segments, 2 small ones located mesad of body margin (fig. 123) .... *cynodontis*, p. 301

No small marginal setae mesad of abdominal body margin 5

5. With 5 marginal setae on each side of abdominal segments 1–6; conical setae on dorsal derm absent or very short, usually as long as wide at base .... *insignis*, p. 303

With 3 or 4 marginal setae on each side of abdominal segments 1–6; most conical setae on dorsal derm slender, 1.5 to 2 times as long as wide at base 6

6. With 3 marginal setae of different sizes on each side of abdominal segments 1–6; dorsal conical setae larger on prosoma and on abdomen 20–45 μ long .... *herbaceus*, p. 301

With 4 (rarely 3–5) marginal conical setae on each side of abdominal segments 1–6, 2 setae smaller than others; dorsal conical setae on prosoma 15–30 μ, on abdomen only 7–20 μ long .... *pseudinsignis*, p. 304

174. **Rhizococcus agropyri** Borchsenius – Borchsenius' Felt Scale
Figs 22, 118d

Syn. & Lit.: *Rhizococcus agropyri* Borchsenius, 1949:359; Ossiannilsson, 1959:194; Danzig, *1962b:852; Rogojanu, 1966b:432; Tsalev, 1968:207; Dziedzicka & Koteja, *1971:575; Koteja, 1974a:104; Dziedzicka, 1977:72. *R. obscurus* Borchsenius, 1949:360. *Eriococcus agropyri*, Hoy, 1963:68. *Acanthococcus agropyri*, Kosztarab & Kozár, 1978:69; Tereznikova, *1981:17.

Ovisac gray, white or yellowish, with almost flat dorsum and carinate submargin, 4 mm long and 1.5 mm wide, they form compact colonies often overlaying each other on leaf sheaths (Danzig, 1962b). Adult female elongate-oval, yellow, 2.3–3.3 mm long, 1.2–1.7 mm wide; antenna 7-, rarely 6-segmented, 260–279 μ long; labium ca 125 μ long, 80 μ wide; stylet loop ca 1.5 times as long as labium; hind legs 570–645 μ long, hind coxae and femur with a group of translucent pores; hind femur ca 137–165, tibia

125–143, tarsus 150 μ long, claw with denticle; oval disc pores in a submarginal band on venter; quinquelocular pores scattered on abdominal sternites and at spiracles; few multilocular disc pores on venter of prosoma, others form rows and bands on abdominal sternites; macrotubular ducts numerous, of two sizes, form rows and bands on both body surfaces; microtubular ducts scattered on entire dorsum; conical setae with truncate or rounded apex, form a row along body margin, with 2 setae (1 somewhat larger than other) on each margin of abdominal sternites 1–7; small setae, 4.5–10 μ long, form rows on dorsum and along body margin; ventral hairlike setae of various lengths; anal ring partly with double pore row, with 8 setae, each 100–129 μ long; anal lobe apex and inner margin distinctly sclerotized, apical seta 234–315 μ long; subapical seta 148 μ long; lobes with 2 dorsal medial setae, ca 30 and 52 μ long, and 1 lateral seta 25 μ long.

Distribution. A common steppe-inhabiting mesophilous species. In CE: Bu, Hu, Po, Ro, USSR (European Part and Middle Asia); also known from Sweden.

Biology. Infests grasses of the genera: *Agropyron, Agrostis, Aneurolepidium, Elymus* and *Lolium*. Often becomes a pest. Females collected from May through October, each layed 98–211 eggs (Matesova, 1968a). Probably has more than 1 generation per year. No males known.

Natural enemies. Parasites: *Aphycus nitens, Coccidencyrtus breviventris, Marietta zebra, Xana kurdjumovi* (Kosztarab, 1956b); *Eunotus acutus, Parasyrphophagus lindus* (Trjapitzin, 1978); *Enargopelte* sp. (Matesova, 1968a).

175. **Rhizococcus cingulatus** (Kiritchenko) – Kiritchenko's Felt Scale

Syn. & Lit.: *Eriococcus cingulatus* Kiritchenko, 1940b:131. *E. longispinus* Borchsenius, 1937:184. *E. cingulatus*, Hoy, 1963:79. *Acanthococcus cingulatus*, Danzig, 1975a:62; Kosztarab & Kozár, 1978:69; Tereznikova, *1981:22. *A. c. orientalis*, Danzig, 1975:79, 1980a:223. *A. c. terrestris* Matesova, in Danzig, 1975a:80. *R. cingulatus*, Borchsenius, 1949:356; Danzig, *1962b:849; 1964:633.

Ovisac oval, white or cream colored, 3.6 mm long, 2.2 mm wide. Adult female oval, yellow or rusty-red, 2.3 mm long, 1.3 mm wide; antennae 7-segmented, 232–251 μ long; stylet loop extends to mesothoracic coxae; hind leg 525–550 μ long, its coxa with many translucent pores, claw thick, with denticle; disc pores with 5 or 7-loculi scattered on entire venter; macrotubular ducts of 2 sizes found scattered in large number on both body surfaces, most of those on venter more slender; microtubular ducts few on entire dorsum; dorsal conical setae slender, sharp-pointed on prosoma, with rounded apex on abdomen, form a row on body margin with 2 large setae and sometimes 1 small on each margin of abdominal segments; spinelike setae thin, 7 μ long, form 2–3 transverse rows on most tergites, a submarginal row of thick setae on venter of prosoma; other ventral setae short, hairlike; pre-anal setae ca 55 μ long; anal ring partly of a double row of pores with 8 setae, each 105–122 μ long; anal lobe with 3 thin dorsal setae (one often similar to marginal setae), each 57–62 μ long, apical seta 285–315 μ long and subapical seta ca 118 μ long.

Distribution. A xerophilous species, known only from steppes in the USSR, from Ukraine to the Far East.

Biology. Collected from roots of dicotyledons (except *Koeleria*) in the genera: *Achillea, Artemisia, Astragalus, Centaurea, Clematis, Dianthus, Koeleria, Potentilla, Pyrethrum, Scabiosa, Trigonella* and *Vitis*. Bisexual; females present from May to September and each lays ca 70 eggs.

176. **Rhizococcus cynodontis** (Kiritchenko) – Bermuda-grass Felt Scale
Fig. 123

Syn. & Lit.: *Eriococcus cynodontis* Kiritchenko, 1940b:133. *E. formicicola*, Borchsenius, 1937b:182. *E. insignis*, Borchsenius, 1949:357; Hoy, 1963:96. *R. cynodontis* Danzig, *1962b:844; Tsalev, 1968:207; Koteja, 1974a:104; Kosztarab & Kozár, 1978:69. *Acanthococcus cynodontis* Tereznikova, *1981:26.

Egg sac oval, snow white, 1.5–2.5 mm long, 0.8–1.5 mm wide. Adult female elongate oval, 2–3 mm long, lemon or brownish yellow; antennae 6- or 7-segmented, 247–264 μ long; labium ca 100 μ long, 75 μ wide; hind femur 148–151, tibia 124–130, tarsus 148–150 μ long; oval disc pores present on ventral submargin; most other disc pores with 7 loculi, scattered on entire venter; microtubular ducts scattered on dorsum; macrotubular ducts somewhat more slender on venter of abdomen, on dorsum they form transverse bands on prosoma and a band 2 or 3 rows wide on each abdominal tergite; marginal conical setae blunt or truncate apically on abdominal segments 5–7, pointed on prosoma and first 4 abdominal segments, they form a marginal row that includes 4, rarely 3 setae on each abdominal segmental margin, 1 or 2 of 4 setae much smaller than others and normally mesad of body margin; spine-like setae 8–12 μ long, form transverse rows on tergites; preanal setae 30–56 μ long; anal ring partly of dense double row of pores, with 8 setae, each ca 72 μ long; anal lobes sclerotized each with 3 dorsal conical setae.

Distribution. A rare species, occasionally found in high densities. Known only from Bu, Hu, It and USSR (Crimea, Caucasus).

Biology. Found on leaves of grass, *Cynodon dactylon*, only. Females collected from July through September. Males unknown; its biology needs study.

Natural enemy. Parasite *Eucoccidophagus breviventris* (Kozár and Sugonjaev, 1979).

177. **Rhizococcus herbaceus** Danzig – Danzig's Felt Scale

Syn. & Lit.: *Rhizococcus herbaceus* Danzig, *1962a:847, 1964:633; Dziedzicka & Koteja, *1971:557; Koteja, 1974a:104; Dziedzicka, 1977:71; Kosztarab & Kozár, 1978:68. *Acanthococcus herbaceus*, Tereznikova, *1981:29.

Ovisac elongate-oval, white. Adult female elongate-oval, greenish, ca 2.4–3 mm long, 1–1.4 mm wide; antennae 6- or 7-segmented, ca 223–250 μ long, third segment distinctly longest; labium ca 140 μ long, 90 μ wide; hind leg ca 650 μ long, hind coxa with translucent pores; hind femur 153–180, tibia 133–150, tarsus 125 μ long; ventral disc pores with 5 and 7 loculi, numerous, form bands on abdominal sternites, scattered on prosoma; microtubular ducts ca 7 μ long, common on entire body, except median area of venter, form a band along ventral margin; macrotubular ducts of 1 type

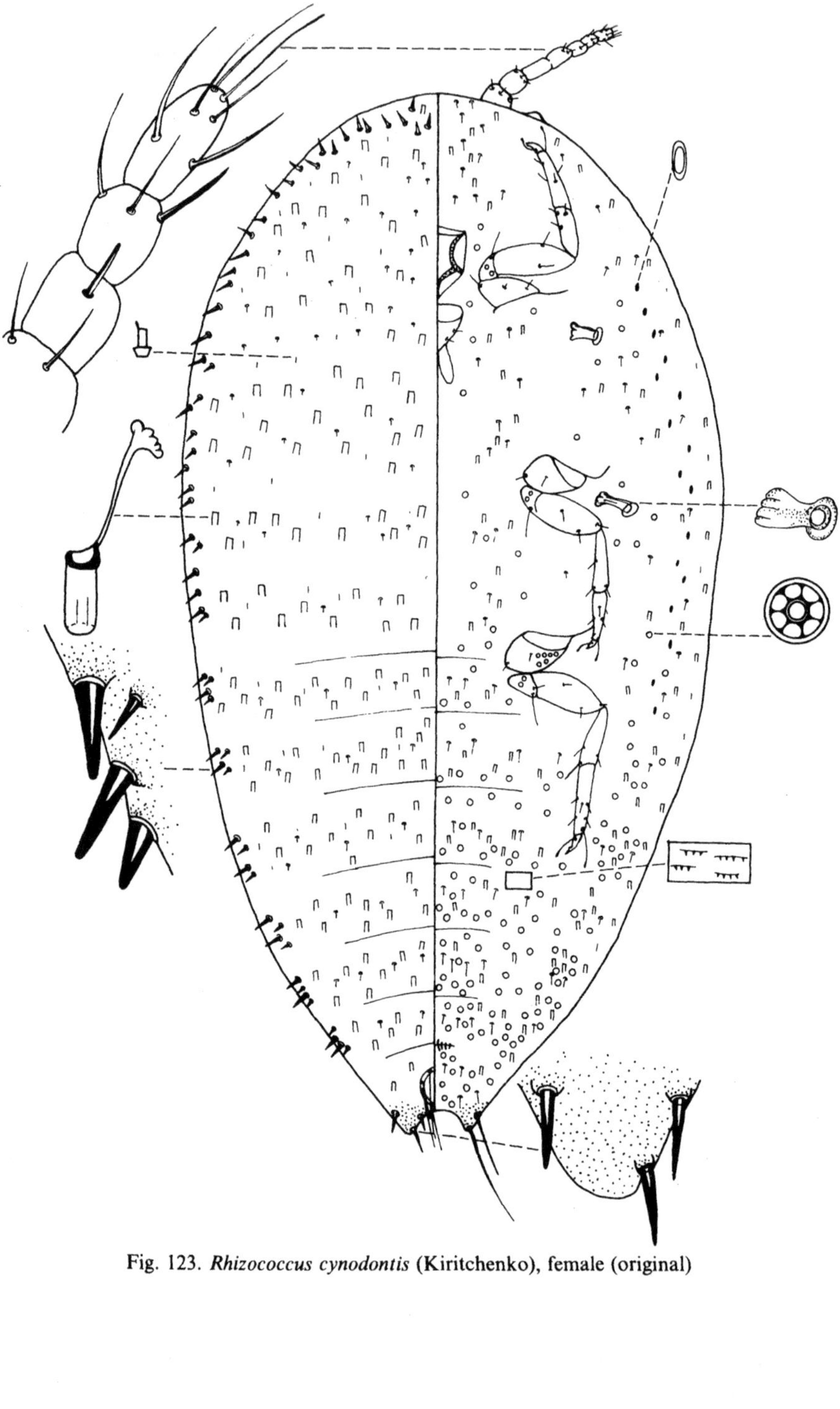

Fig. 123. *Rhizococcus cynodontis* (Kiritchenko), female (original)

scattered on both body surfaces; conical setae of various sizes and shapes: form a row on dorsal margin, these blunt on abdomen, pointed on prosoma, 26–52 μ long, 3 setae on margin of each abdominal tergite, third seta shorter than others; small spinelike setae on dorsum far apart and form transverse rows; most setae 20–45 μ long on dorsum; some on prosoma with bent tips; small setae form a row along ventral margin; ventral hairlike setae form transverse rows on abdominal segments, scattered or in groups on prosoma; anal ring with 1 row of pores and 8 setae, each ca 111–140 μ long; anal lobe with apical seta 250 μ long and 3 dorsal setae, 36–45 μ long.

Distribution. A rare boreal species; known only from Po and USSR (Regions of Leningrad and Transcarpathian Region).

Biology. Feed on leaves of grasses in the genera: *Brachypodium, Calamagrostis;* also on *Carex,* and *Luzula.* Females collected each month from June through September in Poland. Males unknown; biology needs studying.

178. **Rhizococcus insignis** (Newstead) – Conspicuous Felt Scale

Syn. & Lit.: *Eriococcus insignis* Newstead, 1891:164; Green, 1922:20; Schmutterer, 1952b:407; Hoy, 1963:95; Gomez-Menor, Ortega, 1968:553. *E. cynodontis* Kiritchenko, 1940:132. Latter species considered a synonym by Borchsenius (1949) was recognized as a good species by Danzig (1962b) and others. *E. formicicola,* Gavalov, 1928:19 (nomen nudum); *E. saratogensis* Rau, 1937:195. *R. insignis,* Borchsenius 1949:357; Ossiannilsson, 1959:195; Janetschek, 1959:156; Danzig, *1962b:841; 1964:633; Dziedzicka & Koteja, *1971:561; Kozár, 1971:158 (this record should refer to *R. cynodontis*); Koteja, 1974a:104. *Acanthococcus insignis,* Kosztarab & Kozár, 1978:69; Danzig, 1975a:62, 1980a:223; Tereznikova, 1981:29.

Ovisac elongate-oval, white or cream-colored, up to 5 mm long, 2 mm wide. Adult female elongate-oval, tapering toward both ends, yellow, 2.6–3 mm long, 1.1–1.3 mm wide; antennae 6- or 7-segmented, 260–322 μ long; labium ca 120 μ long, 75 μ wide; hind legs 630–680 μ long; hind coxa with a group of translucent pores, hind femur with 1–4 pores, 162–173-, tibia 125–151-, tarsus 150–180 μ long; disc pores with 7-loculi, fewer with 5-loculi and oval disc pores scattered on venter; tubular ducts numerous on both surfaces; microtubular ducts on dorsum only; macroducts of two sizes on venter; conical setae on dorsal margin sharp-pointed on anterior, blunt toward posterior end, of these 5, rarely 4 found on margin of abdominal segments 1–7, 2 always longer than others; small spinelike setae, usually as long as wide at base, form a row on ventral margin and form transverse rows on tergites; ventral hairlike setae of various sizes; preanal setae 40 μ long; anal ring with pores and 8 setae, each 111–138 μ long (only 6 setae on specimens from Poland, 8 on specimens from USSR); anal lobes sclerotized, with 3 dorsal conical setae; apical setae 309 μ long. First instar described by Gomez-Menor Ortega (1968).

Distribution. A common steppe inhabiting mesophilous species. In CE: Au, Bu, Cz, Fr, Ge, Hu, It, Po, USSR (from Ukraine to Far East); also known from Denmark (Kozarzhevskaya and Reitzel, 1975), England, Sweden, Iraq and USA.

Biology. Primarily feeds on leaves of grasses; host genera: *Agropyron, Agrostis, Anthoxanthum, Arrhenatherum, Avenastrum, Bromus, Calamagrostis, Carex, Cichorium, Cynodon, Dactylis, Deschampsia, Elymus, Festuca, Hieracium, Holcus, Koeleria, Luzula, Nardus, Phleum, Poa, Polytrichum, Pteridium, Pteris, Rumex, Saxifraga,*

Silene, Ulex and *Urtica.* Kosztarab and Kozár (1978) noted adults in June; first instars appeared during first part of June; adult females were present again in August and September; based on these records, it is assumed that this species has 2 generations in the area. Schmutterer (1952c) noted in S Germany 1 yearly generation, and eggs overwintering. Males known in USA. It damages Timothy grass in NW USA.

Natural enemies. Parasites: *Beocharis pascuorum, Metaphycus nitens,* and *Scutellista obscura* (Kozár & Sugonjaev, 1979); *Trichomasthus rhizococci* (Trjapitzin, 1978).

179. **Rhizococcus palustris** Dziedzicka & Koteja – Polish Felt Scale

Syn. & Lit.: *Rhizococcus palustris* Dziedzicka & Koteja, *1971:573; Koteja, 1974a:104; Dziedzicka, 1977:71; Kosztarab & Kozár, 1978:68.

Ovisac white or cream-colored, encloses female completely. Adult female elongate-oval, tapering posteriorly, according to Dziedzicka and Koteja (1971) 2.1–2.3 mm long, 1.2–1.3 mm wide; antennae 7-segmented, 250–270 μ long, fourth segment longest; labium 100–120 μ long, ca 75 μ wide; hind leg longest, 570–760 μ long, hind coxa with 7–13 translucent pores on both surfaces; claw and tarsal digitules dilated apically; oval disc pores form a narrow band along ventral margin; numerous disc pores with 3 to 7 loculi concentrated on posterior sternites; microducts form transverse bands on dorsum; macroducts of large size form transverse bands on tergites and a band along ventral margin; smaller macroducts scattered on rest of venter; short dorsal conical setae form transverse bands and rows on tergites; large conical setae form a marginal row, on anterior body margin these sharp-pointed and on posterior margin with blunt apex, all setae on abdominal margin relatively short, less than 2 1/2 times as long as wide at base, usually only 2 large conical setae on each abdominal segmental margin; ventral hairlike setae form groups or transverse rows; anal ring with 8 setae, each 100–150 μ long; anal lobes conical, partly sclerotized, with apical seta 260–280 μ long and dorsum of each lobe with 3 conical setae.

Distribution. Known from Poland only.

Biology. Infested leaves of *Carex* sp. and *Eriophorum vaginatum* in a peat-bog, apparently host specific to Cyperaceae. Females completed egg laying by August. Males not known.

180. **Rhizococcus pseudinsignis** (Green) – Boreal Felt Scale

Syn. & Lit.: *Eriococcus pseudinsignis* Green, 1921:149, 1922a:20; Schmutterer, 1952b:407; Hoy, 1963:110. *R. graminicola,* Ossiannilsson, 1959:195 (a misidentification according to Danzig, 1962b). *R. pseudinsignis,* Borchsenius, 1949:354; Danzig, *1962b:845 (she suggested that Borchsenius' 1949 description and illustration apply to *R. kondarensis,* also that Schmutterer's (1952b) description and illustration probably apply to *R. herbaceus*), 1964:633; Dziedzicka & Koteja, *1971:569; Koteja, 1974a:104 Danzig & Kozár, 1974:11; Kosztarab & Kozár, 1978:68. *Acanthococcus pseudinsignis* Tereznikova, *1981:35.

Ovisac. Elongate-oval, white to light yellow, 2.5–3 mm long, 1.2–1.5 mm wide. Adult female. Elongate-oval, yellow, ca 2.1–3.4 mm long, 1–1.4 mm wide; antennae 7-segmented, 186–254 μ long; labium ca 125 μ long, 75 μ wide; stylet loop extends to line between mid-coxae; hind coxa with a large group of translucent pores; hind leg ca 720 μ long, hind femur 138–187, tibia 130–160, tarsus 135–160 μ long; claw with a minute denticle, tarsal and claw digitules dilated apically; quinquelocular and 7-locular disc pores few, scattered on entire venter; no oval disc pores; microtubular ducts form rows scattered on entire dorsum and along body margin; macrotubular ducts of 2 sizes, larger type numerous on both body surfaces, smaller type on ventral sternites; conical setae form a row on dorsal margin, these blunt on abdomen, pointed toward head, each 20–42 μ long, row includes 4-, rarely 3 setae, on each side of abdominal segments 1–6; small spinelike setae, 6–10 μ long, form a longitudinal row on ventral margin and transverse rows on prosomal dorsum; preanal setae ca 47 μ long; hairlike setae of venter form transverse rows on abdominal sternites; anal ring with partly double pore row and 8 setae, each ca 130 μ long; anal lobes weakly sclerotized with 3 dorsal setae similar to marginal conical setae. First instar described by Schmutterer (1952b).

Distribution. A common boreal species in some places. In CE: Bu, Cz, Ge, Hu, It, Po, USSR (Transcarpathian Region) and Yu; also known from England, and Sweden.

Biology. Primarily on grasses; infests leaves of the plants in the host genera: *Agrostis, Ammophila, Anthoxanthum, Arrhenatherum, Brachypodium, Calamagrostis, Carex, Dactylis, Deschampsia, Festuca, Holcus, Luzula, Phragmites* and *Taraxacum.* According to Schmutterer (1952b) it has 1 yearly generation; eggs overwinter; adults appear during first part of June, start laying 29–87 eggs per female from end of June. Dziedzicka and Koteja (1971) collected adult females each month from April through September. Green (1921) found most ovisacs concealed at base of upper surface of leaves.

Natural enemies. Parasites: *Adelencyrtus breviventris, Anaphycus subapterus, Coccophagus* sp., *Enargopelte obscura, Marietta picta,* and *Rhopus* sp. (Tereznikova, 1981).

FAMILY **LECANODIASPIDIDAE** – FALSE PIT SCALES, LECANODIASPIDIDS

Syn. & Lit.: Asterolecanini (in Lecanoidea: Asterolecaniinae) Balachowsky, 1948:256, Lecaniodiaspididae Borchsenius, 1959:840; 1960:216. Lecanodiaspididae, Afifi & Kosztarab, 1969:1; Williams & Kosztarab, 1970:1; Howell & Kosztarab, 1972:2; Lambdin & Kosztarab, 1973:1, 1977:14; Koteja, 1974b:306; Koteja & Liniowska, 1976:671. Lecanodiaspidinae (in Coccidae), Koteja, 1974a:55; Danzig, 1980a:62.

Females and their eggs completely covered by a papery test; test oval to subcircular and convex in most species. Adult female body oval or subcircular, often with spiracular indentations; antennae normally 8 or 9-segmented, occasionally reduced to 2–7 segments, rarely 10-segmented, terminal segments with several fleshy slender setae; clypeolabral shield (Koteja & Liniowska, 1976) pentagonal, strongly sclerotized and sculptured, with a pair of clypeal and a pair of labral setae; labium hemispherical, appears 1-segmented, with 6–10 setae; legs vary from unsegmented stubs to well developed 5-segmented legs, rarely entirely absent; spiracles often with marginal sclerotization; spiracular furrows present, often bifurcate in *Lecanodiaspis,* each furrow lined with 3- to 10-locular pores, but most quinquelocular; cribriform plates of various shape and size, normally in 2 submedial longitudinal rows, but rarely absent or form 2 large groups on both sides of dorsum; anal ring oval to subcircular, bordered anteriorly by arched plate and posteriolaterally by 2 anal plates, latter often medially fused, ring with 6 to 18 setae (normally with 10) and 0 to 3 irregular rows of pores; anal cleft extends to base of anal plates; anal lobes prominent, broadly rounded at apex, with or without apical setae; simple disc pores distributed on both derms, but fewer on venter; discoidal pores present on dorsum in *Psoraleococcus* Borchsenius; bilocular pores (called square-shaped pores in earlier literature) in a submarginal band on venter, or rarely irregularly spaced; trilocular pores occasionally on dorsum, multilocular pores, primarily 10-loculars, arranged in 6 or 7 transverse abdominal rows or bands, except in *Anomalococcus* Green and *Pterococcus* Howell & Kosztarab; 8-shaped pores of 3 types on dorsum: extremely bent pores, slightly bent pores and flat pores; on venter 2 types present, these extremely bent or flat, both normally restricted to marginal and submarginal area; tubular ducts with clavate terminal filament unevenly distributed on both derms; marginal setae vary on dorsum from short and conical to long and hairlike, setae on venter variable in size and shape, these often form well defined submarginal and transverse rows; spiracular setae, when present, terminate at spiracular furrows on body margin in most species, normally 1 or 2 anterior and 1 posterior seta on each margin, latter rarely absent, they vary from needle-like to fleshy setae; microspines common on venter, on inner margins of anal lobes and in medial area of abdominal segments and at bases of coxae, when legs present.

Adult males described by Giliomee (1967a:185), Afifi and Kosztarab (1969); male test by Borchsenius (1960); first instars by Williams and Kosztarab (1970).

Distribution. World-wide; with ca 71 species in 9 genera; 1 genus with 1 species in CE.

Biology. They feed predominantly on woody plants, often producing galls or pits; some species are monophagous, others oligophagous or polyphagous. Only the biology of *Lecanodiaspis prosopidis* (Maskell) was studied. These developed through 3 instars in females and 5 in the males. They overwintered in egg stage; each female laid 71–188 eggs, and had one yearly generation in North America (Howell and Kosztarab, 1972).

Genus **Lecanodiaspis** Targioni-Tozzetti

Syn. & Lit. *Lecaniodiaspis* Targioni-Tozzetti, 1869:261; Cockerell, 1899:276; Ferris, 1955:44; Borchsenius, 1960:233. *Prosopophora* Douglas, 1892:207; Borchsenius, 1960:223. *Birchippia* Green, 1900:448. *Lecanodiaspis*, Afifi & Kosztarab, 1969:1; Williams & Kosztarab, 1970:1; Kosztarab & Howell, 1972:98; Hodgson, 1973:413; Howell & Kosztarab, 1972:2; Lambdin & Kosztarab, 1973:1, 1977:14; Koteja, 1974b:336; 1974a:113; Koteja & Liniowska, 1976:671.

Type-species: *Lecanodiaspis sardoa* Targioni-Tozzetti, 1869.

Test of female waxy, papery, encloses adult and eggs. Adult female oval; antennae normally 7–9-segmented, rarely 3–6-segmented; legs 1–5-segmented; spiracles associated with primarily quinquelocular pores that extend into spiracular furrow, terminating in spiracular setae on body margins; cribriform plates normally in 2 (rarely 4) longitudinal rows; anal ring with 6–18 setae and 2 or 3 rows of cone or nipple-shaped pores; an arched plate above anal ring, normally bilobed; anal plates usually triangular, ridged and with 2 stout setae dorsally, normally connected in the middle;

anal cleft short; anal lobe slightly protruding in young adults, in older specimens not well defined; various disc pores; 8-shaped pores, tubular ducts and setae as given in the family description.

Distribution. World-wide with ca 55 species; 1 in CE.

Biology. As for the family.

181. **Lecanodiaspis sardoa** Targioni-Tozzetti – Mediterranean Pit Scale
Fig. 124

Syn. & Lit.: *Lecaniodiaspis sardoa* Targioni-Tozzetti, 1869:162; Lindinger, 1912:105, 179; Borchsenius, 1960:234. *L. sardoa*, Leonardi, 1920:263; Goux, 1943b:131; Bodenheimer, 1953a:135, 138, 158; Howell & Kosztarab, 1972:89; Koteja, 1974a:113.

Test of female oval, convex, grayish yellow, ca 2.5 mm long, 2.0 mm wide and 1.5 mm high, with ca 10 tubercles in mid-dorsal carina, lateral ridges absent. Adult female oval, anal lobes slightly protruding; 1.4–2.6 mm long, 0.8–2.1 mm wide; antennae 8 or 9-segmented, 136–164 μ long; labium with 8 setae; legs reduced to stubs with claw and 2 digitules; anterior spiracular furrow with a band of 27–67 primarily quinquelocular pores, posterior furrow divided, with 37–102 pores; cribriform plates unevenly areolate, in 2 longitudinal rows, 3–5 per row, each 7–12 μ in diameter; anal ring oval, 55–65 μ long, 28–46 μ wide, with 10 setae, each 81–112 μ long, and 4 rows of cone-shaped pores; arched plate bilobed, 53–77 μ long, 19–40 μ wide; anal plates triangular, with heavy ridges; dorsum with simple, trilocular and 8-shaped pores of 2 sizes, also tubular ducts; venter with simple, bilocular (square-shaped), multilocular (with 9–13 loculi) and 8-shaped pores, also with tubular ducts; spiracular setae similar to other marginal setae on dorsum; on venter various other setae, also microspines. Adult males described by Borchsenius (1960), Afifi and Kosztarab (1969); first instar by Borchsenius (1960), Williams and Kosztarab (1970).

Distribution. Circum-Mediterranean. In CE: Fr, It, Yu, also Portugal, Spain, Turkey, N Africa, Corsica and Sardinia. A rare species.

Biology. Poorly known. Host genera: *Cistus, Helianthemum.*

Natural enemy. Parasite: *Protyndarichus coccidiphagus* (Borchsenius, 1960).

FAMILY **CEROCOCCIDAE** – ORNATE PIT SCALES, CEROCOCCIDS

Syn. & Lit.: Cerococcini (in Lecanoidea: Asterolecaniinae) Balachowsky, 1948:256. Cerococcinae, Borchsenius, 1960:102; Danzig, 1980a:62. Cerococcidae, Koteja, 1974a:49; Koteja & Liniowska, 1976:668; Lambdin & Kosztarab, 1977:11; Kosztarab & Kozár, 1978:131; Hamon & Kosztarab, 1979:8.

Females and their eggs enclosed in papery wax test. Adult female body membranous, pyriform, with prominent anal lobes, 2–5 mm long; antennae reduced to unsegmented stubs, with several terminal setaceous and fleshy setae; labium 3-segmented, with 10–12 setae; legs if present stublike; spiracles large, with associated disc pores, also with lateral projecting spiracular furrows, posterior furrows usually bifid, furrows lined with quinquelocular to 7-locular pores that form a cluster at the apex of furrows; no spiracular spines; normally with oval or subcircular cribriform plates submedially on posterior dorsal abdominal segments, these arranged in 2 clusters or in transverse rows or bands; anal ring in anal tube; with pores and 8–12 setae; a shieldlike anal plate well developed; no arched plate; anal lobes prominent, conical, each with an apical seta; simple disc pores often on both derms; bilocular pores unevenly distributed on venter of prosoma; trilocular pores occasionally scattered on both derms; quinquelocular pores form small groups at bases of antennae, normally in

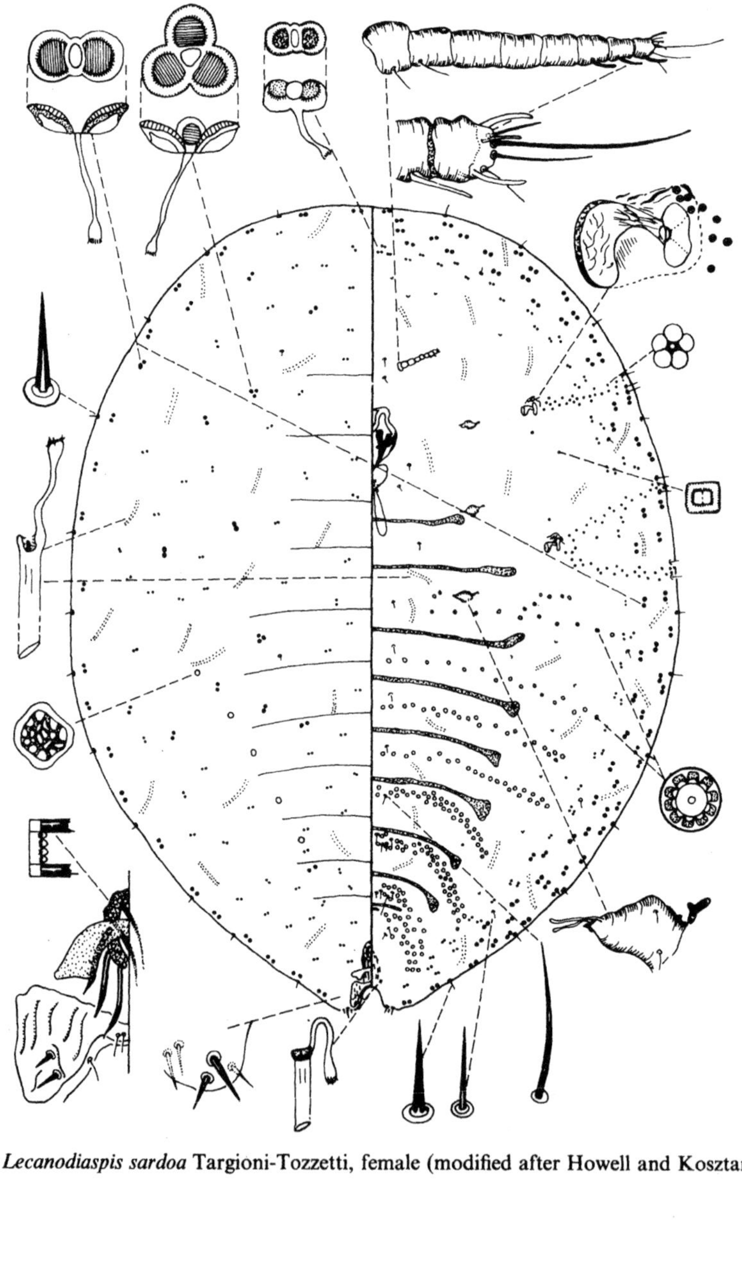

Fig. 124. *Lecanodiaspis sardoa* Targioni-Tozzetti, female (modified after Howell and Kosztarab, 1972)

spiracular furrows mixed with 6 and 7-locular pores; multilocular pores arranged in several transverse abdominal rows or bands on venter; 8-shaped pores normally of two shapes (flat and bent) and sizes on both derms, often form various patterns on dorsum; tubular ducts of various shapes and sizes on both derms, each with a clavate terminal filament, and a cuplike indentation at base of terminal filament, often with small teeth or spines; body setae variable in number, shape and size, usually needle-like or tacklike, seldom fleshy; microspines and macrospines common on venter of abdomen. Adult males described by Patterson (1901); various instars by Howell et al. (1971), first instars by Šulc (1953); Borchsenius (1960), Hamon and Kosztarab (1979).

Distribution. World-wide with ca 64 species in 3 genera. In CE 1 genus with 1 species.

Biology. They develop through 3 instars in females and 5 in males; eggs overwinter; normally 1 yearly generation in temperate climates. Some monophagous, others polyphagous; infest woody and herbaceous plants; often produce pits on twigs, damage to fruit and ornamental trees.

Genus **Cerococcus** Comstock

Syn. & Lit.: *Cerococcus* Comstock, 1882:213; Balachowsky, 1930b:201; Goux, 1932:141; Russell, 1941:3; Ferris, 1955:31; Borchsenius, 1960:101; Lambdin & Kosztarab, 1977:15; Hamon & Kosztarab, 1979:8; Kosztarab & Kozár, 1978:131. *Antecerococcus* Green, 1901:560. *Phenacobryum* 1902:114. *Amelococcus* Marchal, 1904:557. *Cercococcus* Scott, 1907:455. *Coricoccus*. Mahdihassan, 1933:562.

Type-species: *Cerococcus quercus* Comstock, 1882.

The description of the genus same as for the family Cerococcidae. In the related two genera, *Asterococcus* Borchsenius and *Solenophora* Maskell, the 8-shaped pores and tubular ducts form a marginal band, while other 8-shaped pores are rare on dorsum and the test is with shiny exuviae; in *Cerococcus* the 8-shaped pores and tubular ducts are not restricted to the marginal band, also other 8-shaped pores are common on dorsum, while the exuviae are absent from the test (Lambdin and Kosztarab, 1977). First instars described by Hamon and Kosztarab (1979).

Distribution. World-wide, with ca 56 species, 1 in CE.

Biology. As given under the family.

182. **Cerococcus cycliger** Goux – Thyme Pit Scale
Figs 125, 126

Syn. & Lit.: *Cerococcus cycliger* Goux 1932:140; Šulc, 1953:1; Schmutterer, 1955b:99; Zahradnik, 1957:5; Borchsenius, 1960:105; Lambdin & Kosztarab, 1977:93; Kosztarab & Kozár, 1978:114; Hamon & Kosztarab, 1979:42; Kozár et al., 1980:131; Koteja, 1984b:413.

Female test convex, globose, white, with light reddish tint; 1.7–3.0 mm long, 1.5–2.5 mm wide; with 4 marginal and 4 submedial white wax projections. Adult female pyriform 1.0–2.2 mm long, 1.1–1.5 mm wide; antennae 21–45 μ long, with 5–7 fleshy setae; labium triangular, 74–94 μ long, 69–98 μ wide, with 10 setae; legs entirely absent; spiracles located submarginally; anterior spiracular furrow with 72–90 disc pores, band 1–2 pores wide on venter, terminating in 2–4 pores wide furrow on dorsum;

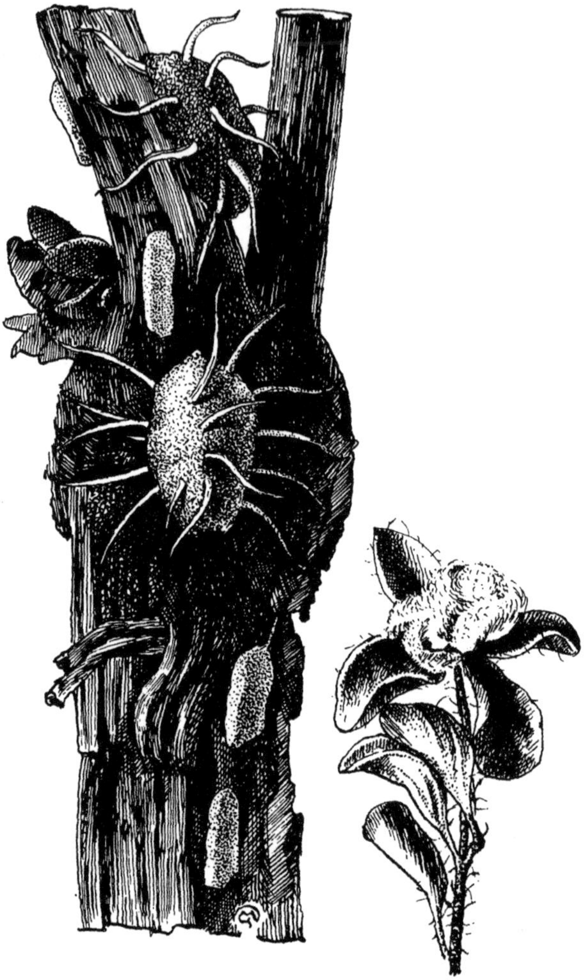

Fig. 125. *Cerococcus cycliger* Goux, females in ovisacs and male cocoons (after Kosztarab and Kozár, 1978)

posterior spiracular furrows bifurcate with 101–135 pores; cribriform plates in 2 submedial clusters with 2–6 plates per cluster; each plate subcircular, unevenly areolate, diameter 6–14 μ; anal ring 44–59 μ long, 30–43 μ wide, with 10 setae, each 78–92 μ long; anal plate posterior to anal ring, triangular, serrated at apex, 31–45 μ long, 43–63 μ wide: anal lobes sclerotized on inner margin, apical seta 160–210 μ long; simple disc pores distributed on both surfaces, diameter 1.7–2.9 μ; bilocular pores numerous on venter; 2–7 quinquelocular pores at base of each antenna, and primary type in spiracular furrows; multilocular pores, primarily 10-locular, form 7 transverse abdominal rows; 8-shaped pores on dorsum of 3 sizes arranged in 4 transverse bands, middle-sized pores dominant and scattered on entire dorsal derm, small pores in spiracular pore clusters; on venter of 2 sizes, these restricted to marginal and submarginal areas; tubular ducts numerous throughout both surfaces, each 29–44 μ long, 1.8–2.6 μ wide, sclerotized cup with 1–3 conical spines; slender setae on both body surfaces, a fleshy pair at apex of anal cleft. Adult male partially described by Goux (1932), first instar by Šulc (1953), also Hamon and Kosztarab (1979).

Distribution. Western Palearctic. In CE: Bu, Cz, Fr, Ge, Hu, It, Po and Yu.

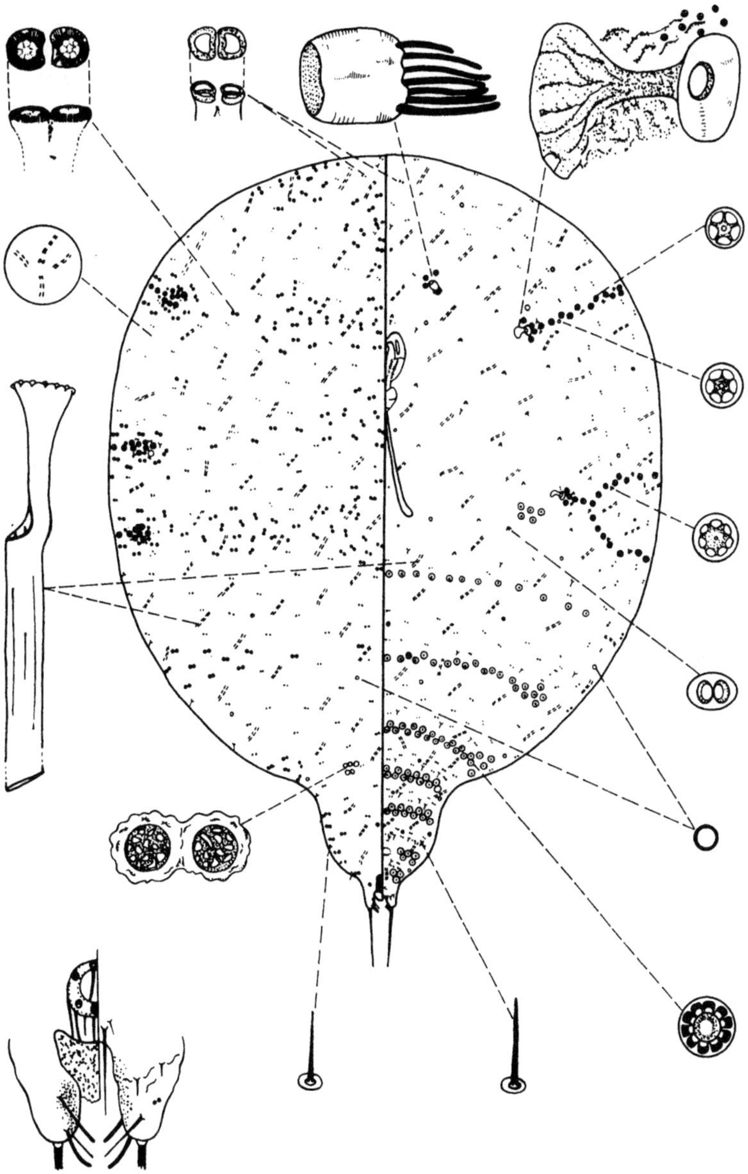

Fig. 126. *Cerococcus cycliger* Goux, female (modified after Lambdin and Kosztarab, 1977)

Biology. A xerophilous species. Occurs on stem of *Helianthemum, Origanum,* and *Thymus.* Most common on latter host. Bisexual; with one yearly generation; apparently eggs overwinter; egg-laying females collected in August in Hungary.

FAMILY **ASTEROLECANIIDAE** – PIT SCALES

Syn. & Lit.: Asterolecaniinae Cockerell, 1896:327; Schmutterer, 1952b:562. Asterolecaninae, Balachowsky, 1948:251. Asterolecaniidae, Russell, *1941:1; Bodenheimer, 1953c:115; Ferris, 1955:8; Borchsenius, *1960:77; Giliomee, 1968:297; Schmutterer, 1972:420; Koteja, 1974b:310; Lambdin & Kosztarab, 1977:12; Kosztarab & Kozár, 1978:134; Danzig, 1980:237.

Test of female transparent or semitransparent, usually greenish or yellowish, normally margin with wax filaments. Adult female after oviposition shrinks and assumes a crescent shape and occupies anterior part of test, remaining space being filled with eggs. Adult female circular, often elongate oval, convex, usually 0.5–3.7 mm long; antenna small, unsegmented, with hairs on apex; labium unsegmented; legs absent or rarely reduced to small tubercles with claw; only thoracic spiracles, these far from body margin, each with a band of quinquelocular pores toward margin, rarely spiracular furrow developed; no spiracular spines; anal ring normally present, small or large, found in a tube, with 0–6 setae. Body margin with 1–2 rows of large 8-shaped pores, and usually with submarginal row of quinquelocular pores. Dorsum normally with tubular ducts, with or without large 8–shaped pores in groups or dispersed randomly, occasionally with disc pores on margin, without anal and cribriform plates. Venter with multilocular pores arranged in transverse rows on abdominal segments; setae form a submarginal row. Anal lobes absent or developed in varying degrees; apical setae present. First instars and adult males described by Russell (1941), Borchsenius (1960), and Giliomee (1968:297).

Distribution. World-wide, with ca 200 species in ca 23 genera. In CE 2 genera with 6 species.

Biology. Live on woody and herbaceous plants. Normally in temperate climates with 1 generation yearly. Some with parthenogenetic, others with biparental reproduction. Some species cause considerable economic injury, many produce pits.

Note. Some authors include in this family the Cerococcidae and Lecanodiaspididae, treated separately by us. Members of this family can be distinguished from the related Cerococcidae and Lecanodiaspididae by the presence of anal plates and cribriform plates in the latter two families.

Key to genera of Asterolecaniidae

Anal ring well developed, with 6 setae, normally with pores; anal ring situated in a tube; with 2–3 rows of large 8-shaped pores on most of body margin; normally on herbaceous plants, especially *Hedera helix* in CE (fig. 128) . . . *Planchonia,* p. 318

Anal ring reduced, usually with 2 setae, without pores; anal ring in or very close to surface of derm; with a single row of large 8-shaped pores on body margin except *viennae* with 2 irregular pore rows; on woody plants, especially *Quercus* in CE (fig. 127) . . . . *Asterodiaspis,* p. 313

Genus **Asterodiaspis** Signoret

Syn. & Lit.: *Asterodiaspis* Signoret, 1877:209. Comstock, 1881:329; Lindinger, 1937:180; Bodenheimer, 1951:328; Borchsenius, *1960:174; Kosztarab & Kozár, 1978:136; Danzig, 1980:240. *Asterolecanium* (in part) Leonardi, 1920:238; Russell, *1941:1; Ferris, 1955:15; Schmutterer, 1972d:420.

Type-species: *Aonidia ilicicola* Targioni-Tozzetti, 1888, misidentified by Signoret as *Asterolecanium quercicola* (Bouché) 1851.

Test of female rigid, circular or oval, normally transparent, 0.6–2.2 mm long and 0.5–2 mm wide. Adult female circular or slightly elongate, with posterior part of body slightly produced; dorsum convex; old females usually brownish; antenna thimble-shaped or conical with 2, rarely 1, 3, or 4 setae; labium unsegmented, with 2 or 3 pairs of setae; quinquelocular pores form spiracular bands toward body margin; anal ring reduced, without pores and with 0–2 short setae; anal lobes absent or weakly developed. Body margin usually with a single row of large 8-shaped pores, but *A. viennae* with an irregular double row; a row of quinquelocular pores parallel with 8-shaped pores. Dorsum with or without large 8-shaped pores, with disc pores and tubular ducts. Venter with quinquelocular or multilocular pores arranged in transverse rows on abdomen, dark-rimmed 8-shaped pores usually present; disc pores absent. Anal lobes not, or weakly developed; apical setae present. First instars described by Russell (1941), Podsiadlo (1970, 1974a, 1974b); also first instars and adult males by Borchsenius (1960).

Distribution. Palearctic and Indo-Malayan Regions, with 18 species (Borchsenius, 1960); in CE 5 species.

Biology. Feed on woody plants of the family Fagaceae: *Lithocarpus, Pasania,* and *Quercus*. Last is the most common host genus.

Key to species of *Asterodiaspis*

(Based on adult females, except last two species where nymphs are also included)

1. Two irregular rows of large 8-shaped pores on margin of prosoma and anterior part of abdomen .... *viennae,* p. 318
   One row of large 8-shaped pores on margin of body 2
2. Dorsum with scattered large 8-shaped pores of two sizes, larger ones slightly smaller than marginal 8-shaped pores 3
   Dorsum with only minute 8-shaped pores, these much smaller than marginal 8-shaped pores 4
3. Row of large marginal 8-shaped pores extends to apex of abdomen, 1 such pore between apical setae; adult nearly circular, ca 1 mm in diameter; 4–9 multilocular pores in 2–3 transverse rows on abdominal venter .... *bella,* p. 314
   Row of large marginal 8-shaped pores extends only to a distance equal to length of 1 to 5 pores from apex of abdomen; adult oval, 1.5–1.7 mm long; 42–47 multilocular pores in 4 transverse rows on abdominal venter .... *roboris,* p. 316

4(2). Adult female usually with 4–41 multilocular pores in 3–4 transverse rows on abdominal venter. First instar with at least 16 dorsal 8-shaped pores (fig. 127) .... *quercicola*, p. 314
Adult female with 40–71 multilocular pores in 4 transverse rows on abdominal venter. First instar usually without, but rarely with 1–3 dorsal 8-shaped pores .... *variolosa*, p. 317

183. **Asterodiaspis bella** (Russell) – Southern Pit Scale

Syn. & Lit.: *Asterolecanium bellum* Russell, *1901:50; Bodenheimer, 1943:25; Borchsenius, 1950:135; Kosztarab, 1959:404; Boratynski, 1961:4; Schmutterer, 1972:420. *Asterodiaspis bella*, Borchsenius, *1960:179; Kosztarab & Kozár, 1978:137; Kozár et al., 1980:131.

Test of female circular, semitransparent, greenish, brownish or clear yellow, 0.9–1 mm in diameter. Adult female nearly circular, diameter 0.8–0.9 mm; antenna with 2 thick setae slightly longer than diameter of antenna; 22–36 quinquelocular pores form row or band between each spiracle and body margin; anal opening ventral, nearly circular, its margin slightly sclerotized, with 2 setae on anterior margin. Body margin with large 8-shaped pores forming a continuous single row, space between them 1–3 times length of a pore; quinquelocular pores absent. Dorsum normally with scattered 8-shaped pores of three sizes; disc pores numerous; tubular ducts 26–28 μ long. Venter with single submarginal row of 8-shaped pores, less numerous than marginal pores; a loose group of 5–7 dark-rimmed 8-shaped pores on each side of labium, some arranged in 2 indefinite transverse rows posterior to labium; 4–9 multilocular pores with 5 to 8 loculi, in 2 or 3 transverse rows of 1–4 pores each. Apical setae 20–22 μ long (Russell, 1941). First instar described by Russell (1941).

Distribution. In CE: Bu, Hu, and It. Also known from Greece, Iraq, Israel, and Morocco.

Biology. On branches of *Quercus* spp. Biology unknown.

184. **Asterodiaspis quercicola** (Bouché) – Small Pit Scale, Golden Oak Scale
Fig. 127

Syn. & Lit.: *Lecanium quercicola* Bouché, 1851:112. *Asterolecanium variolosum minor* Leonardi, 1909:107. *Asterolecanium minus* Lindinger, 1912:360; Russell, *1941:132; Ferris, 1955:25; Boratynski, 1961:4; Schmutterer, 1972:420. *Asterolecanium quercicola*, Leonardi, 1920:253; Ferris, 1955:25, 30; Schmutterer, 1972:421; Podsiadlo, *1974b:487. *Asterodiaspis minus*, Borchsenius, 1960:202; Kosztarab & Kozár, 1978:138. *Asterodiaspis quercicola* Borchsenius, 1960:197; Kosztarab & Kozár, 1978:139; Kozár et al., 1980:131.

Test of female nearly circular, slightly convex dorsally, greenish-yellow or yellow, transparent to translucent; with whitish marginal filaments; ca 0.8–1.6 mm long, 0.6–1.5 mm wide. Adult female almost circular, 0.6–1.2 mm long, 0.5–1.2 mm wide; antenna globular, with 2 thick setae as long as diameter of antenna; each spiracular band with 19–98 quinquelocular pores (Podsiadlo, 1974b); anus well removed from margin, circular with margin slightly sclerotized, with 2 setae on anterior edge. Body

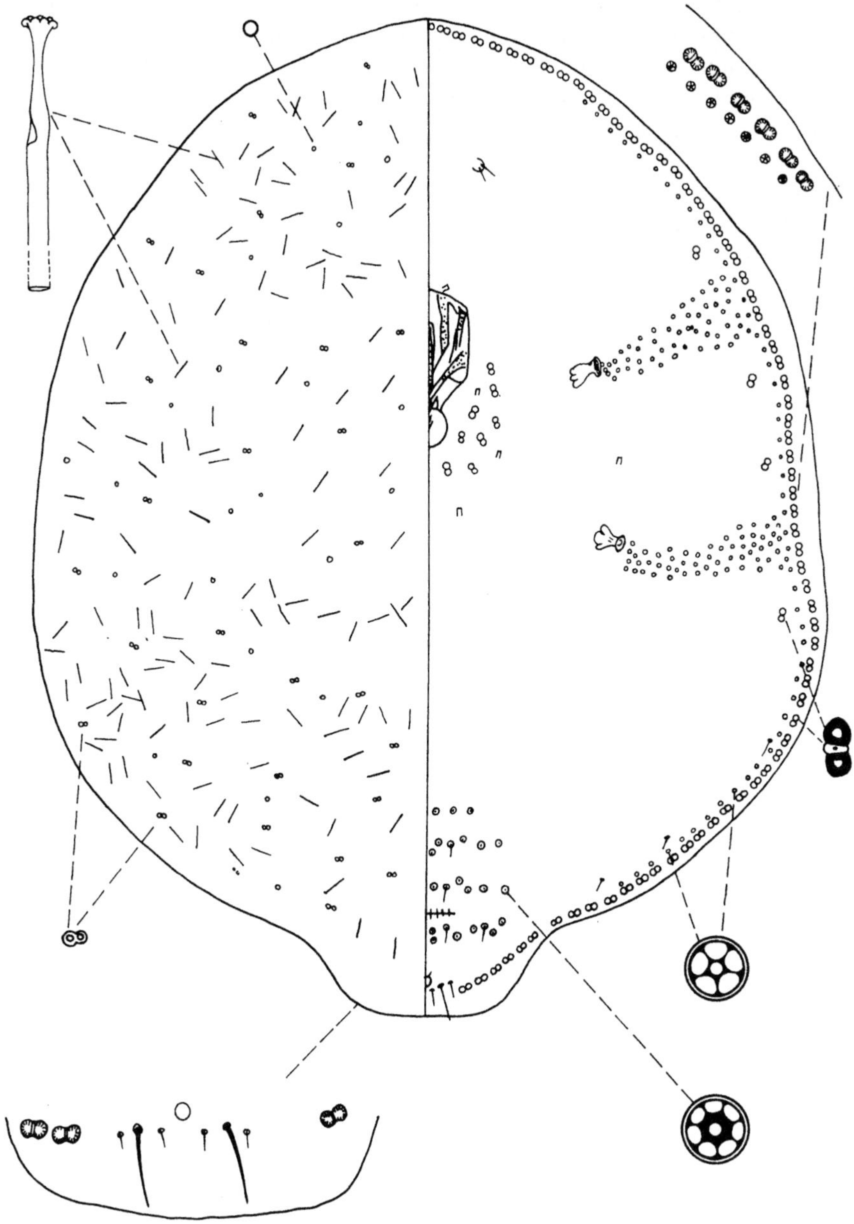

Fig. 127. *Asterodiaspis quercicola* (Bouché), female (modified after Kosztarab and Kozár, 1978)

margin with large 8-shaped pores in a single row, these a pore's width to length apart; quinquelocular pores in a single row, slightly more numerous than 8-shaped pores. Dorsum with numerous minute 8-shaped pores and disc pores; tubular ducts 15–36 μ long. Venter with 4–10 dark-rimmed 8-shaped pores on each side of labium, also some scattered elsewhere; submarginal 8-shaped pores in a single or double row; multilocular pores with 4–10 loculi arranged in 3 or 4 transverse rows, usually total 4–41 pores. Apical setae 16–44 μ long (Podsiadlo, 1974b). First instars described by Russell (1941), Borchsenius (1960), Boratynski (1961) and Podsiadlo (1974b). These with at least 8 pairs of dorsal 8-shaped pores; rarely 1–3 present in *A. variolosa.*

Distribution. A common widely distributed mesophilous species. In CE: Au, Bu, Fr, Ge, Hu, It, Po, Ro and USSR. Also known from Great Britain, Ireland, Turkey and Middle East, S Africa, New Zealand, USA and Mauritania.

Biology. Lives in shallow or deep pits on bark of *Quercus* spp. Occasionally a pest when it destroys young trees, especially in parks. In case of high population density, it infests leaves also. One generation yearly, but in the southern USSR (Caucasus) it can develop a partial second generation (Borchsenius, 1957); adult females overwinter, lay 15–64 eggs per female; eggs hatch after a few hours to 2 days; nymphs usually settle on young branches; molt to second instar at end of July and young adults appear in August. Reproduction parthenogenetic.

Natural enemies. Parasites: *Euaphycus asterolecanii, Habrolepis dalmani* and *H. zetterstedti* (Hadzibejli, 1956; Sugonjaev, 1958).

185. **Asterodiaspis roboris** (Russell) – Oak Pit Scale

Syn. & Lit.: *Asterolecanium roboris* Russell, *1941:177; Borchsenius, 1950:135; Boratynski, 1961:4. *Asterodiaspis roboris,* Borchsenius, *1960:179; Kosztarab & Kozár, 1978:136; Kozár et al., 1980:131.

Test of female oval, sometimes with faint transverse striations, greenish yellow, transparent, thin; marginal filament whitish; 1.5–2 mm long, 1.5–1.75 mm wide. Adult female oval, posterior end slightly produced; 1.5–1.7 mm long, 1.2 mm wide; antenna thimble-shaped, with 2 thick setae; spiracular bar expanded at inner end, each spiracular pore row with 71–99 quinquelocular pores; anal opening ventral, circular, its margin sclerotized, without setae or with 1. Body margin with 8-shaped pores forming a single row, not reaching apex of abdomen, slightly more than a pore's length apart; quinquelocular pores in an irregular single and double row. Dorsum with scattered 8-shaped pores; no minute 8-shaped pores; disc pores numerous; tubular ducts 40 μ long. Venter with ca 25 dark-rimmed 8-shaped pores in a loose row surrounding mouthparts, some scattered on anterior end, others arranged in 4 transverse rows on abdomen; submarginal 8-shaped pores far from margin, arranged in an irregular double row; multilocular pores with 10 loculi, in 4 transverse rows, each row with 6–17 pores, total 42–47; submarginal setae in a complete row. Apical setae 40 μ long (Russell, 1941). First instar described by Russell (1941).

Distribution. A rare species. In CE: Au, Bu, Cz, Hu and It. Also reported from Spain.

Biology. Lives on *Quercus* spp. Biology unknown.

186. **Asterodiaspis variolosa** (Ratzeburg) – Pit-making Oak Scale
Fig. 23

Syn. & Lit.: *Coccus variolosus* Ratzeburg, 1870:187. *Asterolecanium variolosum*, Lindinger, 1912:280; Leonardi, 1920:250; Russell, *1941:218; Schmutterer, 1952b:563, 1972:420; Schmutterer et al., 1957:452; Ferris, 1955:26; Boratynski, 1961:4. *Asterodiaspis variolosa*, Borchsenius, 1960:194; Podsiadlo, *1972:389, 1974a:75, 1974b:487, 1975a:347, 1976:69; Koteja, 1974a:144, 159; Kosztarab & Kozár, 1978:139.

Test of female circular or slightly longer than wide, convex dorsally; greenish, brownish or golden yellow, transparent, with whitish fringe around margin, 1–2.2 mm long, 0.8–2 mm wide. Adult female slightly longer than wide, 0.8–2.1 mm long, 0.7–1.7 mm wide; antenna thimble-shaped, with 2 thick setae, as long as diameter of antenna; labium heart-shaped, as long as wide; outside each spiracular peritreme a group of 7–16 quinquelocular pores, other pores in a row extending to body margin, a total of 33–113 pores; anal opening ventral, near body margin, its margin sclerotized, with 2 setae on anterior edge. Body margin with single row of 8-shaped pores, a pore's width to length apart; quinquelocular pores in a single row, except in double row near spiracular pore bands. Dorsum with numerous minute 8-shaped and disc pores; tubular ducts 19–39 μ long. On venter at each side of beak, with a loose group of 7–10 dark-rimmed 8-shaped pores, some 8-shaped pores scattered on anterior end, others form 2 loose transverse rows posterior to beak; submarginal 8-shaped pores in an irregular double row; multilocular pores with 4 to 13 loculi in 4 transverse rows, each row with 7–23 pores, total number 20–71; submarginal setae in a complete row. Apical setae 22–45 μ long (Russell, 1941; Podsiadlo, 1974a). First instar described by Leonardi (1920), Russell (1941), Borchsenius (1960), Boratynski (1961) and Podsiadlo (1972; 1974a). Last author found only this stage reliable for species determination, these without dorsal submarginal row of 8-shaped pores; rarely with 1–3 pores; ca 8 pairs present in *A. quercicola*.

Distribution. A common cosmopolitan mesophilous species. In CE: Au, Bu, Cz, Fr, Ge, Hu, It, Po, Sw, USSR and Yu. Also known from S, W and N Europe, N and S Africa, Middle East, Japan, S and N America, New Zealand and Australia.

Biology. Live on bark of oaks, *Quercus* spp., in shallow to deep pits, occasionally at high population density, destroys branches and young trees. It has one yearly generation, but according to Schmutterer (1972) two generations develop in New Zealand. Podsiadlo (1975) reported adult females overwintering; lay eggs in Poland from beginning of June through beginning of August; eggs hatch in 3–3½ weeks from mid-June to end of August; second instars develop from mid-July to mid-September; adults appear from beginning of August. According to Schmutterer (1972) each female lays 86–176 eggs. Reproduction parthenogenetic.

Natural enemies. Parasites: *Euaphycus asterolecanii, Habrolepis dalmani, H. italicus, H. pascuorum, Metaphycus variolosus, Microterys chalcostomus. H. dalmani* is an effective biological control agent (Kosztarab, 1959; Schmutterer, 1972; Trjapitzin, 1978). Predator: *Camptobrochis lutescens* (Schmutterer, 1952b).

187. **Asterodiaspis viennae** (Russell) – Viennese Pit Scale

Syn. & Lit.: *Asterolecanium viennae* Russell, *1941:222; Borchsenius, 1950:135; Boratynski, 1961:4. *Asterodiaspis viennae*, Borchsenius, *1960:187. Kosztarab & Kozár, 1978:136.

Test of female nearly circular, dorsally convex with faint transverse striations, bright greenish yellow, semitransparent, shiny, punctate, with whitish marginal filaments, diameter ca 1 mm. Adult female nearly circular, diameter ca 0.95 mm; antenna short, with 2 thick setae as long as diameter of antenna; each spiracular band with 21–45 quinquelocular pores; anal opening ventral, circular, its margin sclerotized with 2 setae on anterior margin. Body margin with 8-shaped pores, most in an irregular double row, ca a pore's length apart; quinquelocular pores in a single row of 4–17 (usually 10–12). Dorsum with numerous minute 8-shaped and disc pores; tubular ducts 28 μ long. Venter with trilocular or quinquelocular pores replacing multiloculars, in 2 rows of 1–3 each; submarginal 8-shaped pores usually in an irregular double row terminating near anterior row of multilocular pores; submarginal setae in a complete row. Apical setae 29 μ long (Russell, 1941). First instar described by Russell (1941).

Distribution. Known only from Austria: Rekawinkel, near Vienna.

Biology. Found on *Quercus cerris*. Biology unknown.

Genus **Planchonia** Signoret

Syn. & Lit.: *Planchonia* Signoret, 1870:282; Maskell, 1894:94, 1895:61; Lindinger, 1937:193; Bodenheimer, 1951:328; Borchsenius, *1960:144; Kosztarab & Kozár, 1978:139. *Asterolecanium* (in part) Leonardi, 1920:238; Russell, *1941:1; Ferris, 1955:15; Schmutterer, 1972:420.

Type-species: *Coccus fimbriatus* Fonscolombe, 1834.

Test of female oval, thin, shiny, transparent, dorsum convex, margin often with wax filaments, 2–4 mm long, 1.5–2.5 mm wide. Adult female oval to almost circular, 1.5–3.7 mm long, 1–2.5 mm wide; antenna with 2 long and 1–7 short hairs; spiracular pore bands extend to body margin; anal ring with 6 setae and normally with pores. Body margin with 8-shaped pores in a band 2–3 pores wide, except on posterior half of abdomen where it narrows to 1 row; quinquelocular pores form submarginal band parallel with marginal 8-shaped pores; band of smaller 8-shaped pores on submargin of prosoma and anterior half of abdomen. Dorsum with scattered large 8-shaped pores, tubular ducts, and disc pores; occasionally minute 8-shaped pores also present. Venter with multilocular pores forming transverse rows on abdomen; minute 8-shaped pores scattered or in groups. Anal lobes developed. First instar described by Borchsenius (1960), first and second instars by Russell (1941).

Distribution. In the Holarctic and Ethiopian Regions, with 7 species. In CE, 1 species.

Biology. Usually infest herbaceous plants belonging to various plant families.

188. **Planchonia arabidis** Signoret – Ivy Pit Scale, Pittosporum Scale
Figs 24, 128

Syn. & Lit.: *Planchonia arabidis* Signoret, 1877a:608; Borchsenius, *1960:145; Rogojanu, 1966b:432; Komosinska & Podsiadlo, 1967:685; Kosztarab & Kozár, 1978:141; Kozár et al., 1980:131. *P. valloti* Lichtenstein, 1882:75. *Asterolecanium massalongianum* Targioni-Tozzetti, 1893:295. *Pollinia thesii* Douglas, 1893:55. *A. fimbriatum*, Leonardi, 1920:241; Lindinger, 1912:175. *A. arabidis*, Newstead, 1903:154; Balachowsky, 1928:141; Russell, *1941:44; Ferris, 1955:18, 21; Schmutterer et al., 1957:450; Kosztarab, 1959:404; Tranfaglia, 1974:54.

Test of female broad pyriform, dorsum strongly convex; from whitish to brownish, translucent to opaque; 2–3.5 mm long, 1.5–2.5 mm wide (Russell, 1941). Adult female oval to pyriform, 1.7–3.2 mm long, 1.2–2.2 mm wide; antenna with 2 large thick and 2–7 short setae; spiracles heavily sclerotized, peritreme with quinquelocular pores and each with a band of 34–70 pores extending to body margin; anal ring with 6 setae, each 80–104 μ long, and with ca 46 pores. Body margin with 2–3 rows of large 8-shaped pores; these joined submarginally by 2–3 parallel rows of quinquelocular pores. Dorsum with numerous large 8-shaped pores, these often form 5–7 transverse groups in median area; minute 8-shaped pores scattered between large 8-shaped pores and margin; disc pores numerous; tubular ducts distributed evenly, ca 40 μ long. Venter with a submarginal band of small 8-shaped pores; multilocular pores in 3 or 4 complete transverse rows or bands, each with 14–52 pores and 2–4 interrupted rows, a total of 55–180 pores. Apical seta 100–128 μ long (Russell, 1941). First and second instars described by Russell (1941) and Borchsenius (1960).

Distribution. A common species, probably of Circum-Mediterranean origin. In CE: Au, Bu, Cz, Fr, Ge, Hu, It, Po, Ro, USSR, and Yu. Also known from S and W Europe, N Africa, Turkey, USA, Philippines and Singapore.

Biology. It infests stems, leaf petioles and leaves of primarily herbaceous perennials, most often *Hedera helix* in CE. Other host genera in Europe: *Achillea, Arabis, Aralia, Centaurea, Fraxinus, Hieracium, Lechea, Ligustrum, Linaria, Lithospermum, Lychnis, Medicago, Origanum, Osmodium, Panax, Penstemon, Phlox, Pittosporum, Teucrium, Thesium, Thymus, Trifolium, Verbascum,* and *Weigelia.* Often produces gall-like deformations on plants. Borchsenius (1960) reported one yearly generation from Caucasus; eggs overwinter and hatch during March and April, first molt during May, second at the beginning of June; adults fully developed by end of June and laid eggs from mid-July to mid-September, 40–200 per female. Reproduction probably parthenogenetic. Landis (1968) reported serious damage by this species in USA on sugar beets, *Beta vulgaris*. There are also 15 additional host genera represented in USA (Russell, personal communications).

Natural enemies. Parasites: *Cheiloneurus kollari, Discodes aeneus, D. cercopiformis, D. differens, Marietta picta* (Kosztarab, 1959; Tranfaglia, 1974; Trjapitzin, 1978).

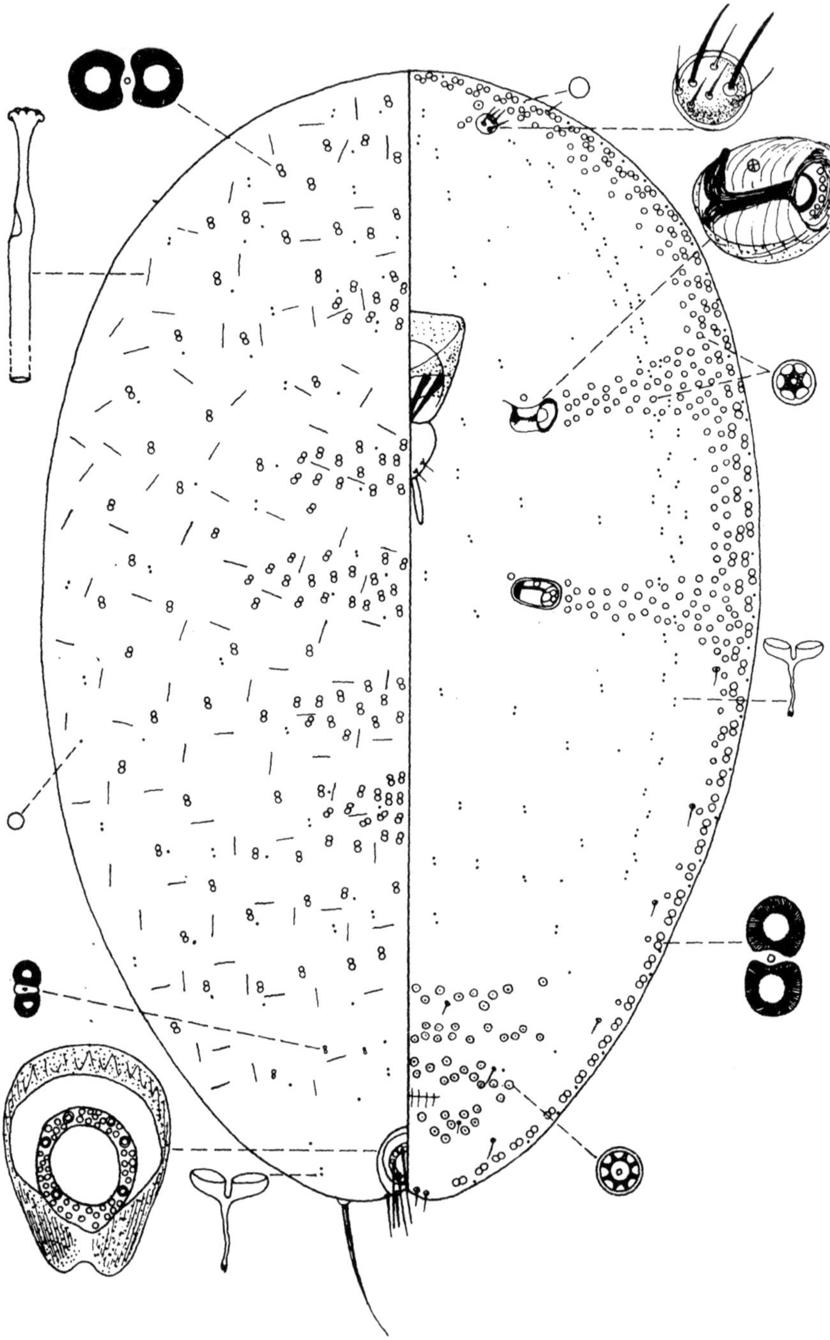

Fig. 128. *Planchonia arabidis* Signoret, female (original)

FAMILY **DIASPIDIDAE** – ARMORED SCALES

Fig. 138

Syn. & Lit.: Lepidosaphidae Shimer, 1868:372. Diaspidae Targioni-Tozzetti, Maskell, 1879:18. Diaspidinae, Balachowsky, (1937–1954b), Zahradnik (1952). Diaspididae, Ferris (1937–1942), McKenzie (1956), Schmutterer (1959), Beardsley & Gonzalez, (1975). Diaspidoidea, Borchsenius (1966), Bazarov & Shmelev (1971), and Nakahara (1982).

The family is the largest and most studied family of the Coccinea, both world-wide and in Central Europe. It includes many destructive pest species. In Central Europe Schmutterer (1959) provided a detailed monograph on the family for Germany, Zahradnik (1951c) for the Czechoslovakian Diaspidinae, and Balachowsky (1937–1954b) comprehensively treated three tribes (Aspidiotini, Odonaspidini, Parlatorini) for the western Palearctic and one tribe (Diaspidini) for the entire Palearctic Region. Other pertinent literature dealing with the Central European armored scales will be included in the citations for each treated species. Diaspididae is a fairly homogeneous group which is considered to be the most highly advanced and specialized family of the Coccinea.

Scale cover or test. Removable, and is composed of two (seldom one) exuviae and a waxy secretion. A ventral cover is secreted by some species (*Targionia vitis*), and when united with the dorsal cover, encloses the body. Male scale cover or test consists normally of one exuviae and waxy secretion. It has little taxonomic significance, therefore we deleted it from our descriptions. In the genus *Leucaspis* the adult females remain within the cast skin of the second molt, and are called pupillarial. Females develop through three stages, males through five. Ultrastructure of scale formation is described in detail by Foldi (1982).

Adult female. Without typical insect appendages, and has a saclike body, usually circular or pear-shaped in Aspidiotinae and spindle-shaped in most Diaspidinae. The head is fused with the thoracic segments, and this fused cephalothorax is referred to as the prosoma. Eight abdominal segments are more or less distinct, while the ninth to eleventh segments have been greatly reduced and are not apparent. The segmental lines are usually visible on the dorsum as furrows, or intersegmental sclerotizations. The terminal segments of the abdomen, starting with segment four, are fused into a pygidium.

The antennae are reduced to unsegmented tubercles, bearing one or more setae (rarely none, e.g. *Chionaspis austriaca*). The eyes are usually reduced to weakly sclerotized spots. The labium is one-segmented with 0–4 pairs of setae. Legs are lacking but in a few species are indicated by small sclerotized tubercles. Associated with the anterior spiracles in the Diaspidinae and Leucaspidinae are one or more disc pores, while in the Aspidiotinae these pores are usually absent. On the venter of the pygidium is located the vulva. Its use in taxonomy is limited; we deleted it from our descriptions. The pores around the vulva are called perivulvar pores which usually are arranged in four or five groups: a median group anterior to the vulva, and an anterior lateral and posterior lateral group at each side. The median group may consist of one or many pores or is entirely absent. In some species the perivulvar pores are absent, while in others, more than five groups or supernumerary pores may be present (e. g. 12–16 groups in *Leucaspis signoreti*). The range in number of pores in each group is expressed for example as 4–6 (9–12) 3–5, where 4–6 refers to the median group, (9–12) to anterior lateral groups, and 3–5 to posterior lateral groups.

The wax for the scale covering is produced by glands that discharge through tubular ducts. Many small ducts (e.g. ventral microducts) found on both surfaces of the body are of limited use in taxonomy at the present. Larger ducts (macroducts) are usually found on the dorsum of the pygidium and abdominal segments, and are important as diagnostic characters. There are two types of macroducts. Two-barred macroducts are relatively short and broad, with two parallel, transverse, sclerotized bars. The presence of these ducts is an important characteristic of the subfamilies Diaspidinae and Leucaspidinae. In species of *Parlatoria* the above dorsal marginal macroducts have a conspicuous sclerotized ring surrounding the orifice. In the tribe Aspidiotini one-barred ducts are present, which are slender, at least six times their width, and bear one sclerotized bar. Dorsal ducts may be arranged in well defined marginal, submarginal or submedial groups on the pygidium or abdominal segments. The macroduct orifices often are in rows or in furrows on the pygidium in the Aspidiotinae, starting at the base of the second and third interlobular space. The anus is a small, circular or oval, opening in the dorsum of the pygidium. The size and relative position of the anus is a key character in separating species, and some genera of Aspidiotinae.

Sclerotized processes are frequently present along the margin of the pygidium. The marginal scleroses, which arise in the spaces between the lobes or from the bases of the lobes, are called paraphyses, and are probably apophyses for muscle attachment. They are slender and usually well developed in Aspidiotinae while in most Diaspidinae the paraphyses are absent. The pygidial segments are in part produced at their lateral margins into flattened, paired, sclerotized processes or lobes. Their size, shape, and number are useful generic and specific characters. The median lobes of most genera are well separated. However, in some genera (i. e. *Chionaspis*) the median lobes are usually close together, and definitely zygotic. The following abbreviations are used for the lobes: L_1, median lobes; L_2, second lobes from center; and L_3, L_4 and L_5, etc. in sequence for the other lobes of the pygidial margin. In Diaspidinae L_2 and often L_3 usually are bilobed, with the inner (mesal) lobule (L_{2a} or L_{3a}) usually being larger than outer (lateral) lobule (L_{2b} or L_{3b}). The area between lobes is the interlobular space.

Plates are present on the margin of the pygidium; they may be confined to the interlobular spaces, or they may occur along the entire margin. Some plates are more or less branched or fimbriated and may bear a single microduct which discharges at the apex of the points or branches. The presence of these fringed plates is usually characteristic in the subfamily Aspidiotinae. Other plates may be elongate, conical, rarely bifid or slightly fimbriate at the tip, and bear one, or rarely more than one, microduct. These plates are called gland spines and occur primarily in the subfamily Diaspidinae. The plates vary between species of *Leucaspis*; these are entirely absent in *L. loewi,* long and pointed in *L. pini,* long and bifurcate at tip in *L. pusilla.* The number of plates or gland spines present in each interlobular space is expressed for example as 2,2,2–3,3 where 2 refers to the number of plates between the median lobes (in median notch); second number, 2, refers to the number of plates in the interlobular space between L_1 and L_2; third and fourth numbers, 2–3, express the range in number of plates between L_2 and L_3; and the fifth number, 3, gives the number of plates anterior to each L_3. When the gland spines are basally swollen, short and sclerotized, they are called gland tubercles. Sclerotized tubercles with a tubular duct but without a spine are known as duct tubercles and are found on the submargins of thorax and anterior abdominal segments of some Diaspidinae species. These are found on the prosomal submargin of *Parlatoria*. The marginal structures, such as lobes, plates, gland spines and gland tubercles are used for the production and manipulation of the scale cover.

Eggs. Oval, color usually same as egg-laying female, ca 0.2–0.3 mm long, 0.1–0.2 mm wide, and are covered by fine waxy powder.

Immature stages. First instar females (N_1) or **crawlers** are elongate-oval, minute, 0.2–0.3 mm long; have 1 pair of simple eyes, 5- or 6-segmented antennae, 3 pairs of well developed legs, 2 pairs of spiracles, few macro- and microducts on derm, anus on pygidial dorsum; on pygidial margin 1 or more pairs of lobes, a few plates or gland spines, and a pair of long apical setae; the test is without exuviae. **Second instar females (N_2)** are usually circular or elongate-oval; similar to adult females in shape, color, body segmentation, antennae, legs, spiracles and anus, but are smaller, lack perivulvar pores and vulva, have fewer pygidial lobes, plates, gland spines; fewer macro- and microducts (except pupillarial forms) than adults and second instar males; the test has only one exuviae. **First instar males** or **crawlers** are very similar to female crawlers, but the pattern and number of dorsal setae are somewhat different, as shown by Stoetzel and Davidson (1974), at least at tribal level. For example, Aspidiotini males have 1 pair of submedial setae on the dorsum of first and second abdominal segments in all stages, which are absent in the females. **Second instar males** elongate-oval; have 2 pairs of ventral eyespots and antennae vestigial; Aspidiotini have 1 short submedial dorsal seta on each side of abdominal segments 1 and 2; more macro- and microducts than females, unless females are pupillarial; pygidium has lobes, plates or gland spines; legs absent; test is elongate with 1 exuviae. **Third instar males** or **prepupae** have an elongate oval well-segmented body, 2 pairs of eyespots, antennae and legs as short reduced limb buds; penial sheath and pygidium absent. **Fourth instar males** or **pupae** elongate oval, similar to prepupa, but all appendages more fully developed than in prepupa and still unsegmented; have wing buds in alate species, 2 pairs of eyespots, and penial sheath partly developed. **Fifth instars** or **adult males** are spindle-shaped, with 3 body regions, head broadly united with the thorax, without constriction; 1 dorsal pair and 1 ventral pair of simple eyes, seldom with 1 lateral pair; antennae 8- to 10-segmented, with long hairlike setae; mouthparts vestigial and non-functional; no ducts or pores on derm; penial sheath as long as or longer than half length of body, with a ventral slit containing aedeagus; test elongate-oval, often parallel-sided, 0.8–1.5 mm long, white or grayish, with 1 exuviae. For comprehensive studies on adult males refer to Jancke (1955), Bustshik (1958) and Ghauri (1962).

The immature stages and adult males have been described in less than 5% of the known species. Therefore, their use at present in classification is limited. References for more detailed descriptions of the immature stages and adult males will be given for each species, when available.

Distribution. World-wide there are about 1,700 species belonging to about 350 genera. From Central Europe (CE) we are discussing 18 genera and 45 species, which are permanent members of the CE fauna. The following 45 introduced species are known only in greenhouses in CE and are not treated here. Because of space limitations, these will only be listed: *Abgrallaspis cyanophylli* (Signoret), *A. gliwicensis*

Komosinska, *Acutaspis perseae* (Comstock), *Aonidia lauri* Bouché, *Aonidiella citrina* (Coquillett), *Aspidiotus destructor* Signoret, *A. nerii* (Bouché), *A. spinosus* Comstock, *Chrysomphalus bifasciculatus* Ferris, *C. dictyospermi* (Morgan), *C. aonidum* (L.), *Diaspis barrancorum* Lindinger, *D. boisduvalii* Signoret, *D. bromeliae* (Kerner), *D. coccois* Lichtenstein, *D. echinocacti* (Bouché), *Dynaspidiotus britannicus* (Newstead), *Fiorinia fioriniae* (Targioni-Tozzetti), *Furchadaspis zamiae* (Morgan), *Genaparlatoria pseudaspidiotus* (Lindinger), *Gymnaspis aechmeae* Newstead, *Hemiberlesia lataniae* (Signoret), *H. palmae* (Cockerell), *H. rapax* (Comstock), *Howardia biclavis* (Comstock), *Ischnaspis longirostris* (Signoret), *Kuwanaspis pseudoleucaspis* Kuwana, *Lopholeucaspis cockerelli* (Grandpré & Charmoy), *Mycetaspis personata* (Comstock), *Lepidosaphes machili* (Maskell), *Odonaspis secreta* Cockerell, *Parlatoria crotonis* Douglas, *P. proteus* (Curtis), *P. ziziphi* (Lucas), *Pinnaspis aspidistrae* (Signoret), *P. buxi* (Bouché), *P. strachani* (Cooley), *Pseudaonidia tricuspidata* Lindinger, *P. trilobitiformis* (Green), *Pseudoparlatoria parlatorioides* (Comstock), *P. ostreata* Cockerell, *Selenaspidus albus* McKenzie, *S. pertusus* Brain, *S. pumilus* Brain, *S. rubidus* McKenzie. For further information on the aforementioned species, consult the works of Balachowsky (1937–1954b), Zahradnik (1968) and Schmutterer (1959).

Biology. Most species feed on woody plants and often are found on the trunk and branches, less frequently on leaves and roots. Members of the genera *Chortinaspis* and *Rhizaspidiotus* are restricted to herbaceous plants. Some species of *Chortinaspis* feed exclusively on roots of grasses. Different instars of San José scale infest all parts of the plants with the exception of roots. The majority of the species are oligophagous, but many are polyphagous, as the San José scale, which is listed from over 700 plant species. True monophagous species are very rare.

The number of yearly generations depends on climatic conditions. In Central Europe most species have one generation, except for the San José scale which has two. They overwinter in different stages: eggs in *Lepidosaphes ulmi,* first instars (N_1) in *Quadraspidiotus perniciosus,* second instars (N_2) in *Q. ostreaeformis,* adult females in *Epidiaspis leperii,* or as eggs and adult females in *Aulacaspis rosae*. Reproduction is usually biparental (*Pseudaulacaspis pentagona*), and seldom parthenogenetic (*Lepidosaphes ulmi*). Egg production per female varies between species and individuals in each taxon from 12 to 200, most often is between 50 to 100. A few species give birth to young ovoviviparously as *Q. perniciosus,* but most lay eggs. Embryonic development lasts from a few hours to three weeks, except for species that overwinter in the egg stage (*L. ulmi*) where it lasts ca eight months. The hatching nymphs settle after a few hours of crawling and produce a protective test. Because "crawlers" can't move farther than a few meters, they are distributed passively with the aid of wind, birds, mammals and with the infested plants.

Natural enemies. usually reduce the population size. The most common ones are chalcids, ladybird beetles, predatory mites, birds, and fungi. The natural enemies are listed for each scale species, because they are often important as biological control agents.

Key to genera of Diaspididae

1. Dorsal macroducts of one-barred type, these at least 6 times as long as their diameter; gland spines and plates only on pygidial margin; pygidial lobes unilobular; without gland tubercles on prosomal margin; anterior spiracles usually without associated pores; body circular or pyriform (Subfamily Aspidiotinae) 2

 Dorsal macroducts of two-barred type, these rarely longer than 3 times their diameter; gland spines often extend to prepygidial segments; L_2 most often bilobulate; gland tubercles present or absent on prosomal margin; anterior spiracles commonly with associated pores; body usually elongate spindle-shaped, rarely pyriform (Subfamilies Diaspidinae, Leucaspidinae) 7

2. Plates on pygidial margin much reduced, absent or glandspine-like; dorsal and ventral pygidial ducts about the same size and numerous; ventral test well-developed and thick (Tribe Targionini) 3

 Plates on pygidial margin well-developed, usually wide and fimbriated at apex; dorsal pygidial ducts definitely larger than ventral ducts and in moderate number; ventral test usually thin (Tribe Aspidiotini) 4

3. Perivulvar pores absent; gland spines or plates absent; dorsal pygidial duct orifices not thickened (fig. 153)..................... *Targionia*, p. 382

 Perivulvar pores usually in 4 small groups; gland spines or reduced plates present; dorsal pygidial duct orifices thickened (fig. 151) .... *Rhizaspidiotus*, p. 380

4 (2). Marginal scleroses or paraphyses absent or only weakly developed on dorsum of pygidial margin 5

 Marginal scleroses or paraphyses well developed on dorsum of pygidial margin 6

5. Perivulvar pores absent; plates between lobes long, at least 3 times as long as wide; L_3 reduced or absent (fig. 134) .... *Chortinaspis*, p. 336

 Pervivulvar pores present; plates between lobes wide, not more than 2 times as long as wide; L_3 fully developed (fig. 142)........... *Nuculaspis*, p. 361

6 (4). L_1 fully developed; L_2 and L_3 reduced to small points or absent (fig. 135) . ... *Diaspidiotus*, p. 337

 L_1 and L_2 usually well developed; L_3 only indicated by a small prominence (fig. 149) .... *Quadraspidiotus*, p. 367

7 (1). Only L_1 developed; L_2 and L_3 entirely absent; from *Acer campestre* in Hungary .... *Ferreroaspis*, p. 345

 L_2, also L_3 normally developed, at least as small lobelike tubercular projections, or if L_2 absent (*Epidiaspis*) most gland spines large and with bent tip; on a variety of host plants and in different parts of CE 8

8. L_2 unilobulate; gland plates usually present, short and wide, most with fimbriated apex (Subfamily Leucaspidinae) 9

 L_2 almost always bilobulate; gland spines present, only absent in *Mohelnaspis* (Subfamily Diaspidinae) 11

9. Adult female enclosed in the exuviae of N_2 (pupillarial); lobes small, triangular, without notches on margin; dorsal macroducts long and

narrow, without sclerotized ring surrounding their orifices; plates narrow, spiniform, rarely wide and fimbriated at apex (fig. 140) .... *Leucaspis* p. 355

Body of adult female free, covered by test only; lobes wide, notched on both margins; dorsal macroducts short and wide, with conspicuous sclerotized ring surrounding their orifices; plates wide, and with multiple fimbriation at apex (Tribe Parlatorini) 10

10. Plates between L_1 and L_2 as wide as or wider than L_1 or L_2; body red when live; mesophilous; feed on conifers only; a monotypic Palearctic genus (fig. 152) .... *Syngenaspis,* p. 380

Plates between L_1 and L_2 much narrower than L_1 or L_2; body usually not red when live; thermophilous; feed on a variety of plants; genus world-wide with about 50 species .... *Parlatoria,* p. 362

11 (8). Median lobes parallel or only slightly divergent, their bases not united by scleroses; two gland spines between median lobes (except *Mohelnaspis* where all gland spines absent and median lobes far apart); test elongate oystershell-shaped, usually brownish, seldom black (Tribe Lepidosaphidini) 12

Median lobes divergent or parallel, their bases usually united by scleroses; no gland spines between median lobes (except *Carulaspis*); test circular, rarely elongate oystershell-shaped, usually light in color (Tribes Chionaspidini and Diaspidini) 13

12. Median lobes reduced, and far apart, space between them more than 3 times their diameter; L_2 and L_3 reduced to lobe-like tubercular projections; marginal dorsal macroducts and gland spines absent; feed on grasses (fig. 141) .... *Mohelnaspis,* p. 360

Median lobes fully developed, close together, space between them less than 2 times their diameter; L_2 developed, bilobular, L_3 often reduced or absent; marginal macroducts and gland spines present; feed on a variety of hosts, usually woody plants (fig. 138) .... *Lepidosaphes,* p. 346

13(11). Female elongate oval or spindle-shaped; prosoma narrower than the first 3 abdominal segments; test of female elongate, usually oystershell-shaped, exuviae at pointed end 14

Female usually circular, oval, or broad pyriform (elongate pyriform, with mushroom-like swollen prosoma in *Aulacaspis*); prosoma wider than the first 3 abdominal segments; test of female circular, rarely oval, exuviae central or subcentral 15

14. Median lobes not or only slightly divergent; submarginal and submedian macroducts form rows on anterior of pygidium and on prepygidial segments; no wide furrow on pygidial dorsum between bases of L_1 and anus; test of female white; on various trees and shrubs (fig. 133)............ .... *Chionaspis,* p. 332

Median lobes divergent; most submarginal and submedian macroducts scattered on pygidium, only a few form rows; a wide furrow on pygidial dorsum between bases of L_1 and anus; test of female brown; preferred host *Euonymus* (fig. 154) .... *Unaspis,* p. 385

15(13). L_2 absent or reduced to small sclerotized point; most gland spines large and with bent tip (fig. 137) .... *Epidiaspis*, p. 343
L_2 present, well developed, often bilobulate; gland spines not bent at tip 16

16. Median notch with 2 gland spines; anus wider than length of L_1; feed on Cupressaceae and *Viscum album* (fig. 130) .... *Carulaspis*, p. 329
Median notch without gland spines (except 2 small setae in *Pseudaulacaspis*); anus not wider than length of L_1; feed on other hosts than Cupressaceae and *Viscum album* 17

17. Body elongate-pyriform with mushroom-like swollen prosoma; gland spines not branched at apex; L_1 narrow, invaginated, strongly divergent, without notches on margin; with submedian dorsal ducts on abdominal segment VI; preferred hosts *Rosa* and *Rubus* (fig. 129) .... *Aulacaspis*, p. 327
Body broadly oval without mushroom-like swollen prosoma; some gland spines often branched at apex; L_1 wide, not protruding, divergent, with notches on both margins; no submedian dorsal ducts on abdominal segment VI; polyphagous (fig. 144) .... *Pseudaulacaspis*, p. 364

Genus **Aulacaspis** Cockerell

Syn. & Lit.: *Aulacaspis* Cockerell, 1893:180; Ferris, 1937:9; Zahradnik, 1952:179; Balachowsky, 1954b8:240; Schmutterer, 1959:197; Borchsenius, 1966:135; Takagi, 1970:82; Kosztarab & Kozár, 1978:147; Danzig, 1980a:322.

Type-species: *Aspidiotus rosae* Bouché, 1833.

Test of female circular to slightly oval, white, or grayish white, with exuviae located centrally or subcentrally. Adult female elongate pear-shaped with swollen prosoma; anterior abdominal segments and metathorax lobed laterally; anterior spiracles always with associated disc pore group. Pygidium with 3 or 4 pairs of lobes; L_1 divergent, invaginated at base by sclerotized yoke; other lobes bilobed; gland spines large, pointed; dorsal macroducts in submarginal and median groups; perivulvar pores in 5 groups.

Distribution. World-wide with about 50 species; 1 species in CE.

Biology. Although members of this genus are found on a wide variety of plants, including trees and shrubs; only a few species are of economic importance.

189. **Aulacaspis rosae** (Bouché) – Rose scale, Blackberry scale
Figs 25, 129

Syn. & Lit.: *Aspidiotus rosae* Bouché, 1833:53. *Aulacaspis rosae spinosa* Maskell, 1897a:241. *A. rosae*, Lupo, 1938:157; Zahradnik, 1951:180; Balachowsky, 1954b:242; Schmutterer, 1959:197; Săvescu, 1961:19; Kosztarab, 1963:57; Borchsenius, 1966:139; Bazarov & Smelev, 1971:110; Komosinska, 1974b:24; Kosztarab & Kozár, 1978:147; Danzig, 1980b:322.

Test of female circular, flat, white to grayish white, diameter about 1.5–2 mm; exuviae subcentral. Adult female elongate, pear- or mushroom-shaped with swollen

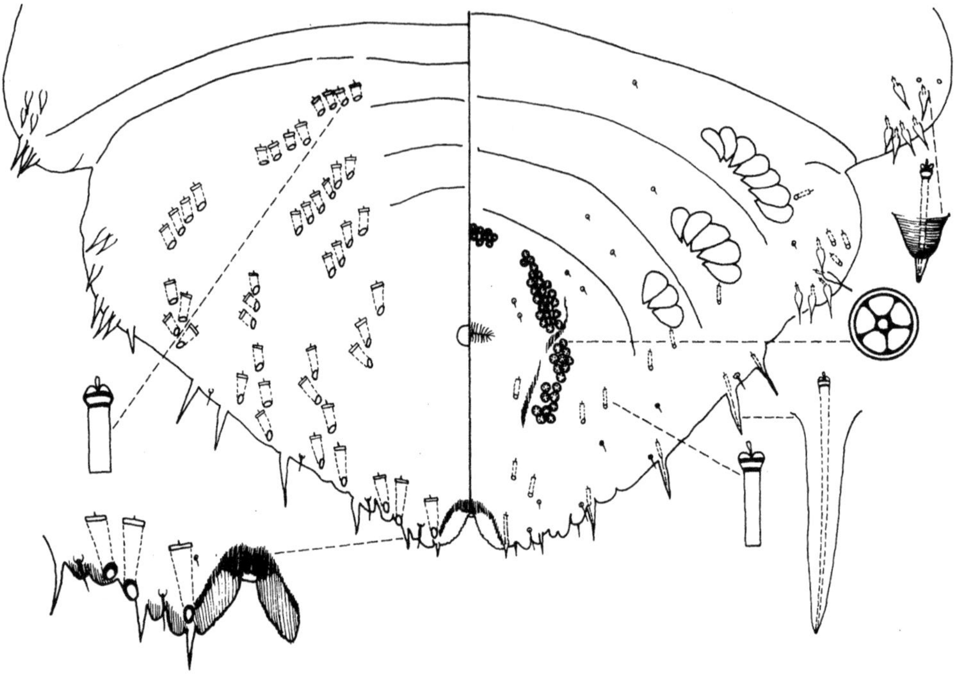

Fig. 129. *Aulacaspis rosae* (Bouché), female pygidium (modified after Kosztarab, 1963)

prosoma, brownish dark-red, about 1–1.1 mm long, 0.6–0.7 mm wide on slide; with 14–30 disc pores associated with each anterior spiracle, 5–17 pores at each posterior spiracle; abdominal segments 1–3 strongly lobed. Pygidium with L_1 large, divergent, with finely serrate inner margins, enclosing a triangular median notch; L_2 and L_3 bilobed, rounded at apex; gland spines long; dorsal macroducts short, arranged in rows composed of submarginal and submedian series; few ventral microducts in submarginal area; perivulvar pores: 8–33 (15–40) 14–40; vulva and anus located in middle of pygidium. Adult male described by Jancke (1955), Bustshik (1958) and Ghauri (1962); second instar by Boratynski (1953).

Distribution. A common species with world-wide distribution, probably present in most European countries. In CE: Au, Bu, Cz, Fr, Ge, Hu, It, Po, Ro, Sw, USSR, and Yu.

Biology. Common host genera: *Rosa* and *Rubus*. Also found on other Rosaceae, *Agrimonia* and *Hydrangea*. Other hosts listed by Fernald (1903) are doubtful. It infests the trunk and branches. Often heavy infestations on roses result in death of plants. In CE there is one yearly generation; but more generations develop in tropical climates. Species reproduces bisexually, overwinters as eggs or adult females, and lays 50–150 eggs per female (Bazarov and Smelev, 1971).

Natural enemies. Parasites: *Adelencyrtus aulacaspidis, Alloxysta erythrothorax* v. *dubia, Aphytis longiclavae, A. mytilaspidis, A. proclia, Archenomus bicolor, Arrhenophagus chionaspidis, Asaphes vulgaris, Blastotrix sericea, Cheiloneurus microphagus,*

Coccidencyrtus berlesei, Coccobius notatus, Habrolepis mayri, Metaphycus pretiosus, Microterys herbidus, Prospaltella fasciata, Pteroptrix macaropedicellata, Thomsonisca amathus, Xystus erythrocephalus (Balachowsky, 1954b; Schmutterer et al., 1957; Trjapitzin, 1978).

Genus **Carulaspis** MacGillivray

Syn. & Lit.: *Carulaspis* MacGillivray, 1921:305; Ferris, 1937:31; Balachowsky, 1954b:202; Boratynski, 1957:246; Schmutterer, 1959:188; Goidanich, 1960:1; Lupo, 1966:34; Borchsenius, 1966:160; Kosztarab & Kozár, 1978:148.

Type-species: *Diaspis juniperi* Bouché, 1851.

Female test circular, slightly convex, white, exuviae central or subcentral. Adult female circular to slightly oval; pygidium with well developed nonzygotic L_1, L_2 bilobed, L_3 usually reduced; glandspines present; dorsal macroducts only slightly smaller than marginal ones; perivulvar pores usually in 5 compact groups; anal opening large, circular, slightly posterior of center of pygidium.

Distribution. World-wide with 7 species; 3 in CE.

Biology. Most species with host preference for plants in the family Cupressaceae.

Key to species of *Carulaspis*

1. A large submarginal dorsal macroduct on both sides of 6th abdominal segment, at base of L_3 area; a sclerotized marginal spur on each side of 4th abdominal segment, this sharp, spinelike; on *Viscum album* . *visci*, p. 331
   No submarginal macroduct on 6th abdominal segment, at base of L_3 area; sclerotized marginal spur on each side of 4th abdominal segment obtuse or with more or less rounded apex; on Cupressaceae 2
2. Without marginal macroduct between median lobes (L_1); L_1 longer than wide; inner lobe of L_2 similar to L_1 long and narrow, outer lobe long and wide, parallel-sided, rounded at apex; inner lobe of L_3 present, but reduced . . . . *carueli*, p. 329
   With 1 marginal macroduct between median lobes; L_1 about as long as wide; inner lobe of L_2 wider than L_1, outer lobe small triangular, pointed at apex; entire L_3 indicated by sclerotized teeth only (fig. 130). . . . *juniperi*, p. 330

190. **Carulaspis carueli** (Signoret) – Minute Cypress Scale

Syn. & Lit.: *Diaspis caruelii* Signoret, 1869:436; Targioni-Tozzetti, 1868:736 (nomen nudum). *D. minima* Targioni-Tozzetti, 1868:736 (nomen nudum). *C. minima*, Balachowsky, 1954b:210; Boratynski, 1957:249, 1968:3; Baccetti, 1960:23; Lupo, 1966:34; Kosztarab & Kozár, 1978:151; Kozár et al., 1979. *C. carueli* Goidanich, 1960:35; Borchsenius 1966:160; Bazarov & Smelev, 1971:144.

Test of female circular, white, diameter 1.3 mm; exuviae yellow, central. Adult female almost circular, 0.5–0.7 mm long, 0.4–0.6 wide; anterior spiracle with 2 associated disc pores. Pygidium triangular with very thin long paraphyses; L_1 longer than wide, with rounded apex; L_2 bilobed, inner lobule more or less quadrangular, outer lobule smaller and with rounded apex; L_3 bilobed, reduced; gland spines large and thick; without marginal duct between bases of median lobes; perivulvar pores: 4–14 (7–17) 0–13. Adult male described by Bustshik (1958), and Ghauri (1962); nymphs by Baccetti (1960).

Distribution. In CE: Au, Bu, Fr, It, Ro, and USSR. Also from England, S Europe, Africa, Middle East; S and N America. Species has a more southern distribution than *juniperi.*

Biology. Host genera: *Biota, Callitris, Chamaecyparis, Cryptomeria, Cupressus, Juniperus, Sequoia* and *Thuja.* A bisexual species with one generation per year that overwinters as fertilized adult females. It is often a pest of ornamental plantings.

Natural enemies. Parasites: *Aphytis aonidiae, A. mytilaspidis* (DeBach and Rosen, 1976). Predators: *Aleuropterix juniperi* (Henry, 1976).

191. **Carulaspis juniperi** (Bouché) – Juniper Scale
Figs 26, 130

Syn. & Lit.: *Aspidiotus juniperi* Bouché, 1851:112. *Diaspis carueli* Targioni-Tozzetti, 1868:756, *D. visci,* Lupo, 1938:129. *Carulaspis visci* (Schrank), Ferris 1937:18, Balachowsky, 1954b:206; Schmutterer, 1959:189. *C. juniperi,* Boratynski, 1957:247; Goidanich, 1960:1; Baccetti, 1960:64; Ghauri, 1962:136; Kosztarab, 1963:59; Lupo, 1966:34; Borchsenius, 1966:161; Kuznetsov, 1970:1649; Bazarov & Shmelev, 1971:142; Komosinska, 1974b:23; Kosztarab & Kozár, 1978:151; Kozár et al., 1979:131.

Test of female circular, slightly convex, white, about 1–1.5 mm in diameter, exuviae subcentral or central. Adult female circular, yellow, darker with eggs, 0.4–0.9 mm long, 0.4–0.8 mm wide; anterior spiracles with 1 or 2 associated disc pores. Pygidium broadly triangular, rounded at apex, with small paraphyses; L_1 about as long as wide; L_2 bilobed, inner lobule wider than L_1 with posteriolateral margin oblique, outer lobule much smaller, L_3 vestigial; gland spines well developed; marginal dorsal ducts larger than other ducts; 1 marginal duct between bases of median lobes; only few ventral ducts present; perivulvar pores: 6–16 (6–18) 6–12. Adult male described by Ghauri (1962), nymphs by Baccetti (1960).

Distribution. In CE: Au, Bu, Cz, Fr, Ge, Hu, It, Po, Ro, Sw, USSR, and Yu. Also W and S Europe, N Africa, Middle East, N and America and Australian Region. A common species, especially in urban environments.

Biology. It infests leaves and cones. Host genera: *Biota, Chamaecyparis, Cryptomeria, Cupressocyparis, Cupressus, Juniperus, Libocedrus, Picea, Pinus, Sequoia, Taxodium, Taxus,* and *Thuja.* Because of past nomenclatural problems with this species, the host list and the available biological records should be re-evaluated. According to Schmutterer (1952b, 1959) the mated females overwinter; start laying eggs in mid-May, and lay 9–56 eggs per female. The eggs hatch in 3–4 weeks and adults appear at end of July. There is one yearly generation. A xerophilous species of more northern distribution than *carueli.*

Natural enemies. Parasites: *Aphytis mytilaspidis, Aspidiotiphagus citrinus, Euaphycus botanicus, Prospaltella serviniellus* (Balachowsky, 1954). Predators: *Chilocorus*

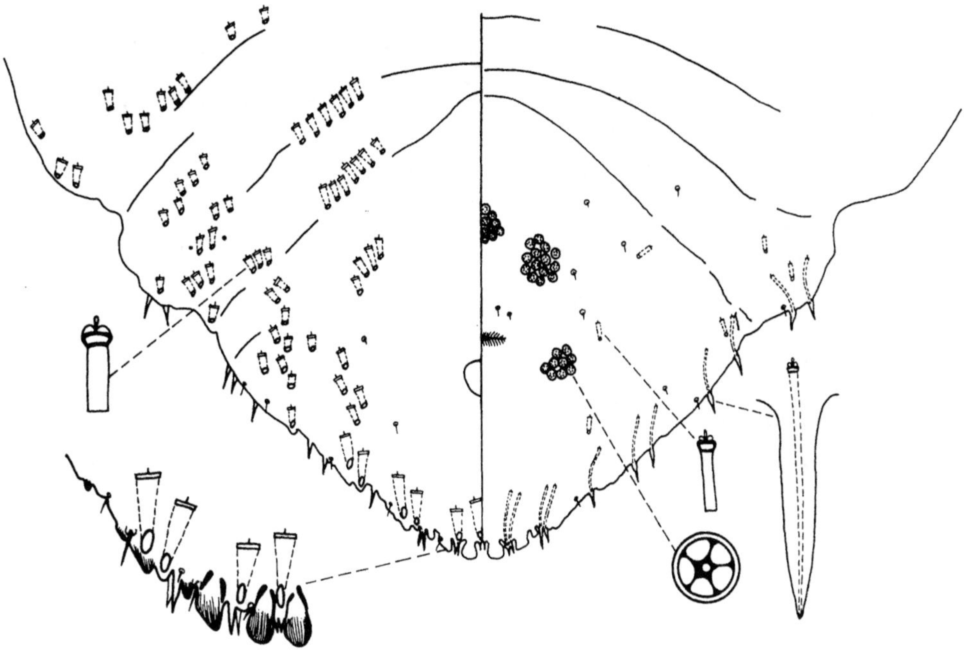

Fig. 130. *Carulaspis juniperi* (Bouché), female pygidium (modified after Kosztarab, 1963)

bipustulatus (Schmutterer, 1952b); *Exochomus quadripustulatus* (Schmutterer et al., 1957); *Scapheremaeus marginalis* (Kosztarab, 1963); *Aleuropteris juniperi* (Henry, 1976). Kuznetsov (1970) also reported *Chilocorus renipustulatus, Exochomus flavipes, Mirrha octodecimguttata* and *Vibidia duodecimguttata.*

192. **Carulaspis visci** (Schrank) – Mistletoe Scale
Fig. 131

Syn. & Lit.: *Coccus visci* Schrank, 1781:296. *Diaspis juniperi* v. *visci* Lindinger, 1909:222. *D. visci* Löw, 1872:273. *C. visci,* Baccetti, 1960:23; Goidanich, 1960:1; Lupo, 1966:34; Borchsenius, 1966:161; Boratynski, 1968:33; Kosztarab & Kozár, 1978:150.

Test of female circular, white, with two yellowish exuviae subcentral. Adult female short oval, widest at mesothorax; antenna with 1 large curved seta; anterior spiracles each with 1, rarely 2 trilocular disc pores; margin of 4th abdominal segment with a sharp, spinelike sclerotized spur on each side. Pygidium with L_1 as wide as long; L_2 well developed, about the same size as L_1, outer lobe small; L_3 reduced to 2 sclerotized points; gland spines well developed; a macroduct in submarginal area on both sides of 6th abdominal segment, at base of L_3; perivulvar pores: 7–13 (8–16) 8–13.

Distribution. In CE: Au, Cz, Ge, Hu, It (Lindinger, 1912; Goidanich, 1960). Specimens from Switzerland were received through USA Plant Quarantine in 1973.

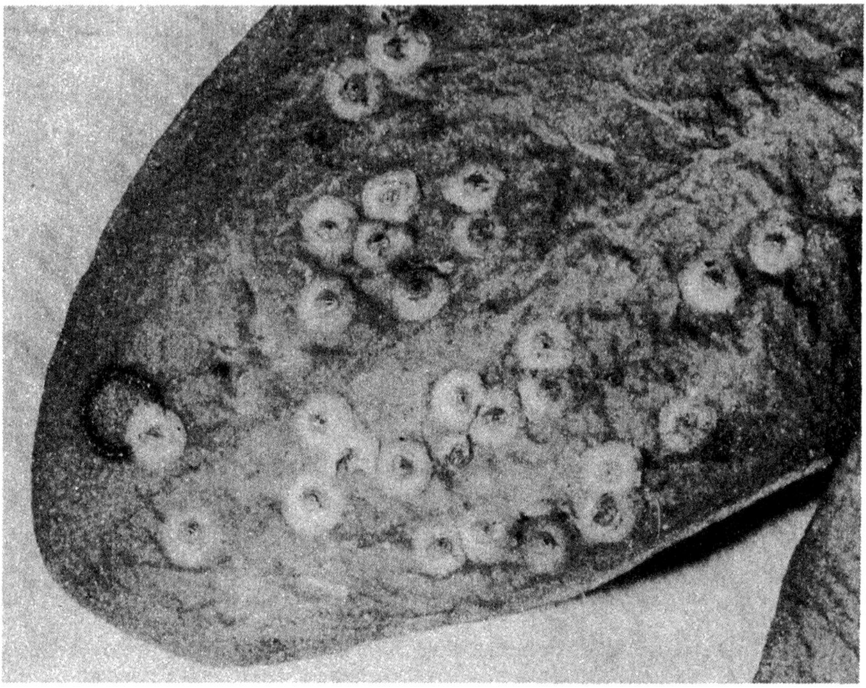

Fig. 131. *Carulaspis visci* (Schrank), female tests on leaf of *Viscum album* (Photo A. Kim and M. Kosztarab)

Biology. A monophagous species; on leaves and stem of *Viscum album*. It produces pits like Asterolecaniidae.

Note. Many earlier host, distribution and biological records of this species found in the literature are based on misidentifications and should be re-evaluated.

Genus **Chionaspis** Signoret – Scurfy Scales

Syn. & Lit.: *Chionaspis* Signoret, 1869:442. *Fundaspis* MacGillivray, 1921:307. *Marchaliella* Bodenheimer, 1951:331. *Phenacaspis* Cooley & Cockerell, Cockerell, 1899:398; *Chionaspis,* Ferris, 1937:613; Balachowsky, 1954b:317; Schmutterer 1959:228; Borchsenius, 1966:94; Takagi & Kawai, 1967:29; Danzig, 1980a:311.

Type-species: *Coccus salicis* Linnaeus, 1758.

Test of female elongate oystershell-shaped, white, with 2 exuviae at pointed end. Adult female elongate spindle-shaped, lobed laterally, most reddish; antennae usually with 1 seta; disc pores associated with anterior spiracles, seldom with posterior spiracles; dorsal submedian, submarginal and marginal macroducts, and ventral marginal microducts present on derm. Pygidium with well-developed median lobes (L_1) close together; L_2 always present, usually bilobed, rounded apically; L_3 similar to

L_2, but smaller, sometimes entirely lacking; gland spines well developed; marginal macroducts or gland spines absent between median lobes; perivulvar pores in five groups. Immatures of both sexes and adult males were described by Willoughby and Kosztarab (1974), also by Knipscher et al. (1976), 1st instars and 2nd instar females by Balachowsky (1954b), and 2nd instar males by Takagi and Kawai (1967), also by Howell (1980).

Distribution. World-wide with about 80 species, most in the Holarctic Region; only 3 in CE.

Biology. It feeds principally on woody plants. Most species have 1 yearly generation and overwinter either as eggs or adult females. Several species are of economic importance.

Key to species of *Chionaspis*

1. L_2 unilobular; a large, clavate paraphysis present at outer basal angle of L_2; basal one half of median lobes (L_1 fused; gland spines with thickened base (fig. 132) . . . . *lepineyi*, p. 334
   L_2 bilobular; large, clavate paraphysis absent from basal angle of L_2; L_1 not fused; gland spines without thickened base 2
2. L_1 small, symmetrical, without notch on outer margin; L_3 reduced to tooth-like structure; antenna without seta; only 1–3 dorsal marginal macroducts on each segment 4–6; on needles of conifers only . . . . *austriaca,* p. 333
   L_1 large, asymmetrical, with notch on outer margin; L_3 well-developed, lobelike; antenna with 1 seta; 4–8 dorsal marginal macroducts on each segment 4–6; polyphagous, on deciduous trees and shrubs (fig. 133). . . . . *salicis,* p. 335

193. **Chionaspis austriaca** Lindinger – Austrian Pine Scale

Syn. & Lit.: *Chionaspis austriaca* Lindinger, 1912:252. *C. barbeyi* Balachowsky, 1930c:267. *C. austriaca* Balachowsky, 1954b:336; Schmutterer, 1959:233; Borchsenius, 1966:96; Kosztarab & Kozár, 1978:153.

Test of female wide oystershell-shaped, white with yellowish exuviae, 1.5–3 mm long, 0.7–1 mm wide. Adult female oval, red, about 1 mm long, 0.5 mm wide; antenna without setae; each anterior spiracle with 5–7 associated disc pores, posterior with 1–2. Pygidium with L_1 relatively small, symmetrical without notches, rounded at apex; L_2 bilobed, inner lobe almost as large as L_1, outer lobe small; L_3 reduced to sclerotized teeth-like structure; gland spines small, 1 in each interlobular space 2 and 3; a much larger one on each side of segments 5 and 6; dorsal median macroducts only 1 or 2 on each side of segment 6, 3 or 4 on segments 4 & 5; few ventral microducts in submarginal band; perivulvar pores: 4–7 (8–18) 7–12; vulva and anal opening at about same distance from posterior margin.

Distribution. In CE: Au, Fr, Hu, It, Sw and Yu (Zivojinovic and Petrovic, 1955). Also in Corsica and Spain. Rare.

Biology. Unknown. On needles of conifers in genera *Abies* and *Pinus*.

194. **Chionaspis lepineyi** Balachowsky – Oak Scurfy Scale
Fig. 132

Syn. & Lit.: *Chionaspis lepineyi* Balachowsky, 1928: 273, 1954b:325; Borchsenius, 1966:98; Zahradnik 1972:430; Kosztarab & Kozár, 1978:152.

Test of female pear-shaped, grayish white, 1.6–2.1 mm long. Adult female spindle-shaped, red; antenna with curved seta; anterior spiracles with 3–5 associated disc pores, posterior with 1–2 pores; short gland spines in submarginal groups on venter of metathorax and abdominal segments 1–4. Pygidium with median lobes large, basal half fused, notched on outer margin, rounded at apex; L_2 unilobular, notched on outer margin and with a large clavate paraphysis arising from basal angle; L_3 reduced to several low tubercles; gland spines all large, with thickened base, 1 in each interlobular space 2 and 3; submedian dorsal macroducts missing on segment 6, only 2 on each side of segment 5, and 3–4 on segments 3 and 4; 3–4 submarginal macroducts on segment 5, 5–7 on segments 3 and 4; perivulvar pores: 19–50 (27–44) 13–22. Male tests commonly found in large white patches on infested trees.

Fig. 132. *Chionaspis lepineyi* Balachowsky, female and male tests on *Quercus* sp. twig (after Kosztarab and Kozár, 1978)

Distribution. In CE: Cz, Hu, Sw and USSR. Also in Spain, Morocco.

Biology. Host genera: *Castanea* and *Quercus,* on trunk and branches. A bisexual species, with one generation per year that overwinters as adult females. Each female lays about 60 eggs during second half of May, and crawlers emerge in June (Kosztarab and Kozár, 1978). A Mediterranean mesophilous element.

Natural enemies. Parasites: *Aspidiotiphagus citrinus, Coccophagoides similis, Thomsonisca amathus, Thysanus subaeneus* (Kosztarab, 1956b; Zahradnik, 1972; Trjapitzin, 1978).

195. **Chionaspis salicis** (Linnaeus) – Willow Scale

Figs 27, 133

Syn. & Lit.: *Coccus salicis* Linnaeus, 1958:456. *Aspidiotus populi* Bärensprung 1849:167. *A vaccini* Bouché, 1851:111. *Chionaspis fraxini* Signoret, 1869:445. *C. sorbi* Douglas, 1893b:130. *C. salicis* f. *salicis* Danzig, 1970:1018. *C. salicis* f. *polypora* Danzig, 1970:1018. *C. salicis* f. *micropori* Danzig, 1970: 1018. *C. salicis,* Lupo, 1938:277; Dinther, *1950:171; Balachowsky, 1954b:321; Schmutterer, 1959:230; Săvescu, 1961:32; Borchsenius, 1966:99; Zahradnik, 1972:430; Komosinska, 1974b:21; Kosztarab & Kozár, 1978:153; Danzig, 1980a:311. There are more synonyms in the early literature.

Test of female broad oystershell-shaped, convex, white, about 2–3 mm long, with yellow exuviae. Adult female spindle-shaped, 1.3–1.7 mm long, widest at 1st or 2nd abdominal segment, red; antenna with 1 seta; anterior spiracles each with 4–12 associated disc pores, posterior with 0–2. Pygidium with L_1 asymmetrical, with notch on inner margin close to tip; L_2 bilobed, with rounded tip, L_{2b} much smaller than L_{2a}; L_3 well developed, lobelike; 1–2 large pointed gland spines in interlobular spaces 2 and 3;

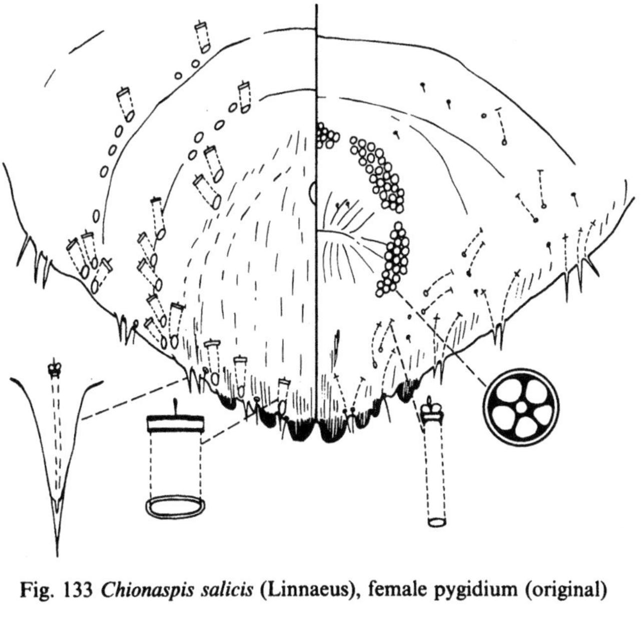

Fig. 133 *Chionaspis salicis* (Linnaeus), female pygidium (original)

dorsal submedian macroducts 1–3 on each side of segment 6; about 4–8 submarginal macroducts on each side on segments 3–5; few ventral microducts in submarginal area; perivulvar pores in 5 groups: 8–19 (19–37) 16–32; vulva in middle of pygidium; anus circular, located close to median group of perivulvar pores. Adult male was described by Jancke (1955), Ghauri (1962) and others. Boratynski (1953) described the second nymphs, while Dinther (1950) all stages.

Distribution. A common boreo-montane Trans-Palearctic species. In CE: Au, Bu, Cz, Fr, Ge, Hu, It, Po, Ro, Sw, USSR, and Yu. Also known from Denmark, England, Greece, Holland, Portugal, Spain, Sweden, Turkey, Corsica, Malta, Middle East, N Africa, and Far East.

Biology. A polyphagous species, known from plants belonging to over 17 families. Host Genera: *Acer, Alnus, Andromeda, Arctostaphylos, Betula, Cornus, Corylus, Cotoneaster, Cytisus, Erica, Euonymus, Fraxinus, Genista, Grossularia, Helianthemum, Hippophae, Jasminum, Laburnum, Ledum, Ligustrum, Loranthus, Lyonia, Myrtus, Oxycoccus, Paeonia, Populus, Pyrus, Quercus, Rhamnus, Rhododendron, Ribes, Rosa, Salix, Sarothamnus, Sophora, Sorbus, Syringa, Tamarix, Tilia, Ulmus, Vaccinium* and *Viburnum* (Borchsenius, 1966; Bazarov and Smelev, 1971; Zahradnik, 1972).

It has 1 yearly generation and reproduces bisexually or parthenogenetically. Red eggs overwinter and hatch in Germany at end of April (Schmutterer, 1952b). Some nymphs settle under tests of females, a few on leaves; first molt at end of May. Winged or wingless males are active from mid-June through mid-July. Eggs are laid from end of August; egg production varies on different hosts between 8–15 per female (Schmutterer, 1952b; Balachowsky, 1954b; Tsalev, 1968). Balachowsky (1954b) found it in France at 2,000 m elevation.

Natural enemies. Parasites: *Anabrolepis zetterstedti, Aphelinus fuscipennis, Aphytis mytilaspidis, A. proclia, A. sugonjaevi, Archenomus longicornis, Archenophagus chionaspidis, Azotus celsus, Cheiloneurus microphagus, Prospaltella gigas, Pteroptrix dimidiata,* and *Thomsonisca amathus.* (Dinther, 1950; Zahradnik, 1972; Trjapitzin, 1978). Predators: *Anthocoris minki, Brysoptera rufifrons, Chilocorus bipustulatus, C. renispustulatus, Cybocephalus politus, Hemisarcoptes malus, Histiogaster entomophagus, Novius suturalis, Pilophorus clavatus, Platybunus pinetorum* and *Tyrophagus dimidiatus* (Dinther, 1950; Schmutterer, 1952b).

Note. Often a pest on forest trees in its northern range of distribution, and causes significant damage to currant, *Ribes* sp. (Danzig, 1972c).

Genus **Chortinaspis** Ferris

Syn. & Lit.: *Chortinaspis* Ferris 1938:125. Balachowsky, 1948:382; Borchsenius, 1966:279; Danzig, 1964:651; Kosztarab & Kozár, 1978:154.

Type-species: *Aspidiotus chortinus* Ferris.

Test of female oval, thick, usually brown, exuviae terminal, seldom central; ventral test well developed, white, often adheres to body of female. Adult female oval or pear-shaped. Pygidium broadly rounded; L_1 usually wide, with crenulate, or rounded apex;

L_2 short, wide, or triangular; L_3 and L_4 often reduced to sclerotized points or often absent; plates usually small, fimbriated or pointed; 2 plates in median interlobular space as long as L_1; no paraphyses; dorsal ducts slender, form definite rows in longitudinal furrows; ventral ducts smaller and fewer than dorsal ducts; perivulvar pores absent; anus relatively large, wider than length of L_1, removed from apex of pygidium about 2 or 3-times its diameter.

Distribution. There are 13 species in the Holarctic Region, and 1 in Rhodesia; only 1 in CE.

Biology. Little is known on their biology. They feed on roots and stems of Gramineae, except, 1 species *C. iridis*, in the Middle East, that feeds on *Iris*.

196. **Chortinaspis subterraneus** (Lindinger) – Rhizome Armored Scale
Fig. 134

Syn. & Lit.: *Epidiaspis subterranea* Lindinger, 1912:174. *Hemiberlesia provincialis* Vayssière, 1914:207. *H. subterranea*, Leonardi, 1920:94. *Separaspis subterraneus*, Lupo, 1954:2. *C. subterraneus*, Balachowsky, 1948:143; Borchsenius, 1966:280; Tereznikova, 1967:474; Danzig, 1972c:208; Kosztarab & Kozár, 1978:154.

Test of female elongate oval, 1.4–1.6 mm long, seldom circular, thick, with ventral test. Adult female oval or pear-shaped, widest at mesothorax; antenna with 1 long seta. Pygidium with 4 pairs of lobes: L_1 large, wider than long, notched slightly on both sides; L_2, L_3 and L_4 of similar small size and triangular; plates narrow, longer than L_1 and fimbriated at apex, their formula 2, 2, 3–4, 3; dorsal ducts with small sclerotized head, form definite bands in the pygidial furrows, 1 at base af median interlobular space reaches anus; ventral ducts smaller, shorter, usually in groups along pygidial margin; perivulvar pores absent; anus large, circular, removed from apex of pygidium about 2½ times its diameter.

Distribution. In CE: Fr, Hu, It, USSR, and Yu.

Biology. A rare thermophilous species. On rhizomes of grasses in the genera *Agropyron, Ammophila* and *Festuca.* Biology is unknown.

Natural enemy. Parasite: *Physcus subterraneus* (Trjapitzin, 1978).

Genus **Diaspidiotus** Berlese & Leonardi

Syn. & Lit.: *Diaspidiotus* Berlese & Leonardi, 1896:350; Gomez-Menor, 1937:80; Ferris, 1938:164; Lupo, 1948:194; Balachowsky, 1950:488; Zahradnik, 1952c:142; Gomez-Menor, 1956:13; Schmutterer, 1959:100; Kosztarab, 1963:22; DeLotto, 1963:144; Borchsenius, 1966:320; Morrison & Morrison, 1966:58; Bazarov & Smelev, 1971:197; Kosztarab & Kozár, 1978:154; Danzig, 1980a:339.

Type-species: *Diaspis ancylus* Putnam, 1877.

Test of female circular to wide oval; exuviae central or subcentral. Adult female pear-shaped, widest at mesothorax; antenna with 1 seta. Only L_1 fully developed; L_2 or L_3 indicated by small points; plates rarely absent (*D. distinctus*), most plates or gland spines fringed, usual formula: 2,2,3; well-developed paired intersegmental scleroses between lobes; dorsal ducts arranged in definite rows or series; perivulvar pores present or absent; anus small, well removed from apex of pygidium.

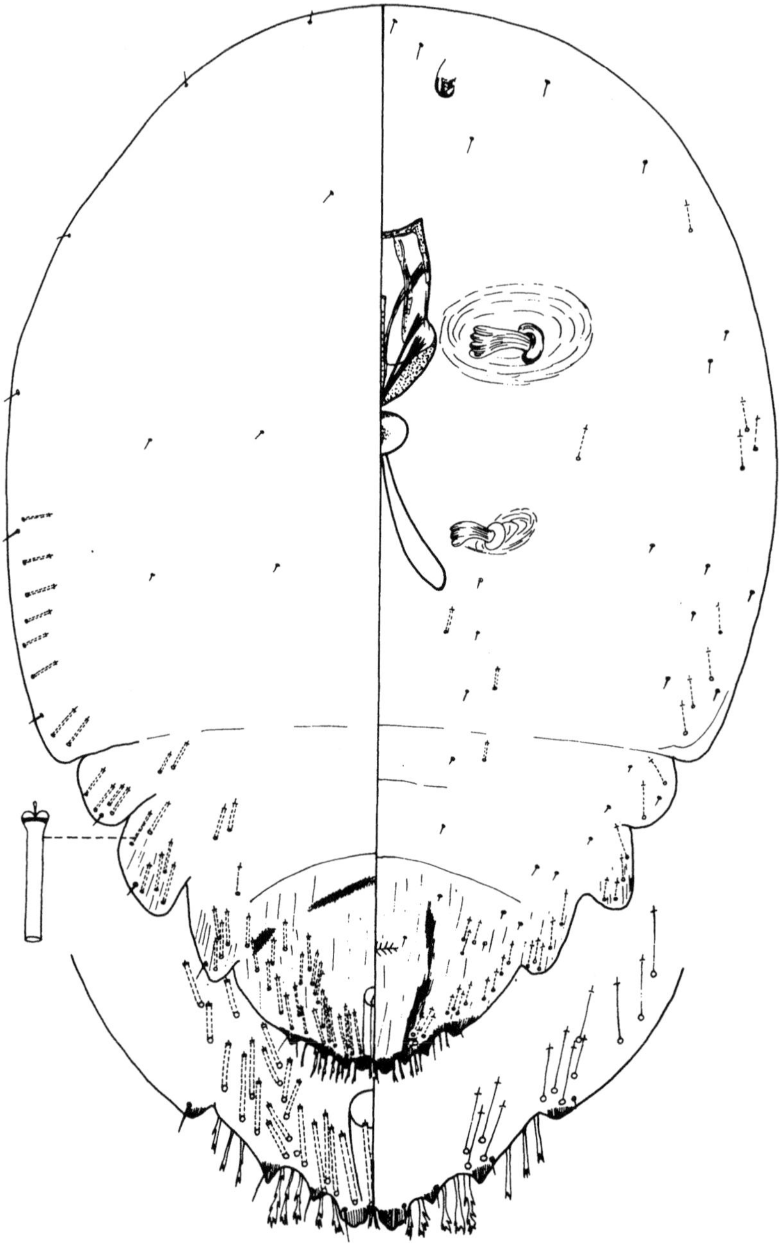

Fig. 134. *Chortinaspis subterraneus* (Lindinger), female (original)

Distribution. World-wide, with about 36 species; 6 in CE.

Biology. Live on trees and shrubs. Some are pests of cultivated plants.

Note. *D. slovenicus* (Bachmann, 1955) was left out from our work, because it is known only from the Adriatic coast in Yugoslavia.

Key to species of *Diaspidiotus*

1. Perivulvar pores present 2
   Perivulvar pores absent 4
2. Median lobes notched on both margins; dorsal ducts form 4 rows on each side of pygidium; duct between bases of median lobes does not extend beyond anal opening; plates between median lobes almost as long as median lobes; feed only on plants belonging to Ericaceae . . . . *bavaricus*, p. 340
   Median lobes notched on outer margin only; dorsal ducts form 2 or 3 rows on each side of pygidium; duct between bases of median lobes extends beyond anal opening; plates between median lobes absent or shorter than 1/2 length of median lobes; feed on non-ericaceous plants 3
3. Dorsal ducts form 3 rows on each side of pygidium; plates between bases of median lobes short, pointed; other plates wide and fimbriated at apex; prosoma at maturity not especially sclerotized and without lateral constrictions . . . . *hungaricus*, p. 341
   Dorsal ducts form 2 rows on each side of pygidium; plates between median lobes absent; other plates daggerlike, pointed; prosoma at maturity heavily sclerotized and with lateral constrictions (fig. 135a) . . . . . . *osborni*, p. 341

4(1). Median lobes longer than wide, notched on both margins; duct between bases of median lobes absent; marginal plates absent; derm heavily sclerotized and covered with fine wrinkles, except pygidium . . . . *distinctus*, p. 340
   Median lobes not longer than wide, notched on outer margin only; duct between bases of median lobes present; marginal plates present; derm not sclerotized and not covered with wrinkles 5

5. Dorsal ducts less than 32; median lobes deeply notched on outer margin; gland spines small daggerlike or absent . . . . *alni*, p. 339
   Dorsal ducts more than 32; median lobes only shallowly notched on outer margin; gland spines moderately wide and usually fimbriated at apex (fig. 135b) . . . . *wuenni*, p. 343

197. **Diaspidiotus alni** (Marchal) – Alder Armored Scale

Syn. & Lit.: *Targionia alni* Marchal, 1909:871. *Aspidiotus alni*, Thiem & Gerneck, 1934b:130. *D. alni*, Balachowsky, 1950:529; Schmutterer, 1959:103; Borchsenius, 1966:32; Danzig, 1967:849; Zahradnik, 1972; 439; Kosztarab & Kozár, 1978:156.

Test of female circular, slightly convex, gray, diameter 1.5–2 mm. Adult female pear-shaped; antenna with 1 seta. Pygidium with only L_1 developed, large, deeply notched on outer margin, rounded at apex; paired intersegmental scleroses well developed; 2

short gland spines between median lobes, longer gland spines laterad of L_1; up to 32 dorsal ducts on entire surface in 3 rows on each side of pygidium, 1 dorsal duct between bases of median lobes extends to anus; ventral ducts in a submarginal band; perivulvar pores absent; anus small, circular, removed about 3-times the length of L_1 from apex of pygidium. Adult male described by Bustshik (1958); second instar (N_2) by Borchsenius (1939).

Distribution. Common in boreal environment on trees growing on creek and river banks. In CE: Au, Fr, Ge, Hu, Po(?), and USSR.

Biology. It lives on trunks and branches of *Alnus, Carpinus, Fagus* and *Quercus*. On *Alnus* it is often found under the thin outer bark layer. The biology is poorly known.

Natural enemy. Parasite: *Pteroptrix dimidiata* (Kosztarab, 1959).

198. **Diaspidiotus bavaricus** (Lindinger) – Blueberry Armored Scale

Syn. & Lit.: *Aspidiotus bavaricus*, Lindinger, 1912:31; Balachowsky 1932a:102; Thiem & Gerneck, 1934b:130. *D. bavarius*, Balachowsky, 1950:500; Zahradnik, 1952c:143; Boratynski, 1953:461; Gomez-Menor, 1957:40; Schmutterer, 1959:101; Kosztarab & Kozár, 1978:155. *Quadraspidiotus bavaricus*, Borchsenius, 1966:328.

Test of female circular or broad-oval, gray, diameter 1.4–2 mm; exuviae central, brownish-yellow. Adult female pyriform, widest at mesothorax, when alive lemon yellow; antenna triangular, with a long seta. Pygidium: L_1 developed, wide, rounded at apex, deeply notched on both margins; L_2 and L_3 indicated only by sclerotized points; intersegmental scleroses elongate and narrow; gland spines between median lobes pointed, about 3/4 length of L_1, others wide with well fimbriated apex; their formula 2, 2, 2–4; dorsal ducts numerous, in definite longitudinal rows, duct between median lobes not reaching anus; few ventral ducts along margin of pygidium; perivulvar pores: 0 (3–8) 3–9. Adult males described by Theron (1958) and Ghauri (1962). Second instars (N_2) of both sexes were described by Boratynski (1953); N_1 was described by Gomez-Menor (1957).

Distribution. Found in most European countries; in CE: Au, Cz, Fr, Ge, Hu, It, Po, Sw, USSR and Yu. Also known from Corsica.

Biology. A boreo-montane species that is oligophagous on plants in the family Ericaceae, including the genera: *Arbutus, Calluna, Erica, Rhododendron* and *Vaccinium*. According to Schmutterer (1959), a biparental species with 1 yearly generation in S Germany, that overwintered as N_2, became adults in May. Each female laid an average of 110 eggs during July and August, which hatched in 2–4 days.

Natural enemies. Parasites: *Anabrolepis zetterstedti*, **Anagyrietta pantherina*, **Eusemion corniger, Metaphycus callunae, Trichomasthus bavarici* (Schmutterer, 1959; Trjapitzin, 1978). *Possibly hyperparasites.

199. **Diaspidiotus distinctus** (Leonardi) – Leonardi Armored Scale

Syn. & Lit.: *Targionia distincta*, Leonardi, 1900:305, 1920:114. *Chorizaspidiotus distinctus*, Lupo 1954:29. *D. distinctus*. Balachowsky, 1950:523; Gomez-Menor, 1960:160; Zahradnik, 1962:92, 1972:339; Borchsenius, 1966:323.

Test of female circular, diameter 1–1.3 mm, white or light grey; exuviae central or subcentral, orange; ventral test white. Adult female almost circular, 0.9 mm long, yellow or brownish-yellow; derm surface heavily sclerotized and covered with fine wrinkles, except pygidium. Only L_1 developed, large, longer than wide, notched on both sides; plates and perivulvar pores absent; dorsal ducts form 3 rows, no duct between bases of L_1, nor on abdominal segment IV; very few small ventral ducts along margin; anus large, circular; pygidial end with 7 long setae on each side of margin.

Distribution: A rare Mediterranean species. In CE: Cz, Fr, It, Ro and Sw; also known from Spain, Morocco and Israel.

Biology: Little is known on its biology. It lives on plants in the genera: *Corylus, Matricaria* (in Spain) and *Quercus*. In its northern range adult females overwinter (Zahradnik, 1972).

200. **Diaspidiotus hungaricus** Kosztarab – Hungarian Oak Scale

Syn. & Lit.: *Diaspidiotus hungaricus* Kosztarab, 1956a:435; Lindinger, 1957:548; Kosztarab, 1959:416; Borchsenius, 1966:324; Zahradnik, 1972:439; Kosztarab & Kozár, 1978:155.

Test of female circular, ash-gray with light gray margin, diameter 1.8 mm; exuviae subcentral or central, lst exuviae orange yellow, 2nd yellow. Adult female pear-shaped, about 1.2 mm long; when live, lemon yellow with pygidial margin dark yellow; antenna with 1 bent seta. Pygidium triangular; L_1 fully developed, deeply notched on outer margin, rounded at apex; L_2 and L_3 indicated by sclerotized points only; plates wide and fringed, those between median lobes short and pointed at tip, plate formula: 2, 2, 2; paired intersegmental scleroses short and wide; total of 48–56 dorsal ducts, slender, long, in definite rows, 1 between bases of L_1 extends beyond anal opening; 8/12 ventral ducts located in submarginal area; perivulvar pores: 1–3 (3–7) 2–4.

Distribution. Rare, species known from Hungary.

Biology. Found at base of trunk and branches of *Quercus petraea,* in a mixed oak forest. There is one yearly generation and the adult females overwinter. Males or their tests have not been found (Kosztarab and Kozár, 1978).

Natural enemies. Parasites: *Aphytis bovelli, Azotus pinifoliae, Coccophagoides similis* and *Pteroptrix dimidiata* (Kosztarab, 1956b).

201. **Diaspidiotus osborni** (Newell & Cockerell) – Osborn's Scale
Fig. 135a

Syn. & Lit.: Aspidiotus osborni Newell & Cockerell, 1898:229. *Diaspis snowii* Hunter, 1899:14. *Aspidiotus yulupae* Bremner, 1907:367. *D. osborni,* Ferris, 1938:186; Kosztarab, 1963:30; Borchsenius, 1966:325; Munting, 1971:134; Stoetzel & Davidson, 1974:500; Kozár, et al., 1979:129.

Test of female circular or slightly elongate, light gray, diameter 1.5 mm; exuviae subcentral, orange yellow. Adult female pear-shaped, ca 1 mm long, at full maturity prosoma sclerotized, with moderate constrictions between pro- and mesothorax, and between meso- and metathorax; orange yellow. Pygidium with L_1 large, protruding,

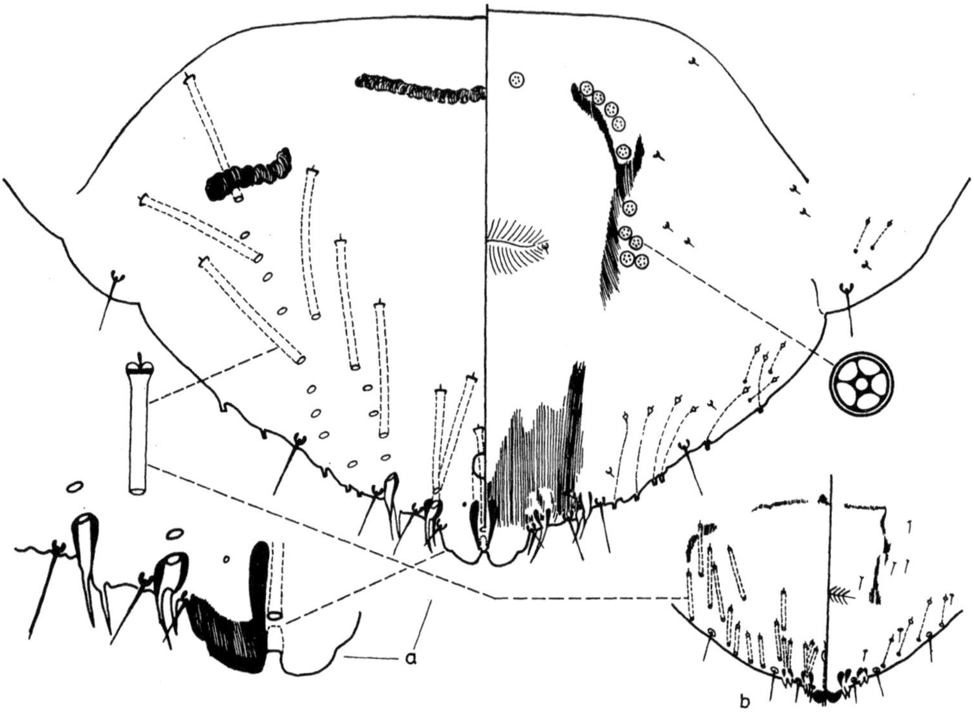

Fig. 135. Pygidium of female, (a) *Diaspidiotus osborni* (Newell & Cockerell) (modified after Kosztarab 1963) (b) *Diaspidiotus wuenni* (Lindinger) (modified after Kosztarab and Kozár, 1978)

mesal margin 1/2 as long as outer margins; outer margins sloping toward apex, notched halfway, rounded at apex; L_2 and L_3 absent; plates small, daggerlike, none between median lobes; paired, elongate scleroses in median and interlobular spaces; dorsal ducts few, slender, form 2 rows on each side of pygidium, 1 duct between median lobes, 2–4 in interlobular space between L_1 and L_2; few ventral ducts in marginal area; perivulvar pores: 0–2 (3–10) 3–10; anus small, removed 2 or 3 times its diameter from base of L_1. Nymphal stages described by Stoetzel and Davidson (1974).

Distribution. Probably of N American origin. Known from Canada, USA, Mexico, and S Africa. It was reported for the first time from Europe in Bulgaria by Kozár et al. (1979) and Italy.

Biology. Infests trunk and branches of trees in the genera: *Betula, Carpinus, Castanea, Cornus, Crataegus, Diospyros, Fagus, Fraxinus, Hicoria, Juglans, Morus, Ostrya, Platanus, Prunus, Quercus, Robinia, Salix, Tilia* and *Vitis*. Stoetzel and Davidson (1974) reported that in Maryland (USA) 2 yearly generations occurred. Mated females overwintered, and each laid 4–7 eggs per day the following spring. Winged males were present during late June and early July, while wingless males were active during the last half of August.

Natural enemy. Parasite: *Ablerus clisiocampae.*

202. **Diaspidiotus wuenni** Lindinger – Oak Armored Scale
Fig. 135b

Syn. & Lit.: *Aspidiotus alni* Lindinger, 1911:245 (not Marchal, 1909). *A. wünni* Lindinger, 1923:147, 152. *D. wünni,* Bachmann & Geier, 1950:118; Balachowsky, 1950:526; Zahradnik, 1952c:147; 1972:440; Schmutterer, 1959:105; Lellakova-Duskova, 1959:338. *D. wuenni,* Borchsenius, 1966:327; Kosztarab & Kozár, 1978:156.

Test of female circular, slightly convex, gray, diameter 2–2.5 mm; exuviae central or subcentral. Adult female pear-shaped, widest at pro- and mesothorax, diameter 1 mm; live females light yellow with darker to brownish pygidium (Zahradnik, 1951); antenna has 1 sharp seta. Pygidium with only L_1 developed, shallowly notched on outer margin; gland spines moderately wide, usually fimbriated at apex, 2 gland spines and 2 scleroses in each interlobular space; over 32 dorsal macroducts, on entire surface 1 duct between bases of L_1 extends past anus; ventral ducts few, in a submarginal band; perivulvar pores absent. Adult male described by Lellakova-Duskova (1959).

Distribution. Only moderately frequent. In CE: Au, Bu, Cz, Ge, Hu, Sw, USSR and Yu.

Biology. Oligophagous on Fagaceae. It is found on trunk and branches of *Castanea, Fagus* and *Quercus.* In Czechoslovakia adult females overwinter (Zahradnik, 1952).

Natural enemy. Parasite: *Aspidiotiphagus citrinus* (Zahradnik, 1972).

Genus **Epidiaspis** Cockerell

Syn. & Lit.: *Epidiaspis* Cockerell, 1899:398b. *Cockerellaspis* MacGillivray, 1921:318. *Epidiaspis,* Ferris, 1937:99; Zahradnik, 1951:188; Balachowsky, 1954b:217; Schmutterer, 1959:194; Borchsenius, 1966:179; Kosztarab & Kozár, 1978:157.

Type-species: *Diaspis leperii* Signoret, 1869.

Test of female circular to oval, convex, usually white. Adult female pear-shaped to circular; disc pores associated with anterior spiracles, rarely lacking at posterior ones. Pygidium with well developed L_1; L_2 and L_3 usually reduced or lacking; gland spines large, often with forked or curved tip; dorsal ducts scattered, of different sizes; perivulvar pores in 5 (rarely 7) groups.

Distribution. World-wide, except Australian and Oriental Regions, with 13 species, most in N and C America, 1 in CE.

Biology. Members of this genus feed on a variety of trees and shrubs. The life cycle of only *E. leperii* was studied in detail.

203. **Epidiaspis leperii** (Signoret) – Italian Pear Scale; Grey Pear Scale
Figs 28, 136, 137

Syn. & Lit.: *Diaspis leperii* Signoret, 1869:437; Balachowsky & Mesnil, 1935:387. *D. ostreaformis* Signoret, 1869:439. *D. pyri* Colvée, 1881:31. *A. piricola* Del Guercio, 1894:142. *D. fallax* Horváth, 1897:95. *E. piricola,* Frank & Krüger, 1898:80. *E. betulae,* Lindinger, 1912:388; Thiem & Gerneck, 1934a:529;

Zahradnik, 1952c:189; Săvescu, 1960:125; *E. pyri*, MacGillivray, 1921:318. *E. leperii*, Ferris, 1937:51; Geier, 1949:177; Balachowsky, 1954b:220; Schmutterer, 1959:195; Borchsenius, 1966:179; Zahradnik, 1972:429; Kosztarab & Kozár, 1978:157.

Test of female circular, slightly convex, white, yellowish- or grayish-white, diameter 1.1–1.6 mm, exuviae yellow, subcentral or central. Adult female pyriform, red or orange-yellow, ca 0.7–1 mm long; antenna with 1 curved seta; few dorsal ducts on margin of prepygidial segments; 3–5 disc pores associated with each anterior spiracle, 1 or 2 at posterior ones. Pygidium with prominent L_1 close together, rounded at apex, notched on outer margin; L_2 and L_3 reduced to small sclerotized points; gland spines large, usually with tips bent toward L_1; dorsal macroducts scattered, 1 between bases of median lobes; few ventral ducts in marginal and submarginal area; perivulvar pores in 5 (rarely 7) compact groups: 7–12 (12–17) 7–12; anus slightly posterior of middle of pygidium. Wingless adult males, N_1 and N_2, were described by Geier (1949).

Distribution. In CE: Au, Bu, Cz, Fr, Ge, Hu, It, Po(?), Ro, Sw, USSR, and Yu. Also S and W Europe, Middle East, N Africa, N and S America. A common species that apparently prefers warm climates.

Biology. Geier (1949) provided detailed life history in Switzerland. It infests the bark of trees in the genera: *Aesculus, Amygdalus, Berberis, Cerasus, Cornus, Crataegus, Cydonia, Gleditsia, Heteromeles, Juglans, Malus, Mespilus, Olea, Photinia, Prunus, Pyrus,* and *Ribes* (Borchsenius, 1966). In CE it is one of the most important pests of pear, and a common pest of plums. As a result of feeding, plant tissue stops growing at the feeding sites and the infested branches become deformed and break easily at these sites. In Hungary, there is one yearly generation. Mated females overwinter and start laying eggs at end of May; egg laying and hatching of nymphs can last up to 2 months. Most nymphs emerge between June 5–15, usually in 6 days after egg laying, and settle near the female tests, that often results in several layers of tests on top of each other. First nymphal stage lasts 1 month, followed by 20–30 days for 2nd stage. Adults were

Fig. 136. Deformation on twig caused by *Epidiaspis leperii* (Signoret) on *Pyrus communis* (Photo F. Kozár)

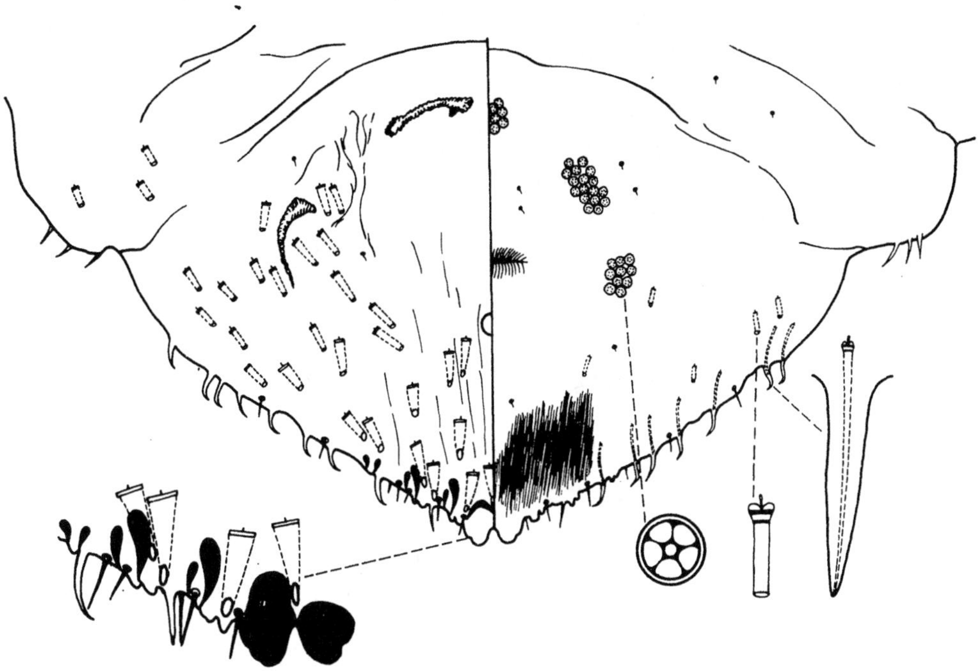

Fig. 137. *Epidiaspis leperii* (Signoret), female pygidium (modified after Kosztarab, 1963)

observed at end of July and first part of August; adult males were present for 20–30 days (Kozár, 1975). In USSR, each female produces 20–90 eggs (Danzig, 1972c).

Natural enemies. Parasites: *Aphytis maculicornis, A. mytilaspidis, A. proclia, Archenomus maritimus, Azotus marchali, Coccophagoides similis, Pteroptrix dimidiata* (Schmutterer et al., 1957; Zahradnik, 1972; Trjapitzin, 1978). Predators: *Chilocorus bipustulatus, Exochomus quadripustulatus; Cymbaeremeus* sp. (Geier, 1949, Schmutterer et al., 1957).

Notes. Live specimens of *E. leperii* are easily separated from other species on fruit trees in CE by the red or orange-yellow colored body and white tests. Slide-mounted females are differentiated from species in the related genera by their prominent L_1, the absence of L_2 and L_3, and the large gland spines with curved tips.

Genus **Ferreroaspis** Kozár

Syn. & Lit.: *Ferreroaspis* Kozár, 1983:92.

Type-species: *Acanthomytilus hungaricus* Vinis 1981. A monotypic genus.

Test of female convex, oystershell-shaped, with 2 exuviae at anterior end. Adult female spindle-shaped; antennae sclerotized, reduced to one-segmented stubs, each with 2 thick setae; labium one-segmented; trilocular disc pores associated with each spiracle; duct tubercles on submargin of metathorax and anterior abdominal segments. Pygidium with only 1 pair of lobes, these well-developed; L_2 and L_3 entirely absent; paraphyses absent; median pair of gland spines as long as lobes; other gland spines reduced, short, spine-like; dorsal macroducts two-barred type, few, small to medium in size, arranged in short rows on pygidial and posterior abdominal segments; marginal macroducts largest; microducts in submarginal area of venter; perivulvar pores in 5 small groups. For more complete description, distribution and biology see *F. hungaricus,* the only included species.

Note. This genus is separated from other Lepidosaphidina in CE by the presence of only 1 pair of large pygidial lobes.

204. **Ferreroaspis hungaricus** (Vinis) – Vinis' Armored Scale

Syn. & Lit.: *Acanthomytilus hungaricus* Vinis, 1981:201. *Ferreroaspis hungaricus,* Kozár, 1983:92.

Test of female 0.5–1 mm long, 0.2–0.3 mm wide. Live female pyriform, 0.45–0.75 mm long, lilac color. Slide-mounted females widest at second abdominal segment; stylet loop reaches line between hind spiracles; anterior spiracles with 1–2 trilocular pores, posteriors each with 3 pores. Pygidial margin semicircular, with only L_1 developed, these large, broadly notched one on outer margin, rounded apically; gland spines between L_1 strong spine-like; other marginal gland spines very small; dorsal macroducts few, of various sizes; marginal macroducts 5 pairs, each ca 8 μ in diameter, 17 μ long; microducts in small groups along body margin; perivulvar pores: 2–4 (1–6) 2–5; anal opening circular, close to anterior margin of pygidium. First and second instars described and illustrated by Vinis (1981).

Distribution. Known from Hungary (Budapest) only.

Biology. In bark crevices of twigs and trunk of *Acer campestre*; one generation per year; mature females overwinter, lay 16–18 eggs per female from end of June to end of July, nymphs hatch in 6 to 8 days after egg laying; nymphs settle in bark crevices and molt twice, mature by end of September; adult males rare and present in low numbers (Vinis, 1981).

Genus **Lepidosaphes** Shimer – Oystershell Scales
Fig. 29

Syn. & Lit.: *Lepidosaphes* Shimer, 1868:372. *Mytilococcus* Amerling, 1858:27 (nomen nudum); Lupo, 1939:69; Zahradnik, 1951:161, 1972:424. *Mytilaspis* Targioni-Tozzetti, 1868:737. *Cornuaspis* MacGillivray, 1921:274. *Cephalaspis* MacGillivray, 1921:274. *Insulaspis* Mamet, 1950:32. *Lepidosaphes,* Balachowsky, 1954b:28; Takahashi, 1955:67; Schmutterer, 1959:156; Borchsenius, 1963:1161, 1966:43; Morrison & Morrison, 1966:107; Kosztarab & Kozár, 1978:159.

Type-species: *Coccus ulmi* Linnaeus, 1758. Williams (1971) proposed *L. conchiformis* (Gmelin) as type.

Test of female strongly convex, oystershell-shaped, usually brownish or gray with 2 exuviae at anterior end. Adult female spindle-shaped; disc pores at anterior spiracles only. Pygidium with 2 pairs of lobes, L_2 bilobed; L_3 reduced or absent; without elongate paraphyses; gland spines thick; dorsal macroducts arranged in rows on segments 1 to 6; marginal macroducts largest; microducts in submarginal area of venter; perivulvar pores in 5 groups.

Distribution. World-wide with about 150 species, 13 of these in the Palearctic Region, 8 in CE.

Biology. Members of the genus feed on a variety of plants, primarily woody. They overwinter in egg stage or as adult females.

Note. Some of the species assigned to this genus were placed in recently resurrected genera, but are treated here as synonyms. Until the entire subtribe of Lepidosaphidina is revised, we see no reason to place the few species included here into several genera.

Key to species of *Lepidosaphes*

1. Over 12 median dorsal ducts on each half of 6th abdominal segment, forming a continuous band, extending from margin of pygidium almost to anus; segments 2–4 bear sclerotized spurs on margin 2
   - Less than 12 median dorsal ducts on each half of 6th abdominal segment; segments 2–4 without sclerotized spurs on margin 4
2. 24–30 dorsal ducts on each half of 5th abdominal segment; L_1 with 1–3 well developed notches on each margin; total perivulvar pores 98–130; no males known . . . . *populi*, p. 351
   - Over 30 dorsal ducts on each half of 5th abdominal segment; L_1 without or with 1–2 very small or irregular notches on each margin; total perivulvar pores less than 80; with males 3
3. L_1 irregularly notched on margins; usually 31–38 dorsal ducts on each half of 5th abdominal segment; known from *Tilia europaea*, in Romania . . . . *tiliae*, p. 352
   - L_1 without or with only 1 or 2 very small notches on each margin; usually 32–50 dorsal ducts on each half of 5th abdominal segment; polyphagous and cosmopolitan (fig. 138) . . . . *ulmi*, p. 353

4 (1). 2 (rarely 3 or 4) dorsal submedian macroducts on each half of abdominal segment 6; on conifers only 5
   - 4 or more dorsal submedian macroducts on each half of abdominal segment 6; on deciduous trees and shrubs 6

5. Space between median lobes (L_1) wider than L_1; 2 gland spines in each interlobular space much longer than L_1 . . . . *newsteadi*, p. 351
   - Space between median lobes equal or smaller than L_1; 1 (rarely 2) gland spines in interlobular spaces 2 and 3 shorter than L_1 . . . . *juniperi*, p. 349

6 (4). L_1 twice notched on outer margin; median gland spines shorter than L_1 . . . . *conchiformis*, p. 348
   - L_1 once notched on outer margin; median gland spines longer than L_1 . . 7

7. L_{2b} well developed, elongate tongue-like; usually 4 dorsal submedian macroducts on each half of segment 6; dorsal ducts absent from median area of prepygidial abdominal segment; posterior perivulvar pore groups each

with 3–4 pores; abdominal segments 2 & 3 sclerotized in old females ... . *minima*, p. 350

L_{2b} small, triangular; 5–6 dorsal submedian macroducts on each half of segment 6; dorsal ducts in median area of most prepygidial abdominal segments; posterior perivulvar pore groups each with 5–10 pores; abdominal segments 2 and 3 not sclerotized .... *granati*, p. 349

205. **Lepidosaphes conchiformis** (Gmelin) – Red Oystershell Scale

Syn. & Lit.: *Coccus conchiformis* Gmelin 1789:2221. *Mytilaspis linearis* Targioni-Tozzetti, 1868:737. *M. ficus* Signoret, 1870:94. *L. rubri* Thiem, 1931:557. *L. conchiformis-ulmi* Koroneos, 1934:81. *Mytilococcus saliceti* Lindinger, 1936:158. *M. conchiformis* Lupo 1939:96, 1943:196. *L. conchiformis*, Balachowsky & Mesnil, 1935:593; Borchsenius 1966:53; *L. conchiformis* forma *conchiformis*, Lupo, 1943:196; Balachowsky, 1954b:65; Schmutterer, 1959:167; Săvescu, 1961:30; Rafiq & Ghani, 1971:69; Zahradnik, 1972:424; Komosinska, 1974a:343, 1974b:15, 1975:127; Kosztarab & Kozár, 1978:162.

Female test oystershell-shaped, convex; usually dark brown, 1.2–2.8 mm long, 0.5–0.6 mm wide, exuviae yellow; ventral test incomplete, white. Adult female spindle-shaped, with sclerotized pygidium; pale pink to reddish violet, about 1.0–1.3 mm long, 0.4–0.6 mm wide; antenna with 2–4 thick setae; anterior spiracles with 1–7 associated disc pores; posterior spiracles without disc pores; numerous dorsal ducts present from mesothorax through all abdominal segments; ventral glandular tubercles present in groups in the submarginal area of abdominal segments 1–4. Pygidium with L_1 wide, slightly longer than gland spines, with a single inner notch and double outer notch; L_2 bilobed, inner lobe much larger than outer lobe, both triangular; L_3 reduced to 3–4 small teeth; gland spines well developed; 4 to 12 dorsal submedian ducts on segment 6; perivulvar pores: 4–7 (7–12) 5–11; anal opening close to anterior margin of pygidium (Balachowsky, 1954b; Komosinska, 1974a). Descriptions of apterous male, also of 1st instar and 2nd instars, by Komosinska (1974a).

Distribution. In CE: Bu, Cz, Fr, Ge, Hu, It, Po, Ro, Sw, USSR, and Yu. Also in England, Greece, Spain, Portugal, Turkey, in Middle and Far East, N Africa, Central Asia, S and N America (Balachowsky, 1954b; Borchsenius, 1966). A common species.

Biology. This polyphagous species is known from the following host genera: *Acer, Betula, Carpinus, Celtis, Corylus, Diospyros, Ficus, Fraxinus, Juglans, Malus, Pistacia, Pyrus, Syringa, Tilia, Ulmus* and *Zelkova*. It is biparental, has 1 yearly generation in CE, and overwinters as adult females. Eggs are laid in Poland during May, 5–8 per female, which hatch at the end of May. Second instars develop by the end of June, and adults appear by the end of July (Komosinska, 1975). An average of 26 eggs are laid in Germany according to Schmutterer (1959).

Natural enemies. Parasites: *Anaphes gracilis, Aphytis mytilaspidis, A. proclia, Apterencyrtus microphagus, Azotus atomon, A. pinifoliae, Cheiloneurus microphagus, Coccophagoides similis, Coccophagus insidiator, Hispaniella lauri, Pteroptrix dimidiata, Prospaltella leucaspidis* (Schmutterer, 1959; Zahradnik, 1972; Komosinska, 1975). *C. similis* was found to be the dominant parasite. Predators: *Pilophorus cinnamopterus, Hemisarcoptes malus* (Schmutterer, 1952b).

Note. This species is related to *L. minima*; and the distinguishing characters are discussed at *L. minima*. Until recently the two were treated as two forms of *L. conchiformis*.

206. **Lepidosaphes granati** Koronéos – Hawthorn Oystershell Scale

Syn. & Lit.: *L. conchiformis-granati* Koronéos, 1934:72, 81. *Mytilococcus granati*, Lupo, 1939:101. *L. crataegi* Borchsenius, 1949:343. *L. granati*, Balachowsky, 1954b:72; Danzig, 1972c:215; Kosztarab & Kozár 1978:162. *Mytilaspis granati*, Borchsenius, 1966:54.

Test of female brown, 1.5–2.6 mm long. Adult female elongate, oval; antenna with 2 short setae; anterior spiracle with 3 associated disc pores, posterior with 1–2; a large group of microducts also associated with each posterior spiracle; a row of dorsal ducts on each abdominal segment. Pygidium partly sclerotized; L_1 wide, notched on both sides; L_2 bilobed, outer lobe small, pointed; L_3 lacking, in its place a thickening of marginal macroduct rims is found; median gland spines longer than lobes; a pair of long gland spines on margin of each pygidial segment; 5–6 dorsal submedian macroducts on each half of segment 6; few microducts in marginal band on venter; perivulvar pores: 2–6 (7–9) 5–10; anus close to anterior margin of pygidium.

Distribution. In CE: Bu, Hu, It, and USSR. Also known from Greece, Iran, Morocco (Balachowsky, 1954b, Kosztarab and Kozár, 1978). A rare species in CE.

Biology. Host genera: *Crataegus, Elaeagnus, Punica, Rhamnus,* and *Zelkova* (Balachowsky, 1954b; Borchsenius, 1966). According to Danzig (1972c) 2 yearly generations develop in southern USSR. Mated females overwintered and began laying eggs in mid-April. About 25 eggs per female were laid on *Punica granatum*. It was present at 1,700 m elevation in Morocco (Balachowsky, 1954b).

207. **Lepidosaphes juniperi** Lindinger – Juniper Oystershell Scale

Syn. & Lit.: *Lepidosaphes juniperi* Lindinger, 1912:188; Mesnil, 1941:82; Balachowsky, 1954b:86; Bustshik, 1960:176; Borchsenius, 1966:63; Bazarov & Smelev, 1971:65; Zahradnik, 1972:425; Komosinska, 1974:20; Kosztarab & Kozár, 1978:161. This species is often included in the genus *Insulaspis* or *Mytilococcus*.

Test of female brown, elongate oystershell-shaped, 1.5–2 mm long, exuviae light brown. Adult female oval, widest at metathorax, light violet; 2 setae on antenna of unequal size; 1–3 disc pores associated with each anterior spiracle; 2–5 cone-shaped gland spines associated with each posterior spiracle. Pygidium wide, with square cut tip; L_1 wide, notched on both sides; distance between L_1's smaller than their width; L_2 bilobed, outer lobe smaller than inner lobe; both rounded at tip; only 1 (rarely 2) gland spines in interlobular spaces 2 and 3, these shorter than lobes; 6 marginal macroducts on each side; 2–4 dorsal submedian ducts on abdominal segment 6; other dorsal ducts form a band in submarginal area of pygidium; only few scattered microducts on venter of pygidium; perivulvar pores: 2–4 (4–11) 2–8; anal opening circular, located near anterior margin of pygidium.

Distribution. In CE: Bu, Po, Sw, and USSR. Also S Europe, Corsica, Greece, Turkey, and Middle East, Central Asia (Borchsenius, 1966; Tsalev, 1968; Komosinska, 1969). A rare species in CE.

Biology. Oligophagous on conifers in the families Abietaceae and Cupressaceae. Known host genera: *Biota, Cupressus, Juniperus, Pinus,* and *Thuja* (Borchsenius, 1966). It is found at up to 2,200 m elevation. According to Bazarov and Smelev (1971) it is bisexual and has 1 yearly generation in Central Asia. It is seldom a pest.

Natural enemy. Parasite: *Archenomus opacus* (Trjapitzin, 1978).

Note. Bustshik (1960) suggested the possibility of 2 "races" of this species.

208. **Lepidosaphes minima** (Newstead) – Fig Oystershell Scale

Syn. & Lit.: *Mytilaspis minima* Newstead, 1897:95. *M. ficifolii* Berlese, 1904:3. *L. ficifoliae* v. *ulmicola* Leonardi, 1907:36. *Mytilococcus ficifoliae,* Lupo, 1939:124. *L. ficus* Stafford and Barnes, 1948:567 (misidentification). *L. conchiformis* forma *minima* Balachowsky, 1954:69. *L. turkmenica* Borchsenius & Bustshik, 1955:304. *Insulaspis minima,* Borchsenius, 1966:64.

Test of female small, oystershell-shaped, often irregularly S-shaped, usually pale brown with lighter posterior end. Adult female spindle-shaped, differs from *L. conchiformis* in the following morphological characters: abdominal segments 2 and 3 sclerotized (not in *conchiformis*); dorsal submedian ducts greatly reduced in number on each segment, 4 (rarely 5) on segment 6 (4–12 in *conchiformis*); median gland spines longer than L_1 (shorter in *conchiformis*); L_1 only once notched on outer margin (twice notched in *conchiformis*); perivulvar pores: 2–4 (4–6) 3–4 (more in each group in *conchiformis*). Anal opening large, circular, close to anterior margin of pygidium.

Distribution. In CE: Fr, It. Also the Mediterranean area, including Malta, Turkey, Algeria and Egypt; Middle East, Central Asia, California and Chile (Borchsenius, 1966). Apparently of Mediterranean origin.

Biology. It is known only from leaves and twigs of *Ficus* and *Ulmus*. Its biology in Fresno, California, was studied by Stafford and Barnes (1948) who considered this species (*"L. ficifoliae"*) to be a summer form or leaf form of *L. ficus* Signoret, latter being the stem infesting form. Mated females overwintered, and started laying eggs in February. Each female laid about 30 eggs which started hatching around the first of April. Egg laying of the summer generation started in early June. Each female laid about 12 eggs which started hatching on June 17 and continued through July. Adults mated in the fall. Female tests found on leaves during the summer were shorter, narrower and lighter colored than those of overwintering generation. The female morphology was also different.

Natural enemies. Stafford and Barnes (1948) listed as parasites *Aphytis mytilaspidis* and *A. chrysomphali*; as predators *Chilocorus bivulnerus, Lindorus lophanthae* and lacewing larvae.

209. **Lepidosaphes newsteadi** (Šulc) – Pine Oystershell Scale

Syn. & Lit.: *Mytilaspis newsteadi* Šulc, 1895:8. *Mytilococcus newsteadi*, Lupo, 1939: 119. *L. newsteadi*, Mesnil, 1949:74; Zahradnik, 1951:166, 1972:425; Schmutterer, 1952b:575, 1959:164; Balachowsky, 1954b:83; Borchsenius, 1966:65; Kosztarab & Kozár, 1978:161. It was treated in the earlier literature in the genera *Insulaspis*, *Mytilaspis* and *Mytilococcus*.

Test of female elongate, almost parallel-sided, brown or light brown, 1.6–3 mm long. Ventral test grayish-white. Adult female spindle-shaped, widest at metathorax, whitish-pink to violet; antenna with 2 setae; each anterior spiracle with 1–6 associated disc pores, none at posterior spiracles. Pygidium trapezoidal, L_1 rounded apically, usually notched on both sides; the space between the 2 L_1's wider than L_1; L_2 bilobed, outer lobe much smaller than inner lobe; L_3 lacking; gland spines well developed, all longer than lobes; 6 dorsal marginal macroducts on each side, 2–4 submedial macroducts on each half of segment 6; very few microducts on venter; perivulvar pores: 0–4 (6–11) 4–12; anal opening circular, close to anterior pygidial margin.

Distribution. Palearctic. In CE: Au, Bu, Cz, Fr, Ge, Hu, It, Po, Ro, Sw, and USSR. Also known from Holland, Turkey, Lebanon (Rogojanu, 1962; Tereznikova, 1963; Borchsenius, 1966; Tsalev, 1968). A common mesophilous species in semi-dry coniferous forests.

Biology. Host genera: *Abies, Cedrus, Juniperus (?), Picea,* and *Pinus*. Occurs on needles and often is a pest. According to Schmutterer (1952b) it has 1 yearly generation. Fertilized females overwinter and lay about 7–56 eggs per female in Germany, from end of April to end of May. Eggs hatch in about 40–50 days; adults develop by beginning of August.

Natural enemies. Parasites: *Aphytis mytilaspidis, Prospaltella aurantii, P. intermedia, P. leucaspidis, Pteroptrix dimidiata* (Zahradnik, 1972; Trjapitzin, 1978).

Note. According to S. Nakahara (personal communication) the records from Bermuda Isl., and N America are misidentifications of *L. pallida* (Maskell) (= *L. maskelli*) which occur on Juniper and other Cupressaceae.

210. **Lepidosaphes populi** Săvescu – Poplar Oystershell Scale

Syn. & Lit.: *Lepidosaphes populi* Săvescu, 1955:951, 1961:27; Borchsenius, 1966:44; Tsalev, 1967:59, 1968:215; Kosztarab & Kozár, 1978:160.

Test of female oystershell-shaped, brown with posterior end whitish, 2.2–3 mm long, 0.75–1.2 mm wide. Adult female spindle-shaped, 1.2–1.8 mm long, 0.68–0.82 wide; antenna with 1–2 (rarely 3) short setae; anterior spiracles with 5–8 associated disc pores, none at posterior spiracle. Pygidium yellow with 2 pairs of lobes; L_1 large with rounded margin, 1 or 2 pronounced notches on each side; L_2 bilobed, inner lobe larger, with notch on outer margin; total of 130–140 dorsal macroducts; perivulvar pores: 13–21 (18–30) 18–30, total 100–130 pores, rarely 98–99; anal opening close to posterior margin of pygidium. First instars 320–368 μ long, 160–174 μ wide, antennae 82 μ long; second instars 420–500 μ long, 240–280 μ wide, antennae reduced, spiracles small, anterior with 1 pore.

Distribution. In Bu, Ro, and USSR. It has been confused with *L. ulmi* and probably is more widely distributed.

Biology. It is found on trunk and branches of *Fraxinus, Gleditsia, Populus, Robinia* and *Salix*; often causes damage. This species is apparently parthenogenetic, has one generation per year and overwinters as eggs. In Bulgaria the crawlers appear from mid-May. The adult females were present from end of June and started laying eggs at the end of July. Each female laid 26–124 eggs (Tsalev, 1967).

Natural enemies. Records of Tsalev (1967). Parasites: *Aphytis mytilaspidis, A. proclia,* and *Habrolepis zetterstedti*.The first chalcid parasitized 16.4 to 54.2% of scales. Predator: *Chilocorus bipustulatus.*

Note. The taxonomic status of this species requires further study. It should be compared with *L. salicina* Borchsenius and with known biotypes of *L. ulmi.* Apparently its biology differs from both.

211. **Lepidosaphes tiliae** Săvescu – Linden Oystershell Scale

Syn. & Lit.: *Lepidosaphes tiliae* Săvescu, 1957:197, 1961:25; Borchsenius, 1966:44; Kosztarab & Kozár, 1978:160.

Test of female oystershell-shaped, brownish-gray or gray-dark-brown, 3–4 mm long. Adult female spindle-shaped, yellowish-white, 1.2–1.8 mm long, 0.68–0.82 wide; antennae reduced with 2, seldom 3–4 curved setae; anterior spiracles with 5–7 associated disc pores; none at posterior spiracles; lateral sclerotized spurs on margin of abdominal segments 2–3. Pygidium yellow to orange-yellow; L_1 wide, margin seldom with notches; L_2 bilobed, inner lobe larger than outer and with an outer notch; with 100–120 total dorsal ducts (seldom under 100) on fifth abdominal segment and 60–76 total (seldom under 60) on fourth segment; perivulvar pores in 5 groups: 8–12 (10–21) 9–19, total 62–80 pores, seldom 54–56. Adult male and other instars described by Săvescu (1957). First instar elongate, narrowing toward posterior end, yellowish-white, 260–320 μ long, 140–160 μ wide, the 6-segmented antenna 72 μ long. Second instar pear-shaped, white transparent, with yellow pygidium; 420–500 μ long, 200–240 μ wide.

Distribution. In Ro only. Balás (1966) reported its presence from Hungary, based on male tests only.

Biology. Monophagous species, known from *Tilia europea.* According to Săvescu (1957) the males infest the leaves in large numbers. This species has one generation per year and overwinters as eggs. Crawlers appear during first third of May; adult females appear during first third of June and lay 50–110 eggs per female.

Note. Although *L. tiliae* is treated here, this species is so closely related morphologically to *L. ulmi* that further research is necessary to ascertain that they are indeed two different species.

212. **Lepidosaphes ulmi** (Linnaeus) – Oystershell Scale, Mussel Scale
Figs 29, 138

Syn. & Lit.: *Coccus ulmi* Linnaeus, 1758:455. More than 30 junior synonyms are listed for this species by Borchsenius (1966). Only the more common ones used earlier in the European literature are listed here. *Chermes arborum linearis* Geoffroy 1762:509. *Coccus linearis* Modeer, 1778:22. *C. amygdali* Schrank, 1801:143. *C. berberidis* Schrank, 1801:143. *Diaspis linearis* Costa, 1835:21. *Aspidiotus falciformis* Baerensprung, 1849:168. *A. pomorum* Bouché, 1851:110. *Mytilaspis pomorum* v. *vitis* Goethe, 1884:118. *M. ulicis* Douglas, 1886:249. *M. ceratoniae* Gennadius, 1895:277. *M. confusus* Horváth, 1897:81. *M. pomorum* v. *candidus* Newstead, 1901:82. *M. pomorum* v. *ulicis* Newstead 1901:200. *Mytilococcus communis* Amerling, 1858:99. *M. oleae* Leonardi, 1908:150. *Lepidosaphes abietis* Signoret 1870:91. *L. ulmi bisexualis* Thiem, 1933:638. *L. ulmi cotini* Koroneos, 1934:80. *L. ulmi rosae* Koroneos, 1934:82. *L. ulmi* v. *oleae* Koroneos, 1934:61. *L. ulmi*, Balachowsky, 1934:37; Balachowsky & Mesnil, 1935:380; Danzig, 1959:879; Schmutterer, 1952:20, 1959:158; Săvescu, 1961:31; Kosztarab, 1963:87; Borchsenius, 1966:44; Bazarov & Smelev, 1971:54; Zahradnik, 1972:426; Kosztarab & Kozár, 1978:160.

Test of female variable, usually elongate oystershell-shaped, convex, brownish or gray, about 1.8–3.5 mm long; ventral test white. Adult female elongate spindle-shaped, distinctly lobed laterally, white to reddish-yellow; 1.2–1.5 mm long; abdominal segments 2–4 each with a sclerotized spur on margin; each anterior spiracle with 3–7 associated quinquelocular disc pores, none at posterior spiracles. Pygidium sclerotized at apex; median lobes large, wide, notched on both sides, rounded apically; L_2 bilobed; L_3 lacking; gland spines large and slender; 2 between median lobes, about as long as lobes; 6 large marginal macroducts; other dorsal macroducts fairly small, in a continuous group on segment 6, extending from margin to anus; ventral microducts few and small; perivulvar pores: 7–17 (12–27) 7–24; anal opening circular, close to anterior margin of pygidium; vulva in the middle of pygidium. Adult males were described by Bustshik (1958), first instars by Howell and Tippins (1977) and second instars by Boratynski (1953).

Distribution. In CE known from Au, Bu, Cz, Fr, Ge, Hu, It, Po, Ro, Sw, USSR, and Yu; also from other parts of Europe, Africa, Middle East, Orient, North and South America, Australian Region (Borchsenius, 1966). This cosmopolitan species is commonly collected.

Biology. A polyphagous species, known from the following host genera: *Abies, Acer, Aesculus, Ailanthus, Alnus, Amelanchier, Amygdalus, Andromeda, Arctostaphylos, Armeniaca, Berberis, Betula, Buxus, Caesalpinia, Cajanus, Calluna, Carpinus, Castanea, Ceanothus, Cerasus, Ceratonia, Chimaphila, Clematis, Cornus, Corylus, Cotinus, Cotoneaster, Crataegus, Cyanotis, Cydonia, Cytisus, Elaeagnus, Erica, Euphorbia, Euonymus, Fagus, Fraxinus, Genista, Geranium, Ginkgo, Hippophae, Hypericum, Juglans, Laburnum, Ledum, Ligustrum, Lonicera, Lycium, Maackia, Malus, Matsudana, Melica, Mespilus, Myrica, Myotus, Olea, Oxycoccus, Pachysandra, Padus, Pavia, Persica, Philadelphus, Platanus, Populus, Potentilla, Prunus, Ptelea, Pyrus, Quercus, Quinaria, Rhamnus, Ribes, Ricinus, Robinia, Rosa, Rubus, Salix, Sarothamus, Sassafras, Sesbania, Sorbaria, Sorbus, Spartium, Spartocytisus, Spiraea, Staphylea, Symphoricarpus, Syringa, Thelycrania, Tilia, Ulex, Ulmus, Vaccinium, Viburnum, Vincetoxicum, Viscum, Vitis* and *Ziziphus* (Danzig, 1959b; Kosztarab, 1963; Borchsenius, 1966). Parthenogenetic and biparental forms differ in their host preferences. The first prefer hosts of the Rosaceae family, while latter is found on a

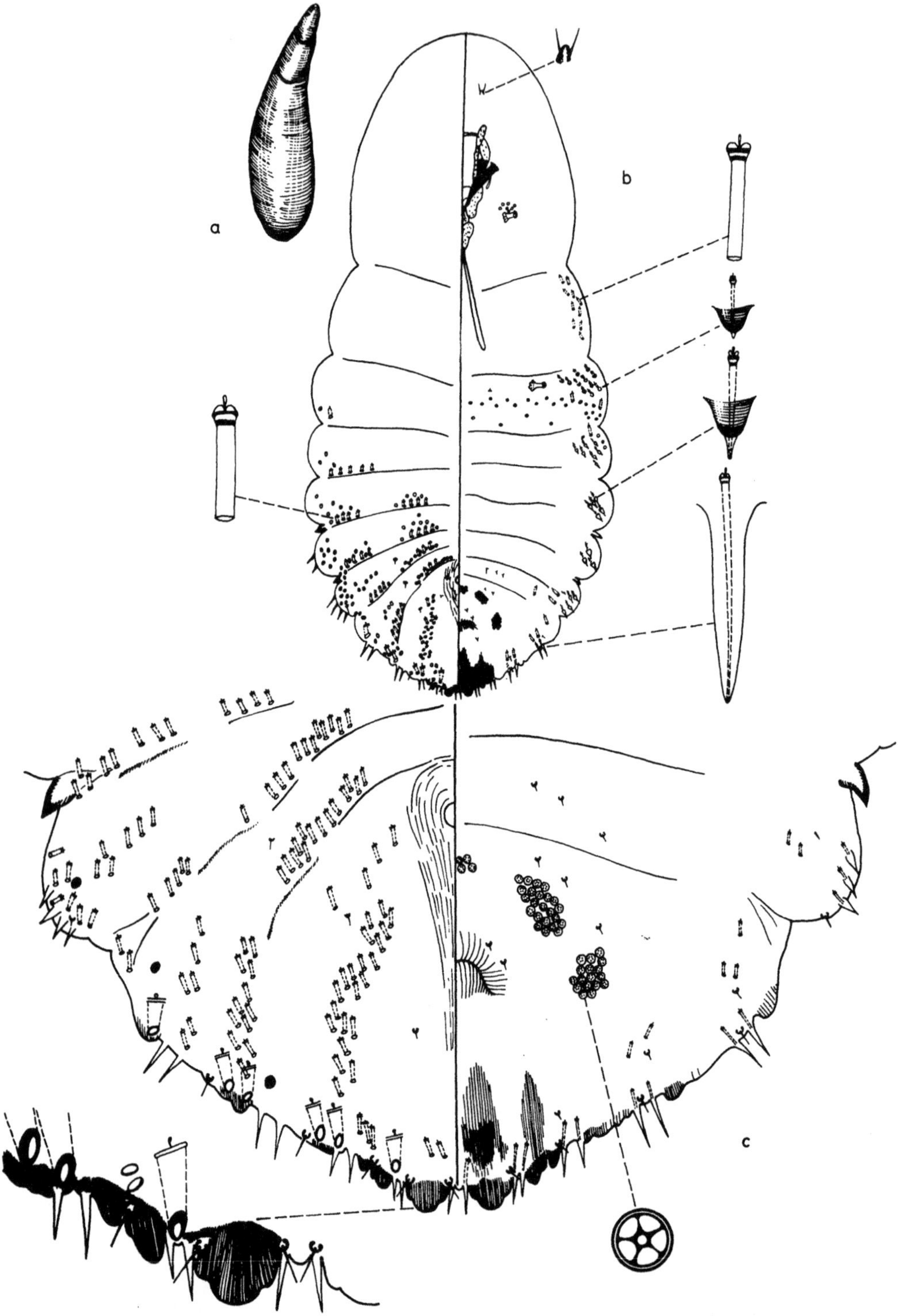

Fig. 138. *Lepidosaphes ulmi* (Linnaeus) (a) female scale, (b) female, (c) enlarged pygidium (modified after Kosztarab, 1963)

wide variety of host plants. Danzig (1959) noted that the host preference is flexible enough allowing for transfer of specimens to a variety of hosts, but there is a reduction in their biopotential. The flexibility of the bisexual form is noted in their adaptation to more extreme climatic conditions. The parthenogenetic or biparental condition did not change even with host alteration. It was possible only with some difficulty to mate males of the bisexual form with females of the parthenogenetic form, thus proving the differentiation of the two forms. It overwinters in the egg stage and usually has 1 yearly generation. However, in Romania Săvescu (1955) reported the possibility of 2 yearly generations. Eggs start hatching during second half of May, after the flowering of apples. The crawlers first appear in sunny areas and settle close to the adult females. They then produce the white test. According to Schmutterer (1952b) the first molt occurs at the beginning of June, and the adults appear at end of July or in August. Eggs are laid in August or September, depending on the geographic location. Egg production is dependent on the host plant. Danzig (1959b) found 40–47 eggs per female on rosaceous trees in the parthenogenetic form, and each biparental female laid 20 eggs on *Padus* and 68 eggs on *Tilia*. Bazarov & Smelev (1971) reported 10 to 120 eggs per female. Occasionally this species becomes an important pest in neglected orchards, on forest trees, and on ornamental trees, especially under urban conditions.

Natural enemies. Parasites: *Anabrolepis zetterstedti, Anaphes gracilis, Aphytis abnormis, A. caucasicus, A. diaspidis, A. moldavicus, A. mytilaspidis, A. proclia, Apteroptrix longiclava, Archenomus maritimus, A. longicornis, Aspidiotiphagus citrinus, Cheiloneurus microphagus, Coccophagus niger, C. parvipennis, C. similis, Eusemion cornigerum, Gyranus matritensis, Habrolepis dalmani, Hispaniella lauri, Metallonoidea britannica, Microterys macharus, Physcus testaceus* and *Pteroptrix opaca* (Balachowsky, 1954b; Schmutterer, 1959; Kosztarab, 1963; Zahradnik, 1972). Predators: lady beetle, *Chilocorus bivulnerus; Dometorina* sp., *Hemisarcoptes malus, Octobdellodes* sp., *Oripoda elongata, Scapheremaeus marginalis* and *Tyrophagus entomophagus* (Schmutterer, 1952b; Kosztarab, 1963).

Note. There are indications from their morphology and biology that a species complex exists in *L. ulmi,* as currently recognized. Thiem (1933) has separated this species into two forms, *L. ulmi ulmi* Linnaeus, 1758 and *L. ulmi bisexualis* Thiem, 1933. Moreover, *L. populi* Săvescu, 1955, and *L. tiliae* Săvescu, 1957, were considered forms of *L. ulmi,* but in recent years they have become accepted as good species by various European workers. A thorough study of the morphology and biology of *L. ulmi* and related species is required to clarify their taxonomic status.

Genus **Leucaspis** Signoret

Syn. & Lit.: *Leucaspis* Targioni-Tozzetti, 1868:734, (nomen nudum); Signoret, 1869:99; *Leucodiaspis,* Zahradnik, 1951:99. *Rhopaloaspis* Del Guercio, 1903:188. *Euleucaspis* Lindinger, 1905b:252. *Anamaspis* Leonardi, 1906:22; Borchsenius, 1966:218. *Actenaspis* Leonardi 1906:25. *Pusillaspis* Lindinger, 1906a:27. *Leucaspis,* Green, 1915a:459; Leonardi, 1920:117; Ferris, 1936:25 & 68; Gomez-Menor, 1937:133; Lupo, 1944:206; Balachowsky, 1953:844; Schmutterer, 1959:142; Borchsenius, 1966:211; Takagi, 1969:25; Kosztarab & Kozár, 1978:162.

Type-species: *Aspidiotus pini* Hartig, 1839 (= *Leucaspis candida* Targioni-Tozzetti).

Female pupillarial; entirely enclosed in the exuviae of second instar; only one exuviae on elongate white or grayish-white test. Adult female elongate spindle-shaped, widest at metathorax or 1st abdominal segment; antenna tubercle-like with few setae; gland tubercles common along body margins; usually anterior spiracles with associated pores. Pygidial lobes small, in 1–4 pairs, L_1's separated from each other by much space; plates rarely absent, often spiniform, rarely wide and fimbriate at apex; dorsal macroducts only along submargin; perivulvar pores normally in 5 groups, usually arranged in a semicircle, often with supplementary pore groups. Second instars described by Balachowsky (1953) and Zahradnik (1952c); 1st and 2nd instars by Lupo (1944).

Distribution. In Palearctic region with 9 species, 1 in S America; 4 in CE. A number of additional species were formerly assigned to *Leucaspis*.

Biology. Most species infest needles of *Pinus* spp. They usually have 1 yearly generation, and overwinter as adults or nymphs. A study on the parasites of 4 species of *Leucaspis* in Corsica was published by Benassy (1968).

Key to species of *Leucaspis*

1. Pygidial plates absent; lobes very small and with rounded apex; distance between median lobes twice the diameter of anus .... *loewi,* p. 356
   Pygidial plates present; lobes variable in size, usually pointed at apex; distance between median lobes about same as diameter of anus or less . 2
2. Perivulvar pores in 12–16 groups, but some groups often combined; gland spines narrow, pointed, 2–3 times longer than lobes, never fimbriate at tip; 4–5 pairs of lobes .... *signoreti,* p. 359
   Perivulvar pores in 5 groups, or combined in a semicircular band; some gland spines wide and with fimbriate tip; 2 to 4 pairs of lobes 3
3. Perivulvar pores in 5 groups; a group of 12–22 gland tubercles laterad of each anterior spiracle; 28–52 dorsal macroducts in a submarginal band; 2–4 pairs of well developed lobes .... *pini,* p. 357
   Perivulvar pores arranged in a semicircular band; no gland tubercle groups laterad of anterior spiracles; only about 6 dorsal macroducts in pygidial submargin; 3 pairs of faintly developed lobes (fig. 140)... *pusilla,* p. 358

213. **Leucaspis loewi** Colvée – Löw's Pine Scale

Syn. & Lit.: *Leucaspis loewi* Colvée, 1882:10. *L. pini* Morgan, 1892:13. *Fiorinia sulci* Newstead 1894:232. *Anamaspis loewi* Leonardi, 1906:23. *A. loewi,* Borchsenius, 1966:219. *L. loewi,* Leonardi, 1920:132; Gomez-Menor, 1937:135; Lupo, 1944:237; Zahradnik, 1951:106; Balachowsky, 1953:868; Schmutterer, 1952:582, 1959:147; Rogojanu, 1962:295; Kuznetsov, 1970:1644; Zahradnik, 1972:442; Kosztarab & Kozár, 1978:163.

Test of female pear-shaped, or wide oystershell-shaped, white, 1.4–2.2 mm long, 0.8–1.2 mm wide. Adult female elongate oval, violet to gray in color; antenna with 3–5 setae; anterior spiracle with 2–7 quinquelocular pores, no pores at posterior one. Pygidial margin semicircular, without plates; 3 pairs of lobes, small, with rounded apex; distance between median lobes twice diameter of anus; 28–46 dorsal macroducts in a submarginal band; perivulvar pores in 5 groups: 6–15 (8–23) 8–16, groups often form a continuous band. Second instar described by Zahradnik (1952c) and Balachowsky (1953), 1st and 2nd instars by Lupo (1944).

Distribution. Western Palearctic. In CE: Au, Bu, Cz, Fr, Ge, Hu, It, Po, Ro, Sw, USSR, and Yu. Also from S and W Europe and N Africa.

Biology. A common species on pine, *Pinus* spp., needles. It is seldom a pest but will cause yellowing and premature dropping of needles. One yearly generation. In S Germany they overwinter as N_2 and as mated or unmated adult females. Adults develop from overwintering N_2 by first half of May and after mating lay ca 26 eggs per female by end of May or in June; eggs hatch in 18–20 days. Overwintering fertilized females often lay some eggs during fall and complete their egg laying in May. From their eggs adult males and females develop by August, and the ensuing mated females overwinter (Schmutterer, 1959). Kuznetsov (1970) described the life cycle for Crimea.

Natural enemies. Parasites: *Anthemus funicularis, A. leucaspidis, A. pini, Aspidiotiphagus citrinus, Azotus pinifoliae, Parasauleia trjapitzini, Prospaltella aurantii, P. intermedia, P. leucaspidis* (Schmutterer, 1959; Kuznetsov, 1970; Trjapitzin, 1978). Predators: *Chilocorus bipustulatus, C. renipustulatus, Exochomus flavipes, Myrrha octodecimguttata, Vibidia duodecimguttata* (Kuznetsov, 1970).

214. **Leucaspis pini** (Hartig) – Hartig's Pine Scale
Fig. 30

Syn. & Lit.: *Coccus pini* Hartig, 1839:642. *Aspidiotus pini* Bouché, 1851:111. *Leucaspis candida* Targioni-Tozzetti, 1868:734; Schmutterer, 1952b:583. *Aspidiotus flavus* Signoret 1870:108. *L. affinis* Leonardi, 1906:4. *Leucodiaspis pini,* Zahradnik, 1951:100; *Leucaspis pini,* Leonardi, 1920:122; Gomez-Menor, 1937:142; Lupo, 1944:208; Balachowsky, 1953:849; Schmutterer, 1959:144; Borchsenius, 1966:212; Zahradnik, 1972:443; Kosztarab & Kozár, 1978:163.

Test of female elongate, almost parallel-sided, white or grayish-white, 2.4–2.9 mm long, 0.7–0.8 mm wide. Adult female elongate spindle-shaped, antenna with 4–6 setae; 3–9 disc pores associated with each anterior spiracle, none at posterior spiracles; a group of 12–22 gland tubercles laterad of each anterior spiracle. Pygidial margin semicircular, with 2–4 pairs of triangular lobes; 2–10 gland spines in each interlobular space, these seldom fimbriated at tip, plates total 36–48; 28–52 dorsal ducts forming a submarginal band; perivulvar pores in 5 groups: 8–12 (9–18) 6–15, few pores cephalad of anterior lateral group; anus close to anterior margin of pygidium. Adult male (*L. candida*) described by Jancke (1955), first and second instars by Lupo (1944), second instar by Zahradnik (1952c) and Balachowsky (1953).

Distribution. A common species. In CE: Au, Bu, Cz, Fr, Ge, Hu, It, Po, Ro, Sw, USSR, and Yu. Also in S and W Europe, Middle East, N Africa, S America (Schmutterer, 1959; Borchsenius, 1966).

Biology. A mesophilous species living on needles of pines, and rarely a pest. According to Schmutterer (1952b, 1959) it occurs at up to 1,100 m elevation in Germany. It has 1 yearly generation and overwinters as N_2 and as mated and unmated females. Each overwintered female lays between 17–51 eggs in May. The overwintered

N_2 develop into adults by June and start laying eggs after that time. Both biotypes are biparental.

Natural enemies. Parasites: *Anthemus leucaspidis, A. pini, Aphytis abnormis, A. mytilaspidis, Aspidiotiphagus citrinus, Azotus atomon, A. pinifoliae, Centrobia walkeri, Debachiella pini, Parasauleia trjapitzini, Prospaltella aurantii, P. intermedia, P. leucaspidis, Tetrastichus clavicornis, Thysanus subaeneus* (Fulmek, 1943; Balachowsky, 1953; Benassy, 1963, 1968; Trjapitzin, 1978; Kosztarab and Kozár, 1978).

215. **Leucaspis pusilla** Löw – Small Pine Scale
Figs 139, 140

Syn. & Lit.: *Leucaspis pusilla* Löw, 1883b:3. *L. leonardii* Cockerell, 1903:84. *L. perezii* Green, 1915:463. *Leucodiaspis pusilla*, Zahradnik, 1952c:104. *L. pusilla*, Leonardi, 1920:118; Gomez-Menor, 1937:138; Lupo, 1944:216; Balachowsky, 1953:860; Schmutterer, 1959:149; Rogojanu, 1966a:325; Borchsenius, 1966:213; Kuznetsov 1966:46; Zahradnik, 1972:444; Kosztarab & Kozár, 1978:164.

Test of female oval, oystershell-shaped, white, 1.5–2 mm long. Adult female spindle-shaped, antenna with 2–4 setae; 2–3 disc pores associated with each anterior spiracle, none at posterior ones. Pygidium with 3 pairs of faintly developed lobes, sometimes lacking; 20–32 very long gland spines, most with fringed apex; about 6 dorsal ducts in submarginal area on each side, none on other parts of dorsum; 18–32 perivulvar pores arranged in a semicircle; venter of derm with semicircular wrinkles in median area, and

Fig. 139. Damage of *Leucaspis pusilla* Löw by reducing in size the twig and needles (Photo F. Kozár)

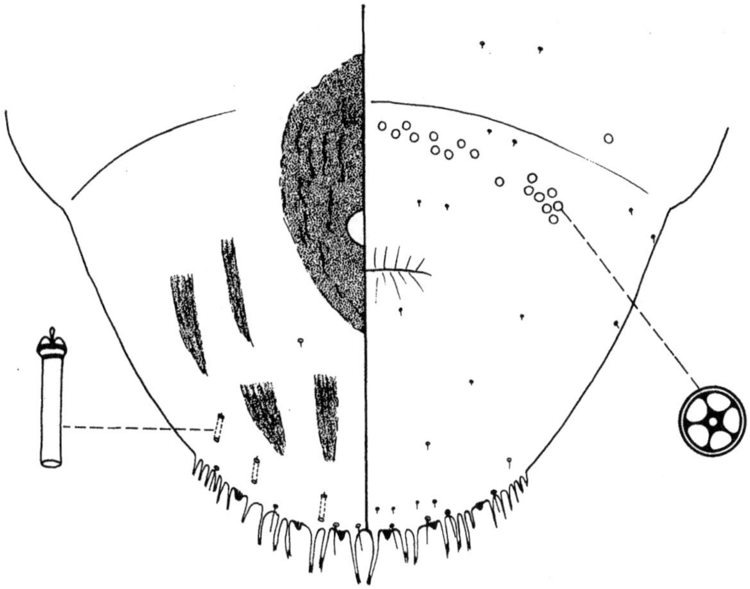

Fig. 140. *Leucaspis pusilla* Löw, female pygidium (original)

with a submarginal line bordering the lateral extensions of intersegmental membrane; anus anterior from center of pygidium, vulva in center. Adult male described by Bustshik (1958); female 2nd instar by Balachowsky (1953), and Schmutterer (1959); 1st and 2nd instars by Lupo (1944).

Distribution. In CE: Bu, Cz, Fr, Hu, It, Ro, USSR, and Yu. Also S Europe, Middle East, N Africa and Argentina.

Biology. A Mediterranean mesophilous species. Found on needles of *Pinus* and *Cedrus;* occasionally a pest. Kozár (1969) reported high populations in Hungary that reduced the growth of shoots and needles. A bisexual species with one generation per year that overwinters as adult females and N_2 in Hungary. Tereznikova's (1963) report of 2 yearly generations with 2 hatching dates, probably was due to the species overwintering in 2 different stages. Kuznetsov (1966) also noted 1 yearly generation.

Natural enemies. Parasites: *Anthemus leucaspidis, A. pini, Aspidiotiphagus citrinus, Azotus atomon, Coccophagoides similis, Parasauleia trjapitzini, Prospaltella intermedia, P. leucaspidis, P. pusilla* (Balachowsky, 1953; Bénassy, 1968; Trjapitzin, 1978).

216. **Leucaspis signoreti** Targioni-Tozzetti – Signoret's Pine Scale

Syn. & Lit.: *Leucaspis signoreti* Targioni-Tozzetti, 1868:735. *L. corsa* Lindinger, 1905b:252. *L. signoreti* v. *mauretanica* Balachowsky, 1928a:135. *L. signoreti,* Leonardi, 1920:125; Lupo, 1944:223; Balachowsky, 1953:853; Rogojanu, 1966a:325; Borchsenius, 1966:214; Zahradnik, 1972:445; Kosztarab & Kozár, 1978:163.

Test of female elongate oystershell-shaped, 2.5–3.2 mm long, with 1 exuviae only. Adult female spindle shaped, antenna with 5–6 setae; 6–16 quinquelocular pores associated with each anterior spiracle, none at posterior ones; a group of 20–28 gland tubercles laterad of each anterior spiracle. Pygidial margin semicircular; with 4–5 pairs of triangular pointed lobes, some rarely bilobed; in each interlobular space 2–19 gland spines, long, narrow, pointed, 2–3 times longer than lobes; a total of 36–50 dorsal macroducts in a narrow submarginal band starting at base of L_4 and extending anteriorly on prepygidial segments; perivulvar pores in 12–16 groups organized in 3–4 transverse rows; vulva located in center of pygidium, anus at anterior 1/3 of pygidium. Second instar described by Balachowsky (1953), 1st and 2nd instars by Lupo (1944).

Distribution. A rare Mediterranean species. In CE: Fr, It, Ro, and USSR (Moscow area and the Transcarpathian Region). Also the islands of Corsica, Sardinia and Cyprus, N Africa (Tereznikova, 1959; Borchsenius 1966).

Biology. A thermophilous species, on *Pinus* spp. According to Balachowsky (1953) it occurs in Corsica at up to 1,400 m elevation. It is bisexual, but otherwise very little is known about its biology.

Natural enemies. Parasites: *Anthemus leucaspidis, Aphytis abnormis, Aspidiotiphagus citrinus, Azotus atomon, Parasauleia trjapitzini, Prospaltella intermedia, P. leucaspidis* (Malenotti, 1918; Fulmek, 1943; Bénassy, 1968; Trjapitzin, 1978).

Genus **Mohelnaspis** Šulc

Syn. & Lit.: *Mohelnaspis* Šulc, 1937:29; Ferris, 1938a:58; Zahradnik, 1951:170; Balachowsky, 1954b:152; Schmutterer, 1959:172; Kosztarab & Kozár, 1978:164.

Type-species: *Mohelnaspis massiliensis* Goux, 1937.

Female test elongate oystershell-shaped, brown, with two oval exuviae at pointed end. Adult female spindle-shaped; antenna with 2 setae. Pygidium reduced, only median lobes developed; gland spines and marginal macroducts lacking; perivulvar pores in 5 small groups.

Distribution. Western Palearctic. It includes 1 species.

Biology. All host plants are grasses, Gramineae.

Note. This monotypic genus is related to *Berlesaspis* MacGillivray and *Evallaspis* Lupo.

217. **Mohelnaspis massiliensis** (Goux) – Grass Scale
Figs 31, 141

Syn. & Lit.: *Poliaspidoides massiliensis* Goux, 1937:35. *Mohelnaspis moravica* Šulc, 1937:30. *M. recepi* Bodenheimer, 1941:66. *M. massiliensis*, Zahradnik, 1951:171; Balachowsky, 1954b:154; Schmutterer, 1959:173; Borchsenius, 1966:31; Kosztarab & Kozár, 1978:165.

Test of female very narrow and elongate oystershell-shaped; brownish gray; about 1.5–3.0 mm long, 0.5–0.6 wide; the small exuviae extend out from the test margin; white ventral test developed. Female elongate spindle-shaped, orange-yellow, with sclerotized pygidial end; antenna with 2 setae; spiracles usually each with 3–6 disc pores. Pygidium with median lobes (L_1) developed, far apart, rounded on outer margin; L_2 and L_3 represented by double teeth only; no gland spines on pygidial

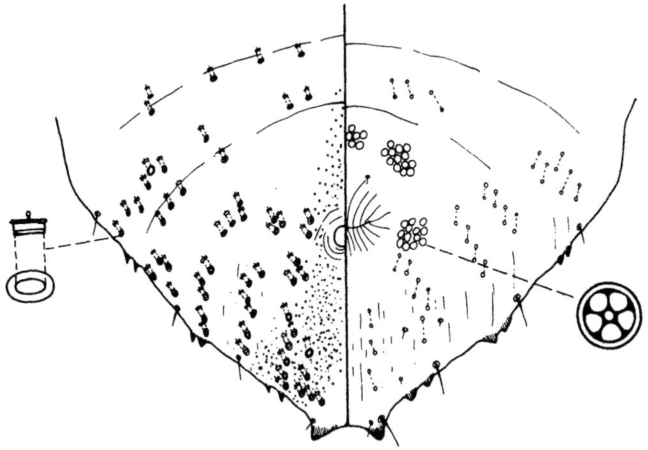

Fig. 141. *Mohelnaspis massiliensis* (Goux), female pygidium (original)

margin; many macroducts present on dorsum, marginal macroducts absent; microducts scattered on venter of pygidium; perivulvar pores: 5–11 (7–16) 6–12; 3 large setae present on each side in submarginal row of venter; anal opening and vulva located at about same distance from posterior end of pygidium, closer to anterior pygidial margin.

Distribution. Only known from Cz, Fr, Hu and Turkey.

Biology. It infests grass blades of *Alopecurus, Brachypodium, Festuca* and *Sesleria* (Zahradnik, 1962; Borchsenius, 1966). A thermophilous steppe-inhabiting species that overwinters as adult female (Schmutterer, 1959) and has one generation per year.

Natural enemy. Parasite: *Physcus* sp. (not *testaceus* Masi) was reported from S France by Parker (1960).

Note. The extreme reduction of pygidial lobes, absence of marginal macroducts and absence of pygidial plates or gland spines will distinguish this from other armored scales in CE, except *Leucaspis loewi,* which occurs on pines.

Genus **Nuculaspis** Ferris

Syn. & Lit.: *Nuculaspis* Ferris, 1938:241; Lupo, 1948:203; Kosztarab, 1963:39; Borchsenius, 1950:215, 1966:274; Danzig, 1964:651, 1980a:338.

Type-species: *Aspidiotus californicus* Coleman, 1903.

Test of female elongate oval, black or gray, 1.7–2.3 mm long; exuviae yellow, central. Adult female circular, at full maturity heavily sclerotized and with pygidium retracted into apex of body. Lobes 3 or 4 pairs; L_1 largest; plates small, fringed at apex; paraphyses absent; most dorsal and ventral ducts along pygidial margin; perivulvar pores usually in 4 groups; anus relatively large, oval.

Distribution. Holarctic, with 6 species; 1 in CE.

Biology. Feed on needles of conifers only.

218. **Nuculaspis abietis** (Schrank) – Conifer-needle Scale
Figs 142, 143

Syn. & Lit.: *Coccus abietis* Schrank, 1776:48. *C. pineti* Schrank, 1801:146. *Aspidiotus flavus* Hartig, 1839:642. *A. abietis* Comstock, 1883:57. *A. pini* Targioni-Tozzetti, 1884:383. *A. abietis* (Schrank), Leonardi, 1920:48; Thiem & Gerneck 1934b:130; Gomez-Menor, 1937:55. *Dynaspidiotus abietis* (Schrank), Balachowsky, 1948:336; Zahradnik, 1952c:152; Goidanich, 1956:207; Schmutterer, 1959:60; Rogojanu, 1962:299; Kosztarab & Kozár, 1978:157. *N. abietis*, Lupo, 1948:204; Borchsenius, 1966:274; Kuznetsov, 1970:1647; Zahradnik, 1972:433.

Test of female wide oval, 1.7–2.3 mm long, central part black, margins light gray, with yellow exuviae almost in center. Teneral adult female pear-shaped, yellowish; prosoma heavily sclerotized in old specimens; antenna small, with 1 bent seta. Pygidium triangular, rounded at apex; L_1 large and wide, only slightly notched on both sides; L_2 rounded at apex, notched on outer margin; L_3 small, triangular, plates as long as lobes, fimbriated at apex, their formula: 2, 2, 3–4, 3–5; dorsal ducts in rows of about equal lengths, forming dense submarginal band; ventral ducts few, along margin; perivulvar pore formula: 0–6 (4–11) 4–8; anus about as large as width of L_1, removed twice its diameter from apex of pygidium. Second instar described by Borchsenius (1939).

Distribution. A common species. In CE: Au, Bu, Cz, Fr, Ge, Hu, It, Po, Ro, Sw, USSR and Yu. Also known from W Europe and the Mediterranean area.

Biology. Occasional pest that infests needles of conifers in the genera *Abies, Juniperus, Picea, Pinus* and *Pseudotsuga.* A bisexual species with one yearly generation; N_2 overwinter in S Germany (Schmutterer, 1959) and become adults in April. Each female lays an average of 68 eggs in June and July which hatch usually in a few days, or occasionally in a few hours. Kuznetsov (1970) reported 11 days for eclosion.

Natural enemies. Parasites: *Aphytis abnormis, A. mytilaspidis, Aspidiotiphagus citrinus, Coccophagus similis, Prospaltella aspidioticola, P. aurantii, P. gigas, P. leucaspidis, Pteroptrix dimidiata* (Zahradnik, 1972; Kuznetsov, 1970; Trjapitzin, 1978). Predator: *Chilocorus bipustulatus* (Schmutterer, 1952b).

Genus **Parlatoria** Targioni-Tozzetti

Syn. & Lit.: *Parlatoria* Targioni-Tozzetti, 1868:735. *Apteronidia* Berlese, 1895:80. *Websteriella* Leonardi, 1899:208. *Euparlatoria* Leonardi, 1903:15. *Parlatorea* Lindinger, 1905:131. *Archangelskaia* Bodenheimer, 1951:331. *Parlatoria,* Leonardi, 1920:136; Ferris, 1937:171; McKenzie, 1945:47; Hall, 1946:527; Balachowsky, 1953:773; Schmutterer, 1959:131; Borchsenius, 1966:186.

Type-species: *Aspidiotus proteus* Curtis, 1843.

Test of female elongate oval, or broad oystershell-shaped, often with 2nd exuviae large, covering most of the test. Adult female somewhat broad-ovoid. Prosoma and prepygidial abdominal segments often with duct tubercles on submargin of venter; disc pores always associated with anterior spiracles. Pygidium normally with 3–4 pairs of unilobular lobes; plates usually shorter than lobes and wide, with apical margin fimbriated; dorsal macroducts short, two-barred, with a conspicuous sclerotized ring

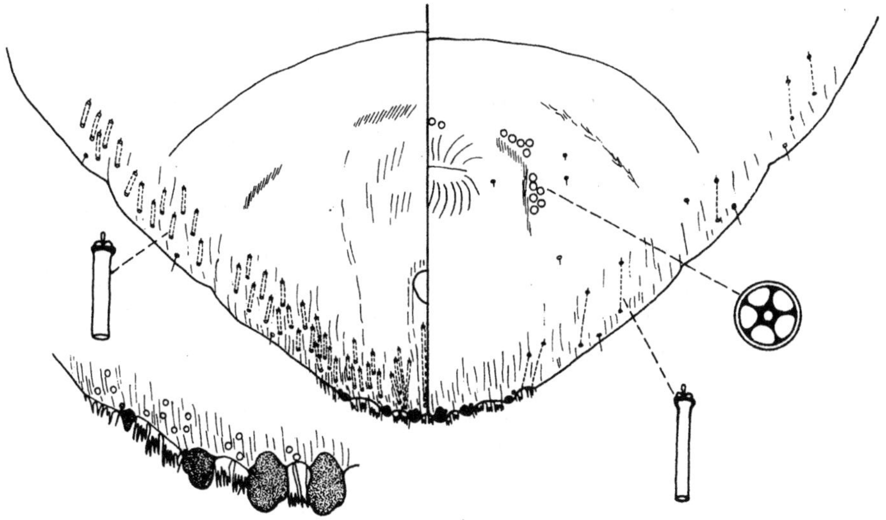

Fig. 142. *Nuculaspis abietis* (Schrank), female pygidium (original)

surrounding their orifice; perivulvar pores, when present, usually in 4 groups; vulva in middle of pygidium.

Distribution. World-wide, with about 50 species, most in SE Asia, 1 in CE.

Biology. Members of the genus are thermophilous and feed on a variety of host plants. A number of species are important pests in the tropics and subtropics.

Fig. 143. *Nuculaspis abietis* (Schrank), female and male tests on needles (after Kosztarab and Kozár, 1978)

219. **Parlatoria oleae** (Colvée) – Olive Parlatoria Scale
Fig. 32

Syn. & Lit.: *Diaspis oleae* Colvée, 1880:40. *P. calianthina* Berlese & Leonardi, 1896:346. *P. affinis* Newstead, 1897:97. *Syngenaspis oleae*, MacGillivray, 1921:248. *Parlatoria judaica* Bodenheimer, 1924:59. *P. pergandii* Bodenheimer, 1943:10. *P. morrisoni* Bodenheimer, 1944:82. *P. oleae*, Leonardi, 1920:137; Balachowsky & Mesnil, 1935:386; Ferris, 1937:176; McKenzie, 1945:69; Balachowsky, 1953:794; Ghauri, 1962:37; Borchsenius, 1966:193; Hafez et al., 1967:347; Bazarov & Smelev, 1971:152; Zahradnik, 1972:441.

Test of female wide oval, moderately convex, white or gray, ca 1.5–2.5 mm long, 1–1.6 mm wide. Adult female oval, about 1 mm long, 0.9 wide, widest at metathorax; antenna with 1 long seta; 2–8 disc pores associated with each anterior spiracle; 1 to 10 duct tubercles in 3–5 groups on each side of ventral submargin of prosoma. Pygidium with well developed 3 pairs of lobes, each once notched on outer margin; L_4 short; plates wide, somewhat shorter than lobes, with fringed apex, their number starting from middle: 2, 2, 3, 4, also laterally 3–4 groups, each with 3–6 plates; dorsal macroducts short, 1 in each interlobular space, 1–4 in area between perivulvar pores, smaller submedial macroducts on abdominal segments 4 and 5, others scattered; 4–7 ventral microducts on submargin of each segment; perivulvar pores usually in 5 groups: 0–8 (14–30) 10–22; anus at posterior 2/3, vulva at middle of pygidium.

Distribution. A very common species of Mediterranean origin. In CE: Bu, Fr, Hu(?), It, Ro, USSR, and Yu. Also S Europe and N Africa, Middle East, Oriental Region (none from SE Asia, Far East), S and N America and Australia (Borchsenius, 1966). One unconfirmed record from pear, originating from Hungary, was found in USA Plant Quarantine files of 1936. Considering the strict orchard inspections in Hungary, it is doubtful that this material originated from that country.

Biology. A mesophilous species with extensive host list (Borchsenius, 1966) that includes over 80 plant genera. An important pest of fruit trees and ornamental plants, which lives on trunk, branches, fruits, and leaves of *Amygdalus, Armeniaca, Buxus, Cerasus, Fraxinus, Mahonia, Malus, Olea, Persica, Prunus, Pyrus, Ribes, Rosa, Rubus, Sorbus, Syringa, Vitis,* etc. It produces red discoloration on apples and other fruits. Danzig (1972c) found 2–3 yearly generations in USSR. According to Bazarov and Smelev (1971) mated females overwinter; and in the 1st generation each female lays a maximum of 62 eggs. The 2nd generation females lay a maximum of 43 eggs. Hafez et al. (1967) found the highest level of infestation on the lower parts of pear trees in Egypt.

Natural enemies. Parasites: *Aphytis hispanicus, A. maculicornis, A. mytilaspidis, A. proclia, Aspidiotiphagus citrinus, Azotus chrysomphali, Coccophagoides similis, Habrolepis pascuorum, Hispaniella lauri, Pachyneuron coccorum, Prospaltella inquirenda* (Balachowsky, 1953; Bazarov and Smelev, 1971; Trjapitzin, 1978). Predator: *Chilocorus bipustulatus* (Bazarov and Smelev, 1971).

Genus **Pseudaulacaspis** MacGillivray

Syn. & Lit.: *Pseudaulacaspis* MacGillivray, 1921:305. *Sasakiaspis* Kuwana, 1926:7. *Pseudaulacaspis* Ferris, 1937:108; Balachowsky, 1954b:235; Schmutterer, 1959:200; Borchsenius, 1966:174; Kosztarab & Kozár, 1978:165; Danzig, 1980a:318.

Type-species: *Aulacaspis pentagona* Targioni-Tozzetti, 1885.

Test of female circular, convex, white; exuviae subcentral or marginal. Adult female broadly oval or pyriform, with strongly lobed margins, yellow or orange, 0.9–1.1 mm long on slide. Pygidium strongly sclerotized; L_1 wide, large, with several notches; L_2 bilobed; L_3 reduced to sclerotized points; gland spines large, often bifurcate or trifurcate at apex; all dorsal ducts large, about the same size, in definite rows; perivulvar pores in 5 large groups; anal opening at about the middle of pygidium.

Distribution. World-wide, most of the 9 species in S and E Asia, 1 in CE.

Biology. Members of this genus usually infest woody plants. Some species are important pests.

220. **Pseudaulacaspis pentagona** (Targioni-Tozzetti) – White Peach Scale, West Indian Peach Scale, Mulberry Scale

Figs 33, 144

Syn. & Lit.: *Diaspis pentagona* Targioni-Tozzetti, 1886:1; Balachowsky & Mesnil, 1935:390. *D. amygdali* Tryon, 1897:118. *D. lanatus* Cockerell, 1892:137. *D. patelliformis* Sasaki, 1894:107. *Aspidiotus vitiensis* Maskell, 1895:40. *Diaspis geranii* Maskell, 1897b:299. *D. amygdali* v. *rubra* Maskell, (in part) 1898:228. *Aulacaspis pentagona auranticolor* Carnes, 1907:160. *P. pentagona,* Ferris, 1937:220; Balachowsky, 1954b:236; Schmutterer, 1959:200; Ghauri 1962:168; Kosztarab, 1963:99; Borchsenius, 1966:175; Takagi & Kawai, 1967:29; Varga, 1971:424; Danzig, 1972:220c; Zahradnik, 1972:429; Kosztarab & Kozár, 1978:165; Kawai, 1980:276; Davidson et al., 1983:753. There are over 300 articles dealing with this species. We only listed here some of the more recent ones.

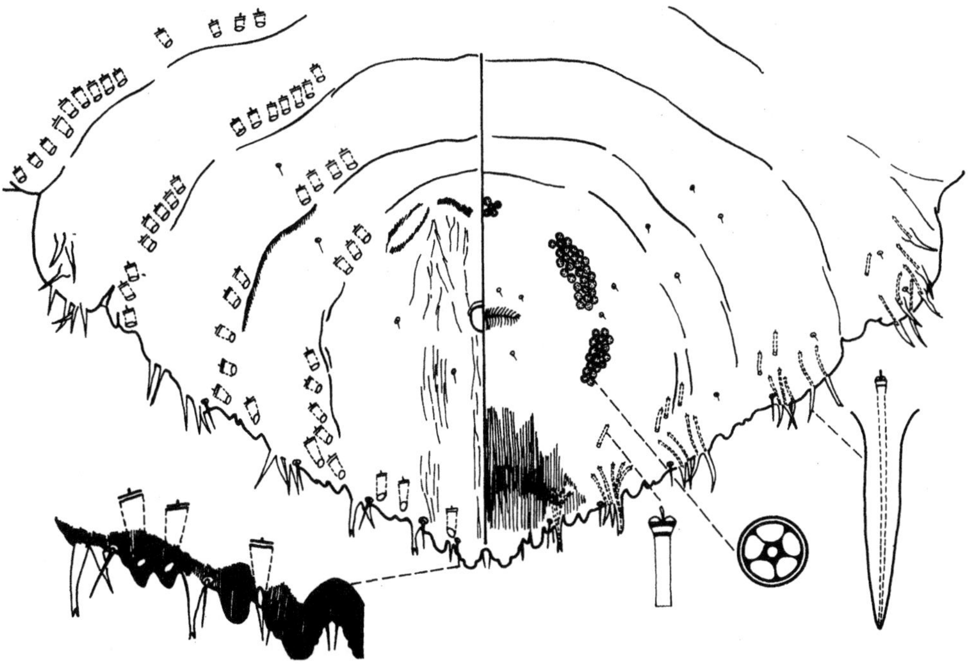

Fig. 144. *Pseudaulacaspis pentagona* (Targioni-Tozzetti), female pygidium (modified after Kosztarab, 1963)

Test of female circular or subcircular, white to yellowish white, ca 1.5–2.8 mm in diameter; exuviae marginal or submarginal. Adult female broadly oval or pyriform, with strongly lobed margins, orange-yellow, 0.9–1.1 mm long; 12–18 disc pores associated with each anterior spiracle, none at posterior spiracles; ducts absent from dorsum or venter of prosoma. Pygidium wide, strongly sclerotized at apex; L_1 large and protruding, notched on both sides; L_2 bilobed, outer lobule small, both rounded at apex; L_3 only moderately developed; usually some of the large gland spines forked at apex; dorsal ducts uniform in size, forming definite rows; perivulvar pores: 6–20 (24–45) 18–38; anal opening circular anterior to middle of pygidium. Adult male described by Lobdell (1937), Theron (1958) and Ghauri (1962); nymphs by Lupo (1957), Takagi and Kawai (1967). Bennett and Brown (1958) provided morphological characters for separating the sexes in the egg and first instar stages.

Distribution. Cosmopolitan. In CE: Bu, Fr, Ge, Hu, It, Sw, USSR and Yu. Also in almost all tropical and subtropical countries, and in the southern part of temperate zone. An introduced Oriental species that is spreading northward in Europe, and reached Budapest only recently.

Biology. A polyphagous species with an extensive host list. For a complete listing see Borchsenius (1966). In CE it is a common pest on *Morus* spp. and on stonefruit trees, especially peach and plum. In Hungary 2 generations per year occur and elsewhere there are 2–4 generations. Adult females overwinter, and each female lays about 140 eggs (Danzig, 1972c). Varga (1971) noted egg laying (80–100 per female) from mid-May in Hungary, adult males appeared from beginning July to mid-August, and again from mid-September to end of October. Murakami (1970) summarized the biology in Japan where it has 3 yearly generations. Males form large obvious colonies. Its biology was often confused with that of *P. prunicola* (Maskell) (Kawai, 1980; Davidson et al., 1983).

Natural enemies. Parasites: *Ablerus clisiocampae, A. perspeciosus, Adelencyrtus aulacaspidis, Aphytis diaspidis, A. fuscipennis, Apterencyrtus microphagus, Archenomus orientalis, Arrhenophagus chionaspidis, Aspidiotiphagus citrinus, A. lounsburyi, Azotus capensis, Cales noaki, Coccophagus kuwanae, Dimacrocerus platensis, Marietta mexicana, Prospaltella aurantii, P. berlesei, P. diaspidicola, P. murtfeldti, P. niigatae, Psilomirinus flabicaudus, Pteroptrix dimidiata, Tetrastichus canadensis, Thomsonisca amathus, Thysanus aspidioti, T. carideі, T. platensis* (Balachowsky, 1954b; Trjapitzin, 1978). Predators: *Callicaria superba, Chilocorus bipustulatus, C. kuwanae, C. rubidus, Chrysopa boninensis, Cybocephalus gibbulus, C. rufifrons, Eilema griseola aegrota, Exochomus quadripustulatus, Hyperaspis japonica, Lindorus lophanthe, Lithosia quadradives, Milthochrista calamina, Paramysia oblongoguttata, Pseudoscymnus hareja, Rodolia concolor, Stethorus japonicus, Sticholotis punctata Stigmatophora flava, Telsimia nigra.* Entomophagous fungi: *Nectria coccophila, Sphaerostilbe coccophila* (Murakami, 1970; Varga, 1971; Danzig, 1972c; Zahradnik, 1972). Among the parasites, *P. berlesei* was the most effective biological control agent in S Europe and reduced the scale population to below economic level. Danzig (1972c) reported the effective use of *L. lophanthe*.

Note. Apparently there are two forms of this species. In warmer climates the gland spines normally are trifurcate while in the temperate zone they are usually bifurcate or simple (Davidson et al., 1983).

Genus **Quadraspidiotus** MacGillivray

Syn. & Lit.: *Quadraspidiotus* MacGillivray, 1921:388. *Forbesaspis* MacGillivray, 1921:388. *Comstockaspis* MacGillivray, 1921:391. *Affirmaspis* MacGillivray, 1921:393. *Euraspidiotus* Thiem & Gerneck, 1934b:231. *Quadraspidiotus*, Ferris, 1938:251; Lupo, 1948:174; Balachowsky, 1950:397; Zahradnik, 1952c:113; Schmutterer, 1959:76; Gomez-Menor, 1959:151; Kosztarab, 1963:41; DeLotto, 1963:144; Borchsenius, 1966:327; Bazarov & Smelev, 1971:211; Kosztarab & Kozár, 1978:166.

Type-species: *Aspidiotus ostreaeformis* Curtis, 1843.

Test of female circular, or slightly oval, moderately convex, usually gray; exuviae central or subcentral. Adult female pear-shaped or circular with large, broad pygidium. Pygidium with 2 pairs of lobes developed, usually close together, L_2 smaller than L_1 and usually notched on outer margin; L_3 sometimes represented by a small prominence; usually 5 pairs of intersegmental scleroses developed; plates usually small and simple, occasionally fimbriate at apex; dorsal ducts small, slender; ventral ducts few, small; perivulvar pores present or absent; anus relatively small, its diameter less than length of L_1.

Distribution. World-wide with ca 45 species; 9 in CE.

Biology. Members of this genus feed on woody and herbaceous plants. Some species are important pests of cultivated trees.

Key to species of *Quadraspidiotus*

1. Perivulvar pores absent 2
   Perivulvar pores present 3
2. L_1 asymmetrical, notched on outer margin; 7 groups of plates on pygidial margin; less than 10 dorsal ducts on each side of segment V; overwinter as first instars; polyphagous (fig. 150) .... *perniciosus*, p. 375
   L_1 symmetrical, notched on both margins; 5 groups of plates on pygidial margin; more than 10 dorsal ducts on each side of segment V; overwinter as adult females; feed on Leguminosae .... *sulci*, p. 378
3. L_3 entirely absent 4
   L_3 present, but sometimes reduced in size 5
4. Plates between median lobes as long as lobes, other plates also well developed; about 8 dorsal ducts on each side of segment IV; overwinter as second instars; usually on herbaceous plants .... *labiatarum*, p. 370
   Plates between median lobes much shorter than lobes, all other plates greatly reduced or absent; no dorsal ducts on segment IV; overwinter as adult females; on deciduous trees; (some specimens from Morocco without perivulvar pores) .... *lenticularis*, p. 371
5. Fourth abdominal segment with dorsal submarginal ducts 6
   Fourth abdominal segment without dorsal submarginal ducts 7
6. Prosoma heavily sclerotized; 4th abdominal segment with 8–24 dorsal submarginal ducts on each side; ventral microducts form evident groups on prosomal margin; 5 or more duct orifices laterad of L_1; prefers *Salix*, *Populus* (fig. 146) .... *gigas*, p. 368

Prosoma membranous; 4th abdominal segment with 3–7 dorsal submarginal ducts on each side; ventral microducts do not form groups on prosomal margin; less than 5 duct orifices laterad of L_1; polyphagous (fig. 149) ... . *ostreaeformis*, p. 372

7. Dorsal ducts present on margin of prepygidial segments; less than 70 total dorsal ducts on pygidium, in 6 rows or groups; L_1 wider than long; duct at base of median notch extends anterior to anal opening .. *zonatus*, p. 379

Without dorsal ducts on margin of prepygidial segments; more than 70 total dorsal ducts on pygidium, in more than 6 rows or groups; L_1 not wider than long; duct at base of median notch extends only to anal opening 8

8. Plates between median lobes shorter than lobes; mesal plates between lobes pointed; 80–90 total dorsal ducts on pygidium; L_1 not or only shallowly notched on inner margin; overwinter as adult females ... *marani*, p. 372

Plates between median lobes as long as lobes; mesal plates between lobes fimbriated at apex; 100–120 total dorsal ducts on pygidium; L_1 well notched on both margins; overwinter as second instars ....*pyri*, p. 377

221. **Quadraspidiotus gigas** (Thiem & Gerneck) – Poplar Armored Scale
Figs 145, 146

Syn. & Lit.: *Aspidiotus gigas* Thiem & Gerneck, 1934:131. *A. multiglandulatus* Borchsenius, 1935:132. *Q. ostreaeformis*, Ferris, 1938:256 (a misidentification). *Q. gigas*, Balachowsky, 1950:465; Zahradnik, 1952c:137, 1972:434; Schmutterer, *1952:571, 1959:97; Gomez-Menor, 1960:166; Săvescu, 1961:21; Kosztarab, 1963:44; Lellakova-Duskova, *1963:611, 1966:297; Borchsenius, 1966:330; Komosinska, 1968:51; Bazarov & Smelev, 1971:213; Kosztarab & Kozár, 1978:172. *Diaspidiotus gigas*, Danzig, 1980a:340.

Test of female circular, moderately convex, gray, diameter 1.7–2.8 mm; exuviae central or subcentral orange-yellow. Adult female pear-shaped, prosoma strongly sclerotized, with marked constriction between meso- and metathorax, prepygidial segments distinctly indicated; live light yellow; 1.3–1.6 mm long. Pygidium wide, acute at apex; L_1 wide, notched on lateral margin, L_2 much smaller; L_3 small, tooth-like; plates between median lobes small and pointed, 2 plates laterad of L_1 pointed or slightly fimbriated; 3 plates laterad of L_2 of variable shape and size; 2 pairs of intersegmental scleroses well developed, each median one encloses 4–5 duct orifices, each lateral one 3; dorsal ducts in several rows or groups, 8 to 20 ducts on each side of abdominal segment IV, 1 duct at base of median notch extends to anus; numerous ventral ducts in a submarginal band; both dorsal and ventral ducts extend into submargin of prepygidial segments; perivulvar pores: 4–11 (10–23) 9–18; anus circular, removed about twice its diameter from base of L_1. Adult male described by Bustshik (1958); males and all immature stages described by Lellakova-Duskova (1963).

Distribution. A common Trans-Palearctic species. In CE: Bu, Cz, Fr, Ge, Hu, It, Po, Ro, Sw, USSR, and Yu. Also in other parts of Europe, Middle East, Algeria, Canada, and USA.

Biology. A mesophilous species which lives on trunk and branches of *Populus* and *Salix* species, and also known from *Tilia* sp. An occasional pest that causes deformations

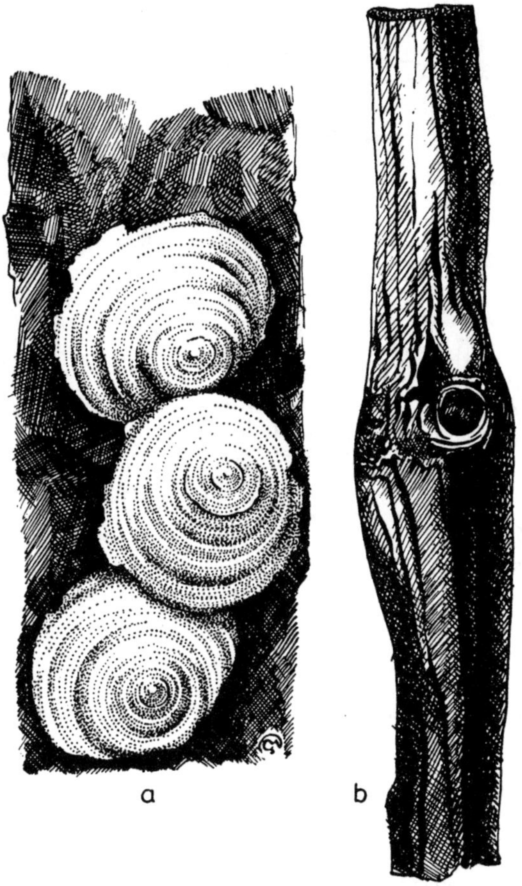

Fig. 145. *Quadraspidiotus gigas* (Thiem & Gerneck), (a) female tests on bark, (b) deformation on the twig (after Kosztarab and Kozár, 1978)

and dying of branches. It has 1 yearly generation and overwinters usually in N_2 stage, rarely as N_1. According to Schmutterer (1959) and Lellakova-Duskova (1963) adults develop by end of April. Egg laying starts in June and continues through September; each female lays 86–215 eggs that hatch the next day after being laid. Males are common.

Natural enemies. Parasites: *Aphytis mytilaspidis, Archenomus caucasicus, A. longicornis, A. maritimus, Aspidiotiphagus citrinus, Azotus atomon, A. celsus, A. matritensis, Coccophagoides similis, Comperiella bifasciata, Hispaniella lauri, Metaphycus duplus, Prospaltella gigas, Pteroptrix dimidiata, P. maritima, Trichomasthus dissimilis* (Kosztarab, 1956b; Zahradnik, 1972; Trjapitzin, 1978). Predators: *Chilocorus bipustulatus, Coccinella bipunctata, Ectemnus nigriceps, Loricula elegantula, L. pselaphiformis, Temnostethus longirostris* (Zahradnik, 1972).

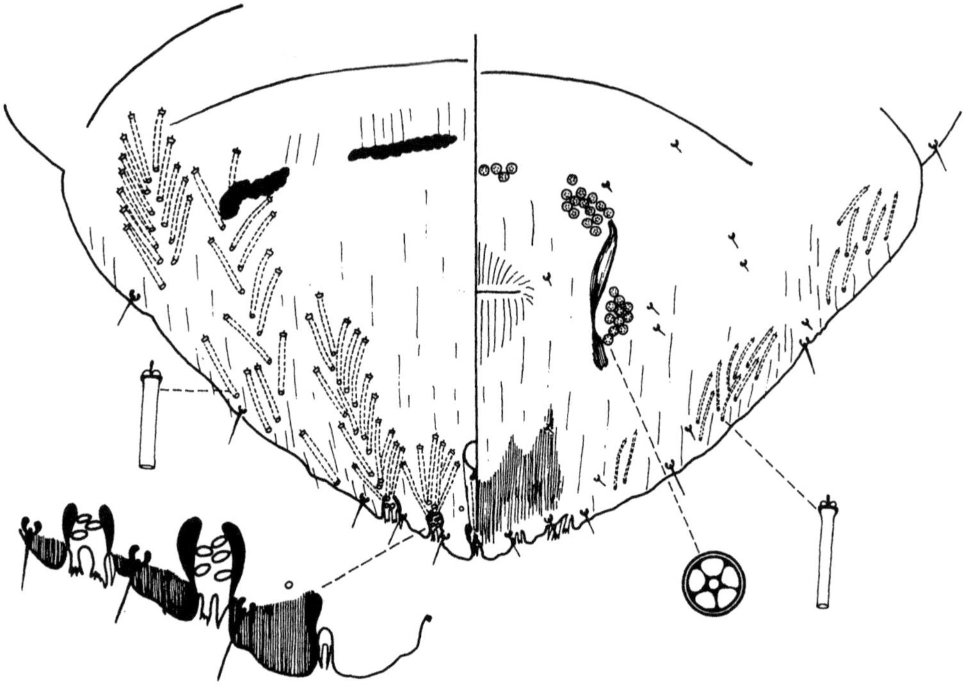

Fig. 146. *Quadraspidiotus gigas* (Thiem & Gerneck), female pygidium (modified after Kosztarab, 1963)

222. **Quadraspidiotus labiatarum** (Marchal) – Marchal's Armored Scale
Fig. 147

Syn. & Lit.: *Aspidiotus labiatarum* Marchal, 1909:872. *A. privignus* Lindinger, 1909:151. *A. thymi* Šulc, 1934:19. *A. baudysi* Šulc, 1934:20. *A. viticola* Leonardi, 1913:61. *A. labiatarum*, Leonardi, 1920:59; Koronéos, 1934:10; Thiem & Gerneck, 1934a:535, 1934b:130. *Rhizaspidiotus labiatarum*, Lupo, 1948:199. *Q. labiatarum*, Balachowsky, 1950:478; Zahradnik, 1951c:139; Gomez-Menor, 1957:46; Schmutterer, 1959:95; Borchsenius, 1966:332; Kosztarab & Kozár, 1978:171.

Test of female oval, convex, white or yellowish white, rarely grayish, 1.4–2.1 mm long; exuviae subcentral. Adult female circular or oval, yellow, antenna with 1 seta. Pygidium with L_1 protruding, close together, narrowed at base, deeply notched on outer margin; L_2 small, usually triangular, rarely parallel-sided; L_3 absent or much reduced; plates usually wide and fimbriate, formula 2,2,2–3,3; plates between median lobes longer or as long as L_1 and fimbriated at apex; paired intersegmental scleroses developed; dorsal ducts form rows on pygidium, a few ducts on each side of segment IV, and on margin of prepygidial segments; a duct at base of median notch extends to anus; ventral ducts small and few; perivulvar pores in small groups: 1–6 (2–5) 1–3; anus removed twice its diameter from base of L_1.

Distribution. A rare Mediterranean xerophilous species. In CE: Cz, Fr, Ge, Hu, It, Sw, and Yu. Also Greece, Spain, Corsica Isl., Algeria and Morocco. It was collected at 1,700 m elevation in N Africa (Balachowsky, 1950).

Biology. On stems and leaves of plants in the genera: *Cytisus, Euphorbia, Fumana, Genista, Globularia, Helianthemum, Hypericum, Prasium, Rhus, Salicornia, Salvia, Satureia, Sedum, Spergula, Stachys, Teucrium, Thymelea* and *Thymus*. According to Schmutterer (1959) it overwinters as N_2 and probably is biparental.

Natural enemy. Parasite: *Metaphycus hubai* (Trjapitzin, 1978).

223. **Quadraspidiotus lenticularis** (Lindinger) – Lindinger's Lenticular Scale

Syn. & Lit.: *Aspidiotus lenticularis* Lindinger, 1912:149; Leonardi, 1920:61; Koronéos, 1934:13. *Q. lenticularis* Ferris, 1941:45, 1943:96; Lupo, 1948:189; Bachmann & Geier, 1950:118; Balachowsky, 1950:433; Schmutterer, 1959:92; Gomez-Menor, 1960:163; Borchsenius, 1966:332; Zahradnik, 1972:435; Kosztarab & Kozár, 1978:171; Kozár et al., 1979:132.

Test of female circular, gray or dark gray, diameter 2–2.4 mm; exuviae central or subcentral. Adult female pear-shaped. Pygidium triangular; L_1 elongate, notched on outer margin; L_2 similar but much smaller; L_3 absent; plates very small, not fringed at apex, often absent; intersegmental scleroses only between L_1 and L_2 well developed; dorsal ducts form 4 elongate rows on pygidium, none on segment IV, 1 duct at base of median notch extends well beyond anus; ventral ducts small and few; perivulvar pores

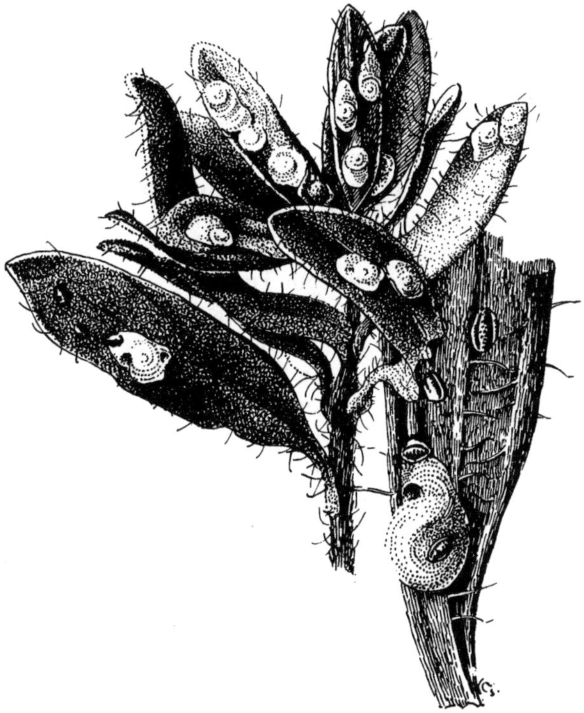

Fig. 147. *Quadraspidiotus labiatarum* (Marchal), female, male and nymphal tests on leaves of *Thymus* (after Kosztarab and Kozár, 1978)

usually in 4–5 loose groups but sometimes absent (in Morocco): 0–2 (0–6) 0–5; anus removed 2 or 3 times its diameter from base of L_1.

Distribution. A Mediterranean species infrequently found in CE. In CE: Bu, Fr, Hu, It, Sw, USSR, and Yu. Also Denmark, Greece, Spain, Canary Isl., Morocco and Australia.

Biology. A mesophilous species infesting bark of trunk and branches of *Betula, Euphorbia, Ficus, Fraxinus, Olea, Pistacia, Populus, Prunus, Quercus* and *Rhamnus.* Little is known about the biology. It probably reproduces bisexually. According to Zahradnik (1972) adult females overwinter.

Natural enemy. Parasite: *Prospaltella fasciata* (Trjapitzin, 1978).

224. **Quadraspidiotus marani** Zahradnik – Zahradnik's Pear Scale

Syn. & Lit.: *Quadraspidiotus marani* Zahradnik, 1952a:449. *Q. schneideri* Bachmann, 1952:357, 1953:357. *Q. marani* f. *olivicola* Pegazzano, 1955:314. *Q. marani,* Duskova, 1952:452, 1953b:229; Zahradnik, 1955a:125, 1955b:88, 1972:435; Schmutterer, 1952b:570, 1959:88; Müller & Eisenschmidt, 1954:151; Săvescu, 1960:119; Lellakova-Duskova, 1965:202; Tsalev & Valcheva, 1965:22; Borchsenius, 1966:333; Kosztarab & Kozár, 1978:170; Kozár et al., 1979:535.

Test of female circular, moderately convex, dark gray, diameter 1.8–2.1 mm; exuviae central or subcentral. Adult female pear-shaped, dark yellow, ca 1.5 mm long; antenna with 1 seta. Pygidium with L_1 usually notched on both sides; L_2 smaller, shallowly notched on outer margin; L_3 with rounded apex, without notch; plates between median lobes pointed and shorter than L_1, other plates wider, most fimbriated at apex, but plate laterad of L_1 daggerlike, never fimbriated; 80–90 total dorsal ducts, 1 between median lobes extends to anus, none on abdominal segment IV; 50–60 small ventral ducts along margin of pygidium; perivulvar pores in 4–5 groups; anus circular, removed 3–4 times its diameter from bases of L_1. Adult male described by Lellakova-Duskova (1965), N_2 by Bachmann (1953).

Distribution. A common species in CE reported from: Au, Bu, Cz, Fr, Ge, Hu, It, Po, Ro, Sw, USSR, and Yu. Also known from Turkey.

Biology. The species infests trunk and branches of *Crataegus, Fraxinus, Malus, Prunus, Pyrus* and *Sorbus*; often a pest in orchards. One yearly generation. Adult females overwinter, each lays ca 115 eggs in June–July (Schmutterer, 1959), that hatch in 1–3 days. Adult males appear in September.

Natural enemies. Parasites: *Aphytis mytilaspidis, A. proclia, Archenomus bicolor, Cheiloneurus microphagus, Metaphycus notatus.* Predators: *Chilocorus bipustulatus, Cybocephalus politus* (Zahradnik, 1972).

225. **Quadraspidiotus ostreaeformis** (Curtis) – Yellow Plum Scale, European Fruit Scale

Figs 148, 149

Syn. & Lit.: *Aspidiotus ostreaeformis* Curtis, 1843b:805. *A. betulae* Bärensprung, 1849:165. *A. tiliae* Bouché, 1851:111. *A. hippocastani* Signoret, 1869:136. *A. oxyacanthae* Signoret, 1869:137. *A. scutiformis* Göthe, 1899:18. *A. ostreaeformis,* Newstead 1901:99; Thiem & Gerneck, 1934a:532. 1934b:142;

Fig. 148. *Quadraspidiotus ostreaeformis* (Curtis), female and male tests on tree bark (after Kosztarab and Kozár, 1978)

Balachowsky & Mesnil, 1935:361. *Q. ostreaeformis,* Ferris, 1938:256 (a misidentification of *Q. gigas*); Rogojanu, 1939:1; Lupo, 1948:175; Balachowsky, 1950:405; Zahradnik, 1952c:115, 1972:435; Gomez-Menor, 1956:21; Schmutterer, 1959:77; Kosztarab, 1963:48; Borchsenius, 1966:334; Bazarov & Smelev, 1971:211; Kosztarab & Kozár, 1978:174. *Diaspidiotus hunteri* (Newell), Ferris, 1938:182 (misidentification). *Diaspidiotus ostreaeformis,* Danzig, 1980a:341.

Test of female circular, convex, dark gray, diameter 1.4–1.9 mm; exuviae subcentral, rarely central, orange-yellow. Adult female pear-shaped, 1.2–1.5 mm long, live lemon yellow. Pygidium with 3 pairs of lobes; L_1 not or only slightly notched on lateral margin, rounded apically; L_2 much smaller, usually notched on outer margin; L_3 reduced to a triangular sclerotized point; plates between median lobes needle-like, half as long as L_1, 1st plate laterad of L_1 daggerlike, 2nd wide and fimbriated at apex; 2 plates laterad of L_2 usually wide and fimbriate; 5 sets of paired intersegmental scleroses well developed; dorsal macroducts form groups or rows, extend as far anteriorly as abdominal segment II; ventral ducts form a band along margin; perivulvar pores in 5 groups: 2–8 (5–14) 5–17; anus circular, removed about 3 times its diameter from base of L_1. Adult male described by Bustshik (1958) and Ghauri (1962). Nymphs described by Borchsenius (1939), Boratynski (1953) and MacFarlane (1961).

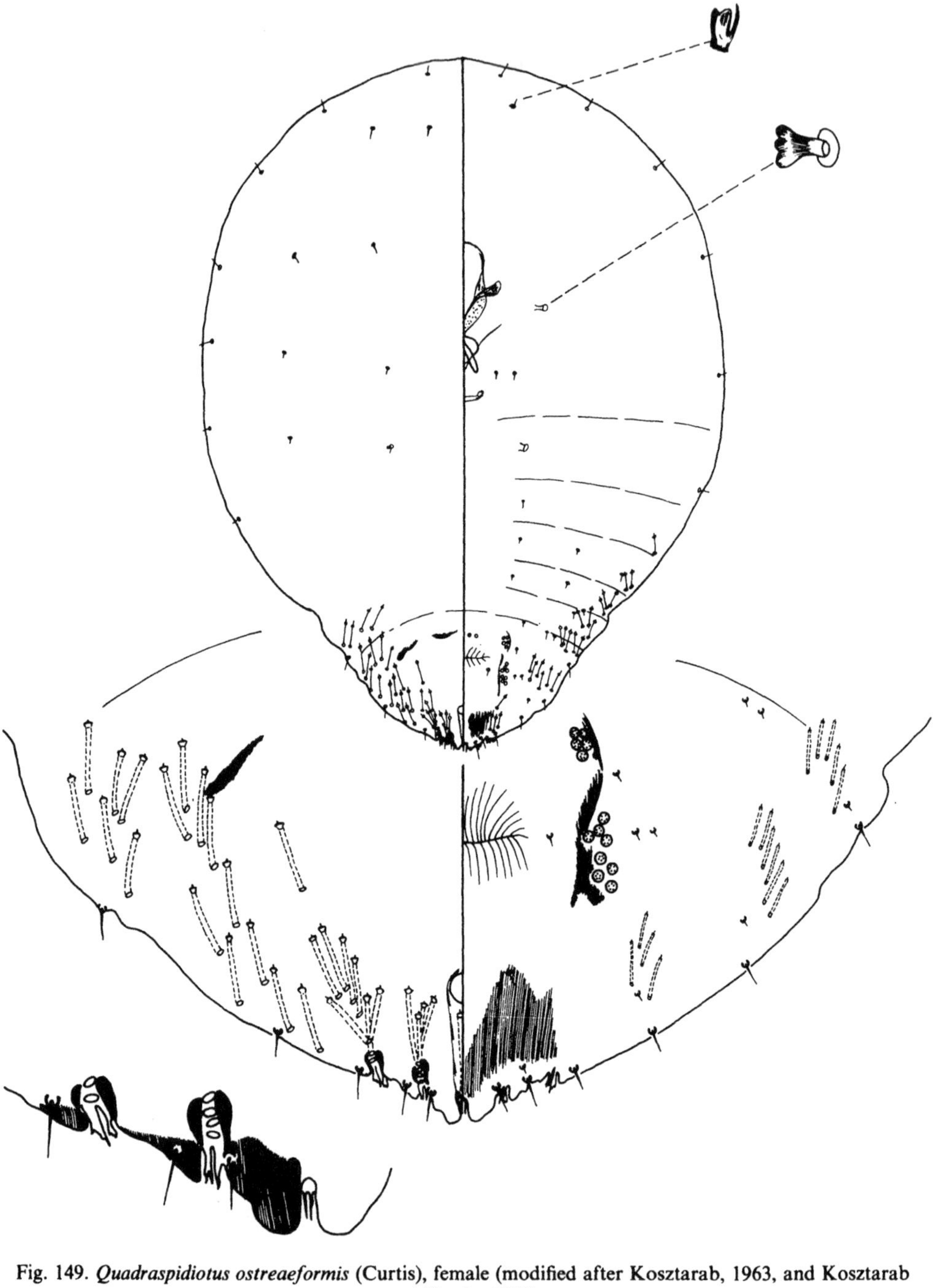

Fig. 149. *Quadraspidiotus ostreaeformis* (Curtis), female (modified after Kosztarab, 1963, and Kosztarab and Kozár, 1978)

Distribution. A very common species with almost world-wide distribution. In CE: Au, Bu, Cz, Fr, Ge, Hu, It, Po, Ro, Sw, USSR and Yu. Also known from other parts of Europe, N Africa, Middle East, Far East, S and N America, New Zealand, and Australia.

Biology. Host genera: *Abies, Acer, Aesculus, Alnus, Amygdalus, Betula, Calluna, Caragana, Carpinus, Catalpa, Cerasus, Chaenomeles, Cornus, Corylus, Cotoneaster, Crataegus, Cydonia, Cytisus, Fagus, Ficus, Fraxinus, Gleditsia, Juglans, Ledum, Malus, Morus, Myrica, Olea, Ostrya, Pavia, Persica, Phoenix, Platanus, Populus, Prunus, Pyrus, Quercus, Rhamnus, Ribes, Rosa, Salix, Sorbaria, Sorbus, Spiraea, Syringa, Tilia, Ulmus* and *Vaccinium.*

It has 1 yearly generation in CE and overwinters as N_2. The two adults appear at end of April in Hungary, and lay 72–203 eggs per female (Schmutterer, 1952). N_1 stage lasts 45–80 days. It often reaches high population density, and becomes a pest on fruit trees, especially on plum in CE.

Natural enemies. Parasites: *Anabrolepis zetterstedti, Anagyrus schmuttereri, A. schoenherri, Aphytis bovelli, A. mytilaspidis, A. testaceus, Archenomus bicolor, A. caucasicus, A. longicornis, A. maritimus, Azotus atomon, A. celsus, A. marchali, A. matritensis, A. pinifoliae, A. pulcherrimus, Cheiloneurus microphagus, Coccophagoides similis, Hispaniella lauri, Prospaltella aurantii, P. gigas, Pteroptrix dimidiata* and *Thysanus ater* (Schmutterer, 1959; Zahradnik, 1972; Trjapitzin, 1978). Predators: *Chilocorus renipustulatus, Cybocephalus politus* and *Hemisarcoptes malus.*

226. **Quadraspidiotus perniciosus** (Comstock) – San José Scale
Figs 34, 150

Syn. & Lit.: *Aspidiotus perniciosus* Comstock, 1881:304. *Aonidia fusca* Maskell, 1895:43. *Aspidiotus andromelas* Cockerell, 1897:5. *Aonidiella perniciosa* Balachowsky & Mesnil, 1935:369. *A. perniciosus*, Thiem & Gerneck, 1934b:130; Melis, *1943:1, 1951:1. *Q. perniciosus*, *Kiritchenko, 1937b:1; Ferris, 1938:258; Balachowsky, 1950:424; Kawecki, 1950:1, 1985:64; Zahradnik, 1952c:128, 1972:436; Gomez-Menor, 1956:24; Schmutterer, et al. 1957:485; Schmutterer, 1959:80; Săvescu, 1961:123; Kosztarab, 1963:50; Borchsenius, 1966:336; Bazarov & Smelev, 1971:220; Stoetzel & Davidson, 1975:492; Kosztarab & Kozár, 1978:169. *Diaspidiotus perniciosus*, Danzig, 1980a:342. There are hundreds of articles that deal with the classification and/or biology of this species.

Test of female almost circular, slightly convex, light to dark gray, diameter ca 1.5–2.2 mm; exuviae central or subcentral; the 1st instar exuviae gives a craterlike appearance to test. Young adult female pear-shaped, yellow, egg laying females almost circular, reddish-yellow, 0.8–1.2 mm long. Pygidium with 2 pairs of well-developed lobes; L_1 prominent, notched on outer margin, L_2 often notched more than once; L_3 scarcely indicated by a small prominence; plates between median lobes pointed, as long as lobes; 1st plate laterad of L_1 usually pointed, 2nd plate fimbriated; 3 plates laterad of L_2 wide and well fimbriated at apex; 3 short and broad, variously fringed plates beyond L_3, each bearing a ventral microduct; 4 pairs of intersegmental scleroses; dorsal ducts few, slender, 1 between bases of median lobes extends beyond anus; ventral ducts very small, along margin of pygidium; perivulvar pores absent; anus circular, removed ca 2

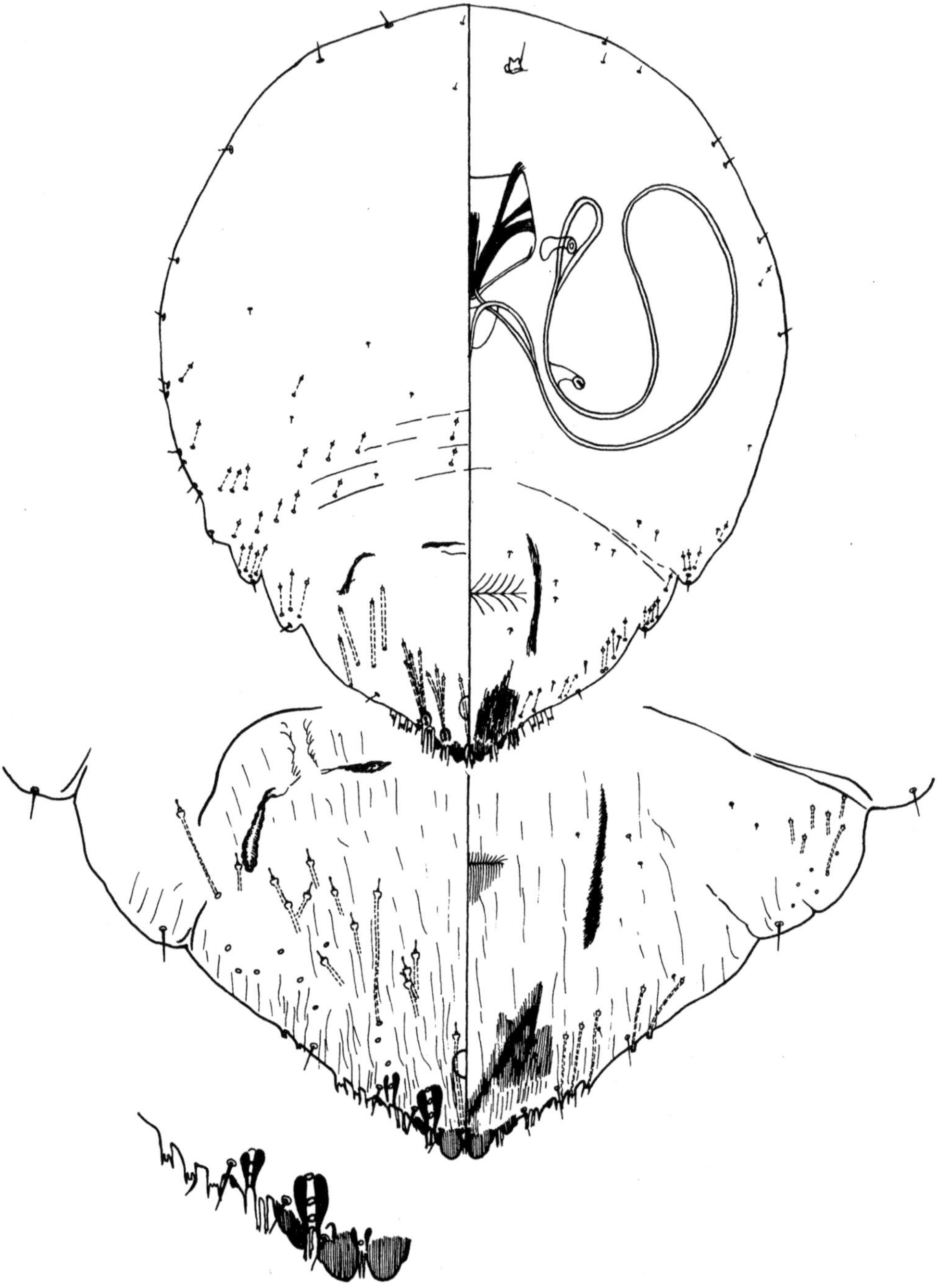

Fig. 150. *Quadraspidiotus perniciosus* (Comstock), female (modified after Kosztarab, 1963, and Kosztarab and Kozár, 1978)

or 3 times its diameter from bases of L_1. Adult males described by Melis (1943), Jancke (1955) and Bustshik (1958), etc.; nymphs described by Bonnemaison (1936), Borchsenius (1939), Melis (1943), and others.

Distribution. A very common cosmopolitan species of Far East origin. In CE: Au, Bu, Cz, Fr, Ge, Hu, It, Po (presumably no longer exist), Ro, Sw, USSR and Yu. Also on all continents.

Biology. Borchsenius (1966) listed 239 host plant genera belonging to many families. Most host plants belong in the Rosaceae. It is considered an important fruit and ornamental tree pest in different parts of the world. It is a primary pest on apple, pear, plum, peach, red currant, gooseberry, rose, hawthorn, and others. In case of heavy infestation the trees die. Red discoloration is produced on fruits. There are as many as 5 generations per year depending on the geographic area. In CE mostly 2 generations are known. This species overwinters in N_1 stage under a black test. Males of the 1st generation are active for about 10 days at the beginning of May in Hungary; and 2nd generation crawlers are active from beginning of August. It was introduced into CE in the 1920's where it became established in a short time and spread into most of Europe. Control with chemicals is possible, but the chemicals are expensive, and several treatments are required. Therefore, natural enemies are gradually becoming more important in control programs. Population dynamics and forecasting methods of this species were intensively studied by Kozár (1975, 1979).

Natural enemies. Parasites: *Anabrolepis zetterstedti, Aphytis mytilaspidis, A. proclia, A. testaceus, Archenomus bicolor, A. longicornis, A. maritimus, Aspidiotiphagus citrinus, Hispaniella lauri, Prospaltella fasciata, P. perniciosi, Pteroptrix dimidiata* (Trjapitzin, 1978). These were the common parasites in CE. Many additional parasites are listed, some possibly erroneously, in the world literature. Predators: *Chilocorus bijugus, C. bipustulatus, C. bivulnerus, C. quadripustulatus, C. renipustulatus, Coccinella bipunctata, Cybocephalus politus, Deraeocoris ruber, Exochomus quadrimaculatus, Hemisarcoptes malus, Microweisea miscella, Scapheremaeus marginalis* and *Scymnillus aterrimus* (Kosztarab, 1963; Zahradnik, 1972). Bichina and Kischina (1976) noted in Moldavia the predation by *Cepea vindobonensis, Chrysopa carnea, C. prasina,* and *C. septempunctata,* and by ants: *Formica pratensis, F. rufa,* and *Tetramorium caespitum.* Folkina (1978) observed the effective predator, *Crematogaster subdentata.* Entomophagous fungi: *Sphaerostilbe* sp. and *Fusarium* sp. (Kosztarab, 1963; Kosztarab and Kozár, 1978); *Alternaria* sp. and *Coniothyrium* sp. (Smolyannikov, 1964).

227. **Quadraspidiotus pyri** (Lichtenstein) – Yellow Pear Scale

Syn. & Lit.: *Aspidiotus pyri* Lichtenstein, 1881:51. *A. patavinus* Berlese, 1896:350. *A. sinaiticus.* Bodenheimer, Bodenheimer & Theodor, 1929:106. *Aspidiotus pyri,* Leonardi, 1920:57; Thiem & Gerneck, 1934b:130; Szelényi, 1936:159. *Diaspidiotus spurcatus* Borchsenius, 1949:242. *Q. pyri (=piri),* Lupo, 1948:185, Balachowsky, 1950:409; Zahradnik, 1951:119, 1972:438; Duskova, 1952:453, *1953a:8; Schmutterer, *1952b:570 (biol.), 1959:85; Bachmann, *1953a:357; Borchsenius, 1966:339; Kosztarab & Kozár, 1978:170.

Test of female circular, moderately convex, dark brown, diameter 1.8–2.1 mm; exuviae central or subcentral. Adult female pear-shaped, yellow when live. Pygidium wide; L_1 notched on both margins; L_2 smaller, notched on outer margin only; L_3 smallest; plates between median lobes as long as L_1 and fimbriated at apex, other plates also fimbriated; 100–120 total dorsal ducts, 1 at base of median notch extends to anus; intersegmental scleroses present; ventral ducts few, in submarginal area and prepygidial segments; perivulvar pores in 4–5 groups: 0–4 (8–14) 7–12; anus removed twice its diameter from median notch. Adult male described by Duskova (1953a) and Bustshik (1958). Nymphs (N_1 and N_2) described by Duskova (1953a).

Distribution. A common species, probably of Mediterranean origin. In CE: Bu, Cz, Fr, Ge, Hu, It, Po, Ro, Sw, USSR, and Yu. Also Spain, Belgium, Holland, Malta, Middle East, N Africa and Australia.

Biology. A mesophilous species infesting trunk and branches of: *Aesculus, Amygdalus, Betula, Carpinus, Cerasus, Crataegus, Ficus, Fraxinus, Juglans, Malus, Mespilus, Olea, Platanus, Populus, Prunus, Pyrus, Salix, Sorbus, Spiraea,* and *Tilia.* It is a pest on fruit trees. There is one yearly generation. Overwinters as N_2 and develops into adult by end of May. According to Schmutterer (1952b) egg laying in S Germany is during June and July; each female lays 41–165 eggs that hatch in a few days. Reproduction is biparental or parthenogenetic (Bachmann, 1953a).

Natural enemies. Parasites: *Aphytis aonidiae, A. mytilaspidis, A. proclia, Archenomus bicolor, A. longicornis, Aspidiotiphagus citrinus, Azotus atomon, A. marchali, Habrolepis dalmani, Hispaniella lauri,* and *Pteroptrix dimidiata.* The listed parasites are important in reducing the scale populations in CE. (Zahradnik 1972; Trjapitzin, 1978). Predators: *Chilocorus bipustulatus, C. renipustulatus, Cybocephalus politus* (Schmutterer, 1952b; Zahradnik, 1972).

228. **Quadraspidiotus sulci** Balachowsky – Šulc Armored Scale

Syn. & Lit.: *Quadraspidiotus sulci* Balachowsky, 1950:52. *Gonaspidiotus sulci,* Lupo, 1954:14. *Q. sulci,* Zahradnik, 1952c:133; Schmutterer, 1959:83; Borchsenius, 1966:340; Kosztarab & Kozár, 1978:169.

Test of female oval, strongly convex, black or blackish-brown, 1.2–1.8 mm long, exuviae submarginal or subcentral, brown; ventral test thick and white. Adult female pear-shaped, lemon-yellow, antenna with straight pointed seta. Pygidium with 2 pairs of lobes, L_1 quadrate, notched symmetrically on both sides; L_2 much smaller; L_3 entirely absent; plate formula: 2, 2, 2–3, plates between median lobes slightly shorter than lobes and with 2 teeth at apex, others fimbriated; dorsal ducts in large number, forming a longitudinal group on each side of pygidium, 1 duct at base of median notch extends to anus; ventral ducts relatively long and numerous on submargin, extending anteriorly into prepygidial segments; intersegmental scleroses short; perivulvar pores absent; anus circular, removed twice its diameter from base of L_1.

Distribution. A rare species known from Cz, Hu and It.

Biology. Prefers xerophilous steppe habitats and lives on bark of *Cytisus* and *Genista.* In Czechoslovakia adults overwinter (Zahradnik, 1952c). Reproduction probably is biparental.

229. **Quadraspidiotus zonatus** (Frauenfeld) – Zonate Armored Scale

Syn. & Lit.: *Aspidiotus zonatus* Frauenfeld, 1868:888; *A. quercus* Signoret, 1869:132. *A. zonatus* Leonardi, 1920:54; Koronéos, 1934:12; Thiem & Gerneck, 1934b:130; Balachowsky & Mesnil, 1935:366. *Q. zonatus*, Lupo, 1948:181; Balachowsky, 1950:418; Zahradnik, 1952c:124, 1972:439; Schmutterer, 1952b:571, 1959:90; Bodenheimer, 1952:339; Gomez-Menor, 1959:159; Ghauri, 1962:103; Borchsenius, 1966:341; Rogojanu, 1966a:325; Kosztarab & Kozár, 1978:171.

Test of female circular, gray, or grayish-black, diameter 1.7–2.8 mm; exuviae central or subcentral. Adult female pear-shaped, slightly convex; yellow, pygidium dark yellow to brown, apex strongly sclerotized; sclerotized spur on each side of prothorax. Pygidium usually with 3 pairs of lobes; L_1 broad, notched on outer margin; L_2 smaller and also notched on outer margin, L_3 reduced or absent; plates between median lobes usually shorter than L_1 and not fringed; other plates wide and fringed, formula: 2,2,3; intersegmental scleroses developed; dorsal ducts form groups, some ducts on margin of prepygidial segments but none on segment IV, 1 duct at base of median notch extends to or beyond anus; ventral ducts in submarginal area of pygidium, and prepygidial segments; perivulvar pores in 4–5 groups: 0–2 (5–18) 4–11; anus removed 2 or 3 times its diameter from base of L_1. Adult male described by Bustshik (1958) and Ghauri (1962); N_2 described by Boratynski (1953).

Distribution. A common Palearctic species. In CE: Au, Bu, Cz, Fr, Ge, Hu, It, Po, Ro, Sw, USSR, and Yu. Also in other parts of Europe, Middle East and N Africa.

Biology. Mesophilous species that lives on trunk and branches (males on leaves) of *Betula, Fagus, Ficus, Fraxinus, Juglans, Loranthus, Quercus, Sorbus,* and *Vaccinium.* Primarily a pest on *Fagus* and *Quercus,* and causes young twigs to die. One yearly generation. Adult females overwinter, and start laying 67–163 eggs per female in June in S. Germany (Schmutterer, 1952). Eggs hatch in 1–3 days, and the crawlers migrate and develop on leaves. Adults appear in August or early September.

Natural enemies. Parasites: *Aphytis mytilaspidis, Azotus pinifoliae, Pteroptrix dimidiata* (Schmutterer, 1959). Predators: *Camptonotus lutescens, Chilocorus bipustulatus, C. renipustulatus* (Schmutterer, 1952b; Tsalev, 1972).

Genus **Rhizaspidiotus** MacGillivray

Syn. & Lit.: *Rhizaspidiotus* MacGillivray, 1921:390. *Chorizaspidiotus* MacGillivray, 1921:391. *Remotaspidiotus* MacGillivray, 1921:391. *Hemiberlesiella* Thiem & Gerneck, 1934:132. *Thymaspis* Šulc, 1934:2. *Rhizaspidiotus*, Ferris, 1937a:34, 1938:265, 1943:99; Zahradnik, 1952c:157; *Balachowsky, 1951:92; Schmutterer, 1959:121; Borchsenius, 1966:245; Bazarov & Smelev, 1971:171; Kosztarab & Kozár, 1978:174.

Type-species: *Aspidiotus dearnessi* Cockerell, 1898.

Test of female circular or broad oval, strongly convex, thick, exuviae central or subcentral; ventral test usually well developed. Adult female pear-shaped or circular, antenna with 1 seta. Pygidial margin sclerotized, often with fine teeth or crenulate; L_1 well developed, other lobes with different degrees of reduction; plates, when present, very small, often slightly fimbriate at tip; dorsal ducts abundant on all pygidial segments; ventral ducts form a marginal band; perivulval pores usually absent or in 24 small groups; anus close to center of pygidium.

Distribution. There are 7 species in the Palearctic Region, 1 in N America, and 1 in Malaya; only 1 in CE.

Biology. Only the biology of *R. dearnessi* has been studied in great detail (Lacroiz, 1926). Members of this genus live on a variety of herbaceous and woody plants, often on roots.

230. **Rhizaspidiotus canariensis** (Lindinger) – Variable Armored Scale

Fig. 151

Syn. & Lit.: *Aspidiotus canariensis* Lindinger, 1911:12. Danzig, (1970) synonymized the following 4 species: *Aspidiotus artemisiae* Hall, 1926:20; *A. kiritschenkoi* Laing, 1929:487; *Thymaspis fusca* Šulc, 1934:3, 18; *R. pavlovskii* Borchsenius, 1955b:249. *Hemiberlesia canariensis*, Leonardi, 1920:102. *A. artemisiae*, Gomez-Menor, 1937:75. *R. artemisiae*, Balachowsky, 1951:655; Zahradnik, 1952c:157; Schmutterer, 1959:122; Bazarov & Smelev, 1971:172; Kosztarab & Kozár, 1978:175. *Thymaspis artemisiae*, Bodenheimer, 1952:344. *R. canariensis*, Borchsenius 1966:246; Danzig, 1970:1021, 1980a:334.

Test of female broad oval, strongly convex, brownish-yellow or brownish-gray, diameter 1.4–2.2 mm, exuviae subcentral, seldom central, usually pale green; ventral test thick, usually remains attached to plant surface after removal of upper test. Adult female pear-shaped, oval, widest at metathorax, pale green or yellowish when live; antenna with 1 or 2 setae; spiracles wide, without associated pores. Pygidial margin crenulate, L_1 prominent, occasionally semicircular or square, notched at apex, with parallel inner margins; L_2 and L_3 triangular; L_4 seldom developed; plates elongate, pointed or fimbriated at apex; tubular ducts abundant on both body surfaces; perivulvar pores variable in number: 0–1 (0–9) 0–9; anus elongate oval, at posterior 1/3 of pygidium. First instar described by Leonardi (1920).

Distribution. A rare species in CE, apparently of Middle Eastern or Mediterranean origin. In CE: Cz, Fr, Ge, It, and USSR. Also known from S Europe, N Africa, Canary Islands, Middle East and Middle Asia. Balachowsky (1951) reported it from 2,000 m elevation in N Africa.

Biology. A xerophilous, steppe-inhabiting species that usually infests stems, seldom leaves, but in dry habitats is found on roots and root collars. Most often found on *Artemisia* spp., but in CE the common host is *Thymus* spp. Other host genera: *Achillea, Argyranthemum, Aster, Bupleurum, Centaurea, Chrysanthemum, Convolvulus, Euphorbia, Helichrysum, Santolina, Tanacetum* and *Teucrium*. Apparently prefers Compositae. A biparental species with one yearly generation in S Germany (Schmutterer, 1959), and two in Middle Asia (Bazarov and Smelev, 1971). The overwintering adult females start laying eggs in April in S Germany. Adults develop during July–August.

Note. Danzig (1970) found much host-induced and geographic variations in the morphological structures, often even in the same population, for this reason four species were synonymized by her.

Genus **Syngenaspis** Šulc

Syn. & Lit.: *Syngenaspis* Šulc, 1895b:2; Ferris, 1937b:105; McKenzie, 1945:85; Zahradnik, 1952c:110; Balachowsky, 1953:820; Schmutterer, 1959:134; Borchsenius, 1966:200; Kosztarab & Kozár, 1978:175; Pellizzari, 1978:1.

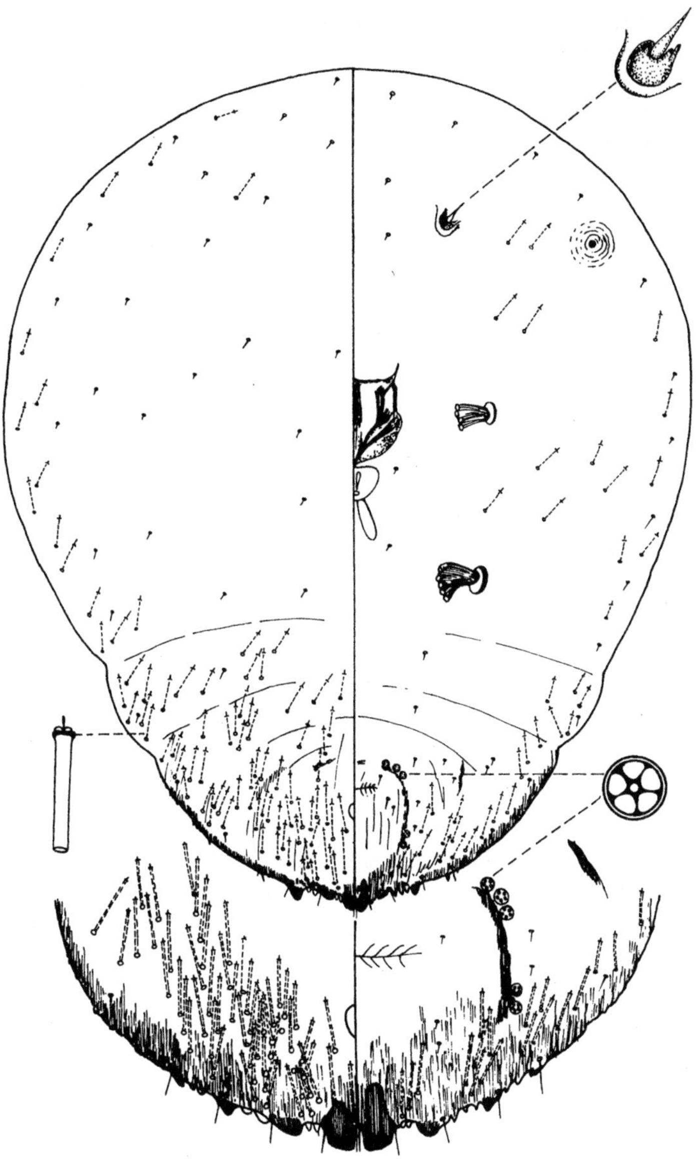

Fig. 151. *Rhizaspidiotus canariensis* (Lindinger), female (original)

Type-species: *Syngenaspis parlatoriae* Šulc, 1895.

Test of female elongate oystershell-shaped, grayish white. Adult female wide oval, with disc pores associated with anterior spiracles only. Pygidium with L_1, L_2, L_3 well developed, elongate, parallel-sided, notched once on each side; L_4 occasionally somewhat reduced; plates wide, long, with comblike apex; most dorsal ducts in submarginal band; perivulvar pores in 5 (rarely 4) groups.

Distribution. Palearctic. A monotypic genus.

Biology. Occurs on needles of conifers.

Note. Several species were mistakenly included in this genus. It is related to *Parlatoria* Targioni-Tozzetti, 1868, but differs from latter by its body shape that is narrowed anteriorly, and by the less sclerotized dorsal macroduct rim orifice.

231. **Syngenaspis parlatoriae** Šulc – Šulc's Conifer Scale
Fig. 152

Syn. & Lit.: *Syngenaspis parlatoriae* Šulc, 1895b:15. *S. sulci* Bodenheimer, 1941:69, 1953b:41. *S. parlatoriae*, McKenzie, 1945:85; Zahradnik, 1952c:110; Schmutterer, 1952:581, 1959:134; Balachowsky, 1953:821; Bachmann, 1953b:183; Danzig, 1959:401; Rogojanu, 1962:298; Zak-Ogaza & Koteja, 1964:431; Borchsenius, 1966:200; Kuznetsov, 1967:247; Kosztarab & Kozár, 1978:176; Vinis, 1981:206.

Test of female elongate oystershell-shaped, 1.2–1.5 mm long, 0.4 mm wide, posterior half white or grayish white, with 2 yellow-orange exuviae. Adult female body oval, widest at 2nd abdominal segment; yellow, or reddish; antenna with 1 long seta; 2 (rarely 3 or 4) disc pores associated with each anterior spiracle; ventral prosoma with duct tubercles on submargin. Pygidium very broad; 3 or often 4 pairs of well developed lobes, similar in size and shape, notched once on each margin; plates large, wide, and with comblike apex; few microducts in a submarginal band on venter, also on prosoma; dorsal macroducts concentrated in a wide submarginal band; perivulvar pores in 5 (rarely 4) groups: 3–5 (7–12, 7–14); anal opening at middle of pygidium.

Distribution. In CE: Au, Bu, Cz, Ge, Hu, It, Po, Ro, USSR, and Yu. Also in Turkey. It was recently found in Hungary by Vinis (unpublished record), and is rare in CE.

Biology. Apparently a S European mesophilous species that lives at the bases and underside of conifer needles in the genera *Abies, Cedrus, Picea, Pinus* and *Tsuga.* Schmutterer (1952b) found 1 yearly generation in S Germany. It overwintered primarily as first instars, but also as adult females. Adults from overwintering nymphs developed by mid-May and each female started laying 5–18 eggs at end of June, which hatched in about 2½ weeks. The overwintering females layed a few eggs in the fall and resumed egg laying in May.

Natural enemies. Parasites: *Archenomus caucasicus, Prospaltella aurantii* (Schmutterer, 1959; Trjapitzin, 1978).

Genus **Targionia** Signoret

Syn. & Lit.: *Targionia* Signoret, 1870:105. *Targaspidiotus* MacGillivray, 1921:392. *Targionia,* Leonardi, 1920:104; Gomez-Menor, 1937:121; Ferris, 1938:273, *1943:81; Balachowsky, 1948:268, 1951:632; Borchsenius, 1966:248; Kosztarab & Kozár, 1978:176.

Type-species: *Targionia nigra* Signoret, 1870.

Test of female circular or oval, dark colored, exuviae subcentral or submarginal; ventral test thick, partly united with dorsal test to form a protective capsule for the insect. Adult female pear-shaped often with the prosoma variously sclerotized; antenna with 1 seta; no disc pores associated with spiracles. Pygidium with 3 pairs of

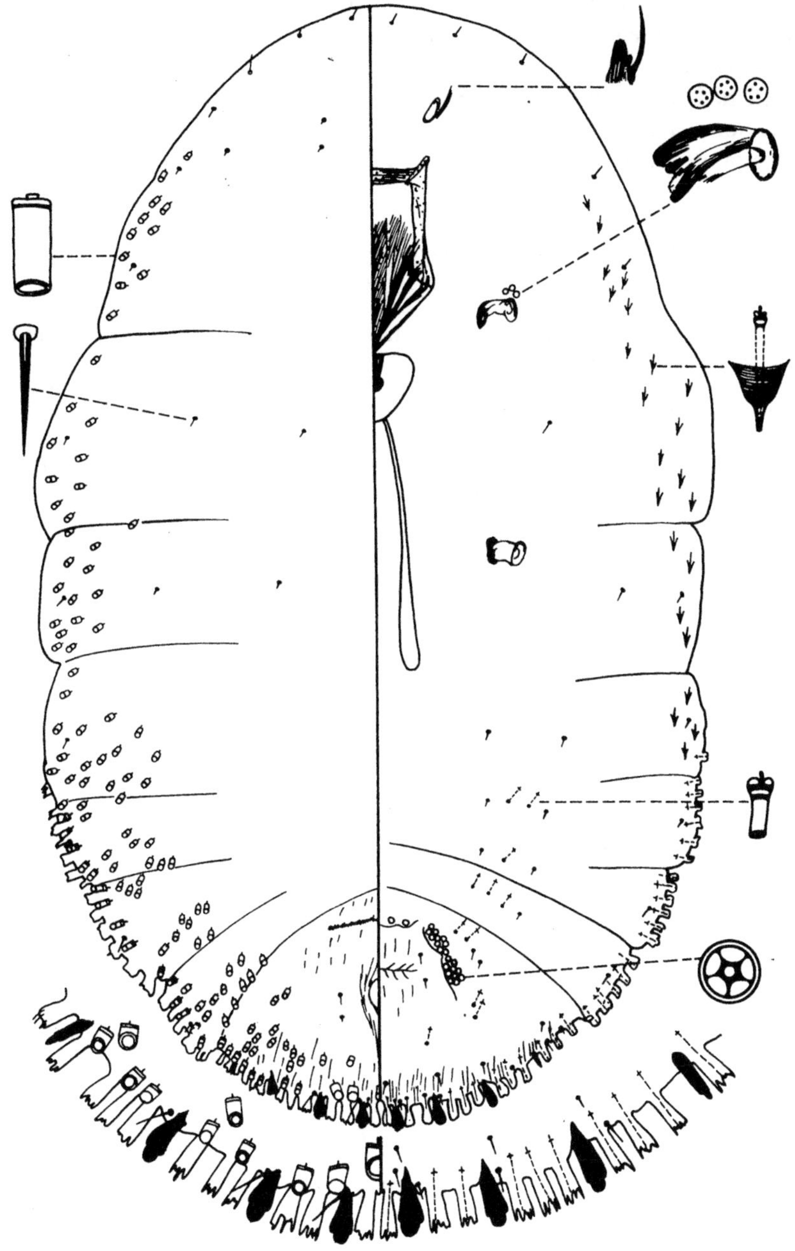

Fig. 152. *Syngenaspis parlatoriae* Šulc, female (original)

lobes, often with an indication of a 4th pair; L_1 well developed, others tooth-like, plates and perivulvar pores entirely absent; dorsal ducts very slender, their orifices often in longitudinal furrows.

Distribution. Five species in SW Palearctic, S America and E Africa; 1 in CE.

Biology. Some are pest on cultivated plants, and feed on a variety of woody and herbaceous plants. The biology of only *T. vitis* was studied.

232. **Targionia vitis** (Signoret) – Black Oak Scale
Fig. 153

Syn. & Lit.: *Aspidiotus vitis* Signoret, 1876a:52. 1876b:439. *Diaspis blankenhorni* Targioni-Tozzetti, 1879:17, 32. *Targionia vitis* v. *suberi* Leonardi, 1907:166. *T. vitis* v. *arbutus* Leonardi, 1909:123. *T. vitis,* Gomez-Menor, 1937:124; Lombardi, 1938:117; Balachowsky, *1951:640; Săvescu, 1961:17; Borchsenius, 1966:250; Bytinski-Salz & Sternlicht, 1967:126; Zahradnik, 1972:440; Kosztarab & Kozár, 1978:177.

Test of female circular or oval, almost black, rarely dark gray, highly convex, covered with concentric circles, diameter 1.8–2.4 mm; ventral test thick, partially united with dorsal cover, thus enclosing female, and adheres well to plant surface. Adult female pear-shaped; color violet when alive; about 1.7 mm long; antenna with 1 seta. Pygidial margin heavily sclerotized, median lobes large, once notched on outer

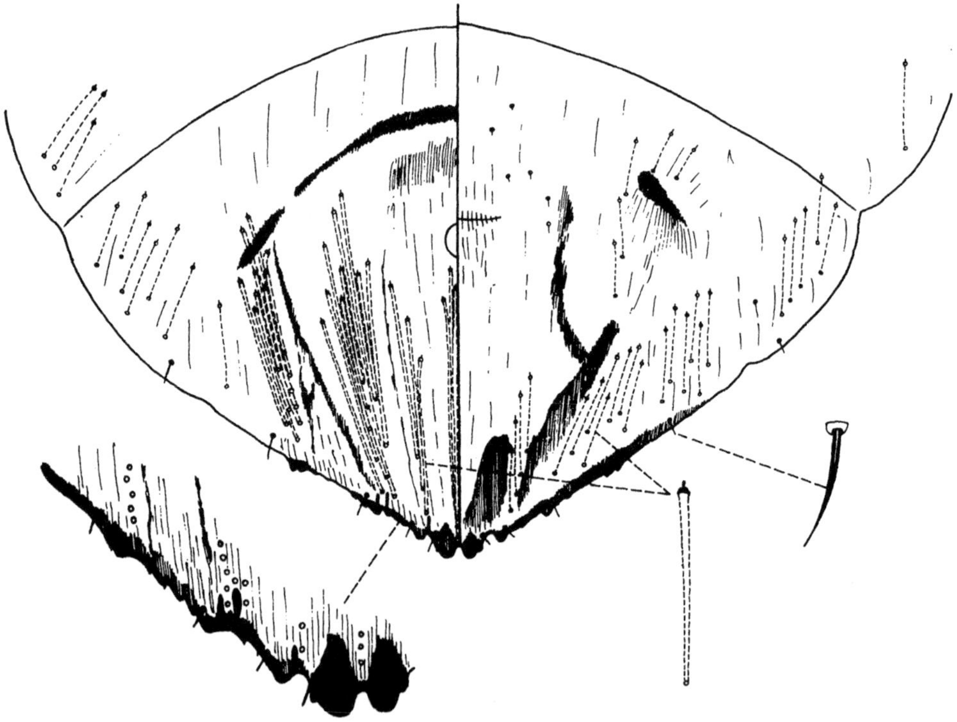

Fig. 153. *Targionia vitis* (Signoret), female pygidium (modified after Kosztarab and Kozár, 1978)

margin, set close together; L_2, L_3 and L_4 reduced to triangular tooth-like structures; plates and perivulvar pores absent; dorsal ducts long and slender, orifices small and in definite rows in longitudinal furrows; a few stout or elongate paraphyses arise from 2nd and 3rd interlobular spaces; venter with many ducts, especially in the submarginal areas, a few on prepygidial segments; both anus and vulva located close to center of pygidium. Adult male and 1st instar described by Lombardi (1938); N_1 by Leonardi (1920).

Distribution. A common circum-Mediterranean species. In CE: Bu, Cz, Fr, Hu, It, Ro, USSR, and Yu.

Biology. A mesophilous species that lives on trunks and branches of trees in the genera: *Arbutus, Castanea, Fagus, Platanus* and *Quercus*. Occasionally produces damage at high population density. One yearly generation in Hungary, where adult females overwinter, give birth to living young from mid-May and males and females develop by end of July. Development is similar in Bulgaria (Tsalev, 1968).

Natural enemies. Parasites: *Aphytis abnormis, Azotus matritensis, Chrysocharidia fimbriata, Coccophagus similis, Habrolepis pascuorum, Ooencyrtus azureus, Pteroptrix dimidiata*. Some records are questionable.

Genus **Unaspis** MacGillivray

Syn. & Lit.: *Unaspis* MacGillivray, 1921:308. *Graphaspis* MacGillivray, 1921:310; *Amerotrachaspis* MacGillivray, 1921:311. *Prontaspis* MacGillivray, 1921:311. *Unaspis,* Ferris, 1937:128; Rao 1949:59; Balachowsky, 1954b:288; Schmutterer, 1959:222; Takagi, 1961:14; Borchsenius, 1966:104; Kosztarab & Kozár, 1978:177.

Type-species: *Chionaspis acuminata* Green, 1896.

Test of female elongate, oystershell- or pear-shaped, usually brown or blackish, exuviae at anterior end. Adult female spindle-shaped, tapering anteriorly; orange-yellow. Pygidium with well-developed 3 pairs of lobes; L_1 divergent, L_2 and L_3 deeply bilobed; gland spines slender and pointed, none between median lobes; dorsal ducts numerous; perivulvar pores in 5 groups, sparse or absent; anal opening circular, located in the anterior part of pygidium.

Distribution. Mostly tropical, with 11 species; 1 species in CE.

Biology. Members of this genus live on a wide variety of trees and shrubs, and some species are important pests.

233. **Unaspis euonymi** (Comstock) – Euonymus Scale
Figs 35, 154

Syn. & Lit.: *Chionaspis euonymi* Comstock, 1881:313. *C. nemausensis* Signoret, 1886:6. *C. eunonymi* Lupo, 1938:272. *Unaspis nakayamai* Takahashi & Kanda, 1939:185; *U. euonymi* Ferris, 1937:130; Rao, 1949:62; Balachowsky, 1954b:294; Schmutterer, 1959:223; Kosztarab, 1963:102; Borchsenius, 1966:106; Rogojanu, 1966a:325; Kosztarab & Kozár, 1978:177.

Test of female oystershell-shaped, broad, dark brown with gray margin, 1.8–2.2 mm long; exuviae yellow at anterior end. Adult female spindle-shaped, lobed laterally,

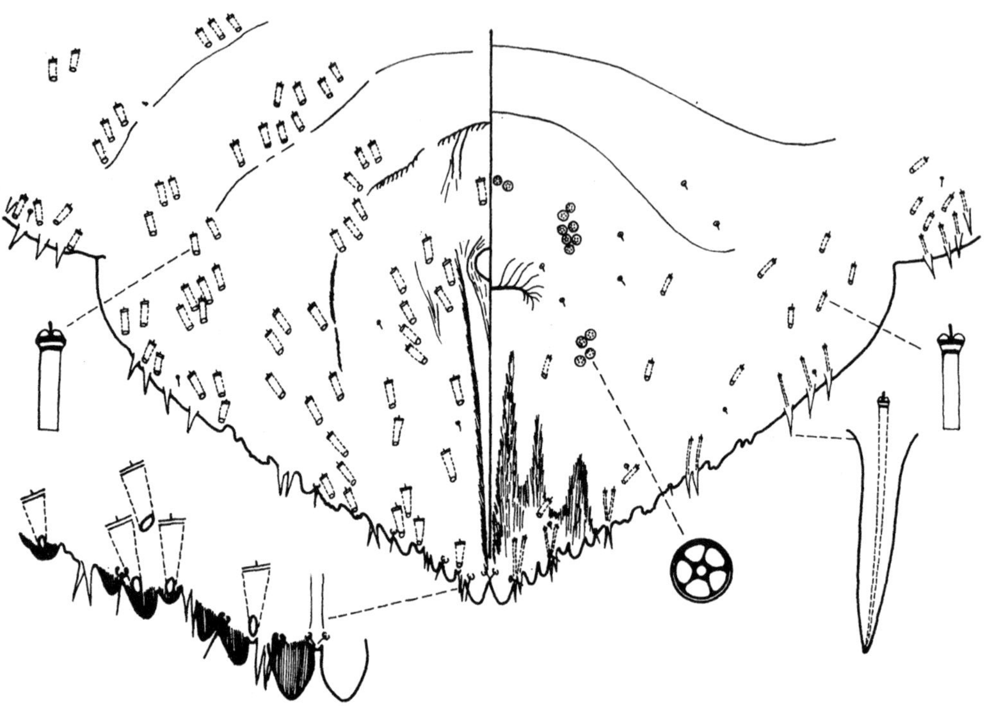

Fig. 154. *Unaspis euonymi* (Comstock), female pygidium (modified after Kosztarab, 1963)

widest at second abdominal segment, orange yellow, 1.2–1.4 mm long; antenna with 2–3 setae; anterior spiracles each with 8–22 disc pores, posteriors each with 2–6 pores; dorsum and venter with a band of ducts on margin: a group of cone-shaped gland spines posterior of hind spiracles. Pygidium broad, triangular, sclerotized at apex, with a median furrow extending from anus to median lobes; L_1 close together, nonzygotic, elongate, tapering toward apex; L_2 and L_3 well developed, deeply bilobed; L_4 indicated by sclerotized area; gland spines slender, usually 3–5 on margin of 4th segment, others in pairs; dorsal macroducts numerous, all of about equal size, most scattered, some arranged in more or less definite rows; few ventral microducts present; perivulvar pores: 2–6 (4–8) 2–8; anal opening circular, its diameter about equal with length of L_1. Male descriptions: Lobdell (1937), Modugno (1953) and Bustshik (1958). Nymphs described by Modugno (1953).

Distribution. Almost world-wide; missing from Australian region. In CE: Au, Bu, Fr, Hu, It, Ro, Sw, USSR, and Yu.

Biology. Host genera: *Celastrus, Daphne, Euonymus, Hibiscus, Ilex, Jasminum, Ligustrum, Lonicera, Olea, Pachistima, Pachysandra, Perychymenum, Prunus* and *Syringa* (Borchsenius, 1966). It is common in CE on *Euonymus* where it infests the trunk, branches and leaves. It is an important and common pest, causing the destruction of trees and shrubs. The number of yearly generations within its range varies, but in Hungary Vinis (1977) found 2 and a partial 3rd generation. It overwinters primarily as adult females, but also many overwinter as nymphs. Females lay 30–50 eggs, and crawlers hatch in 1–2 days. Biparental reproduction.

Natural enemies. Parasites: *Aenasioidea hispanica, Allaptus excisus, Aphytis proclia, Aspidiotiphagus citrinus, Coccidencyrtus poutiersi, Coccophagoides similis, Dicopus citri, Prospaltella gigas, Thysanus merceti* (Balachowsky, 1954b; Schmutterer and Kloft 1957; Trjapitzin, 1978). Predator: *Tyrophagus entomophagus* (Kosztarab, 1963).

III. APPENDIX

After the completion of the manuscript for this book, in 1983, a number of new species have been and are being described from CE. Because it was too late to include them, the following list of these species is provided here.

Acanthococcus centaurea Săvescu (1986, in print) from Ro.
Acanthomytilus jablonowskii Kozár & Matile-Ferrero, 1983 from Hu and It (Kozár, Tranfaglia and Pellizzari, 1984).
Atrococcus gouxi Matile-Ferrero, 1983 from Sw.
Diaspidiotus quercus Săvescu (1986, in print) from Ro.
Dysmicoccus balticus Koteja & Łagowska (1986, in print) from Po.
Eupeliococcus drabae Săvescu (1986, in print) from Ro.
Eupeliococcus tragopogoni Săvescu (1986, in print) from Ro.
Gregoporia istriensis Kozár, 1983 from Yu.
Heterococcus agropyri Săvescu (1986, in print) from Ro.
Heterococcus avenae Săvescu (1986, in print) from Ro.
Heterococcus dethieri Matile-Ferrero, 1983 from Sw.
Lepidosaphes crataegicola Săvescu (1986, in print) from Ro.
Luzulaspis macrospinus Săvescu (1986, in print) from Ro.
Paroudablis brachipodi Săvescu (1986, in print) from Ro.
Paroudablis ulmi Săvescu (1986, in print) from Ro.
Peliococcus schmuttereri Săvescu, 1984 from Ro.
Phenacoccus balachowskyi Săvescu, 1984 from Ro.
Phenacoccus cerasi Săvescu (1986, in print) from Ro.
Phenacoccus convolvuli Săvescu (1986, in print) from Ro.
Phenacoccus matricariae Săvescu, 1984 from Ro.
Phenacoccus prunispinosi Săvescu, 1984 from Ro.
Phenacoccus rehaceki Săvescu, 1984 from Ro.
Pseudococcus agropyri Săvescu (1986, in print) from Ro.
Pseudococcus borchsenii Săvescu, 1984 from Ro.
Pseudococcus brevicornis Săvescu, 1984 from Ro.
Pseudococcus carthami Săvescu (1986, in print) from Ro.
Pseudococcus gouxi Săvescu, 1984 from Ro.
Pseudococcus graminivorus Săvescu (1986, in print) from Ro.
Pseudococcus kaweckii Săvescu, 1984 from Ro.
Pseudococcus kosztarabi Săvescu, 1984 from Ro.
Pseudococcus kozari Săvescu, 1984 from Ro.
Pseudococcus microosteoli Săvescu (1986, in print) from Ro.
Pseudococcus moldavicus Săvescu (1986, in print) from Ro.

Pseudococcus multiporus Săvescu, 1984 from Ro.
Pseudococcus transilvanicus Săvescu (1986, in print) from Ro.
Pseudococcus zahradniki Săvescu, 1984 from Ro.
Pulvinaria ampelopsidis Săvescu, 1983 from Ro.
Pulvinaria brachiungualis Săvescu (1986, in print) from Ro.
Pulvinaria corni Săvescu (1986, in print) from Ro.
Pulvinaria euonymicola Săvescu, 1983 from Ro.
Rhizaspidiotus balachowskyi Kozár & Matile-Ferrero, 1983 from Hu.
Volvicoccus alpinus Matile-Ferrero, 1983 from Sw.

In addition there were several new species described from amber (fossils) by Koteja (1984, and 1986 in print), also Koteja and Zak-Ogaza (1986a, b, in print).

IV. CITED REFERENCES

Afifi, S. 1968. Morphology and taxonomy of the adult males of the families Pseudococcidae and Eriococcidae (Homoptera: Coccoidea). Bull. Brit. Mus. (Nat. Hist.) Entomol. Suppl. 13:210 pp.

Afifi, S., & M. Kosztarab. 1967. Studies on the morphology and taxonomy of the males of *Antonina* and one related genus (Homoptera: Coccoidea, Pseudococcidae). Va. Polytechnic Inst. & State Univ. Res. Div. Bull. 15:43 pp.

Afifi, S., & M. Kosztarab. 1969. Morphological and systematic studies on adult males of some species of *Lecanodiaspis* (Homoptera: Coccoidea: Lecanodiaspididae). Va. Polytech. Inst. & State Univ. Res. Div. Bull. 36:1–23.

Ahmad, R., & M. A. Ghani. 1971. Laboratory studies on the biology of *Lepidosaphes conchiformis* (Gmel.) (Hem., Diaspididae) and its parasite *Aphytis maculicornis* (Masi) (Hym., Aphelinidae). Bull. Entomol. Res. 61:69–74.

Ali, S. M. 1971. A catalogue of the Oriental Coccoidea (Part V, with an index) Indian Museum Bull. 6(2):7–82.

Amerling, C. 1858. Einige physiokratische Beobachtungen an den Obstbäumen in der Umgegend von Prag. Lotos 8:27–30.

Amyot, C. J. B., & A. Serville. 1843. Histoire Naturelle des Insectes. Hémiptères. p. 1–76, 1–676, 1–6. Paris.

Archangelskaja, A. D. 1935. The carmine-producing coccids (Margarodes) of Middle Asia and species of allied genus *Neomargarodes*. (In Russian). Pub. Com. Sci. Uzbek. SSR. Tashkent. 38 pp.

Argyriou, L. C. & S. S. Paloukis. 1976. Some data on biology and parasitization of *Sphaerolecanium prunastri* Fonscolombe (Homoptera: Coccidae) in Greece. Ann. Inst. Phytopathol. Benaki, (n. s.) 11(3):230–240.

Atanasov, E. P. 1959. A contribution to the knowledge of scale insects in the People's Republic of Macedonia. Ybk. Faculty of Agr. & Forest, Univ. Skopje, Yugoslavia 12 (1958/59): 425–435.

Babaev, T. 1970. The peach scale and its natural enemies in Tadzhikistan (In Russian). Mat. III. Respubl. Konf. Molod. Uchenich Tadz. Dushanbe. p. 208–209.

Baccetti, B. 1960. Le Cocciniglie Italiane delle Cupressaceae. Redia 45(Ser. 2):23–111.

Bachmaier, F. 1963. Zur Kenntnis der parasitischen Hymenopteren von *Spilococcus nanae* Schmutterer in Oberbayern. Beitr. Entomol. 13:553–565.

Bachmaier, F. 1965. Untersuchungen über die Insekten-und Milbenfauna der Zwergbirke (*Betula nana* L.) in süddeutschen und österreichischen Mooren, unter besonderer Berücksichtigung der phytophagen Arten und ihrer Parasiten. Veröff. Zool. Staatssamml. München 9:55–158.

Bachmann, F. 1952. *Quadraspidiotus schneideri* n. sp. eine neue Schildlausart. Mitteil. Schweiz. Entomol. Gesell. 25:357.

Bachmann, F. 1953a. Untersuchungen an den gelben Obstbaumschildläusen *Quadraspidiotus piri* Licht. und *Quadraspidiotus schneideri* n. sp. Zeitschr. f. angew. Entomol. 34(3):357–404.

Bachmann, F. 1953b. Beitrag zur Kenntnis der Jugoslavischen Schildlausfauna. Belgrade. Srpska Akad. Nauk Zbornik Radova 31; Inst. Ekol. Biogeog. No. 4 (1952–1953):175–184.

Bachmann, F. 1955. Beitrag zur Kenntnis der Jugoslavischen Schildlausfauna. 2. Mitteil. Biol. Glasnik 8:99–102.

Bachmann, F., & P. Geier. 1950. Einige für die Schweiz neue oder wenig bekannte Cocciden aus der Unterfamilie der Diaspidinae. Mitteil. Schweiz. Entomol. Gesell. 23:117–119.

Bailey, S. F. 1964. A study of the European fruit lecanium scale, *Lecanium corni*, on prune. J. Econ. Entomol. 57:934–938.

Balachowsky, A. 1928a. Contribution à l'étude des Coccides de l'Afrique mineure (2e note). Soc. Hist. Nat. Afrique Nord 19:121–144.

Balachowsky, A. 1928b. Contribution à l'étude des Coccides d l'Afrique mineure (4e note). Nouvelle liste de Coccides Nord-Africains avec description d'èspèces nouvelles. Bull. Soc. Entomol. France 17:273–279.

Balachowsky, A. 1930a. Contribution à l'étude des Coccides de France (3e note). Coccides nouveaux ou peu connus de la Faune de France. Bull. Soc. Entomol. France. 1130 (10):178–184.

Balachowsky, A. 1930b. Contribution à l'étude des Coccides l'Afrique mineure (7e note). Les *Cerococcus* (Asterolecaninae) du Nord Africain. Eos 6:201–219.

Balachowsky, A. 1930c. Deux *Chionaspis* (Hem. Coccidae) nouveaux de l'*Abies pinsapo* et du Cèdre. Bull. Soc. Entomol. France 17:266–273.

Balachowsky, A. 1932a. Sur la présence d'*Aspidiotus bavaricus* Ldgr. (Hem. Coccidae) dans la forêt de Fontainebleau. Bull. Ass. Naturalistes Vallée Loing 15:102–107.

Balachowsky, A. 1932b. Étude biologique des Coccides du bassin occidental de la Méditerranée. Encycl. Entomol (Ser. A) v. 15, Paris, Lechevalier. 214 + LXXI pp.

Balachowsky, A. 1932c. Contribution à l'étude des Coccides de France (9e note). Sur un *Luzulaspis* nouveau des Iles d'Hyéres. Bull. Soc. Entomol. France 37:197–200.

Balachowsky, A. 1934a. Recherches biogéographiques sur la faune des Coccides de Corse. Soc. Biogeog. (Paris) Compt. Rend. 11:34–39.

Balachowsky, A. 1934b. Contribution à l'étude des Coccides de France (17e note). Bull. Soc. Entomol. France 39:67–72.

Balachowsky, A. 1934c. Contribution à l'étude des Coccides de France (19e note). Deux espèces nouvelles pour la faune de France récoltées dans la région Parisienne. Bull. Soc. Entomol. France. 39:130–131.

Balachowsky, A. 1936. Contribution à l'étude des Coccides du Nord Africain (15e note). Sur deux remarquables Pseudococcinae découverts dans le Sahara marocain par Ch. Rungs. Bull. Soc. Entomol. France 10:157–165.

Balachowsky, A. 1937–1953. Les Cochenilles de France, d'Europe, du Nord de l'Afrique, et du Bassin Méditerranéen. I–1937, No. 526:1–66; II–1937, No. 564:73–130; III–1939, No. 784:131–241; IV–1948, No. 1054:243–394; V–1950, No. 1087:397–557; VI–1951, No. 1127:561–720; VII–1953, No. 1202:725–929. Actualités Ci. et Indus. Entomol. Appl. Hermann & Cie, Paris.

Balachowsky, A. 1939. Contribution à l'étude des coccides du Nord-Africain (21e note). II. Remarques sur les *Aclerda* paléarctiques. Ann. Grignon École Natl. Agr. p. 55–57.

Balachowsky, A. 1942. Essai sur la classification des Cochenilles (Homoptera-Coccoidea). Ann. Grignon École Natl. Agr. (Ser. 3) 3:34–48.

Balachowsky, A. 1950a. Remarques biogéographiques sur l'aire de répartition Mondiale du G. *Kermes* (Hom. Coccoidea). Proc. Internatl. Cong. Entomol. (Stockholm) 8:342–346.

Balachowsky, A. 1950b. Les *Kermes* (Hom. Coccoidea) des Chênes en Europe et dans le Bassin Méditerranéen. Proc. Internatl. Cong. Entomol. (Stockholm) 8:139–754, p. 1–16 (reprint).

Balachowsky, A. 1953. I. Sur un Pseudococcini (Hom. Coccoidea) nouveau du Cap Bon (Tunisie). *Parapedronia mathisi* n. sp. Bull. Soc. Sci. Nat. Tunisie 6(1952–53):227–230.

Balachowsky, A. 1953a. *Arctorthezia cataphracta* Olafsen relique de la toundra boréo-arctique des hauts sommets alpins (1). (Coccoidea Ortheziidae). Entomologiste 9(5–6):93–95.

Balachowsky, A. 1953b. Sur deux Pseudococcini Hom. (Coccoidea) nouveaux appartenant à la faune alpine d'altitude. Mitteil. Schweiz. Entomol. Gesell. 26(4):295–300.

Balachowsky, A. 1953c. Sur le statut des genres *Puto* Signoret, *Ceroputo* Šulc et *Macrocerococcus* Leonardi. Mitteil. Schweiz. Entomol. Gesell. 26(4):301–304.

Balachowsky, A. 1953d. Sur les *Kermes* Boitard (Hom. Coccoidea) des chênes du bassin Oriental de la Méditerranée. Rev. Pathol. Vég. Entomol. Agr. France 32:181–189.

Balachowsky, A. 1953e. Sur un *Heliococcus* Šulc (Coccoidea-Pseudococcini) nouveau vivant à 3600 mètres d'altidude dans les Alpes du Dauphiné et du Tyrol. Mitteil. Schweiz. Entomol. Gesell. 26(3):236–240.

Balachowsky, A. 1954a. Sur une nouvelle espèce d'Eriococcini de la forêt de Fontainebleau avec création d'un genre nouveau. Bull. Soc. Entomol. France 59:61–64.

Balachowsky, A. 1954b. Les Cochenilles Paléarctiques de la Tribu des Diaspidini. Mem. Sci. Institut Pasteur. 450 pp.

Balachowsky, A., & L. Mesnil. 1935, 1936. Les Insectes Nuisibles aux Plantes Cultivées. Paris. Part I (1935):1–1137; Part II (1936):1141–1901.

Balás, G. 1966. Animal pests of horticultural plants. 2nd ed. (In Hungarian). Mezőgazd. Kiadó, Budapest. 525 pp.

Bärensprung, F. v. 1849. Beobachtungen über einige einheimische Arten aus der Familie der Coccinen. Ztg. f. Zool. Zoot. und Palaeozool. 1:165–170, 173–176.

Bareth, C., & M. Liefferinge. 1976. Complements à la diagnose d'une Cochenille du hêtre: *Cryptococcus fagi* Baer. (Insecte Homoptère). Nouv. Rev. Entomol. 6(2):209–221.

Bartlett, B. R. 1961. The influence of ants upon parasites, predators, and scale insects. Ann. Entomol. Soc. Am. 54:543–551.

Bartlett, B. R. 1978. Asterolecaniidae, (p. 46–48), Coccidae (p. 57–74), Eriococcidae (p. 129–131), Margarodidae (p. 132–136), Ortheziidae (p. 136–137), Pseudococcidae (p. 137–170). *In:* Introduced parasites and predators of arthropod pests and weeds: A world review, by Clausen, C. P. (ed). USDA-ARS, Agricultural Handbook No. 480:545 pp.

Bazarov, B. B. 1968. Contribution to the fauna and ecology of scale insects and mealybugs (Homoptera, Coccoidea) on the southern slopes of the Gissar mountains. (In Russian). *In* Narzikulova, M. N. The Gorge of Kondara. Akad. Nauk Tadzhikskoi SSR Inst. Zool. i. Parazitol. p. 63–99.

Bazarov, B. B. 1974a. Revision of the genus *Heliococcus* Šulc (Homoptera, Coccoidea) in the Soviet fauna I. Entomol. Obozr. 53(2): 346–361. Also in Entomol. Rev. 32(2): 78–88.

Bazarov, B. B. 1974b. Revision of the genus *Heliococcus* Šulc (Homoptera, Coccoidea) in the Soviet fauna II. Entomol. Obozr. 53(3):603–644. Also in Entomol. Rev. 53(3):97–114.

Bazarov, B. B. & A. M. Nurmamatov. 1975. Mealybugs of the genus *Phenacoccus* Ckll. in the Fauna of Tadzhikistan. (In Russian). Entomol. Tadzhikistan. Sbornik Statii 1975:54–61.

Bazarov, B. B. & G. P. Shmelev. 1971. Armoured scales of Tadzhikistan and the adjoining regions of Middle Asia. (In Russ.) Fauna of Tadzhikistan SSR 11(1):238 pp.

Beardsley, J. W. 1960. A preliminary study of the males of some Hawaiian mealybugs (Homoptera; Pseudococcidae). Proc. Hawaii. Entomol. Soc. 17 (1959):199–243.

Beardsley, J. W. 1962a. Descriptions and notes on male mealybugs (Homoptera: Pseudococcidae). Proc. Hawaii. Entomol. Soc. 28(1961):81–98.

Beardsley, J. W. 1962b. Notes on the biology of the pink sugar cane mealybug. *Saccharicoccus sacchari* (Cockerell), in Hawaii (Homoptera: Pseudococcidae). Proc. Hawaiian Entomol. Soc. 18(1):55–59.

Beardsley, J. W. 1968. External morphology of the adult male of *Matsucoccus bisetosus*. Ann. Entomol. Soc. Amer. 61:1449–1459.

Beardsley, J. W. 1969. The *Anagyrina* of the Hawaiian Islands (Hymenoptera: Encyrtidae) with descriptions of two new species. Proc. Hawaii. Entomol. Soc. 20(2):287–310.

Beardsley, J. W. 1969a. A new fossil scale insect from Canadian amber. Psyche 76:270–279.

Beardsley, J. W. & R. H. Gonzalez. 1975. The biology and ecology of armored scales. Ann. Rev. Entomol. 20:47–73.

Beckmann, J. 1970. I. *Kermes*. Cochenille. Beitr. z. Gesch. Erfindungen 3:1–46.

Bénassy, C. 1963. Observations écologiques sur le parasitisme chez les coccides. Cas de *Leucaspis pini* Hartig. Rev. Pathol. Vég. Entomol. Agr. France 42(1):35–45.

Bénassy, C. 1968. Le parasitisme des Cochenilles du genre *Leucaspis* dans les Alpes-Maritimes et la Corse. I.–Inventaire et répartition des *Leucaspis* et de leurs parasites (Hymenoptères, Chalcidoidea). Ann. Épiphyties 19(4):659–681.

Ben-Dov, Y. 1968. Occurrence of *Sphaerolecanium prunastri* (Fonscolombe) in Israel and description of its hitherto unknown third larval instar. Ann. Épiphyties 19(4):615–621.

Ben-Dov, Y. 1975. On the identity of *Filippia* Targioni Tozzetti, 1868 and *Lichtensia* Signoret, 1873 (Homoptera: Coccidae). J. Entomol. Soc. So. Africa 38(1):109–121.

Ben-Dov, Y. 1976. The identity of two species from the "Lecanium series" of Signoret (Homoptera, Coccidae). Entomol. Monthly Mag. 111 (1975):115–116.

Bennett, F. D. & S. W. Brown. 1958. Life history and sex determination in the diaspine scale, *Pseudaulacaspis pentagona* (Targ.) (Coccoidea). Canad. Entomol. 90:317–324.

Berg, C. 1899. Substitución de 'nombres genericos. III. Buenos Aires Mus. Nac. Comunicaciónes 1:77–80(78).

Berlese, A. 1895. Review of G. del Guercio–Cocciniglie nuovo note e poco note–(Il Nat. Siciliano). Riv. Patol. Veg. 4(1896):185.

Berlese, A. 1896. La Cocciniglie Italiane viventi sugli agrumi. Parte III. I. Diaspiti (Cap. II. Osservazioni anatomiche. Cap. III. Cenni di biologia e danni che i Diaspiti recano arli agrumi.) Riv. Patol. Veg. 4 (1895–1896):195–292; 5:3–73.

Berlese, A. 1904. Sopra una nuova species di Cocciniglia *(Mytilaspis ficifolii)* che attacca le foglie del fico. Atti. R. Inst. Incoragg. Napoli, (5), 12:1–5.

Berlese, A. & G. Leonardi. 1896. Diagnosi di Cocciniglie nuove. Riv. Patol. Veg. 4:345–352.

Bichina, T. I. & E. P. Kiskina. 1976. San José entomophages in Moldavia. (In Russian, English summary). *In:* Biol. Zashch. Rast. Izdat. Stiinta. Kishinev. p. 13–18.

Blahutiak, A. 1977. Natural enemies of European Fruit Lecanium, *Parthenolecanium corni* Bouché (Homoptera: Coccidae), in CSSR. (In Czech). Forest & Food Sci. Ser. A, 5:130 p.

Blanchard, R. 1897. Sur les Cochenilles du genre *Orthezia* Bosc, 1784. Ann. Soc. Entomol. France 65(1896):678–683.

Bodenheimer, F. S. 1924. The Coccidae of Palestine.First report on this family. Palestine Zion. Exec. Agric. Expt. Sta., Inst. Agr. Nat. Hist. Bull. 100 pp.

Bodenheimer, F. S. 1928. Eine Kleine Cocciden-Ausbeute aus Griechenland. Konowia 7:(191)–192.

Bodenheimer, F. S. 1931. Zur Kenntnis der Palearktischen Kermesarten (Rhy. Cocc.). Konowia 10:241–247.

Bodenheimer, F. S. 1941. Seven new species of Coccidae from Anatolia. Istanbul. Univ. Fac. Sci. Rev., Ser. B, 6:65–84.

Bodenheimer, F. S. 1943. A first survey of the Coccoidea of Iraq. Iraq Dir.-Gen. Agr. Bull. 28:33 p.

Bodenheimer, F. S. 1944. Additions to the Coccoidea of Iraq, with descriptions of two new species (Hemiptera-Homoptera). Bull. Soc. Fouad Ier d'Entomol. 28:81–84.

Bodenheimer, F. S. 1944a. Note on the Coccoidea of Iran, with descriptions of new species (Hemiptera-Homoptera). Bull. Soc. Fouad Ier Entomol. 28:85–100.

Bodenheimer, F. S. 1951. Description of some new genera of Coccidae. Entomol. Ber. 13(315):328–331.

Bodenheimer, F. S. 1953a, b, c. The Coccoidea of Turkey I, II, III. Istanbul. Univ. Facult. Sci. Rev., Ser. B., 17(4):315–351; 18(1):1–61; 18:91–164.

Bodenheimer, F. S. & O. Theodor. 1929. Ergebnisse der Sinai-Expedition 1927. Leipzig, Hinrichs. 143 p. Teil IV. Die Coccidenfauna der Sinai-Halbinsel p. 104–117.

Bognár, S. & G. Vinis. 1979. Scale insects in the parks and avenues of Budapest. Acta Agron. Acad. Sci. Hung. 28(1–2):19–26.

Boitard, P. 1828. Les Gallinsectes. *In* Manuel d'Entomologie ou Histoire Naturelle des Insectes 2:169–172.

Bonnemaison, L. 1936. Morphologie comparée du "Pou de San José (*Aonidiella perniciosa* Comst.) et de l'*Aspidiotus* des arbres fruitiers (*Aspidiotus ostreaeformis* Curtis). Rev. Pathol. Vég. Entomol. Agr. France 23:230–243.

Boratynski, K. L. 1952a. *Matsucoccus pini* (Areen, 1925) (Homoptera, Coccoidea: Margarodidae): Bionomics and external anatomy with reference to the variability of some taxonomic characters. Trans. Roy. Entomol. Soc. London 103(9):285–326.

Boratynski, K. L. 1952b. Observations on *Matsucoccus pini* (Green) and a species of *Matsucoccus* Ckll. from Russia (Homoptera, Coccoidea, Margarodidae). Ann. & Mag. Nat. Hist. 5(12):507–508.

Boratynski, K. L. 1953. Sexual dimorphism in the second instar of some Diaspididae (Homoptera: Coccoidea). Trans. Roy. Entomol. Soc. London 104(12):451–479.

Boratynski, K. 1955. Coccoidea (Hem., Homoptera) from the Austrian Tyrol. Entomol. Monthly Mag. 91:66–67.

Boratynski, K. 1957. On the two species of the genus *Carulaspis* MacGillivray (Homoptera: Coccoidea, Diaspidini) in Britain. Entomol. Monthly Mag. 93:246–251.

Boratynski, K. 1960. *Matsucoccus pini* (Green) (Homoptera: Coccoidea, Margarodidae) in Norfolk. Entomol. Monthly Mag. 95:240.

Boratynski, K. 1961. A note on the species of *Asterolecanium* Targioni-Tozzetti, 1869 (Homoptera, Coccoidea, Asterolecaniidae) on oak in Britain. Proc. Entomol. Soc. London (B) 30:4–14.

Boratynski, K. 1962. A new species of *Acanthococcus* Signoret, 1875, (Eriococcidae, Coccoidea, Homoptera) from Britain. Proc. Roy. Entomol. Soc. London (B) 31(5–6):55–60.

Boratynski, K. L. 1964. A note on some British Coccoidea, with new additions to the British Fauna. Proc. Roy. Entomol. Soc. London (B) 33(5–6):103–110.

Boratynski, K. 1968. A note on some species of the genus *Diaspis* Costa, 1828, (Hemiptera, Coccoidea) in the collections of the Naturhistorisches Museum in Vienna; with the description of a new species. Ann. Naturhist. Mus. Wien 72:33–43.

Boratynski, K. 1970. On some species of *"Lecanium"* (Homoptera, Coccidae in the collection of the Naturhistorisches Museum in Vienna; with description and illustration of the immature stages of *Parthenolecanium persicae*. Ann. Naturhist. Mus. Wien 74:63–76.

Boratynski, K. & R. G. Davies. 1971. The taxonomic value of male Coccoidea with an evaluation of some numerical techniques. Biol J. Linn. Soc. 3:57–102.

Boratynski, K., E. Pancer-Koteja & J. Koteja. 1982. The life history of *Lecanopsis formicarum* Newstead (Homoptera, Coccinae). Ann. Zool. (Warszawa) 36:517–537.

Boratynski, K. & D. J. Williams. 1964. A note on some British Coccoidea, with new additions to the British Fauna. Proc. Roy. Entomol. Soc. London (B) 33(5–6):103–110.

Borchsenius, N. S. 1935. Five new species of armoured scales (Coccidae) morphologically allied to San José scale (*Aspidiotus perniciosus* Comst.). (In Russian). Inst. Zashch. Rast. Leningrad 6:127–133.

Borchsenius, N. S. 1937a. Tables for the identification of coccids (Coccidae) injurious to cultivated plants and forests in the USSR. (In Russian). Izd. Leningrad. Obl. Karan. Insp. 147 pp.

Borchsenius, N. S. 1937b. Quarantine and other scale insects (Coccidae) of USSR. (In Russian). Plant Quar. Sta., Tbilisi. 272 pp.

Borchsenius, N. S. 1939. Systematic characteristics of second-instar larvae of the oysterlike Coccidae occurring in the USSR. (In Russian). Leningrad Quar. Lab. 47 pp.

Borchsenius, N. S. 1947. Two new genera of mealybugs and a new species of scale insect (Homoptera-Coccoidea) from Armenia (In Russian) Akad. Nauk. Dokl. Arm. SSR 7:141–143.

Borchsenius, N. S. 1948a. On the revision of the genus *Eriococcus* Sign. (Insecta, Homoptera, Coccoidea). (In Russian) Akad. Nauk Dokl. USSR (n. s.) 60:501–503.

Borchsenius, N. S. 1948b. Revision of the genus *Phenacoccus* Ckll. (Insecta, Homoptera, Coccoidea). (In Russian). Akad. Nauk. Dokl. USSR (n. s.) 61(5):953–956.

Borchsenius, N. S. 1948c. New mealybugs of the family Pseudococcidae from Central Asia (In Russian). Akad. Nauk. Dokl. USSR (n. s.) 63:581–584.

Borchsenius, N. S. 1948. Review of Palearctic mealybugs of the genus *Macrocerococcus* Leon. (Coccoidea, Pseudococcidae). (In Russian). Entomol. Obozr. 30:31–39.

Borchsenius, N. S. 1949. Fauna of USSR. Homoptera, Pseudococcidae. (In Russian). Akad. Nauk. Zool. Inst. (n. s. 38) 7:383 pp.

Borchsenius, N. S. 1949a. A new genus and new species of hard and soft scales (Homoptera, Coccoidea) of USSR fauna. (In Russian). Entomol. Obozr. 39(3–4):334–353.

Borchsenius, N. S. 1950. Mealybugs and scale insects of the USSR (Coccoidea)((In Russian). Akad. Nauk. USSR. Zool. Inst., 32:250 pp.

Borchsenius, N. S. 1950a. Collection and study of hard and soft scales (In Russian). Akad. Nauk. Zool. Inst. Zhivotnyi Mir SSSR. 4:30 pp.

Borchsenius, N. S. 1952. New genera and species of soft-scales of the family Coccidae (= Lecaniidae) of the fauna of USSR and adjacent countries. (Insecta, Homoptera, Coccoidea). (In Russian). Trudy Zool. Inst., Akad. Nauk USSR. 12:269–316.

Borchsenius, N. S. 1953. New genera and species of scale insects of the family Coccidae (Homoptera, Coccoidea). (In Russian.) Entomol. Obozr. 33:281–290.

Borchsenius, N. S. 1955a. New species of coccids of the family Margarodidae of the USSR fauna (Homoptera, Coccoidea). (In Russian). Entomol. Obozr. 34:222–226.

Borchsenius, N. S. 1955b. New species of armored scales from the maritime region of the USSR and northern Korea (Homoptera, Coccoidea). (In Russian). Trudy Zool. Inst. Akad. Nauk. USSR. 21:247–252.

Borchsenius, N. S. 1955c. New species of false-hardscales (Homoptera, Coccoidea, Coccidae) of the fauna of the USSR and adjacent countries. (In Russian). Trudy Zool. Inst. Akad. Nauk USSR. 18:288–303.

Borchsenius, N. S. 1955d. Suborder Coccoidea-scale insects and mealybugs. (In Russian). *In* Handbook of Forest Pests, v. 2, p. 848–885. Akad. Nauk. USSR Moscow.

Borchsenius, N. S. 1956. Notes on the Coccoidea of Korea. (In Russian). Entomol. Obozr. 35:671–679.

Borchsenius, N. S. 1956a. Review of the Palearctic scale-insects of the genus *Eriopeltis* Sign. (Homoptera, Coccoidea). (In Russian). Entomol. Obozr. 35:397–420.

Borchsenius, N. S. 1957. Fauna of USSR, Homoptera, Coccidae (In Russian). Zool. Inst. Akad. Nauk SSSR. (n. s. 66) 9:493 pp.

Borchsenius, N. S. 1958. On the evolution and phylogenetic interrelations of the Coccoidea. (In Russian) Zool. Zhur. 37:765–780.

Borchsenius, N. S. 1959. Notes on Coccid fauna of China. 7. A new family of soft scales–Lecaniodiaspididae, Fam. N. (Homoptera, Coccoidea). Entomol. Obozr. 38(4):840–846, also in Entomol. Rev. 38(4):758–763.

Borchsenius, N. S. 1960. Fauna of USSR, Homoptera, Kermococcidae, Asterolecaniidae, Lecaniodiaspididae, Aclerdidae. (In Russian). Zool. Inst. Akad. Nauk SSSR (n. s. 77) 8:282 pp.

Borchsenius. N. S. 1963. On the revision of the genus *Lepidosaphes* Shimer (Coccoidea, Homoptera, Insecta). (In Russian). Zool. Zhur. 42:1161–1174.
Borchsenius, N. S. 1966. A catalogue of the armoured scale insects (Diaspidoidea) of the world. (In Russian). Akad. Nauk SSSR Zool. Inst. 449 pp.
Borchsenius, N. S., & T. N. Bustshik. 1955. A new species of comma-hardscale (Homoptera, Coccoidea) from Turkmenia. (In Russian). Trudy Zool. Inst. Akad. Nauk SSSR 18:304–307.
Borchsenius, N. S. & E. M. Danzig. 1966. A new species of *Greenisca* Borchs. (Homoptera, Coccoidea, Eriococcidae) from the USSR. (In Russian). Trudy Zool. Inst. Akad. Nauk. SSSR. 37:41–44.
Borchsenius, N. S. & E. M. Tereznikova. 1959a. Two new mealybugs of the genus *Rhizoecus* Künck (Insecta, Homoptera, Coccoidea) from Ukrainian SSR. (In Ukrainian). Dopovidi Akad. Nauk Ukr. RSR. 3:322–325.
Borchsenius, N. S., & E. M. Tereznikova. 1959b. Two new species of mealybugs of the genus *Heliococcus* Šulc (Insecta, Homoptera, Coccoidea). (In Russian). Zool. Zhur. 838:491–494.
Borchsenius, N. S., & M. A. Ter-Grigorian. 1956. Mealybugs parasitizing wheat and other grains in Armenian SSR. (In Russian). Akad. Nauk Armianskoi SSR Izv. Biol. Selsk. Nauk 9:17–27.
Bosc d'Antic, L. A. 1784. Description de l'*Orthezia characias*. J. Phys. 24(1):173.
Bouché, P. F. 1833. Naturgeschichte der schädlichen und nützlichen Garteninsekten. Berlin. Nicolai. 176 pp.
Bouché, P. F. 1844. Beiträge zur Naturgeschichte der Scharlachläuse (Coccina). Stettin. Entomol. Ztg. 5:293–302.
Bouché, P. F. 1851. Neue Arten der Schildlaus-Familie. Stettin. Entomol. Ztg. 12:110–112.
Brandt, J. F., & J. T. C. Ratzeburg. 1833. Getreue Darstellung und Beschreibung der Thiere, die in der Arzneimittellehre in Betracht kommen. Berlin 2: p. 225.
Bremner, O. E. 1907. New Coccidae from California. Canad. Entomol. 39:366–368.
Brown, S. W. 1977. Adaptive status and genetic regulation in major evolutionary changes of coccid chromosome systems. Nucleus. 20:145–157.
Buchner, P. 1965. Endosymbiosis of animals with plant microorganisms. Interscience Pub., John Wiley & Sons, Inc., New York. 909 pp.
Burmeister, H. 1835. Scharlachläuse. Schildläuse. Coccina. (Gallinsecta 1.) In Handb. der Entomol. 2(1):61–83.
Bustshik, T. N. 1958. A contribution to the comparative morphology of the males of the scale insects (Homoptera, Coccoidea, Diaspididae). (In Russian). Trudy Akad Nauk SSSR. Vsesoiuzn. Entomol. Obsh. 46:162–269.
Bustshik, T. N. 1960. Coccid fauna of Western Kopet-Dagh. (In Russian). Trudy Zool. Inst. Akad. Nauk SSSR. 27:167–182.
Bytinski-Salz, H., & M. Sternlicht. 1967. Insects associated with oaks *(Quercus)* in Israel. Israel J. Entomol. 2:107–143.
Cadahia, D. 1971. *Matsucoccus pini* Green (Homoptera: Margarodidae). Su descripcion y bionomia en España. Spain. Bol. Serv. Plag. For. 14(27):81–91.
Canard, M. 1958. Recherches sur la morphologie et la biologie de la cochenille *Eulecanium corni* Bouché (Homoptères-Coccoidea). Ann. École Natl. Super. Argon. Toulouse 6(2):185–271.
Canard, M. 1965. Utilisation des larves néonates pour aider à la détermination des Cochenilles floconneuses. Ann. Soc. Entomol. France (n. s.), 1(2):421–424.
Canard, M. 1965a. Sur quelques Pulvinariini (Homoptera, Coccoidea) du midi de la France. Proc. 12th Internatl. Cong. Entomol., London, 1964 (1965) p. 170.
Canard, M. 1967. Une nouvelle Pulvinaire du littoral Méditerranéen: *Rhizopulvinaria maritima* nov. sp. (Coccoidea-Coccidae). Vie et Milieu, Sér. C: Biol. Terrestre 18:159–167.
Canard, M. 1968. Contribution à l'étude des Rhizopulvinaires méditerranéennes (Hom. Coccidae). Bull. Soc. Entomol. France 73:90–96.
Canard, M. 1980. Présence en Grèce d'*Eulecanium tiliae* (L.) (Homoptera, Coccidae) sur le figuier *Ficus carica* L. et de son parasite *Blastothrix sericea* (Dalm.) (Hymenoptera, Encyrtidae). Biol. Gallo-Hellenica 9(1):157–162.
Carnes, E. K. 1907. The Coccidae of California. Calif. State Comnr. Hort. Bien. Rpt. 2(1905/06):155–222.
Ceianu, I. & I. Teodorescu, 1980. *Leucopis (Leucopomyia) silesiaca* Egg. (Diptera-Chamaemyiidae) predator of *Pulvinaria betulae* L. (In Romanian). Anal. Univ. Bucureşti, Ser. Biol. 29:63–66.

Clausen, C. P. (Ed.) 1978. Introduced parasites and predators of arthropod pests and weeds; a world review. USDA Agric. Res. Serv. Agr. Handb. No. 480:545 pp.
Cockerell, T. D. A. 1892. Museum notes. Inst. Jamaica J. 1:134–137.
Cockerell, T. D. A. 1893a. Museum notes. Coccidae. Inst. Jamaica J. 1:180.
Cockerell, T. D. A. 1893b. Note on the genus *Pseudococcus* Westwood. Entomol. News 4:317–318.
Cockerell, T. D. A. 1893c. A new *Lecanium* from Canada. Canad. Entomol. 25:221–222.
Cockerell, T. D. A. 1894. A check list of Nearctic Coccidae. Canad. Entomol. 26:31–36.
Cockerell, T. D. A. 1895. A new *Pulvinaria* found on orchids. Canad. Entomol. 27:135.
Cockerell, T. D. A. 1896a. A checklist of the Coccidae. Ill. State Lab. Nat. Hist. 4:318–339.
Cockerell, T. D. A. 1896b. Some new species of Japanese Coccidae, with notes. U. S. Dept. Agr. Div. Entomol., Tech. Ser. Bull. 4:47–56.
Cockerell, T. D. A. 1896c. Preliminary diagnoses of new Coccidae. Psyche 7 (1894–1896) (suppl.): 18–21.
Cockerell, T. D. A. 1897. The San José scale and its nearest allies. U. S. Dept. Agr. Div. Entomol. Tech. Ser. 6:31 pp.
Cockerell, T. D. A. 1898. A new scale insect found on bearberry. Canad. Entomol. 30:266–267.
Cockerell, T. D. A. 1899a. Tables for the determination of the genera of Coccidae. Canad. Entomol. 31:273–279, 330–333.
Cockerell, T. D. A. 1899b. Article VII.–First supplement to the check-list of the Coccidae. Bull. Ill. Nat. Hist. Survey 5:389–398.
Cockerell, T. D. A. 1900. Note on the coccid genus *Oudablis*, Signoret. Entomologist 33:85–87.
Cockerell, T. D. A. 1902a. A contribution to the classification of the Coccidae. Entomologist 35:232–233, 257–260.
Cockerell, T. D. A. 1902b. The nomenclature of the Coccidae. Entomologist 35:114.
Cockerell, T. D. A. 1903. Remarks on various Coccidae. *In* Coleman, G. A. Coccidae of the Coniferae, with the descriptions of ten new species from California. J. N. Y. Entomol. Soc. 11:61–85.
Cockerell, T. D. A. 1903a. Some species of *Eulecanium* (Coccidae) from France. Psyche 10:19–22.
Cockerell, T. D. A. 1905. Tables for the indentification of Rocky Mountain Coccidae (Scale insects and mealybugs). Colorado Univ. Studies 2:189–203.
Cockerell, T. D. A. 1909. The Japanese Coccidae. Canad. Entomol. 41:55–56.
Cockerell, T. D. A. & P. J. Parrott. 1899. Contributions to the knowledge of the Coccidae. Industrialist 25:227–37.
Coleman, G. A. 1903. Coccidae of the Coniferae, with the descriptions of ten new species from California. J. N. Y. Entom. Soc. 11:61:85.
Colvée, P. 1880. Ensayo sobre una neuva enfermedad del olivo, producida por una nueva especie del genero *Aspidiotus*. Spain. Min. Fomento. 14:21–41 (23, 30–41), 200–210.
Colvée, P. 1881. Estudios sobre algunos insectos de la familia de los Coccidos. Valencia, Nicasio Ruis Monfort, 40 pp.
Colvée, P. 1882. Nuevos estudios sobre algunos insectos de la familia de los Coccidos. Valencia 16 pp.
Comstock, J. H. 1881. Report of the entomologist. Part II. Report on scale insects. *In* U. S. Dept. Agr., Comnr. Agr. Rpt. 1880:276–349.
Comstock, J. H. 1882. Report of the entomologist, United States Department of Agriculture. Lac Insects. *In* U. S. Dept. Agr., Comnr. Agr. Rpt. 1881–1882:209–214.
Comstock, J. H. 1883. Second report on scale insects including a monograph of the sub-family Diaspinae of the family Coccidae and a list, with notes, of the other species of scale insects found in North America. *In* Cornell Univ. Agr. Exp. Sta., Dept. Entomol. Rpt. 2 (1882–83):47–142.
Comstock, J. H. 1916. Reports on scale insects. Cornell Univ. Agr. Exp. Sta. Bull. 372:425–603. (Reprinted from 1881–3 USDA reports.)
Costa Lima, A. 1942. Insetos do Brasil. 3(23). Homópteros. Esc. Nac. Agron. Ser. Didatica 4:327 pp.
Costa, O. G. 1835. Fauna del Regno di Napoli, famiglia de'Coccinigliferi, o de'Gallinsetti. Emitteri Napoli. 23 pp.
Curtis, J. 1843a. The small brown scale, *Aspidiotus proteus*, Nobis. Gard. Chron. 39:676.
Curtis, J. 1843b. *Aspidiotus ostreaeformis* (the pear-tree oysterscale). Gard. Chron. 46:805.
Dalman, J. W. 1826. Om nagra Svenska Arter af Coccus samt de inuti dem forekommande Parasit-insecter. Kongl. Vetensk.-Akad. Handl. f. 1825:350–374.
Danzig, E. M. 1959a. The scale insect fauna (Homoptera, Coccoidea) of the Leningrad region. Entomol. Rev. 38:395–407.

Danzig, E. M. 1959b. Concerning the biological forms of the apple comma scale *Lepidosaphes ulmi* (L.) (Homoptera, Coccoidea). (In Russian). Zool. Zhur. 38:879–886.
Danzig, E. M. 1960. Some new and little known mealybugs (Homoptera, Coccoidea, Pseudococcidae) of the Leningrad Region. Entomol. Obozr. 39:172–181; also Entomol. Rev. 39(1):117–124.
Danzig, E. M. 1961. Food forms of *Eulecanium franconicum* (Lndgr.) (Homoptera, Coccoidea) Entomol. Obozr. 40(3):571–576, also in Entomol. Rev. 40(3):310–313.
Danzig, E. M. 1962a. Addition to the scale insect fauna (Homoptera, Coccoidea) of the Leningrad region (In Russian). Trudy Zool. Inst. Akad. Nauk SSSR 31:22–24.
Danzig, E. M. 1962b. Revision of the genus *Rhizococcus* Signoret (Homoptera, Coccoidea) in the fauna of the USSR. Entomol. Obozr. 41(4):839–860, also in Entomol. Rev. 41(4):520–535.
Danzig, E. 1964. 5. Suborder Coccinea–coccids or mealybugs and scale insects. (In Russian). Akad. Nauk SSSR Zool. Inst. 1:616–654. (English transl. in Keys to the Insects of the European USSR, ed. by G. Ya. Bei-Bienko, 1967, v. 1. p. 800–850. Publ. for Smithsn. Inst. & Natl. Sci. Found., Washington, DC by Israel Progr. for Sci. Transl.)
Danzig, E. M. 1967. Contributions to the knowledge of the Coccidae (Homoptera) of the Primorye Territory. (In Russian). Trudy Zool. Inst. Akad. Nauk SSSR 41:139–172.
Danzig, E. M. 1968. Contribution to the fauna and biology of scale insects and white-flies (Homoptera, Coccoidea et Aleyrodoidea) from Northern Karelia. Entomol. Obozr. 47:499–504, also in Entomol. Rev. 47:304–306.
Danzig, E. M. 1970. Synonymy of some polymorphous species of coccoids (Homoptera, Coccoidea). (In Russian) Zool. Zhur. 49(7):1015–1024.
Danzig, E. M. 1971a. Three new species of coccids (Homoptera, Coccoidea) from the Far East (In Russian). Zool. Zhur. 50:1414–1417.
Danzig, E. M. 1971b. New and little-known species of mealy-bugs (Homoptera, Coccoidea, Pseudococcidae) from the Far East of USSR. Entomol. Obozr. 50:366–391, also in Entomol. Rev. 50:207–220.
Danzig, E. M. 1972a. Contribution to the fauna of the white flies and scale insects (Homoptera: Aleyrodoidea, Coccoidea) of Mongolia (In Russian). Insect of Mongolia. 1:325–348.
Danzig, E. M. 1972b. New and little-known species of the scale insects (Homoptera, Coccoidea) from Siberia and the Far East of the USSR (In Russian). Trudy Zool. Inst. Acad. Nauk SSSR. 52:261–275.
Danzig, E. M. 1972c. Coccoidea-Scale Insects. *In* Insects and ticks. Pests of Forest (In Russian). Zool. Inst. Akad. Nauk SSSR 1:189–221.
Danzig, E. M. 1974. Contribution to the fauna of the scale insects (Homoptera, Coccoidea) of Mongolia (In Russian). Insects of Mongolia. 2:67–71.
Danzig, E. M. 1975a. New species of mealybugs of the genus *Acanthococcus* Sign. (Homoptera, Coccoidea, Eriococcidae) from the Soviet Far East. Entomol. Obozr. 54(1):62–81, also in Entomol. Rev. 54(1):42–56.
Danzig, E. M. 1975b. Species of mealy-bugs (Homoptera, Coccoidea, Pseudococcidae) new for Mongolia (In Russian). Insects of Mongolia. 3:48–55.
Danzig, E. M. 1975c. Three new species of soft scales (Homoptera, Coccoidea, Coccidae) from southern Siberia and the Far East (In Russian). Zool. Zhur. 54(1):137–138.
Danzig, E. M. 1975d. Review of the cottony grass scales of the genus *Eriopeltis* Sign. (Homoptera, Coccoidea, Coccidae) of the Palearctic. Entomol. Obozr. 54(4):808–813, also in Entomol. Rev. 54(4):69–72.
Danzig, E. M. 1977a. Contributions to the scale insect fauna of North and East Mongolia (Homoptera, Coccoidea) (In Russian). *In* Insects of Mongolia. 5:196–202.
Danzig, E. M. 1977b. On the nomenclature and distribution of some harmful scale insects (Homoptera, Coccoidea). Entomol. Obozr. 56(1):99–109, also in Entomol. Rev. 56(1):73–75.
Danzig, E. M. 1977c. An ecological and geographical review of the scale insects (Homoptera, Coccoidea) of the south of the Far East. *In* Insect Fauna of the Far East (In Russian). Trudy Zool. Inst. Akad. Nauk SSSR. 70:37–60.
Danzig, E. M. 1978. Scale insect fauna of South Sakhalin and Kunashir. *In* New data on the insects of Sakhalin and Kuril Islands (In Russian). Akad. Nauk SSSR, Trudy Biol.–Pochvennogo Inst. (n. s.) 50(153):3–23.
Danzig, E. M. 1980a. Coccoids of the Far East USSR. With a phylogenetic analysis of the coccoid fauna of the world (In Russian). Nauka, Leningrad. 367 pp.

Danzig, E. M. 1980b. On the nomenclature and synonymy of several species of scale insects and white flies (Homoptera: Coccinea, Aleyrodinea). Entomol. Obozr. 59(3):594–596, also in Entomol. Rev. 59(3):73–74.

Danzig, E. M. 1983. New and little-known species of scale insects (Homoptera, Coccinae) of USSR. (In Russian). Entomol. Obozr. 62:514–523.

Danzig, E. M., & S. G. Ivanova. 1976. New species of the genus *Balanococcus* Williams (Homoptera, Coccoidea, Pseudococcidae). Entomol. Obozr. 55(1):76–77, also in Entomol. Rev. 55 (1):53.

Danzig, E. M. & I. M. Kerzhner. 1981. *Coccus* Linnaeus, 1758 and *Parthenolecanium* Šulc, 1908 (Insecta, Homoptera, Coccidae): proposed designation of type species under the plenary powers. Z. N. (S.)2125. Bull. Zool. Nomencl. 38(2):147–152.

Danzig, E. M., & F. Kozár. 1973. A new species of soft scales, *Physokermes inopinatus*, sp. n. (Homoptera, Coccoidea) from Hungary. Entomol. Obozr. 52(4):832–834, also in Entomol. Rev. 52(4):532–533.

Danzig, E. M. & F. Kozár. 1974. Contributions to the coccid fauna of Hungary (In Hungarian). Folia Entomol. Hung. (n. s.) 27(2):9–11.

Davidson, J. A., D. R. Miller & S. Nakahara. 1983. The white peach scale, *Pseudaulacaspis pentagona* (Targioni-Tozzetti) (Homoptera: Diaspididae): Evidence that current concepts include two species. Proc. Entomol. Soc. Washington 85:753–761.

DeBach, P. & D. Rosen. 1976. Armoured scale insects. *In* Studies in biological control (Ed. V. L. Delucchi). Internatl. Biol. Programme 9:139–178.

De Geer, C. 1778. Dixiéme mémoire. Supplément aux mèmoires des volumes précédens, ou description de quelques insectes ailés, suivie de celle de plusieurs insectes du Cap-de-Bonne espérance. *In* Mémoires pour servir à l'histoire des insectes, v. 7:591–605.

Del Guercio, G. 1894. Cocciniglie nuove, note e poco note. Il. Nat. Siciliano 13:141–158.

Del Guercio, G. 1903. Contribuzione allo studio dei Diaspini dell' Olivo. Bull. Soc. Entomol. Ital. 34 (trimes. 3, 1902):179–188.

De Klerk, C. A., Y. Ben-Dov & J. H. Giliomee. 1982. Redescriptions of four vine infesting species of *Margarodes* Guilding (Homoptera: Coccoidea: Margarodidae) from South Africa. Phytophylactica 14:61–76.

De Lotto, G. 1963. The authorship and type of the genus *Diaspidiotus* (Homoptera: Diaspididae). J. Entomol. Soc. S. Africa 24:144–145.

De Lotto, G. 1964. Observations on African mealybugs (Hemiptera: Coccoidea). Bull. Brit. Mus. (Nat. Hist.) Entomol. 14:343–397.

De Lotto, G. 1969. The mealybugs of South Africa (Homoptera: Pseudococcidae), II. Rep. So. Africa Dept. Agr. Tech. Serv. Entomol. Memoirs No. 20:1–30.

De Lotto, G. 1973. Does the Bermuda grass mealybug *Chorizococcus rostellum* (Lobdell, 1930) (Homoptera: Coccoidea: Pseudococcidae) occur in Italy? Entomologica 9:35–36.

Dietz, H. F., & H. Morrison. 1916. The Coccidae or scale insects of Indiana. Office State Entomol., Indianapolis, Indiana. 321 pp.

Dinther, J. B. M. van. 1950. Morphologie en biologie van de Schildluls *Chionaspis salicis* L. Doctoral Diss., Pub. H. Veenman & Zonen. Wageningen. Holland. p. 171–252.

Donisthorpe, H. St. J. K. 1907. Myrmecophilous notes for 1906. Entomol Rec. 19:4–7.

Douglas, J. W. 1881a. 17. Observations on the species of the Homopterous genus *Orthezia*, with a description of a new species. Trans. London Entomol. 1881 (Pt. 3): 297–303.

Douglas, J. W. 1881b. On the species of the genus *Orthezia*. Entomol. Monthly Mag. 17:172–176, 203–205.

Douglas, J. W. 1886. Notes on some British Coccidae. (No. 1). Entomol. Monthly Mag. 157–160.

Douglas, J. W. 1888a. Notes on some British and exotic Coccidae (No. 10). Entomol. Monthly Mag. 25:86–89.

Douglas, J. W. 1888b. Notes on some British and exotic Coccidae (No. 11). Entomol. Monthly Mag. 25:124–145.

Douglas, J. W. 1890a. Notes on some British and exotic Coccidae (No. 16). Entomol. Monthly Mag. 26:153–155.

Douglas, J. W. 1890b. Notes on some British and exotic Coccidae (No. 18). Entomol. Monthly Mag. 26:318–319.

Douglas, J. W. 1891a. Notes on some British and exotic Coccidae (No. 19). Entomol. Monthly Mag. (2nd Ser.) 2:65–69.

Douglas, J. W. 1891b. Notes on some British and exotic Coccidae (No. 20). Entomol. Monthly Mag. 27:95–100.

Douglas, J. W. 1891c. Notes on some British and exotic Coccidae (No. 21). Entomol. Monthly Mag. 27:244–247.

Douglas, J. W. 1892. Notes on some British and exotic Coccidae (No. 23). Entomol. Monthly Mag. 28:207–209.

Douglas, J. W. 1893a. Notes on some British and exotic Coccidae (No. 25). Entomol. Monthly Mag. 29:55–57.

Douglas, J. W. 1893b. Notes on some British and exotic Coccidae (No. 26). Entomol. Monthly Mag. 29:130–131.

Drozdovskiy, E. M. 1966. Chromosome sets in some coccids (Homoptera: Coccoidea). Entomol. Obozr. 45:(4)712–714, also in Entomol. Rev. 45(4):402–403.

Ducasse, G. 1941. Notes sur *Matsucoccus feytaudi*, sp. n. (Homopt. Coccid.). Rev. France Entomol. 8:217–225.

Dufour, L. 1864. Notice sur une nouvelle espèce de Gallinsecte *(Aspidiotus? luzulae)*. Ann. Soc. Entomol. France. (Ser. 4) 4: 207–209.

Dušková, F. 1952. Vergleich der morphologischen Hauptmerkmale der Schildläuse *Quadraspidiotus piri* (Lichtenstein) und *Quadraspidiotus marani* Zahradnik (Coccoidea: Diaspididae). Beitr. z. Entomol. 2(4/5):452–455.

Dušková, F. 1953a. Morphology of *Quadraspidiotus piri* (Lichtenstein) (Homoptera, Coccoidea). (In Czech). Acta Soc. Zool. Bohemoslov. 17:8–36.

Dušková, F. 1953b. Die Morphologischen Merkmale und Ökologische Bemerkungen zu den Weibchen der Schildläuse *Quadraspidiotus piri* (Licht.), *Q. marani* Zahr., *Q. ostreaeformis* (Curt.), und *Q. perniciosus* (Comst.) (Homoptera, Coccoidea). (In Czech). Acta Soc. Zool. Bohemoslov. 17:229–250.

Dziedzicka, A. 1961. Studies on morphology and biology of *Gossyparia spuria* (Mod.) (Homoptera, Coccoidea). (In Polish). Frag. Faun. 9(14):203–219.

Dziedzicka, A. 1968. Studies on the morphology and biology of *Lecanium fletcheri* Ckll. (Homoptera, Coccoidea) and related species. (In Polish). Zool. Polon. 18:125–165.

Dziedzicka, A. 1977. Comparative investigations on the hind legs of scale insects (Coccinea) (In Polish). Sci. Publ. Krakow Sup. Teachers Coll. 20:106 pp.

Dziedzicka, A. & J. Koteja. 1971. A revision of the species of the genus *Rhizococcus* Signoret (Homoptera, Coccoidea) occurring in Poland. Acta Zool. Cracoviensia 16(12):557–579.

Dziedzicka, A., & A. Marchewczyk. 1970. Studies on the morphology and biology of the bitubercular soft scale *(Palaeolecanium bituberculatum* Targ. Homoptera, Coccoidea). (In Polish). Rocznik Nauk. Dud. WSP. w Krakowie 37:28–43.

Dziedzicka, A., & W. Sermak. 1967. The variability of the dorso-marginal glands in larvae II and females of *Lecanium corni* Bouché on *Taxus baccata* L. (In Polish). Rocznik Nauk. Dyd. WSP w Krakowie 29:25–31.

Emeljanov, A. F. 1974. Proposals on the classification and nomenclature of ranges. Entomol. Obozr. 53(3):497–522, also in Entomol. Rev, 53(3):11–26.

Ezzat, Y. M. 1962. The genus *Trionymus* Berg in Egypt, U. A. R. Bull. Soc. Entomol. Egypte 46:67–86.

Ezzat, Y. M. 1965. *Aegyptococcus,* a new genus for *inermis* of Hall (Homoptera: Coccoidea – Pseudococcidae). Bull. Soc. Entomol. Egypte 49:163–167.

Ezzat, Y. M., & N. A. Hussein. 1969. Redescription and classification of the family Coccidae in U. A. R. (Homoptera: Coccoidea). Bull. Soc. Entomol. Egypte 51(1967):359–426.

Ezzat, Y. M., & H. S. McConnell. 1956. The mealybug tribe Planococcini (Pseudococcidae: Homoptera). Bull. Md. Agr. Expt. Sta. A-84:108 pp.

Fabricius, J. C. 1776. Genera Insectorum. Chilonii, Bartsch. 14:310 pp.

Fabricius, J. C. 1794. Entomologia systematica emendata et aucta, Vol. 4:472 pp.

Fallén, C. F. 1814. Specimen Novam Hemiptera Disponendi Methodum Exhibens. Dissert. Lund. Berling. 4:1–26.

Fernald, M. E. 1903. A catalogue of the Coccidae of the world. Mass. Agric. Expt. Sta. Spec. Bull. 88:360 pp.

Ferris, G. F. 1918. The California species of mealy bugs. Stanford Univ. Publs., Univ. Ser. 78 pp.

Ferris, G. F. 1936. Contributions to the knowledge of the Coccoidea (Homoptera). II. Microentomol. 1(2):17–92.

Ferris, G. F. 1937–1942. Atlas of scale insects of North America. Stanford Univ. Press, Stanford, The Diaspididae Ser. I–1937:1–136; II–1938:137–268; III–1941:269–384; IV–1942:385–448.
Ferris, G. F. 1937a. Contributions to the knowledge of the Coccoidea (Homoptera) IV. Microentomol. 2(1):1–45.
Ferris, G. F. 1937b. Contributions to the knowledge of the Coccoidea (Homoptera) VI. Microentomol. 2(3):103–122.
Ferris, G. F. 1938. Contributions to the knowledge of the Coccoidea (Homoptera) VIII. Microentomol. 3(3):57–75.
Ferris, G. F. 1941. The genus *Aspidiotus* (Homoptera: Coccoidea: Diaspididae). Microentomol. 6(2):33–70.
Ferris, G. F. 1943. The genus *Targionia* Signoret and some of its allies (Homoptera: Coccoidea: Diaspididae). Microentomol. 8(3):81–111.
Ferris, G. F. 1950, 1953. Atlas of the scale insects of North America. The Pseudococcidae. Stanford Univ. Press, Calif. Ser. V–1950 (Pt. I):1–278; Ser. VI–1953 (Pt. II):279–506.
Ferris, G. F. 1955. Atlas of the scale insects of North America. VII. The families Aclerdidae, Asterolecaniidae, Conchaspididae, Dactylopiidae and Lacciferidae. Stanford Univ. Press, Calif. 233 pp.
Ferris, G. F. 1957. A review of the family Eriococcidae (Insecta: Coccoidea). Microentomol. 22(4):81–89.
Foldi, I. 1982. Étude structurale et expérimentale de la formation du bouclier chez les Cochenilles Diaspines (Hom. Coccoidea, Diaspididae). Ann. Soc. Entomol. France (N.S.) 18 (3): 317–330.
Folkina, M. Ya. 1978. A case of mass extermination of the San José Scale *Quadraspidiotus perniciosus* by an ant *Crematogaster subdentata* (In Russian). Zool. Zhur. 57(2):301.
Folkina, M. Ya. 1980. The california scale, *Quadraspidiotus perniciosus* (Comst.) (Homoptera, Diaspididae), in the south of the Chimkent Region (In Russian). Trudy Inst. Zool. Akad. Nauk Kazakhskoi SSR. 39:24–31.
Fonscolombe, E. L., & J. H. Boyer de. 1834. Description des *Kermes* qu'on trouve aux environs d'Aix. Ann. Soc. Entomol. France 3:201–218.
Fourcroy, A. F. 1785. Entomologia Parisiensis. Paris, 544 p.
Frank, A. B., & F. Krüger. 1898. Die europäischen Verwandten der San José Schildlaus. Gartenflora 47:393–400.
Frauenfeld, G. von. 1868. Zoologische Miscellen XV. *Aspidiotus zonatus* n. sp. Verh. Zool.-Bot. Gesell., Wien. 18:888–889.
Fulmek, L. 1943. Wirtsindex der Aleyrodiden- und Cocciden-Parasiten. Entomol. Beihefte, 10:1–100.
Gavalov, I. 1928. Matériaux pour servir à l'étude des Coccides de la Tauride. (In Russian). Acta Soc. Entomol. Stavropol 5:17–31.
Geier, P. 1949. Contribution à l'étude de la Cochenille rouge du Poirer (*Epidiaspis leperii* Sign.) en Suisse. Rev. Pathol. Vég. Entomol. Agr. France 28:177–264.
Gennadius, P. 1883. Descriptions de trois nouvelles espèces de Cochenilles. Ann. Soc. Entomol. France (Ser. 6) 3:31–32.
Gennadius, P. 1891. Sur deux Cochenilles du Caroubier dans l'île de Chypre (Hem.). Ann. Soc. Entomol. France (Bull. des Séances) 64:277.
Geoffroy, E. L. 1762. Histoire abrégée des Insectes qui se trouvent aux environs de Paris, Paris, Durand. Vol. 1:523 pp.
Ghauri, M. S. K. 1962. The morphology and taxonomy of male scale insects (Homoptera: Coccoidea). Brit. Mus. (Nat. Hist.), London. 221 pp.
Ghesquière, J. 1934. *Newsteadia collarti* n. sp. Coccide (Ortheziine) radicicole et mycophile de la forêt de Soignes. Bull. & Ann. Soc. Entomol. Belgique 74:27–32.
Ghesquière, J. 1946. Sur la redécouverte en Belgique d'*Ortheziola signoreti* (Hall.). (Hem. Ortheziidae). Bull. & Ann. Soc. Entomol. Belgique 82:236–238.
Giard, A. 1894. Some unusual coccids. Ann. Soc. Entomol. France 62(1893):199–200.
Giliomee, J. H. 1967. Morphology and taxonomy of adult males of the family Coccidae (Homoptera: Coccoidea). Bull. Brit. Mus. (Nat. Hist.) Entomol. Suppl. 7:168 pp.
Giliomee, J. H. 1967a. The morphology and relationships of the male of *Lecaniodiaspis elytropappi* Munting and Giliomee (Homoptera: Coccoidea). J. Ent. Soc. So. Africa 30:185–197.
Giliomee, J. H. 1968. Morphology and relationships of the male of an *Asterolecanium* species (Homoptera: Coccoidea: Asterolecaniidae). J. Ent. Soc. So. Africa. 31:297–308.
Gistel. J. 1848. Naturgeschichte des Thierreichs. Stuttgart, Hoffman. 216 pp.
Gmelin, Ph. F. (Ed.). 1789. Insecta Hemiptera *Coccus*. *In* C. Linnaeus. Systema Nat. (Ed. 13) 1:2215–2222.

Goidanich, A. 1956. Sulla permeabilità del follicolo sericeo in alcuni Diaspididi (Hem. Cocc. Diaspididae). Proc. 10th Internatl. Cong. Entomol. (Montreal, 1956). p. 207–224.
Goidanich, A. 1960. Specializzazione ecologica e nomenclatura delle *Carulaspis* del *Viscum* e elle Cupressaceae (Hemiptera Coccoidea Diaspididae). Bol. Inst. Entomol. Univ. Bologna. 34:1–38.
Goidanich, A. 1962. Abnormal phenological cycle in a Palaearctic naked-bodied lecaniid genus (Hemiptera, Homoptera, Coccoidea). Centro Entomol. alpina e forestale Consiglio Nazionale Ricerche Publ. 50:269–284.
Goidanich. A. 1963. Sulla variabilità ontogenetica nella serie femminile di alcuni Margarodini (Homoptera Coccoidea Margarodidae). Atti Accad. Sci. Torino. 97:223–257.
Gomez-Menor, J. 1959. Homopteros "Sternorryncha" que atacan a la encina. Segunda parte. – Serie Monomera (1). Graelisia. 17:141–201.
Gomez-Menor Ortega, J. 1937. Coccidos de España. Madrid. Inst. Invest. Agron., Estac. Fitopat. Agr. Almeria. 432 pp.
Gomez-Menor Ortega, J. 1956. Cochinillas que atacan a los frutales (Homoptera, Coccoidea: 1. Familia Diaspididae). Bol. Patol. Veg. Entomol. Agr. (Madrid) 22:1–105.
Gomez-Menor Ortega, J. 1956a. Datos sobre Cóccidos de España. Bol. Portici R. Scuola Super. Agr., Lab. Zool. Gen. Agr. "Filippo Silvestri" 33:611–620.
Gomez-Menor Ortega, J. 1957. Adiciones a los Cóccidos de España. Eos 33(1–4):39–86.
Gomez-Menor Ortega, J. 1958. Distribución geográfica y ensayo de la Ecológica de los Cóccidos en España. Simposia Biogeografia Iberica 2. Inst. Biol. Apl. (Barcelona) Pub. 27:5–15.
Gomez-Menor Ortega, J. 1960. Adiciones a los "Cóccidos de España" V. Nota (Superfamilia Coccoidea). Eos 36:157–204.
Gomez-Menor Ortega, J. 1965. Adiciones a los "Cóccidos de España," VI Nota. Especies del género *Evallaspis* con su distributión geográfica en la Peninsula e Islas Baleares (Homoptera). Eos 41:87–114.
Gomez-Menor Ortega, J. 1968. Adiciones a los "Cóccidos de España." VII. Nota. (Hem. Homoptera). Eos 43:541–563.
Göthe, R. 1884. Beobachtungen über Schildläuse und deren Feinde, angestellt an Obstbäumen und Reben im Rheingau. Nassau. Ver. f. Naturk. Jahrb. 37:107–131.
Göthe, R. 1899. Tierische Feinde. (Beobachtungen über Schildläuse). Ber. Kgl. Lehranst. Obqt.-, Wein-Gartenbau, 1898/99:16–22.
Goux, L. 1931a. Notes sur les Coccides (Hem. Coccidae) de la France. II. Contribution à l'étude des *Eriococcus* de la France. Bull. Soc. Zool. France 56:58–73.
Goux, L. 1931b. Notes sur les Coccides (Hem.) de la France. (3e note). Description d'une espèce nouvelle. Bull. Soc. Entomol. France No. 8:113–118.
Goux, L. 1932. Notes sur les Coccides (Hem.) de la France (4e note). Description de *Cerococcus* nouveaux. Bull. Soc. Entomol. France 37:140–144.
Goux, L. 1934. Notes sur les Coccides (Hem.) de la France. (9e note). Contribution à l'étude du genre *Heliococcus* avec description de deux espèces nouvelles. Bull. Soc. Entomol. France 39:164–171.
Goux, L. 1935a. Notes sur les Coccides (Hem.) de la France. (10e) note. Étude d'une espèce nouvelle constituant un genre nouveau. Bull. Soc. Entomol. France 39(20):291–296.
Goux, L. 1935b. Notes sur les Coccides (Hem.) de la France. (11e note). Dix espèces nouvelles pour la France et remarques sur le genre *Antonina* Sign. Bull. Soc. Entomol. France 40:92–96.
Goux, L. 1936a. Notes sur les Coccides (Hem.) de la France (13e note). Étude d'une deuxième *Rhodania* Nouvelle. Bull. Soc. Entomol. France 41:35–40.
Goux, L. 1936b. Notes sur les Coccides (Hem.) de la France. (15e note). Remarques sur les *Eriococcus* du groupe *E. thymi* (Schrank) avec description d'une espèce nouvelle. Bull. Soc. Entomol. France 41:294–299.
Goux, L. 1936c. Notes sur les Coccides (Hem.) de la France. (14e note). Deux *Eriococcus* nouveaux des environs de Marseille. Bull. Soc. Entomol. France 41:344–356.
Goux, L. 1937a. Notes sur les Coccides (Hem.) de la France (16e note). Étude d'une nouvelle du genre *Poliaspidoides* nouveau pour la région Holarctique. Bull. Soc. Entomol. France 42(3):35–42.
Goux, L. 1937b. Notes sur les Coccides (Hem. Coccidae) de la France (17e note). Contribution à la connaissance des *Phenococcus* du groupe *P. cycliger* (Leon.) avec étude de deux espèces nouvelles. Bull. Soc. Zool. France 62:221–236.
Goux, L. 1937c. Notes sur les Coccides (Hem.) de la France. (18e note). Bull. Soc. Entomol. France 42:93–95.

Goux, L. 1938. Notes sur les Coccides (Hem.) de la France (23e note). Description d'un *Trionymus* nouveau. Bull. Mens. Soc. Linn. Lyon 7(6):166–169.
Goux, L. 1940. Remarques sur le genre *Ripersia* Sign. et description d'une *Ripersia* et d'un *Eriococcus* nouveaux (Hem. Coccidae). Notes sur les Coccides de la France (28e note). Bull. Soc. Hist. Nat. Afrique Nord. 318:55–65.
Goux, L. 1941a. Contribution à l'étude d'un faisceau d'espèces constituant un sous-genre nouveau du genre *Pseudococcus* (Hem. Coccidae). Bull. Mus. Hist.-Nat. Marseille 1(1):66–83.
Goux, L. 1941b. Notes sur les *Trionymus* de la France et sur quelques espèces nouvelles pour la faune Française (Hem. Coccidae) (Notes sur les Coccides (Hem. Coccidae) de la France (30e note)). Bull. Soc. Hist. Nat. Afrique Nord 32(1):31–44.
Goux, L. 1942. Notes sur les Coccides (Hem. Coccoidea) de la France (32e note). Description d'un *Phenacoccus* et d'un *Rhizoecus* nouveaux. Bull. Mus. Hist. Nat. Marseille 2(1):33–48.
Goux, L. 1943a. Notes sur les Coccides (Hem. Coccoidea) de la France (33e note). Nouvelle contribution à l'étude des *Pararhizoecus* avec description de deux espèces nouvelles et de leur larve néonate. Bull. Mus. Hist. Nat. Marseille 3(2–3):41–55.
Goux, L. 1943b. Hivernation et diapause hivernale chez les Coccides de la faune Française (Hem. Coccoidea), et remarques sur le déterminisme de la diapause. Bull. Mus. Hist. Nat. Marseille 3:126–145.
Goux, L. 1945. Notes sur les Coccides (Hem. Coccoidea) de la France (34e note). Description d'un *Trionymus* nouveau et de sa larve néonate et remarques sur quelques espèces nouvelles pour la faune Française. Mus. Hist. Nat. Marseille. Bull. 1(1–2):30–38.
Goux, L. 1948. Notes sur les Coccides (Hem. Coccoidea) de la France (39e note)). Les caractères spécifiques chez les *Eriococcus* et la signification des microfilières tubulaires. Bull. Soc. Linn. Provence 16:66–69.
Goux, L. 1953. Notes sur les Coccides (Hem. Coccoidea) de la France (42e note). Étude d'un *Heliococcus* nouveau. Bul. Mens. Soc. Linn. Lyon 22:104–109.
Gozmány, L. (Ed.) 1979. Vocabularium nominus animalium Europae septem linguis redactum. (Septemlingual dictionary of the names of European animals). Akadémiai Kiadó, Budapest, Hungary. Vol. 1, 1171 pp; Vol. 2. 1015 pp.
Green, E. E. 1896. A new British coccid. Entomol. Record & J. Variation 8(11): 260–261.
Green, E. E. 1900. Descriptions of some species of Coccidae collected by Mr. James Lidgett in Victoria, Australia. Ann & Mag. Natl. Hist. (Ser. 7) 6:448–452.
Green, E. E. 1901. On some new species of Coccidae from Australia, collected by W. Froggatt, F. L. S. Proc. Linn. Soc. N. S. Wales, 25(1900):559–562.
Green, E. E. 1902. Notes on a singular coccid from Victoria. Entomol. Monthly Mag. 38:284–285.
Green, E. E. 1914a. A New British Coccid *(Kuwania britannica)*. Entomol. Monthly Mag. 25:197–199.
Green, E. E. 1914b. Ernst Hartert's expedition to the Central Western Sahara XIX. Rhynchota. On a remarkable coccid, with branched antennae, from the Sahara. Novitates Zool. 21: 263–264.
Green, E. E. 1915a. Some remarks on the coccid genus, *Leucaspis*, with descriptions of two new species. Trans. Entomol. Soc. London 1914:459–468.
Green, E. E. 1915b. Observations on British Coccidae in 1914, with descriptions of new species. Entomol. Monthly Mag. 51:175–185.
Green, E. E. 1916. On two new British Coccidae, with notes on some other British species. Entomol. Monthly Mag. 52:23–31.
Green, E. E. 1917. Observations on British Coccidae; with descriptions of new species. Entomol. Monthly Mag. 53:201–210, 260–264, 265–269.
Green, E. E. 1921. Observations on British Coccidae, with descriptions of new species. Entomol. Monthly Mag. 61:146–152, 189–200.
Green, E. E. 1922a. A brief review of the indigenous Coccidae of the British Islands. Proc. So. London Entomol. & Nat. Hist. Soc. 12–25.
Green, E. E. 1922b. The Coccidae of Ceylon. Part V. Dulau, London. 345–472.
Green, E. E. 1923. Observations on British Coccidae–VIII. Entomol. Monthly Mag. 59:212–218.
Green, E. E. 1924. On some new species of Coccidae from various sources. Bull. Entomil. Res. 15:41–48.
Green, E. E. 1925a. Observations on British Coccidae, IX. Entomol. Monthly Mag. 61:34–44.
Green, E. E. 1925b. Notes on the Coccidae of Guernsey (Channel Islands), with descriptions of some new species. Ann. & Mag. Natl. Hist. (Ser. 9) 16:516–527.
Green, E. E. 1926. Observations on British Coccidae–X. Entomol. Monthly Mag. 62:172–183.

Green, E. E. 1928a. Observations on British Coccidae-XI. With descriptions of new species. Entomol. Monthly Mag. 64:20–31.
Green, E. E. 1928b. A brief review of the indigenous Coccidae of the British Islands, with emendations and additions (to February, 1928). Entomologist's Record & Variation (n. s.) 40 (1, suppl.): (5)–(8); (2, suppl.):(9)–(12); (3, suppl.):(13)–(14).
Green, E. E. 1930. Observations on British Coccidae–XII. Entomol. Monthly Mag. 66:9–17.
Green, E. E. 1931a. Notes on some Coccidae from Iceland. Entomol. Tidskr. 52:263–269.
Green, E. E. 1931b. Observations on British Coccidae.–XIII. Entomol. Monthly Mag. 67:99–106.
Green, E. E. 1934. On the type of *Antonina purpurea* Sign., with a description of a new species (*A. sulcii*) from Mid-Europe (Insecta, Coccidae). Proc. Zool. Soc. London, (Pt. 3):509–511.
Habib, A. 1956a. The male *Eulecanium corni* Bouché (Hemiptera-Homoptera: Coccoidea-Coccidae). Bull. Soc. Entomol. Egypte 40:119–126.
Habib, A. 1956b. *Eulecanium taxi* nov. spec. (Homoptera: Coccoidea-Coccidae). Bull. Soc. Entomol. Egypte 40:453–462.
Habib, A. 1957. The morphology and biometry of the *Eulecanium corni*- group, and its relation to host-plants (Hemiptera-Homoptera: Coccoidea). Bull. Soc. Entomol. Egypte 41:381–410.
Hadzibejli, Z. K. 1956. Contribution to the study of *Asterolecanium minus* Lndgr., *Parthenolecanium rufulum* Ckll. and *Diaspidiotus zonatus* Frau. (In Georgian). Trudy Inst. Zashch. Rast. Akad. Nauk Gruzinskoi SSR 11:157–170.
Hadzibejli, Z. K. 1958. A new genus and species of mealybug (Homoptera, Pseudococcidae) from Georgia, Transcaucasia. Entomol. Obozr. 37(4):903–908, also in Entomol. Rev. 37(4):781–784.
Hadzibejli, Z. K. 1963. Coccids of the family Ortheziidae (Homoptera, Coccoidea) of Georgia. (In Russian). Entomol. Obozr. 42:611–620.
Hadzibejli, Z. K. 1966. On the biology and morphology of the genera *Neomargarodes* Green and *Porphyrophora* Brandt (Homoptera, Coccoidea). (In Russian). Entomol. Obozr. 45:693–711.
Hafez, M., H. S. Salama, & A. H. Amin. 1967. Population dynamics of the plum scale, *Parlatoria oleae* (Colvée), in Egypt. Bull. Soc. Entomol. Egypte 51:347–357.
Hall, W. J. 1923. Further observations on the Coccidae of Egypt. Part I. Bull. Egypt Min. Agr. Tech. & Sci. Serv. 36:1–61.
Hall, W. J. 1925. Notes on Egyptian Coccidae with descriptions of new species. Bull. Egypt Min. Agr. Tech. & Sci. Serv. 64:31 p.
Hall, W. J. 1926. Contribution to the knowledge of the Coccidae of Egypt. Bull. Egypt Min. Agr. Tech. & Sci. Serv. 72:41 pp.
Hall, W. J. 1946. On the Ethiopian Diaspidini (Coccoidea). Trans. Roy. Entomol. Soc. London 97(20):497–583.
Haller, E. 1880. Entomologische Notizen. Über die Larve eines noch unbeschriebenen Orthezia-Ähnlichen Thieres. Mitteil. Schweiz. Entomol. Gesell. 6:6–9.
Halmágyi, L. 1974. On the scale insects of spruce trees (*Physokermes* spp.). (In Hungarian, English summary). Növényvédelem 10(10):454–458.
Hambleton, E. J. 1946. Studies of hypogeic mealybugs. Rev. Entomol. (Brasil) 17(1–2):1–77.
Hambleton, E. J. 1976. A revision of the new world mealybugs of the genus *Rhizoecus* (Homoptera: Pseudococcidae). USDA Agr. Res. Serv. Tech. Bull. 1522:88 pp.
Hamon, A. B. & M. Kosztarab. 1979. Morphology and systematics of the first instars of the genus *Cerococcus* (Homoptera: Coccoidea: Cerococcidae). Va. Polytech. Inst. & State Univ. Res. Div. Bull. 146:121 pp.
Hamon, A. B., P. L. Lambdin, & M. Kosztarab. 1976. Life history and morphology of *Kermes kingi* in Virginia (Homoptera: Coccoidea: Kermesidae). Va. Polytech. Inst. & State Univ. Res. Div. Bull. 111:1–31.
Hardy, J. 1868. Notes for 1864. Berwickshire Nat. Club Hist. 1863–1868:135–137.
Hartig, T. 1839. Jahresberichte über die Fortschritte der Forstwissenschaft und forstlichen Naturkunde nebst Original-Abhandlungen aus dem Gebiete dieser Wissenschaften, 1 (4):640–646.
Henriksen, K. L. 1921. Oversigt over de Danske Coccidae. Entomol. Meddel. 13:305–317.
Henry, T. J. 1976. *Aleuropteryx juniperi:* a European scale predator established in North America (Neuroptera: Coniopterygidae). Proc. Wash. Entomol. Soc. 78(2):195–201.
Herberg, M. 1918. Die Schildlaus *Eriopeltis lichtensteini* Sign. Archiv. f. Naturgesch. 82 (Abt. A., No. 10) (1916):1–107.

Herbert, F. B. 1924. The European Elm scale in the West. U. S. Dept. Agr. Bull. 1223:19 pp.
Heyden, C. von. 1860. Aus meinen entomologischen Tagebüchern. (1. Zur Naturgeschichte der Coccinen). Correspondenzbl. f. Sammler von Insecten, insbesondere von Schmetterlingen 1:89–90.
Hodek, I. 1973. Biology of Coccinellidae. Junk Publ. Hague, and Academia Publ., Czechosl. Acad. Sci. 260 pp.
Hodgson, C. J. 1967. Some *Pulvinaria* species (Homoptera: Coccidae) of the Ethiopian Region. Entomol. Soc. So. Africa 30:198–211.
Hodgson, C. J. 1968. Further notes on the genus *Pulvinaria* Targ. (Homoptera: Coccoidea) from the Ethiopian region. J. Entomol. Soc. So. Africa 31(1):140–174.
Hodgson, C. J. 1973. A revision of the *Lecanodiaspis* Targioni-Tozzetti (Homoptera: Coccoidea) of the Ethiopian Region. Bull. Brit. Mus. (Nat. Hist.) Entomol. 27:413–452.
Horváth, G. 1897. Description d'Hemiptères nouveaux et notes diverses. Rev. Entomol. (France) 16:81–97(95).
Howell, J. O. 1980. The value of second-stage males in armoured scale insects (Diaspididae) phyletics. Israel J. Entomol. 14:87–96.
Howell, J. O., & M. Kosztarab. 1972. Morphology and systematics of the adult females of the genus *Lecanodiaspis* (Homoptera: Coccoidea: Lecanodiaspididae). Va. Polytech. Inst. & State Univ. Res. Div. Bull. 70:248 pp.
Howell, J. O., & H. H. Tippins. 1977. Descriptions of first instars of nominal type-species of eight Diaspidid tribes. Ann. Entomol. Soc. Amer. 70(1):119–135.
Howell, J. O., & M. L. Williams. 1976. An annotated key to the families of scale insects (Homoptera: Coccoidea) of America, North of Mexico, based on characteristics of the adult female. Ann. Entomol. Soc. Amer. 69(2):181–189.
Howell, J. O., M. L. Williams, & M. Kosztarab. 1971. Morphology and systematics of *Cerococcus parrotti* (Hunter) with notes on its biology (Homoptera: Coccoidea: Asterolecaniidae). Va. Polytech. Inst. & State Univ. Res. Div. Bull. 64:23 pp.
Hoy, J. M. 1961. *Eriococcus orariensis* Hoy and other Coccoidea (Homoptera) associated with *Leptospermum* Forst. species in New Zealand. New Zeal. Dept. Sci. & Indust. Res. Bull. 141:70 pp.
Hoy, J. M. 1962. Eriococcidae (Homoptera: Coccoidea) of New Zealand. New Zeal. Dept. Sci. & Indust. Res. Bull. 146:219 pp.
Hoy, J. M. 1963. A catalogue of the Eriococcidae (Homoptera: Coccoidea) of the world. New Zeal. Dept. Sci. & Indust. Res. Bull. 150:260 pp.
Huba, A. 1962. The prognosis of the spread and harmfulness of the San José Scale (*Quadraspidiotus perniciosus* Comst.) in Europe (In Czech, English & Russian summary). Polnohospodarstvo (Bratislava) 9 (5–6):415–422.
Hunter, S. J. 1899. The Coccidae of Kansas. Kansas Univ. Quart. Ser. A, 8:1–15.
Imhof, O. E. 1900. Multiocellares geflügeltes Insekt. Biol. Centbl. 20:527.
Ivanova, S. G. 1973. Scale insects of Kaliningrad Region. (In Russian). Zap. Leningrad S–KH. Inst. 212:24–26.
Jablonowski, J. 1916. The scale insects of grapes and other economic plants (In Hungarian). Reprint Kísérletügyi Közl. 19(2nd Pt.):120 pp.
Jakubski, A. W. 1934. The Polish cochineal (*Porphyrophora polonica* (L.)) (In Polish). Warszawa. XX + 502 pp.
Jakubski, A. W. 1965. A critical revision of the families Margarodidae and Termitococcidae (Hemiptera, Coccoidea). Pub. Trustees Brit. Mus. (Nat. Hist.). 187 pp.
Jancke, G. D. 1955. Zur Morphologie der männlichen Cocciden. Zeitschr. f. angew. Entomol. 37(3):265–314.
Jasnosh, V. A. 1967. Biological control methods of the Comstock scale insect in Gruzia. (In Russian). Sbornik Karan. Rast. 19:75–84.
Johnson, W. T. & H. H. Lyon. 1976. Insects that feed on trees and shrubs. Comstock Publ. Associates, Cornell Univ. Press. Ithaca, NY. 464 pp.
Judeich, J. F. & H. Nitsche. 1895. Die Schidläuse. *In* Lehrbuch der Mitteleuropäischen Forstinsektenkunde. Vol. 2:1240–1270.
Kalandra, A. & J. Rozsypal. 1933. Einige Bemerkungen über *Lecanium coryli* L. auf Eschen und über die auf demselben parasitierenden Pilze. (In Czech). Ochr. rostlin 13:153–176.

Kaltenbach, J. H. 1860. Die Deutschen Phytophagen aus der Klasse der Insekten. Naturhist. Ver. Preuss. Rheinlande Verhandl. 17:203–260.

Kaltenbach, J. H. 1874. Die Pflanzenfeinde aus der Klasse der Insekten. Stuttgart, Hoffman. 848 pp.

Kawai, S. 1980. Scale insects of Japan in colors. Publ. Nat. Agr. Ed. Assoc. 455 pp.

Kawecki, Z. 1934. The bituberculate scale (*Lecanium bituberculatum* Targ.) (In Polish). Reprint from "Ogrodnictwa," 30:1–7.

Kawecki, Z. 1935. Die Schildläuse (Coccidae) der Wojewodschaften Kraków und Kielce, die in den Jahren 1933–1934 gesammelt wurden. (In Polish). (Polish Acad. Sci. & Lettres, Physiograph. Comm. Krakow). Ann. Rept. 68:73–90.

Kawecki, Z. 1936. Beiträge zur Kenntnis der Schildlaus-fauna Podoliens, Opoliens und Wolhyniens samt dem wolhynischen Polesie. (In Polish). Kosmos J. Soc. Polon. Nat. Kopernik 61: 319–324.

Kawecki, A. 1948. Some Coccidae from Poland. (In Polish). Polska Akad. Umiej. Mater. Fiz. Kraju 10:1–10.

Kawecki, Z. 1950. San José scale–*Quadraspidiotus (Aspidiotus) perniciosus* Comst. in Europe and its appearance in Poland. (In Polish). Polska Akad. Umiej. Prace Rolniczo-lesne Krakow. 55:54 pp.

Kawecki, Z. 1954. Studies on the genus *Lecanium* Burm. II. The yew scale, *Lecanium pomeranicum* sp. n. and some related species (Homoptera, Coccoidea, Lecaniidae). Ann. Zool. Warszawa 16(2):9–23.

Kawecki, Z. 1955. Studies on the genus *Lecanium* Burm. Part III. *Lecanium sericeum* Ldgr. (Homoptera, Lecaniidae). English transl. 1–10 p. OTS 60–21405, U. S. Dept. Comm., Washington, DC. Reprinted from Ann. Zool. Warszawa 25(14):213–226.

Kawecki, Z. 1957. Notes on scale insects (Homoptera, Coccoidea). (In Polish). Acta Zool. Cracoviensia 2(8):193–205.

Kawecki, Z. 1961. Studies on the genus *Lecanium* Burm. Part V. The nut or thorn scale–*Lecanium coryli* (L.) sensu Marchal nec Šulc (Homoptera, Coccoidea, Lecaniidae). English transl. 1–23 p. OTS 60–21402, U. S. Dept. Comm, Washington, DC. Reprinted from Ann. Zool. Warszawa 27(4):40–69.

Kawecki, Z. 1962a. Studies on the genus *Lecanium* Burm. IV. Materials to a monograph of the brown scale, *Lecanium corni* Bouché, Marchal (♀ nec ♂) (Homoptera, Coccoidea, Lecaniidae). English transl. 1–63 p. OTS 60–21404, U. S. Dept. Comm., Washington, DC). Reprinted from Ann. Zool. (Warszawa) 17:135–245. (1958).

Kawecki, Z. 1962b. The appearance of Coccidae of the genus *Lecanium* Burm. on mistletoe. (Homoptera, Coccoidea, Lecaniidae). Mem. Soc. Entomol. Ital. 41:15–24.

Kawecki, Z. 1967. Studies on the genus *Lecanium* Burm. VI. *Lecanium smreczynskii* sp. n. (Homoptera, Coccoidea, Lecaniidae). Bull. Acad. Polon. Sci. 15(11): 687–689.

Kawecki, Z. 1968. An outline of the biology and the geographical distribution of the globose scale *Sphaerolecanium prunastri* (Fonsc.) (Coccoidea, Lecaniidae). Bull. Acad. Polon. Sci. 16(11):689–693.

Kawecki, Z. 1971. A note on some European Lecaniidae (Coccoidea) with new additions of the Austrian, British, Italian and Polish fauna. Bull. Acad. Polon. Sci. 19(4):255–260.

Kawecki, Z. 1972. *Sphaerolecanium prunastri* (Fonsc.) (Homoptera, Coccoidea, Lecaniidae) and the beginning of its outbreak in Poland. (In Polish). Bull. Entomol. Pologne 42(4):857–863.

Kawecki, Z. 1985. Catalogus faunae Poloniae, Part 21 (No. 3): 107 pp.

Kawecki, Z. & H. Wernerowna. 1975. Description of map by Daniel Zwicker (1650), an inhabitant of Gdansk, giving the distribution of the scale insect *Porphyrophora polonica* (L.) (Coccoidea) in the region of Polesie, supplemented by two but little known studies from the 17th century. (In Polish). Memorabilia Zool. 27:61 pp.

King, G. B. 1900. The genus *Kermes* in North America. Psyche 9:78–84.

King, G. B. 1914. *Kermes lindingeri* King n. sp. Entomol. Rundschau 31:34.

Kiritchenko, A. N. 1928. Fauna of Coccidae of the Ukraine and the Crimea. (Preliminary Report). (In Russ.) J. Plant. Prot. Ukraine 3–4 (1927–28):112–116.

Kiritchenko, A. N. 1931. Second contribution to the coccid fauna of Ukraine and the Crimea. Leningrad Inst. Zashch. Rast. (Lenin Acad. Agr. Sci., USSR, Inst. Plant Prot.) 7(4–6):307–321.

Kiritchenko, A. N. 1932. Description of some new Coccidae (Hemiptera) from Turkestan and Ukraine. Trav. Inst. Zool. Acad. Sci. USSR. 1932:135–142.

Kiritchenko, A. N. 1936. Some new Pseudococcinae of the fauna of USSR (Hemiptera, Coccodea). Rev. Entomol. USSR (1935) 26(1–4):130–159.

Kiritchenko, A. N. 1937a. A new species of coccid from the Ukraine. Ann. & Mag. Nat. Hist. Ser. 10, Vol. 20:395–397.

Kiritchenko, A. N. (Ed.). 1937b. The San José scale in conditions of USSR. (In Russian). Ukrainian Quar. Lab., Moscow. 274 pp.
Kiritchenko, A. N. 1938. Beschreibung von zwei neuen Cocciden aus der Ukraine (USSR). (In Russian). Konowia 16(3&4):229–236.
Kiritchenko, A. N. 1940a. Description of several new species of the genus *Phenacoccus* (Coccoidea, Homoptera, Insecta) in the USSR. Rept. Odessa Univ., Biol. Ser. 4:187–192.
Kiritchenko, A. N. 1940b. Third report on the coccid fauna of USSR (In Russian). Acad. Sci. URSS, Trav. Inst. Zool., (1940):115–137.
Knechtel, W. K. 1929. Zur Kenntnis der Coccidenfauna Rumäniens. Premier Cong. Nat. Roumanie (Cluj, 1928). 1930 (Pt. 2):230–237.
Knipscher, R. C., D. R. Miller, & J. A. Davidson. 1976. Biosystematics of *Chionaspis nyssae* Comstock (Homoptera: Diaspididae), with evidence supporting leaf and bark dimorphism of the scale. Melanderia 25:1–30.
Köhler, G. 1983. Saisonale Einnischung und Dormanz bei der Nesselröhrenschildlaus, *Orthezia urticae* (L.) (Coccina: Ortheziidae). Zool. Jb. Syst. 110:443–454.
Komeili-Birjandi, A. 1981. Biology and ecology of *Parthenolecanium* spp. (Hem. Coccidae). Entomol. Monthly Mag. (1400–1403):47–58.
Komosinska, H. 1968. *Quadraspidiotus gigas* (Thiem et Gerneck) a new species of scale insect (Homoptera, Coccoidea) for the fauna of Poland. (In Polish). Sylwan No. 1:51–54.
Komosinska, H. 1969. Studies of scale insects (Homoptera, Coccoidea, Diaspididae) of Poland, I. (In Polish). Frag. Faun. 15: 267–271.
Komosinska, H. 1974a. Studies on the morphology of *Mytilaspis conchiformis* forma *conchiformis* (Gmelin) (Homoptera, Coccoidea, Diaspididae). Ann. Zool. (Warszawa) 31(2):343–371.
Komosinska, H. 1974b. Physiographical and ecological investigations upon armored scale insects (Homoptera, Coccoidea, Diaspididae) of Poland. (In Polish). Zeszyty Nauk. Akad. Rolniczej Warszawie. Rozprawy Nauk. 43:1–84.
Komosinska, H. 1975. Studies on the biology of *Mytilaspis conchiformis* forma *conchiformis* (Gmelin) (Homoptera, Coccoidea, Diaspididae) in Poland. Ann. Zool. (Warszawa) 33(9):127–148.
Komosinska, H. 1980. *Trionymus tuzensis* sp. n. (Homoptera, Coccoidea, Pseudococcidae) from Poland. Ann. Zool. (Warszawa) 35(19):257–265.
Komosinska, H. & E. Podsiadlo. 1967. Materials to the fauna of scale insects (Homoptera, Coccoidea)–steppe reservations in the Nida Valley (South Poland). I. Bull. Acad. Polon. Sci. Ser. Biol. 25(11):683–686.
Koroneos, J. 1934. Les Coccidae de la Grèce surtout du pélion (Thessalie). I. Diaspinae. Athens, 95 pp.
Kosarzhevskaya E. G. 1968. Technique of preparing slides for coccid (Homoptera: Coccoidea) determination. Entomol. Rev. 47:146–149.
Kosztarab, M. 1950. Contributions to the Hungarian scale insect fauna (In Hungarian). Ybk. Inst. Horticult. Res. (Budapest), 1:115–117.
Kosztarab, M. 1953. The native host plants of the San José scale and their importance from the point of view of protection (In Hungarian). A Növényvédelem Időszerű Kérdései (Budapest), 2(3):16–21.
Kosztarab, M. 1955. Revision und Ergänzung der in der "Fauna Regni Hungariae" angeführten Cocciden. Ann. Hist.-Nat. Mus. Nat. Hung. 6:371–385.
Kosztarab, M. 1956a. Eine neue Schildlaus-Art: *Diaspidiotus hungaricus* n. sp. (Coccoidea, Diaspididae) aus Ungarn. Ann. Hist. Nat. Mus. Nat. Hung. (n.s.) 7:435–438.
Kosztarab, M. 1956b. Parasitologische Untersuchungen an Schildläusen. Acta. Agron. Acad. Sci. Hung. 6(3–4):393–410.
Kosztarab, M. 1959. Biological notes on the scale insects of Hungary, Ann. Entomol. Soc. Amer. 52(4):4:401–420.
Kosztarab, M. 1963. The armored scale insects of Ohio (Homoptera: Coccoidea: Diaspididae). Bull. Ohio Biol. Surv. 2(2):120 pp.
Kosztarab, M. 1968. Cryptococcidae, a new family of the Coccoidea (Homoptera). Va. J. Sci. 19:12.
Kosztarab, M. 1977. The current state of coccoid systematics. Virginia Polytech. Inst. & State Univ. Res. Div. Bull. 127–1–4.
Kosztarab, M. 1981. Recent advances in the study of Coccoidea with special reference to integrated pest management. Acta Phytopathologica 16(1–2):151–156.

Kosztarab, M. 1982. Homoptera. *In* Synopsis and Classification of Living Organisms. McGraw Hill Book Co., New York. 2:447–470.
Kosztarab, M. & D. L. Hale. 1968. A new *Cryptococcus* species from North America, with a key to the species of the genus (Homoptera: Coccoidea) Va. J. Sci. 19:7–11.
Kosztarab, M., & J. O. Howell. 1972. Phylogeny and zoogeography of pit scales of the genus *Lecanodiaspis*. 14th Internat. Cong. Entomol. Abst. p. 98.
Kosztarab, M. & F. Kozár. 1978. Scale Insects–Coccoidea. *In* Fauna Hungariae (In Hungarian). 17(131) Part 22:192 pp.
Kosztarab, M., & F. Kozár. 1983. Introduction of *Anthribus nebulosus* (Coleoptera: Anthribidae) in Virginia for control of scale insects: A review. Va. J. Sci. 34(4):227–236.
Kosztarab, M., & L. M. Russell. 1974. An annotated list of generic names of the scale insects. 2nd Suppl. U. S. Dept. Agr., Agr. Res. Serv. Misc. Pub. 1285:22 pp.
Koteja, J. 1964. Notes on scale insects (Homoptera, Coccoidea) in Poland's fauna. (In Polish). Bull. Entomol. Pologne 34:177–184.
Koteja, J. 1966a. Studies on morphology and biology of *Luzulaspis frontalis* Green (Homoptera, Coccoidea). Bull. Entomol. Pologne. 36(2):17–43.
Koteja, J. 1966b. *Luzulaspis nemorosa* sp. n. (Homoptera, Coccoidea, Coccidae). Bull. Entomol. Pologne. 36(3):45–56.
Koteja, J. 1969. Notes on the Poland's scale insects fauna (Homoptera, Coccoidae). II. Bull. Entomol. Pologne. 39(1):3–15.
Koteja, J. 1969a. *Psilococcus parvus* Borchsenius (Homoptera, Coccoidea)–morphology, biology and taxonomy. Acta Zool. Cracoviensia 14(2):21–41.
Koteja, J. 1970. Systematic position of the genus *Vittacoccus* Borchsenius (Homoptera, Coccoidae). Bull. Entomol. Pologne. 40(2):223–231.
Koteja, J. 1971a. Two new species of *Mirococcus* Borchsenius (Homoptera, Coccoidea, Pseudococcidae). Bull. Entomol. Pologne. 41(1):3–18.
Koteja, J. 1971b. Notes on the Poland's scale-insect fauna (Homoptera, Coccoidea). III. (In Polish). Bull. Entomol. Pologne. 41(2):319–326.
Koteja, J. 1972. Notes on the Polish scale insect fauna (Homoptera, Coccoidea). IV. Bull. Entomol. Pologne. 42(3):565–571.
Koteja, J. 1974a. Comparative studies on the labium in the Coccinea (Homoptera). Sci. Papers Agr. Univ. Krakow 89:160 pp.
Koteja, J. 1974b. On the phylogeny and classification of the scale insects (Homoptera, Coccinea). (Discussion based on the morphology of the mouthparts). Acta Zool. Cracoviensia 19(14):267–325.
Koteja, J. 1978. An introduction to the revision of the *Eriopeltini* Šulc (Homoptera, Coccidae), with establishment of two new genera. Bull. Entomol. Pologne 48:311–327.
Koteja, J. 1979. A revision of the genus *Poaspis* Koteja (Homoptera, Coccidae). Bull. Entomol. Pologne. 49:451–474.
Koteja, J. 1979a. Revision of the genus *Luzulaspis* Cockerell (Homoptera, Coccidae). Bull. Entomol. Pologne. 49:585–638.
Koteja, J. 1980. Campaniform, basiconic, coeloconic and intersegmental sensilla on the antennae in the Coccinea (Homoptera). Acta Biol. Cracoviensia, Ser. Zool. 22:73–88.
Koteja, J. 1981. Frequency of honeydew excretion in relation to circadian activity in scale insects (Homoptera, Coccinea). Bull. Entomol. Pologne. 51:365–376.
Koteja, J. 1984. The Baltic amber Matsucoccidae (Homoptera, Coccinea). Ann. Zool. (Warszawa), 37:437–496.
Koteja, J. 1984a. Notes on the Polish scale insect fauna (Homoptera, Coccinea) V. (In Polish). Bull. Entomol. Pologne. 53:673–677.
Koteja, J. 1984b. *Cerococcus cycliger* (Goux) (Homoptera, Coccinea) new to the Polish fauna. (In Polish). Bull. Entomol. Pologne 54:413–414.
Koteja, J. 1985. Essay on the prehistory of scale insects (Homoptera, Coccinea). Ann. Zool. (Warszawa) 38:461–503.
Koteja, J. 1986. Morphology and taxonomy of male Ortheziidae (Homoptera, Coccinea). Bull. Entomol. Pologne 56:323–374.
Koteja, J. 1986a. Notes on the Polish scale insect fauna (Homoptera, Coccinea) VI. Bull. Entomol. Pologne 56:217–219.

Koteja, J. & J. O. Howell. 1979. *Luzulaspis* Cockerell (Homoptera, Coccidae) in North America. Ann. Entomol. Soc. Amer. 72(3):334–342.

Koteja, J. & F. Kozár. 1979. *Luzulaspis kosztarabi* sp. n. from Hungary (Homoptera, Coccidae). Acta Zool. Acad. Sci. Hung. 25(1–2):121–125.

Koteja, J. & B. Łagowska. 1986. *Dysmicoccus balticus* sp. n. (Homoptera, Coccinea, Pseudococcidae). Bull. Entomol. Pologne (in print).

Koteja, J. & E. Liniowska. 1976. The clypeolabral shield in the taxonomy of the Coccinea (Homoptera). Bull. Entomol. Pologne. 46:653–681.

Koteja, J. & M. Rościszewska. 1970. Revision of the genus *Parafairmairia* Cockerell (Homoptera, Coccoidea). Bull. Entomol. Pologne. 40(2):233–265.

Koteja, J. & B. Zak-Ogaza. 1966. Investigations on scale insects (Homoptera, Coccidae) of the Pieniny Klippen Belt. Acta Zool. Cracoviensia 11:305–332.

Koteja, J. & B. Zak-Ogaza. 1969. The scale-insect fauna (Homoptera, Coccoidea) of the Ojcow National Park in Poland. Acta Zool. Cracoviensia 14(14):351–373.

Koteja, J. & B. Zak-Ogaza. 1972. Morphology and taxonomy of the male *Kermes quercus* (L.) (Homoptera. Coccoidea). Acta Zool. Cracoviensia 17:193–215.

Koteja, J. & B. Zak-Ogaza. 1979. Five species of Pseudococcidae and Eriococcidae (Homoptera) new to the Polish fauna. Polskie Pismo Entomol. 49:671–675.

Koteja, J. & B. Zak-Ogaza. 1981(*Kaweckia* gen. n. in the Eriococcidae (Homoptera, Coccoidea) and notes on related genera. Acta Zool. Cracoviensia. 25(19):501–518.

Koteja, J. & B. Zak-Ogaza. 1981a. The life history of *Steingelia gorodetskia* Nassonov (Homoptera, Coccinea). Ann. Zool. (Warszawa) 36:171–186.

Koteja, J. & B. Zak-Ogaza. 1983. The Coccinea fauna (Homoptera) of the Kraków–Częstochowa Upland (southern Poland) (In Polish). Acta Zool. Cracoviensia 26:465–490.

Koteja, J., B. Zak-Ogaza, & A. El-Nabawi. 1978. Notes on *Heterococcus* Ferris and four mealybugs (Homoptera, Pseudococcidae) new to the Polish fauna. Bull. Entomol. Pologne. 48:501–504.

Kozár, F. 1971. Recent data on Coccoidea (Hom.) damaging Gramineae. (In Hungarian). Folia Entomol. Hung. 24(11):157–162.

Kozár, F. 1972. A new scale insect in our fauna: *Lecanopsis porifera* Borchs. (Homoptera, Coccoidea) that occasionally is a pest of grain crops (In Hungarian). Növénytermelés 21(3):281.

Kozár, F. 1973. An alfalfa pest new for the Hungarian fauna: *Pseudococcus multivorus* Kir. (Homoptera, Coccoidea). (In Hungarian). Növényvédelem 9(12):538–540.

Kozár, F. 1974. Mass infestation and damage of the oak scale *Kermes quercus* L. (Homoptera, Coccoidea). (In Hungarian). Növényvédelem 19(12):534–537.

Kozár, E. 1975. Evidence of sexual attractant in San José scale *Quadraspidiotus perniciosus* Comst. (Homoptera, Coccoidea). (In Hungarian). Növényvédelem 11(8):353–355.

Kozár, F. 1976. The infestation of orchards in Hungary by scales (Homoptera, Coccoidea). (In Hungarian). Ybk. Plant Prot. Res. Inst., Budapest 14:35–47.

Kozár, F. 1978. Some questions on the population dynamics of the San José scale, *Quadraspidiotus perniciosus* Comstock (Homoptera: Coccoidea). Acta Phytopat. Acad. Sci. Hung. 13(1–2):179–195.

Kozár, F. 1979. Comparative studies on density of scale insects in European orchards. (In Hungarian). Agrártud. Közl. 38:135.

Kozár, F. 1981. *Mirococcopsis nagyi* sp. n. and *Luzulaspis rajae* sp. n. from Hungary (Homoptera, Coccoidea). Acta Zool. Acad. Sci. Hung. 27:315–321.

Kozár, F. 1981a. Data on the Coccoidea (Homoptera) fauna of the Hortobágy National Park. *In* The Fauna of the Hortobágy Nat. Park. p. 89–90.

Kozár, F. 1983. *Chorizococcus viktorinae* sp. n. and *Ferreroaspis* gen. n. in the Hungarian fauna (Homoptera: Coccoidea, Pseudococcidae and Diaspididae). Folia Entomol. Hung. 44(1):89–93.

Kozár, F. 1983a. New and little known scale-insect species from Yugoslavia (Homoptera: Coccoidea). Acta Zool. Acad. Sci. Hung. 29(1–3):139–149.

Kozár, F. 1984. Recent data to the knowledge of the scale-insect (Homoptera: Coccoidea) fauna of Hungary. (In Hungarian). Állat. Közl. 71:119–131.

Kozár, F., & E. M. Danzig. 1976. *Atrococcus bejbienkoi* sp. n., and some scale insects new to the Hungarian fauna (Homoptera, Coccoidea). Acta. Zool. Acad. Sci. Hung. 22(1–2):65–67.

Kozár, F. & M. Kosztarab. 1976. A new mealybug, *Chnaurococcus danzigae* sp. n. from Hungary (Homoptera: Coccoidea, Pseudococcidae). Acta Zool. Acad. Sci. Hung. 22(3–4):299–302.

Kozár, F. & M. Kosztarab. 1982. Coccoidea of Central European forests and their host relationships. Acta Mus. Reginaehradecensis S. A. Suppl. (1980):203–211.

Kozár, F. & D. Matile-Ferrero, 1983. Two new species of armoured scale-insects from Hungary (Homoptera, Coccoidea: Diaspididae). Acta Zool. Acad. Sci. Hung. 29(4):389–395.

Kozár, F., G. Ördögh, & M. Kosztarab. 1977. New records to the Hungarian scale insect fauna (Homoptera, Coccoidea). (In Hungarian). Folia Entomol. Hung. (n.s.) 30(1):69–71.

Kozár, F. & E. S. Sugonjaev. 1979. Contribution to the knowledge of parasites of coccoids (Homoptera, Coccoidea). (In Hungarian). Folia Entomol. Hung. 32(2):234–236.

Kozár, F., A. Tranfaglia., & G. Pellizzari. 1984. New records on the scale insect fauna of Italy (Homoptera: Coccoidea). Boll. Lab. Entomol. Agr. Filippo Silvestri, 41:3–10.

Kozár, F., M. Tsalev, R. Viktorin & J. Horváth. 1979. New data to the knowledge of the scale-insects of Bulgaria (Homoptera, Coccoidea). Folia Entomol. Hung. 32(2):129–132.

Kozár, F. & R. A. Viktorin. 1978. Survey of scale insects (Homoptera, Coccoidea) infestations in European orchards. Changes in the scale infestation levels in Hungarian orchards between 1971 and 1976. Acta Phytopathol. Acad. Sci. Hung. 19(3–4):391–402.

Kozár, F. & G. Vinis. 1980. Redescription of *Ritsemia pupifera* Lichtenstein, 1879, reestablishment of the genus and erection of the tribe Ritsemini (Homoptera: Coccoidea, Pseudococcidae). Folia Entomol. Hung. 41(1):97–103.

Kozarzhevskaya, E. & J. Reitzel. 1975. The scale insects (Homoptera: Coccoidea) of Denmark. Stat. Forsogsvirksomhed & Plantekult. 1226:40 pp.

Kozarzhevskaya, E. & J. Reitzel. 1977. Some soil mealybugs (Homoptera: Pseudococcidae) found in Denmark. Stat. Planteavlforsog. Tidsskrift for Planteavl 81:374–380.

Kraatz, G. 1888a. *Signoretia* Kraatz statt *Westwoodia* Sign. Deut. Entomol. Zeitschr. 32:176.

Kraatz, G. 1888b. Synonymische Bemerkung. Deut. Entomol. Zeitschr. 32:360.

Künckel d'Herculais, J. 1878. Histoire de la Aochenille vivant sur les Racines des Palmiers de la Section des Seafortha. Exposé des caractères du genre *Rhizoecus*. Ann. Soc. Entomol. France (Ser. 5) 8:161–164.

Künow, G. 1880. Zwei neue Schildläuse. Entomol. Nachr. 6(5):46–47. (English transl. by J. W. Douglas in Entomol. Monthly Mag. 17:89–90).

Kuwana, I. 1905. A new *Xylococcus* in Japan. (In Japanese). Insect World 9(3):91–95.

Kuwana, I. 1926. The Diaspine Coccoidea of Japan, IV. Genera *Cryptoparlatoria, Howardia, Sasakiaspis* (n. gen., *Diaspis, Aulacaspis, Pinnaspis,* and *Prontaspis*. Japan Dept. Finance, Imp. Plant Quar. Serv. Tech. Bull. 4:44 pp.

Kuwana, I. 1931. The genus *Kermes* of Japan. Japan Min. Agr. Forestry, Bull. Dept. Agr. Sci. 2:15–29.

Kuwana, I. 1932. The genus *Aclerda* in Japan, including Formosa. Philippine J. Sci. 48:57–66.

Kuznetsov. N. N. 1966. Coccid-pests of conifers. (In Russian). Zashch. Rast. 11(7):46–47.

Kuznetsov, N. N. 1967. Coccids (Homoptera, Coccoidea) of the Crimean conifers (Fauna, biology and measures of the control). *In* Pests and Diseases of Fruit and Ornamental Plants. (In Russian). Gosud. Ord. Trud. Krasn. Znameni Nikitskogo Bot. Sad., Trudy 39: 219–304.

Kuznetsov, N. N. 1970. Four little known species of scale insects (Homoptera, Diaspidoidea) damaging conifers. (In Russian). Zool. Zhur. 49(11):1644–1655.

Lacroix, D. S. 1926. Miscellaneous observations on a cranberry scale *Targionia dearnessi* (Ckll.) (Homopt.: Coccoidea). Entomol. News 37:249–251.

La Face, L. 1916. Le metamorfosi dell' *Aclerda berlesii* Buffa (Hemiptera-Coccoidea). Bol. Portici R. Scuola Super. di Agr. Lab. Zool. Gen. e Agr. 11:235–249.

La Face, L. 1921. Osservazioni morfologiche e biologiche sull' *Antonina phragmitis* Marchal. (Hemiptera: Coccoidea). Bol. Portici R. Scuola Super. Agr. Lab. Zool. Gen. Agr. 15:254–267.

Laing, F. 1929. Descriptions of new and some notes on old species of Coccoidea. Ann. & Mag. Nat. Hist. (Ser. 10) 4:465–501.

Lambdin, P. L. & M. Kosztarab. 1973. A revision of the seven genera related to *Lecanodiaspis* (Homoptera: Coccoidea: Lecanodiaspididae). Va. Polytech. Inst. & State Univ. Res. Div. Bull. 83:110 pp.

Lambdin, P. L. & M. Kosztarab. 1977. Morphology and systematics of the adult females of the genus *Cerococcus*. (Homoptera: Coccoidea: Cerococcidae). Va. Polytech. Inst. & State Univ. Res. Div. Bull. 128:251 pp.

Landis, B. J. 1968. *Asterolecanium arabidis*, a scale attacking sugar beets. J. Econ. Entomol. 61(3):871–872.

Leach, W. E. 1815. Entomology. *In* Brewster, D. The Edinburgh Encyclopaedia. 9, Pt. 1:57–172.

Lellakova-Dušková, F. 1959. Männchen der Schildläuse *Diaspidiotus wunni* (Lindinger) (Coccoidea, Diaspididae). Acta. Soc. Entomol. Cechosl. 56(4):338–342.
Lellakova-Dušková, F. 1963. The morphology, metamorphosis, and life-cycle of the scale insect *Quadraspidiotus gigas* (Thiem et Gerneck) (Homoptera, Coccoidea, Diaspididae) Acta. Entomol. Mus. Nat. Pragae 35:611–648.
Lellakova-Dušková, F. 1965. *Quadraspidiotus marani* Zahr., the morphology of developmental stages in the male (Homoptera, Coccoidea, Diaspididae). Acta Entomol. Bohemoslov. 62(3):202–209.
Lellakova-Dušková, F. 1966. Die auf Weidengewächsen (Salicales) lebenden Schildläuse (Coccoidea) in der Tschechoslowakei. Zeitschr. f. angew. Entomol. 57(3):294–309.
Leonardi, G. 1899. Sistema delle "Parlatoriae." Nota preventiva. Riv. Patol. Veg. 8:203–209.
Leonardi, G. 1903. Generi e specie di Diaspiti. Saggio di sistematica delle Parlatoriae. Ann. Portici R. Scuola Super. Agr. (Ser. 2) 5(1904):59 pp.
Leonardi, G. 1906. Generi e specie di Diaspiti. Saggio di sistematica della Leucaspides. Ann. Portici R. Scuola Super. Agr. (Ser. 2) 6:32 pp.
Leonardi, G. 1907. Contribuzione alla conoscenza delle Cocciniglie Italiane. Bol. Portici R. Scuola Super. Agr. Lab. Zool. Gen. Agr. 1:135–169.
Leonardi, G. 1908. Seconda contribuzione alla conoscenza della Cocciniglie Italiane. Bol. Portici R. Scuola Super. Agr. Lab. Zool. Gen. Agr. 3:150–191.
Leonardi, G. 1909. Chermotheca Italica ... Cocciniglie raccolte in Italia. Fasc. 5 (Nos 101–125). Portici, Vesuviano.
Leonardi, G. 1913. Nouve specie di Cocciniglie raccolte in Italia. Bol. Portici R. Scuola Super. Agr. Lab. Zool. Gen. Agr. 7:59–65.
Leonardi, G. 1920. Monografia delle Cocciniglie Italiane. Portici, Della Torre. 555 pp.
Lichtenstein, J. 1879. *Ritsemia pupifera*, eine neue Schildlaus. Entomol. Ztg. 40(1–3):387–389.
Lichtenstein, J. 1881. *Aspidiotus pyri* Licht. Ann. Soc. Entomol. de France (Bull. Entomol.) (Ser. 6) 1:51–52.
Lichtenstein, J. 1882a. Remarks on an aphid and some coccids. Ann. Soc. Entomol. France (Ser. 6) 2:74–75.
Lichtenstein, J. 1882b. Les mâles de quelques Coccidiens. Bull. Soc. Entomol. Ital. 14:329–330.
Lichtenstein, J. 1882c. Bemerkungen zu *Coccus rubi* Schrank. Wien. Entomol. Ztg. 1:124.
Lindeman, K. 1886. Barley louse (*Westwoodia hordei* Lindeman), and injurious insects of South Russia. (In Russian). From No. 8 "Notes" Imp. Obshch. Sel. Khoz., S. Russia, 6 pp.
Lindinger, L. 1905a. Zwei neue Schildläuse aus Asien. Insekten Börse 22:131–132.
Lindinger, L. 1905b. Zwei neue Arten der Coccidengattung *Leucaspis*. Zool. Anz. 29:252–254.
Lindinger, L. 1906a. Die Schildlausgattung *Leucaspis*. Jahrb. Hamburg, Wiss. Anst. 23(1905) Beih. 3:60 p.
Lindinger, L. 1906b: *Lecanium sericeum* n. sp. Insekten Börse 23(37):147.
Lindinger, L. 1907. Fränkische Cocciden. Entomol. Bl. 3:113–117; 136–139. Also in Hamburg Bot. Staatsinst., Abt. f. Pflanzenschutz 9 (1906/07), 8 p.
Lindinger, L. 1909. Beiträge zur Kenntnis der Schildläuse und ihrer Verbreitung. Zeitschr. f. Wiss. Insektenbiol. 5:105–110; 147–152; 220–226.
Lindinger, L. 1911a. Afrikanische Schildläuse. IV. (Kanarische Cocciden, ein Beitrag zur Fauna der Kanarischen Inseln.) Jahrb. Hamburg. Wiss. Anst. 28(1910) Beih. 3:1–38.
Lindinger, L. 1911b. Beiträge zur Kenntnis der Schildläuse und ihrer Verbreitung II. (In part). Zeitschr. f. Wiss. Insektenbiol. 7:9–12, 86–90, 126–130, 172–177, 244–247, 353–358, 378–383.
Lindinger, L. 1912. Die Schildläuse (Coccidae) Europas, Nordafrikas und Vorderasiens, einschliesslich der Azoren, der Kanaren und Madeiras. Mit Anleitung zum Sammeln, Bestimmen und Aufbewahren. Ulmer, Stuttgart 388 pp.
Lindinger, L. 1921. Tätigkeitsbericht der Schädlingsabteilung des Instituts für Angewandte Botanik zu Hamburg für die Zeit vom 14. Februar bis zum 30. Juni 1920. Zeitschr. f. angew. Entomol. 7:424–440.
Lindinger, L. 1923. Einführung in die Kenntnis der Deutschen Schildläuse. Entomol. Jahrb. 32:138–152.
Lindinger, L. 1933. Beiträge zur Kenntnis der Schildläuse (Hemipt.-Homopt. Coccid). Entomol. Anz. 13:143.
Lindinger, L. 1934. Was ist der richtige Name von *Lecanium corni* Marchal? Zeitschr. f. Pflanzenkrank. 44:76–80.
Lindinger, L. 1935. Neue Beiträge zur Kenntnis der Schildläuse (Coccoidea). Entomol. Zeitschr. 49:121–123.
Lindinger, L. 1936. Neue Beiträge zur Kenntnis der Schildläuse (Coccoidea). Entomol. Jahrb. 45:148–167.

Lindinger, L. 1937. Verzeichnis der Schildlaus-Gattungen. (Homoptera-Coccoidea Handlirsch 1903). Entomol. Jahrb. 46:178–198.
Lindinger, L. 1938a. Bemerkenswerte Schildlausfunde. Bombus 6:22–23.
Lindinger, L. 1938b. Verzeichnis der aus Nordwest-Deutschland, insbesondere aus Gross-Hamburg, gemeldeten Schildläuse (Homoptera-Coccoidea). Verhandl. Vereins f. naturw. Heimatforschung, Hamburg. (1937)26:1–15.
Lindinger, L. 1943. Verzeichnis der Schildläus–Gattungen. I. Nachtrag. (Homoptera Coccoidea). Zeitschr. der Wien. Entomol. Gesell. 28:205–208, 217–224, 264–265.
Lindinger, L. 1949. Einige Mitteilungen über Schildläuse (Homopt., Coccoidea). Entomol. Internatl. Zeitschr. f. Gesam. Insektenkunde 1:210–213.
Lindinger, L. 1957. Ein weiterer Beitrag zur Synonymie der Cocciden. Beitr. z. Entomol. 7:543–553.
Linnaeus, C. 1758. Insecta Hemiptera. *Coccus.* Systema Nat. (Ed. 10) 1:455–457.
Linnaeus, C. 1761. Insecta Hemiptera. *Coccus. In* Fauna Suecica, Ed. 2. Stockholmiae, Laur. Salvii. 46 p. (without page numbers), 578 p. (264–266, 548).
List, J. 1886. *Orthezia cataphracta* Shaw. Eine Monographie. Zeitschr. Wiss. Zool. 45:1–85. Reprinted in Zool. Inst. Graz, Arb. (1887)1:201–286.
Lobdell, G. H. 1929. Two new species of *Eriococcus* from Mississippi. (Homoptera: Coccoidea). Ann. Entom. Soc. Amer. 22:762–767.
Lobdell, G. H. 1930. Twelve new mealybugs from Mississippi (Homoptera: Coccoidea). Ann. Entom. Soc. Amer. 23(2):209–236.
Lobdell, G. H. 1937. Two-segmented tarsi in coccids; other notes (Homoptera). Ann. Entom. Soc. Amer. 30:75–80.
Lombardi, D. 1938. Osservazioni sulla morfologia e biologia della *Targionia vitis* Sign. Bol. Inst. Entomol. Univ. Bologna 10:117–138.
Löw, F. 1872. Über *Diaspis visci* Schrank, eine auf der Mistel lebende Schildlaus. Verh. Zool.-Both. Gesell. Wien. 22:273–278.
Löw, F. 1882. Eine neue Cocciden-art *(Xylococcus filiferus).* Verh. Zool.-Bot. Gesell. Wien. 32:271–278.
Löw, F. 1883a. Eine neue Fichten–Schildlaus. Wien. Entomol. Ztg. 2:267–269.
Löw, F. 1883b. Über eine neue Nadelholz-Coccide und den Dimorphismus der Cocciden-Männchen. Wien. Entomol. Ztg. 2:3–7.
Löw, F. 1883c. Eine neue Cocciden. Wien. Entomol. Ztg. 2:115–117.
Lull, R. S. 1899. A new species of *Pulvinaria.* Entomol. News 10:237–242.
Lupo, V. 1938. Revisione delle Cocciniglie Italiane. I. *Diaspis.* II. *Howardia, Adiscodiaspis, Chionaspis, Pinnaspis, Fiorinia, Kuwanaspis.* Bol. Portici R. Ist. Super. Agr. Lab. Zool. Gen. Agr. 30:121–162, 255–322.
Lupo, V. 1939. Revisione delle Cocciniglie Italiane. III. *Mytilococcus.* Bol. Portici R. Ist. Super. Agr. Lab. Zool. Gen. Agr. 31:69–136.
Lupo, V. 1943. Il *Mytilococcus ficifoliae* (Berlese) e una forma estiva del *M. conchiformis* (Gmelin). Bol. Portici R. Ist. Super. Agr. Lab. Entomol. Agr. 5:196–205.
Lupo, V. 1944. Revisione delle Cocciniglie Italiane. IV. Gen. *Leucaspis* Targioni. Bol. Portici R. Ist. Super. Agr. Lab. Entomol. Agr. 5:206–242.
Lupo, V. 1948. Revisione delle Cocciniglie Italiane. VI. *Aspidiotus, Quadraspidiotus, Diaspidiotus, Rhizaspidiotus, Nuculaspis.* Bol. Portici R. Ist. Super. Agr. Lab. Entomol. Agr. 8:137–208.
Lupo, V. 1954. Revisione delle Cocciniglie Italiane. IX. Gen. *Separaspis, Gonaspidiotus, Chorizaspidiotus.* Bol. Portici R. Ist. Super. Agr. Lab. Entomol. Agr. "Filippo Silvestri", 13:1–23.
Lupo, V. 1957 Revisione delle Cocciniglie Italiane. XI. (Gen. *Targionia, Rhizaspidiotus, Aonidia*). Bol. Portici R. Ist. Super. Agr. Lab. Entomol. Agr. "Filippo Silvestri" 15:54–84.
Lupo, V. 1965. Il Genere *Carulaspis* McGill. e le specie Paleartiche ad esso Appartenenti. Accad. Naz. Ital. Entomol. Rend. 13:34–48.
Lupo, V. 1966. Vecchia e nuova sistematica delle Cocciniglie. VI. Cong. Naz. Ital. Entomol. (Padova, 1965) 1966:24–44.
MacGillivray, A. D. 1921. The Coccidae. Urbana, Ill. Scarab, 502 pp.
Mahdihassan, S. 1933. Cryptogamie – sur les differents symbiotes des Cochenilles productrices ou non productrices de Cire. Acad. Sci. (Paris) Compt. Rend. 196:560–562.
Malenotti, E. 1918. I nemici naturali della bianca-rossa (*Chrysomphalus dictyospermi* Morg.). Redia 13:17–51.

Mamet, R. 1950. Notes on the Coccoidea of Madagascar.-I. Inst. Sci. Madagascar Mem., Ser. A. 4:17–38.
Marchal, P. 1904. Sur une Cochenille nouvelle. Ann. Soc. Entomol. France 73:557–561.
Marchal, P. 1908. Notes sur les Cochenilles de l'Europe et du Nord de l'Afrique (1re partie). Ann. Soc. Entomol. France 77:223–309.
Marchal, P. 1909. Sur les Cochenilles du midi de la France et de la Corse. Acad. Sci. (Paris) Compt. Rend. 148:871–872.
Maskell, W. M. 1879. On some Coccidae in New Zealand. Trans. & Proc. New Zeal. Inst. 11(1878):187–228.
Maskell, W. M. 1887. An account of the insects noxious to agriculture and plants in New Zealand, the scale insects (Coccidae). Wellington, Didsbury. 116 pp.
Maskell, W. M. 1892. Further coccid notes: with descriptions of new species, and remarks on Coccidae from New Zealand, Australia, and elsewhere. Trans. & Proc. New Zeal. Inst. 24(1891):1–64.
Maskell, W. M. 1894. Remarks on certain genera of Coccidae. Entomologist 27:93–95.
Maskell, W. I. 1895. Further coccid notes: with descriptions of new species from New Zealand, Australia, Sandwich Island, and elsewhere and remarks upon many species already reported. Trans. & Proc. New Zeal. Inst 27(1894):36–75.
Maskell, W. M. 1897a. On a collection of Coccidae, principally from China and Japan. Entomol. Monthly Mag. 33:239–244.
Maskell, W. M. 1897b. Further coccid notes: with descriptions of new species and discussions of points of interest. Trans. & Proc. New Zeal. Inst. 29(1896):293–331.
Maskell, W. M. 1898. Further coccid notes: with descriptions of new species, and discussion of points of interest. Trans. & Proc. New Zeal. Inst. 30(1897):219–252.
Matesova, G. I. 1955. Contribution to the knowledge of the coccid and diaspine fauna of southeastern Kazakhstan. (In Russian). Trudy Inst. Zool. Akad. Nauk Kazakh. SSR. 4:199–205.
Matesova, G. I. 1957. Nouveaux coccides (Homoptera, Coccoidea) de Kazakhstan Meridio-Orientale. (In Russian & French). Entomol. Obozr. 36(1):163–174.
Matesova, G. I. 1960. New species of soft scales, fam. Pseudococcidae (Homoptera, Coccoidea) of the Kazakhstan fauna. (In Russian). Akad. Nauk Kazakh. SSR Trudy Inst. Zool. (Ent.) 11:205–217.
Matesova, G. I. 1968a. Coccids of Oriental Kazakhstan (Homoptera, Coccoidea). (In Russian). Akad. Nauk Kazakh. SSR Trudy Inst. Zool. 30:102–129.
Matesova, G. I. 1968b. New species of mealybugs (Homoptera, Coccoidea, Pseudococcidae) from East Kazakhstan. (In Russian). Entomol. Obozr. 47(1): 151–159, also in Entomol. Rev. 47(1):85–97.
Matesova, G. I. 1971. An ecological-geographical description of the coccids of the South-East Kazakhstan. (In Russian). Akad. Nauk Kazakh. SSR Trudy Inst. Zool. 32:24–32.
Matile-Ferrero, D. 1970. Hemiptera (Homoptera): Coccoidea. South African Animal Life 14:171–185.
Matile-Ferrero, D. 1983. Espèces nouvelles de Cochenilles associées à une pelouse alpine (Insecta, Homoptera, Coccoidea, Pseudococcidae). Rev. Suisse Zool. 90 (3):629–638.
McClure, M. S. 1980. Foliar nitrogen: A basis for host suitability for elongate hemlock scale, *Fiorinia externa* (Homoptera: Diaspididae). Ecology 61(1):72–79.
McConnell, H. S. 1943. Notes on the anatomy of the coccid genus *Aclerda* and descriptions of three new species. Proc. Wash. Entomol. Soc. 45:99–115.
McKenzie, H. L. 1943a. The seasonal history of *Matsucoccus vexillorum* Morrison (Homoptera: Coccoidea: Margarodidae). Microentomol. 8(2):42–52.
McKenzie, H. L. 1943b. Notes on *Matsucoccus vexillorum* Morrison (Homoptera: Coccoidea: Margarodidae). Microentomol. 8(2):53–57.
McKenzie, H. L. 1945. A revision of *Parlatoria* and closely allied genera (Homoptera: Coccoidea: Diaspididae). Microentomol. 10(2):47–121.
McKenzie, H. L. 1956. The armored scale insects of California. Calif. Insect Surv. Bull. 5:209 pp.
McKenzie, H. L. 1960. Taxonomic study of California mealybugs with descriptions of new species (Homoptera: Coccoidea: Pseudococcidae). Hilgardia 29(15):681–770.
McKenzie, H. L. 1964. Fourth taxonomic study of California mealybugs, with additional species from North America, South America, and Japan (Homoptera: Coccoidea: Pseudococcidae). Hilgardia 35(10):211–272.
McKenzie, H. L. 1967. Mealybugs of California with taxonomy, biology, and control of North American species (Homoptera: Coccoidea: Pseudococcidae). Univ. Calif. Press. 526 pp.
Melis, A. 1943. Contributo alla conoscenza dell' *"Aspidiotus perniciosus"* Comst. Redia 29:1–170.
Melis, A. 1951. Preciasazioni morfo-biologiche sull *'Aspidiotus perniciosus* Comst. Redia (Ser. 2) 36:1–91.

Mesnil, L. P. 1949. Remarques sur quelques Diaspines du Pin silvestre ou du génévrier et leurs parasites. Rev. Pathol. Vég. Entomol. Agr. France 28:73–97.
Miller, D. R. 1969. A systematic study of *Eriococcus* Targioni-Tozzetti, with a discussion of the zoogeography of the Eriococcidae. Ph. D. Thesis, Univ. Calif. Davis. 253 pp.
Miller, D. R. 1970. Review of the mealybug genus *Heterococcus* (Homoptera: Coccoidea: Pseudococcidae) with a description of a new species. Ann. Entomol. Soc. Amer. 63(2):438–453.
Miller, D. R. 1974. Two new *Heliococcus* species, a key to the North American species, and a list of world species (Homoptera: Coccoidea: Pseudococcidae). Pan-Pacific Entomol. 50(2):177–192.
Miller, D. R. 1975. A revision of the genus *Heterococcus* Ferris with a diagnosis of *Brevennia* Goux (Homoptera: Coccoidea: Pseudococcidae). U. S. Dept. Agr., Agr. Res. Serv. Tech. Bull. 1497:61 pp.
Miller, D. R. 1984. Phylogeny and classification of the Margarodidae and related groups (Homoptera: Coccoidea). Verh. SIEEC X. (1983) Budapest. p. 321–324.
Miller, D. R., & J. E. Appleby. 1971. A redescription of *Phenacoccus dearnessi* (Homoptera: Coccoidea: Pseudococcidae). Ann. Entomol. Soc. Amer. 64:1342–1357.
Miller, D. R., & M. Kosztarab. 1979. Recent advances in the study of scale insects. Ann. Rev. Entomol. 24:1–27.
Miller, D. R., & H. L. McKenzie. 1967. A systematic study of *Ovaticoccus* Kloet and its relatives, with a key to North American genera of Eriococcidae (Homoptera: Coccoidea: Eriococcidae). Hilgardia 38(13):471–539.
Miller, D. R. & H. L. McKenzie. 1970. Review of the mealybug genus *Heterococcus* (Homoptera: Coccoidea: Pseudococcidae) with a description of a new species. Ann. Entomol. Soc. Amer. 63:438–453.
Miller, D. R., & H. L. McKenzie. 1971. Sixth taxonomic study of North American mealybugs, with additional species from South America (Homoptera: Coccoidea: Pseudococcidae). Hilgardia 40(17):565–602.
Miller, D. R., & H. L. McKenzie. 1973. Seventh taxonomic study of North American mealybugs (Homoptera: Coccoidea: Pseudococcidae). Hilgardia 41(17):489–542.
Miller, D. R., & D. J. Williams. 1976. Proposed conservation of the family-group name Eriococcidae Cockerell, 1899 (Insecta, Homoptera) and the designation of a type-species for *Eriococcus* Targioni-Tozzetti, 1868 under the plenary powers. Bull. Zool. Nomencl. 33(2):118–123.
Mitic-Muzina, N. 1968. The cycle of developments of the wax louse (*Ceroputo pilosellae* Schultz, Homoptera, Coccoidea). (In Croatian). Zast. Bilja (Beograd) 19:303–317.
Modeer, A. 1778. Om fästflyet, *Coccus*. Goethebrgsk Vetensk. och Vitterhets, Handl. (Ser. 1) 1:11–50.
Modugno., A. 1953. Morfologia e variabilità die caratteri in *Chionsapis euonymi* Comst. Bol. Zool. Agr. Bachicolt. 19:113–135.
Morgan, A. C. F. 1892. Observations on Coccidae (No. 9). Entomol. Monthly Mag. 28:12–16.
Morrison, H. 1925. Classification of scale insects of the subfamily Ortheziinae. J. Agr. Res. 30(2):97–154.
Morrison, H. 1928. A classification of the higher groups and genera of the coccid family Margarodidae. U. S Dept. Agr. Tech. Bull. 52:239 pp.
Morrison, H. 1939. Descriptions of new species of *Matsucoccus* (Hemiptera: Coccidae). Proc. Wash. Entomol. Soc. 41:1–20.
Morrison, H. 1945. The mealybug genus *Heterococcus* Ferris and some of its relatives (Homoptera: Coccoidea). J. Wash. Acad. Sci. 35(2):38–55.
Morisson, H. 1952. Classification of the Ortheziidae. Supplement to "Classification of scale insects of the subfamily Ortheziinae." U. S. Dept. Agr. Tech. Bull. 1052:80 pp.
Morrison, H. 1954. A second species of *Ortheziola* Šulc. Proc. Wash. Entomol. Soc. 56:120–123.
Morrison, H., & E. Morrison. 1922. A redescription of the type species of the genera of Coccidae based on species originally described by Maskell. Proc. U. S. Natl. Mus. 60, Art. 12(No. 2407):120 pp.
Morrison, H., & E. Morrison 1965. A selected bibliography of the Coccoidea, 1st Suppl. U. S. Dept. Agr., Agr. Res. Serv. Misc. Pub. 987:44 pp.
Morrison, H., & E. R. Morrison. 1966. An annotated list of generic names of the scale insects (Homoptera: Coccoidea). U. S. Dept. Agr., Agr. Res. Serv. Misc. Pub. No. 1015:206 pp.
Morrison, H., & A. V. Renk. 1957. A selected bibliography of the Coccoidea. U. S. Dept. Agr., Agr. Res. Serv. Misc. Pub. 734:222 pp.
Müller, R. P., & H. Eisenschmidt. 1954. *Quadraspidiotus schneideri* Bachmann (= *marani* Zahradnik). eine der San José-Schildlaus ähnliche Deckelschildlaus. Nachrichtenbl. Deut. Pflanzenschutzdienst, N. F. 8:151–153.

Munting, J. 1971. On the identity of the grey scale *Diaspidiotus africanus* Marlatt and some related species from southern Africa (Homoptera: Diaspididae). J. Entom. Soc. So. Africa 34(1):119–143.

Murakami, Y. 1970. A review of biology and ecology of diaspine scales in Japan (Homoptera, Coccoidea). Mushi 43 (pt. 7):65–114.

Nakahara, S. 1975. A mealyburg, *Chaetococcus phragmitis* (Marchal). New Jersey Coop. Econ. Insect Rpt. 25(45–48):875.

Nakahara, S. 1981. The proper placements of the Nearctic soft scale species assigned to the genus *Lecanium* Burmeister (Homoptera: Coccidae). Proc. Wash. Entomol. Soc. 83(2):283–286.

Nakahara, S. 1982. Checklist of the armored scales (Homoptera: Diaspididae) of the conterminous United States. U. S Dept. Agr. Animal & Plant Health Insp. Serv., Plant Prot. & Quar. 110 pp.

Nassonov, N. V. 1908a. *Steingelia gorodetskia*, nov. gen. et nov. sp. Nouveau genre et espèce des Coccides du groupe *Xylococcini* (In Russian). Ann. Mus. Zool. Acad. Imp. Sci. St. Petersbourg 13:345–352.

Nassonov, N. V. 1908b. Sur quelques nouvelles Coccides. Ann. Mus. Zool. Acad. Imp. Sci. St. Petersbourg 13:471–499.

Nassonov, N. V. 1910. Sur les transformations de *Kermes quercus* (Linn.) (In Russian). Bull. Acad. Imp. Sci. St. Petersbourg (Ser. 6) 4:47–60.

Neave, S. A. (Ed.) 1950. Nomenclator Zoologicus. Vol. V. 1936–1945. Zool. Soc. London, England. 308 pp.

Newell, W. & T. D. A. Cockerell. 1898. Descriptions of new coccids. *In* Notes on Coccidae occurring in Iowa by Osborn, H. Proc. Iowa Acad. Sci. 5(1897):224–231.

Newstead, R. 1891. On some new or little known Coccidae found in England. Entomol. Monthly Mag. (2nd Ser.) 2:164–166.

Newstead, R. 1892. On new or little known Coccidae, chiefly English (No. 2). Entomol. Monthly Mag. 28: 141–147.

Newstead, R. 1893. New or little known Coccidae, chiefly English (No. 3.) Entomol. Monthly Mag. 29:77–79.

Newstead, R. 1893a. Observations on Coccidae (No. 6). Entomol. Monthly Mag. 4:205–210.

Newstead, R. 1893b. A new coccid in an ant's nest. Entomol. Monthly Mag. 29:138.

Newstead, R. 1894. Observations on Coccidae (No. 10). Entomol. Monthly Mag. 30:232–234.

Newstead, R. 1895a. Observations on Coccidae (No. 11) Entomol. Monthly Mag. 31:165–167.

Newstead, R. 1895b. Observations on Coccidae (No. 13). Entomol. Monthly Mag., (2nd Ser.) 6:233–236.

Newstead, R. 1896. Observations on Coccidae (No. 14). Entomol. Monthly Mag. 32:57–60.

Newstead, R. 1897a. IV.–New Coccidae collected in Algeria by the Rev. Alfred E. Eaton. Trans. Entomol. Soc. London 1897:93–103.

Newstead, R. 1897b. Observations on Coccidae (No. 16). Entomol. Monthly Mag. 33:165–171.

Newstead, R. 1898. Observations on Coccidae (No. 17). Entomol. Monthly Mag. 34:92–99.

Newstead, R. 1901. Monograph of the Coccidae of the British Isles. Vol. 1. London. Ray Society. 220 pp.

Newstead, R. 1903. Monograph of the Coccidae of the British Isles, Vol. 2. London, Ray Society. 270 pp.

Newstead, R. 1907. *Ripersia donisthorpei* and *R. formicarii* n. spp. *In* Donisthorpe, Entomol. Rec. & J. Variation 19:5.

Newstead, R. 1920. Observations on scale-insects (Coccidae).–VI. Bull. Entomol. Res. 10:175–207.

Nikolova, V. 1969. Entomocoenological and biological studies in *Rosa damascena* Mill. plantations. III. Homoptera (In Bulgarian). Bulgar. Akad. Nauk Zool. Inst. Muz. Izv. 29:191–208.

Nur, U. 1962. Sperms, sperm bundles and fertilization in a mealybug, *Pseudococcus obscurus* Essig. J. Morphol. 111:173–199.

Nur, U. 1980. Evolution of unusual chromosome systems in scale insects (Coccoidea: Homoptera). Insect Cytogenetics. Symp. Roy. Entomol. Soc. London: No. 10:97–117. Ed. by R. L. Blackman, G. M. Hewitt & M. Ashburner. Publ. by Blackwell Sci. Publ, Boston.

Olafsen, E. 1772. Reise Igiennem Island–I. Soroe, p. 610–611.

Olmi, M. & A. Sampo. 1975. Scoperta in Val d'Aosta di una Cocciniglia boreo-alpina nuova per l'Italia: *Arctorthezia cataphracta* (Olafsen, 1772) (Homoptera Coccoidea, Ortheziidae). Rev. Valdotaine Hist. Nat. (Aosta) 29: 146–152.

Ossiannilsson, F. 1951. Bidrag till kännedomen om den svenska sköldlusfaunan (Hom. Coccoidea). Opuscula Entomol. 16:1–14.

Ossiannilsson, F. 1959. Bidrag till kännedomen om den svenska sköldlusfaunan (Hom. Coccoidea) II. Opuscula Entomol. 24:193–201.

Ossiannilsson, F. 1971. On the knowledge of Hemiptera on Kullaberg Mtn. (In Swedish). Kullabergs Natur. 14:53.

Ossiannilsson, F. 1977. Mire invertebrate fauna at Eidskog, Norway. V. Auchenorrhyncha, Psylloidea, and Coccoidea (Hem.). Norw. J. Entomol. 24:11–14.

Ossiannilsson, F. 1984. On the distribution of Ortheziidae (Homoptera: Coccoidea) in Sweden. (In Swedish). Entomol. Tidskr. 105:123–127.

Paik, W. H. 1978. Illustrated flora & fauna of Korea. Insecta (VI) (In Korean). Min. Ed. Vol. 22:481 pp.

Panis, A. 1968. Les *Rhizoecus* (Hom. Coccoidea Pseudococcidae) Européens et Méditerranéens d'intérêt économique. Description d'une nouvelle espèce. Ann. Soc. Entomol. France (n. s.) 4(3):549–554.

Panis, A. 1969a. Observations faunistiques et biologiques sur quelques Pseudococcidae (Homoptera, Coccoidea) vivant dans le midi de la France. Ann. Zool. Ecol. Anim. 1(3):211–244.

Panis, A. 1969b. *Aphycus apicalis* (Dalman) et *Leptomastidea bifasciata* (Mayr) (Hym. Encyrtidae), Parasites de *Phenacoccus aceris* (Signoret) (Hom. Pseudococcidae) dans les Alpes-Maritimes. Entomophaga 14(4):383–391.

Parker, H. L. 1960. Parasites and predators associated with some grass scales found in France. Proc. Wash. Entomol. Soc. 62(3):167–179.

Patch, E. M. 1905. The cottony grass scale. Bull. Maine Agr. Exp. Sta. 121:169–179.

Patterson, R. W. 1901. Notes on *Cerococcus*. Proc. Calif. Acad. Sci. (Ser. 3, Zool.) 2:387–398.

Pegazzano, F. 1955. *Quadraspidiotus marani* Zahradnik f. *olivicola*. f. n. (Hem. Hom. Coccidae, Diaspini) nuovo ospite dell'olivo. Redia 40 (Ser. 2):311–324.

Pellizzari, G. 1978. Osservazioni sulla *Syngenaspis parlatoriae* Šulc e su altre cocciniglie infeudate alle Conifere. Bol. Zool. Agr. Bachicolt. Ser. II, 13(1975–76):1–21.

Pflugfelder, O. 1970. Alterung und Tod von *Kermes quercus* L. Weibchen (Coccina). Zool. Anz. 184(5/6):397–403.

Phillips, J. H. H. 1962. Description of the immature stages of *Pulvinaria vitis* (L.) and *P. innumerabilis* (Rathvon) (Homoptera: Coccoidea), with notes on the habits of these species in Ontario, Canada. Canad. Entomol. 94(5):497–502.

Phillips, J. H. H. 1963. Life history and ecology of *Pulvinaria vitis* (L.) (Hemiptera: Coccoidea), the cottony scale attacking peach in Ontario. Canad. Entomol. 95(4):372–407.

Phillips, J. H. H. 1965. Notes on species of *Lecanium* Burmeister (Homoptera: Coccoidea) in the Niagara Peninsula, Ontario, with a description of a new species. Canad. Entomol. 97(3):232–238.

Planchon, G. 1864. Le *Kermes* du chêne aux points de vue zoologique, commercial & pharmaceutique. Thesis (Pharmacist, 1st class) École Supér. Pharmacie Montpellier. 47 pp.

Podsiadlo, E. 1972. A contribution to the explanation of the status of *Asterodiaspis variolosum* (Ratzeburg) (Homoptera, Coccoidea, Asterolecaniidae) on the basis of materials from Poland. Acta Zool. Cracoviensia 17(17):389–404.

Podsiadlo, E. 1974a. Comparative morphological studies on populations of *Asteriodiaspis variolosum* (Ratzeburg) (Homoptera, Coccoidea, Asterolecaniidae) from the Swietokrzyskie Mts. (Poland). Ann. Zool. (Warszawa) 32(7):75–102.

Podsiadło, E. 1974b. Taxonomic studies on the species of *Asterodiaspis* Signoret, 1876 (Homoptera, Coccoidea, Asterolecaniidae) on oak in Poland. Acta Zool. Cracoviensia 19(20):487–530.

Podsiadlo, E. 1975. Studies on the identification and occurrence of species of the genus *Asterodiaspis* Signoret, 1876 (Homoptera, Coccoidea, Asterolecaniidae) in Poland (In Polish). Przeglad Zool. 19(2):211–216.

Podsiadlo, E. 1975a. Life cycle of *Asterodiaspis variolosa* (Ratzeburg, 1870) (Homoptera, Coccoidea, Asterolecaniidae) in Poland. Ann. Zool. (Warszava) 32(15):347–355.

Podsiadlo, E. 1976. Dispersal of individuals of *Asterodiaspis variolosa* (Ratzeburg) (Homoptera, Coccoidea, Asterolecaniidae) on host plants from their hatching place. Ekol. Polska 24(1):69–76.

Podsiadlo, E. 1981a. Chalcidoidea (Hymenoptera) reared from globose scale females, *Sphaerolecanium prunastri* (Fonscolombe) in Warsaw, Poland (Homoptera, Coccidae). Bull. Entomol. de Pologne 51:153–158.

Podsiadlo, E. 1981b. Little known and new Polish Chalcidoidea and Proctotrupoidea (Hymenoptera), parasites of Coccoidea (Homoptera). Bull. Entomol. de Pologne. 51:159–162.

Podsiadlo, E., & H Komosinska. 1976. Further investigations on the scale insect fauna (Homoptera, Coccoidea) in the Nida Valley (Southern Poland). Bull. Acâd. Polon. Sci., Ser. Biol. 24(2):87–91.

Putnam, J. D. 1878. Report on maple bark louse. Annu. Rpt. Iowa State Hort. Soc. 12(1877):317–324.

Rao, V. P. 1949. The genus *Unaspis* MacGillivray (Homoptera: Coccoidea: Diaspididae). Microentomol. 14:59–72.
Rasina, B. P. 1966. New scale species (Coccoidea, Pseudococcidae) from the Latvian SSR. (In Russian). Latv. Entomol. 11:3–29.
Ratkovsky, K. 1899. On the oak scale insect (In Hungarian). Rovartani Lapok 6:70–72.
Ratzeburg, J. T. C. 1843. Bericht über einige neue, den Waldbäumen Schädliche Rhynchoten. Stettin. Entomol. Ztg. 4:201–204.
Ratzeburg, J. T. C. 1844. Fünfte Ordnung. Hemiptera Linn. Die Forst-Insecten oder Abbildung und Beschreibung der in den Wäldern Preussens und der Nachbar Staaten als schädlich oder nützlich bekannt gewordenen Insecten. 3:314 pp.
Ratzeburg, J. T. C. 1870. Eichen Beschädigung durch Schildläuse *(Coccus)*. Tharander Forstl. Jahrb. 20:187–194.
Rau, G. J. 1938. Four more new species of mealybugs from New York State (Hemiptera: Kermidae and Pseudococcidae). Canad. Entomol. 70:157–165.
Rau, G. J. 1942. The Canadian apple mealybug, *Phenacoccus aceris* Signoret, and its allies in Northeastern America. Canad. Entomol. 74:118–125.
Reh, L. 1903–1904. Zur Naturgeschichte Mittel-und Nordeuropäischer Schildläuse. Allg. Zeitschr. f. Entomol. 8:301–308, 351–356, 407–419, 457–469; 9:12–36.
Řehaček, J. 1954. Contribution to the knowledge of the scale insect fauna Fam. Lecaniidae of Czechoslovakia (In Czech). Acta Soc. Entomol. Czechosl. 51:219–223.
Řehaček, J. 1954a. Lecaniinae of Dr. K. Šulc's collection of scale insects. (In Czech). Acta Mus. Moraviae 39:133–145.
Řehaček, J. 1956. Trois espèces nouvelles des Lecaniides pour la faune Tchèchoslovaque. (Hom. Cocc. Lecaniidae) (In Czech). Acta Faun. Entomol. Mus. Nat. Pragae 1(6):47–48.
Řehaček, J. 1957. Neue Schildlausfunde aus der Unterfamilie der Lecaniinae in der Tschechoslowakei. Acta Faun. Entomol. Mus. Nat. Pragae 2:13–18.
Řehaček, J. 1959. A new species of the genus *Luzulaspis* Ckll. (Cocccoidea, Coccidae) from Czechoslovakia. Entomol Obozr. 38(1):176–178, also in Entomol. Rev. 38(1):157–159.
Řehaček, J. 1960. Soft scale (Coccidae) fauna of Slovakia. (In Czech). Biol. Prace 6(12):88 pp.
Reitter, E. 1898. Coleopterologische Notizen. LXIII. Wien Entomol. Ztg. 17:54–56.
Reuter, E. 1904. Two new scale insects for Finland. (In Finnish). Soc. Fauna Flora Fennica, Meddel. 29(1902–03):66–67; 251.
Reyne, A. 1951. A re-description of *Ripersia corynephori* Sign. Tijdschr. Entomol. 94:189–206.
Reyne, A. 1953. *Ripersia wunni* n. sp. (Pseudococcidae). Entomol. Ber. 14(333):235–239.
Reyne, A. 1954. A redescription of *Puto antennatus* Sign. (Homoptera, Coccidea) with notes on *Ceroputo pilosellae* Šulc and *Macrocerococcus superbus* Leon. Leyden Rijks Mus. Natuurlijke Hist. Zool. Meded. 32:291–324.
Reyne, A. 1965. Nederlandse Schildluizen III. Entomol. Ber. 25:96–97.
Rice R. E. & R. A. Jones. 1977. Monitoring flight patterns of male San José scale (Homoptera: Diaspididae). Canad. Entomol. 109(10):1403–1404.
Rice, R. E. & D. S. Moreno. 1970. Flight of male California red scale. Ann. Entomol. Soc. Am. 63(1):91–96.
Richards, W. R. 1958. Identities of species of *Lecanium* Burmeister in Canada (Homoptera: Coccoidea). Canad. Entomol. 90(5):305–313.
Rieux, R. 1975. La spécificité alimentaire dans genre *Matsucoccus* (Homoptera: Margarodidae) avec référence spéciale aux plantes-hôtes de *M. pini* Green. Classement des *Matsucoccus* d'après leurs hôtes. Ann. Sci. Forest. 32(3):157–168.
Rieux, R. 1976. *Matsucoccus pini* Green (1925) (Homoptera, Margarodidae) dans le Sud-Est de la France. Variations intraspécifiques. Comparaison avec des espèces les plus proches. Ann. Zool.-Ecol. Anim. 8(2):231–263.
Robison, W. G., Jr. 1978. Ultrastructure of Coccoidea sperm. Va. Polytech. Inst. & State Univ. Res. Div. Bull. 127:35–51.
Rogojanu, V. 1939. Identification and distribution of some species of scale insect pests of fruit trees in Transylvania, Bucovina and Banat (In Romanian). Cluj (not available).
Rogojanu. V. 1958. Contribution to the systematics and ecology of the family Ortheziidae (Coccidae) of Romania. (In Romanian). Studii Cercet. Biol. (Cluj). 9:301–315.

Rogojanu, V. 1962. Contributions to the knowledge of scale insect (Coccoidea-Homoptera) pests of conifers in green belts. (In Romanian). Studii Cercet. Biol. (Cluj). 13(2):279–307.

Rogojanu, V. 1963. Contributions to the distribution of some spceies of Coccoidea (Lecaniidae) in the Romanian People's Republic (In Romanian). Cercet. Biol., Ser. Zool. (Bucureşti) 14:277–292.

Rogojanu, V. 1966a. New species of coccids for the Romanian fauna (In Romanian). Studii. Cercet. Biol., Ser. Zool. (Bucureşti) 18:323–327.

Rogojanu, V. 1966b. New species of coccids for the fauna of the Romanian Socialist Republic (In Romanian). Studii. Cercet. Biol., Ser. Zool. (Bucureşti) 18:431–433.

Russell, L. M. 1941. A classification of the scale insect genus *Asterolecanium*. U. S. Dept. Agr. Misc. Pub. 424:319 pp.

Russell, L. M. 1970. Additions and corrections to an annotated list of generic names of the scale insects. Suppl. U. S. Dept. Agr., Agr. Res. Serv. Misc. Pub. 1015:13 pp.

Russell, L. M., M. Kosztarab, & M. P. Kosztarab. 1974. A selected bibliography of the Coccoidea, 2nd Suppl. U. S. Dept. Agr., Agr. Res. Serv. Misc. Pub. 1281:122 pp.

Saakyan-Baranova, A. A. & S. S. Muzafarov. 1972. Structure and biology of *Kermococcus quercus* L. (Homoptera, Kermococcidae) and its parasites (Hymenoptera, Chalcidoidea) and interrelationships between them. I & II. (In Russian). Entomol. Obozr. 51(3):498–519, (4):697–714, also in Entomol. Rev. 51(3):296–310, 51(4):411–422.

Saakyan-Baranova, A. A., E. S. Sugonjaev, & G. G. Shel'deshova. 1971. The Brown Scale and its parasites (In Russian). Leningrad, Nauka Press. 165 pp.

Sampo, A. & M. Olmi. 1976. Discovery in Italy of *Orthezia vejdovskyi* Šulc, potentially injurious to the roots of ornamental plants (Coccoidea: Ortheziidae) (In Italian). Atti Accad. Sci. Torino 110:217–219.

Sasaki, C. 1894. On the scale insect of mulberry trees (*Diaspis patelliformis* n. sp.). Bull. Tokyo Imp. Univ., Col. Agr. 2:107–124.

Săvescu, A. 1955. Une nouvelle espèce de Cochenille: La Cochenille virgule du peuplier (*Lepidosaphes populi* Săvescu n. sp.) (Coccoidea-Diaspididae). (In Romanian), Bul. Şti. Biol. Agr. Geol. & Geogr. 7(4):915–925.

Săvescu, A. 1957. Die Linden-Kommaschildlaus (*Lepidosaphes tiliae* sp. n.). Anz. f. Schädlingsk. Schädlingsbekämpf. 33:197–201.

Săvescu, A. 1960. Atlas of Plant Protection. Vol. I. Pests of fruit trees, shrubs, and grapevine (In Romanian). Min. Agr. Bucharest. 270 pp.

Săvescu, A. 1961. Atlas of Plant Protection. Vol. II. Pests of ornamental plants and vegetables. (In Romanian). Min. Agr. Bucharest. 189 pp.

Săvescu, A. 1983. Espèces de coccoïdées nouvelles pour la science signalées en Roumanie. I. *Pulvinaria ampelopsidis* Săvescu sp. n. et *Pulvinaria euonymicola* Săvescu sp. n. (Homoptera-Coccidae). Bull. Acad. des Sci. Agr. et Forest. 12 (Extrait):43–50.

Săvescu, A. 1984. Espèces de coccoïdées nouvelles pour la science, signalées en Roumanie. II. Espèces appartenant aux genres *Pseudococcus* Westw., *Phenacoccus* Ckll. et *Peliococcus* Borchs. (Homoptera, Pseudococcidae). Bull. Acad. des Sci. Agr. et Forest. 13:143–157.

Scheidter, F. 1930. Die Läuse Unserer Nadelhölzer. II. Familie: Coccididae, Schildläuse. Neumann Verlag. 87–119.

Schilling, F. H. 1897. Zottenschildläuse. Prakt. Ratgeber Obst. – Gartenbau, 12:248.

Schimitschek, E. 1939. M. Seitners Bearbeitung der Insektenschädlinge der Zirbe in biozönotischer Darstellung. Zeitschr. f. angew. Entomol. 25:111–124.

Schmidt, L. 1956. Contribution to the entomological fauna of Yugoslavia. Scale Insects (Coccoidea) of Croatia (In Serbo-Croatian). Plant Protect. (Beograd) 36 (sup.), p. 3–11.

Schmutterer, H. 1952a. Zur Kenntnis der Schildlausfauna Bayerns (Homopt., Coccoidea). Nachrichtenbl. Bayer. Entomol. 1:14–15, 18–21.

Schmutterer H. 1952b. Die Ökologie der Cocciden (Homoptera, Coccoidea) Frankens (Pts. 1–3). Zeitschr. f. angew. Entomol. 33:369–420, 544–584; 34:65–100.

Schmutterer, H. 1952c. Über drei neue an Gräsern lebende Schildlausarten der deutschen Fauna (Homoptera: Coccoidea). Beitr. z. Entomol. 2(6):554–566.

Schmutterer, H. 1952d. Die Lebensbaumschildlaus *Eulecanium arion* (Ldgr.) (Homopt., Coccoidea), die Erzeugerin des Lebens-Baumhonigtaues. Zeitschr. f. Bienenforschung 1(7):1–5.

Schmutterer, H. 1953. Ergebnisse von Zehrwespenzuchten aus Schildläusen (Hymenoptera: Chalcidoidea). Part 1. Beitr. z. Entomol. 3(1/2):55–69.

Schmutterer, H. 1954. Zur Kenntnis einiger wirtschaftlich wichtiger Mitteleuropäischer *Eulecanium*-Arten (Homoptera: Coccoidea: Lecaniidae). Zeitschr. f. angew. Entomol. 36:62–83.

Schmutterer, H. 1955a. Neue Beiträge zur deutschen Coccidenfauna: Zur Schildlausfauna der Nördlichen Kalkalpen. *In* Bericht über die 7. Wanderversammlung Deutscher Entomologen Berlin (1954):159–164.

Schmutterer, H. 1955b. Bemerkenswerte Schildlausfunde in Süd-und Südwestdeutschland (Homopt., Coccoidea). Nachrichtenbl. Bayer. Entomol. 4(10):98–102.

Schmutterer, H. 1955c. Ergebnisse von Zehrwespenzuchten aus Schildläusen (Hymenoptera; Chalcidoidea) (Part 2). Beitr. z. Entomol. 5(5/6):510–521.

Schmutterer, H. 1956a. Beitrag zur deutschen Schildlausfauna. Nachrichtenbl. Bayer. Entomol. 5(7):65–67.

Schmutterer, H. 1956b. Neue *Rhizoecus*-Arten aus Mitteleuropa (Homoptera: Coccoidea: Pseudococcidae). Beitr. z. Entomol. 6:516–521.

Schmutterer, H. 1956c. Zur Morphologie, Systematik und Bionomie der *Physokermes*-Arten an Fichte (Homop. Cocc.) Zeitschr. f. angew. Entomol. 39:445–466.

Schmutterer, H. 1957. Eine neue Schildlaus von Zwergbirke (Homopt.: Coccoidea). Nachrichtenbl. Bayer. Entomol. 6(7):70–72.

Schmutterer, H. 1958. *Heterococcus variabilis* n. sp., eine neue Pseudococcide mit bemerkenswerten Variabilität. Acta Faun. Entomol. Mus. Nat. Pragae 3:17–22.

Schmutterer, H. 1959. Schildläuse oder Coccoidea. 1. Deckelschildläuse oder Diaspididae. *In* Die Tierwelt Deutschlands und der Angrenzenden Meeresteile. Fischer, Jena 45:260 pp.

Schmutterer, H. 1965. Zur Ökologie und wirtschaftlichen Bedeutung der *Physokermes*-Arten (Homopt. Coccoidea) an Fichte in Süddeutschland. Zeitschr. f. angew. Entomol. 56(4):300–325.

Schmutterer, H. 1972a. Unterordnung Coccoidea, Schildläuse Allgemeines. *In* Die Forstschädlinge Europas, by W. Schwenke. Paul Parey Publ. Hamburg-Berlin. Vol. 1:387–391.

Schmutterer, H. 1972b. Unterordnung Coccoidea, Schildläuse. Coccidae (Lecaniidae), Napfschildläuse. Soft scales. *In* Die Forstschädlinge Europas, by W. Schwenke. Paul Parey Publ. Hamburg–Berlin. Vol. 1:405–418.

Schmutterer, H. 1972c. Unterordnung Coccoidea, Schildläuse. Kermesidae (Kermidae, Kermococcidae). *In* Die Forstschädlinge Europas, by W. Schwenke. Paul Parey Publ. Hamburg–Berlin, Vol. 1:418–420.

Schmutterer, H. 1972d. Unterordnung Coccoidea, Schildläuse. Asterolecaniidae, Pockenläuse. Star scales. *In* Die Forstschädlinge Europas, by W. Schwenke. Paul Parey Publ. Hamburg–Berlin. Vol. 1:420–422.

Schmutterer, H. 1980. Zum Stand der Erforschung der Schildläuse (Homoptera, Coccoidea) in der Bundesrepublik Deutschland. Mitt. dtsch. Ges. allg. angew. Entomol. 2:49–56.

Schmutterer, H., W. Kloft, & M. Lüdicke. 1957. Coccoidea, Schildläuse, Scale Insects, Cochenilles. Tierische Schädlinge an Nützpflanzen. 2. Teil, Vierte Lieferung. Homoptera II. Teil. *In* Handb. der Pflanzenkrankheiten by Sorauer P. Paul Parey, Berlin. V. 5, pt. 2, 1fg. 4, p. 403–520.

Schrank, F. von P. 1776. Beiträge zur Naturgeschichte. Augsburg, Veith. 137 pp.

Schrank, F. von P. 1781. Coccus, gallinsect. *In* Enumeratio Insectorum Austriae Indigenorum p. 295–296.

Schrank, F. von P. 1801. Fauna Boica, v. 2, pt. 1, 374 p. (Spec. p. 143–148) Nürnberg.

Scott, C. L. 1952. The scale insect genus *Aulacaspis* in Eastern Asia (Homoptera: Coccoidea: Diaspididae). Microentomol. 17: 33–60.

Scott, H. 1907. On *Cercococcus eremobius*, gen. et sp. nov., an aberrant form of Coccidae. Trans. Zool., Linn. Soc. London (Ser. 2) 9:455–464.

Shel'deshova, G. G. 1972. The importance of frost resistance in coccids (Homoptera, Coccoidea). *In* Problems of photoperiodism and diapause in insects by Goryshin, N. I. (In Russian). Leningrad, USSR, Izd. Leningrad Univ. p. 51–54.

Shigo, A. L. 1976. The beech bark disease. J. Arboricult. 2(2):21–25.

Shimer, H. 1868. Notes on the "apple bark-louse" (*Lepidosaphes conchiformis*, Gmelin sp.) with a supposed new *Acarus*. Trans. Amer. Entomol. Soc. 1868: 361–374.

Siewniak, M. 1970. *Matsucoccus mugo* n. sp. (Homoptera, Coccoidea: Margarodidae). Entomol. Nachr. 15: 168–172.

Siewniak, M. 1976. Zur Morphologie und Bionomie der Kiefernborkenschildlaus, *Matsucoccus pini* (Green) (Hom., Coccoidea: Margarodidae). Zeitschr. f. angew. Entomol. 81:337–362.

Signoret, V. 1869. Essai sur les Cochenilles ou Gallinsectes (Homoptères-Coccides). (Parts 3–5). Ann. Soc. Entomol. France (Ser. 4) 9:(75–82); (83–112); (113–134).

Signoret, V. 1870. Essai sur les Cochenilles ou Gallinsectes (Homoptères-Coccides). (Parts 6 and 7). Ann. Soc. Entomol. France (Ser. 4) 10:(135–154); (155–174).

Signoret, V. 1872a. Essai sur les Cochenilles ou Gallinsectes (Homoptères–Coccides). (Part 8). Ann. Soc. Entomol. France (Ser. 5) 1:421–434.
Signoret, V. 1872b. Discussion of coccids found on trip to southern France and Italy. Ann. Soc. Entomol. France (Ser. 5) 2:35–38.
Signoret, V. 1872c. Essai sur les Cochenilles ou Gallinsectes (Homoptères–Coccides). (Part 9). Ann. Soc. Entomol. France. (Ser. 5) 2:33–46.
Signoret, V. 1873. Essai sur les Cochenilles ou Gallinsectes (Homoptères–Coccides). (Parts 10 and 11). Ann. Soc. Entomol. France (Ser. 5). 3:27–48; 395–400; 401–448 (Pt. 11, 401–448, publ. April 22, 1874).
Signoret, V. 1874, 1875, 1876, 1877a. Essai sur les Cochenilles ou Gallinsectes (Homoptères–Coccides). Ann. Soc. Entomol. de France. (Ser. 5) 4. Part 12:(279–298). Part 13:(299–312). (Ser. 5). Part 14:(313–338). Part 15:(339–379). Part 16:(380–407). Part 17:(408–428). Part 18:(429–514). (Ser. 5) 6. Part 18:(591–676).
Signoret, V. 1876a. Note on a new *Aspidiotus* species. Ann. Soc. Entomol. France (Ser. 5) 6:52–64.
Signoret, V. 1876b. Essai sur les Cochenilles ou Gallinsectes. Part 17. Ann. Soc. Entomol. France (Ser. 5) 5:353–394.
Signoret, V. 1877. Note on Coccids. Ann. Soc. Entomol. France. (Ser. 5) 6:209.
Signoret, V. 1886. Exhibition of *Chionaspis euonymi* Comstock. Ann. Soc. Entomol. France (Ser. 6) 6:9.
Silvestri, F. 1911. Dispense di Entomologia Agraria secondo le Lezioni del Prof. F. Silvestri raccolte dal dott. Guido Grandi, parte speciale. Portici, Vesuviano, 573 p.
Silvestri, F. 1919a. Contribuzioni alla Conoscenza degli Insetti Dannosi e dei Loro Simbionti. IV. La Cocciniglia del Prugnio (*Sphaerolecanium prunastri* Fonsc.). Bol. Portici R. Scuola Super. Agr. Lab. Zool. Gen. Agr. 13:70–126(70–82).
Silvestri, F. 1919b. Contribuzioni alla Conoscenza degli Insetti Dannosi e dei Loro Simbionti. V. La Coccinigilia del Nocciuolo (*Eulecanium coryli* L.) Bol. Portici R. Scuola Super. Agr. 13:127–192.
Silvestri, F. 1920. Appendice. Fam. Coccidae-Subf. Lecaniinae. *In* Monografia delle Cocciniglie Italiane, by Leonardi, G. Della Torre, Portici. 501–539.
Silvestri, F. 1939. Fam. Coccidae. *In* Compendio Entomol. Appl. Parte Spec. 1(2):618–860.
Smolyannikov, V. V. 1959. The "hawthorn scale" or the "large apple scale" (*Palaeolecanium bituberculatum* (Targ.)). As a mass pest of the apple in the foothill zone of Stavropol' Territory. (In Russian). 4th Congress of the All-Union Entomol. Soc. Abstract of paper. Pt. 2:80–81.
Smolyannikov, V. V. 1964. The results of studies on San José Scale in Stavropol District (In Russian). *In* Collection of Studies on San José Scale, Kartja Moldovejaske, Kishinev. p. 17–51.
Stafford, E. M. & D. F. Barnes. 1948. Biology of the fig scale in California. Hilgardia 18(16):567–598.
Steinweden, J. B. 1929. Bases for the generic classification of the coccoid family Coccidae. Ann. Entomol. Soc. Amer. 22(2):197–245.
Steinweden, J. B. 1946. The indentity of certain common American species of *Pulvinaria* (Homoptera: Coccoidea: Coccidae). Microentomol. 11(1):1–28.
Stoetzel, M. B., & J. A. Davidson. 1974. Biology, morphology and taxonomy of immature stages of 9 species in the Aspidiotini (Homoptera: Diaspididae). Ann. Entomol. Soc. Amer. 67(3):475–509.
Sugonjaev, E. S. 1958. Some parasites (Hymenoptera: Chalcidoidea) of scale insects in the Leningrad region. Entomol. Obozr. 37(2):308–318, also Entomol. Rev. 37(2):260–267.
Sugonjaev, E. S. 1962. Studies of the fauna and ecology of chalcid parasites of scale insects and mealybugs in the Leningrad region (In Russian). Trudy Zool. Inst. Acad. Nauk SSSR 31:172–196.
Šulc, K. 1895a. On a new genus and species of Coccidae, *Ortheziola vejdovskyi,* n. g., n. sp. (In Czech & English). Prague, K. Böhmisch. Gesell. Wiss. Sitzbar. (1894), No. 44, 9 pp.
Šulc, K. 1895b. Studies on scale insects–I. (In Czech). Prague K. Böhmisch. Gesell. Wiss. Sitzber. (1895), Pt. 2, No. 49:27 pp.
Šulc, K. 1895c. Description of a new species of *Lecanium* from Bohemia. Entomol. Monthly Mag, (Ser. 2) 6:37–38.
Šulc, K. 1898. Studies on scale insects–II. (In Czech). (Prague) K. Böhmisch. Gesell. Wiss. Sitzber (1897), 66:19 pp.
Šulc, K. 1908. New information on scale insects (Coccidae). (In Czech). Klubu Prirod., Prost. Vest. (1907):63–67.
Šulc, K. 1908a. Towards the better knowledge of the genus *Lecanium*. Entomol. Monthly Mag. 44:36.
Šulc, K. 1912a. Coccidae regni Bohemiae, in literatura adhuc Commemoratae. Acta Soc. Entomol. Bohemiae. 9:30–38.

Šulc, K. 1912b. *Heliococcus bohemicus* n. gn., n. sp. (In Czech). Acta Soc. Entomol. Bohemiae. 9:39–48.

Šulc, K. 1932. The Czechoslovakian *Lecanium*-species. (In Czech). Acta. Soc. Sci. Nat. Moravicae 7(5):134 pp.

Šulc, K. 1933. *Metadenopus festucae* n. gn., n. sp. sf. Coccinae, patria Moravia, CSR. (In Czech). Acta Soc. Sci. Nat. Moravicae 8(12):1–17.

Šulc, K. 1934. *Thymaspis fusca* n. gen., n. sp. *Aspidiotus baudysi* n. sp. (Coccidae, SF. Diaspinae), patria Moravia CSR. (In Czech). Acta Soc. Sci. Nat. Moravicae 9(3):21 p.

Šulc, K. 1936. Die weiblichen Geschlechtsorgane von *Xylococcus filiferus* Löw 1882. (In Czech). Československ. Zool. Společ. Věst. 3(1935):60–68.

Šulc, K. 1937. *Mohelnaspis moravica* n. gn. n. sp. (Diaspididae, Coccoidea Hemiptera). (In Czech). Acta Soc. Sci. Nat. Moravicae 10(11):41 p.

Šulc, K. 1941. *Mohelnia festuceti* n. gn., n. sp. (Lecaniidae, Eriopeltini, Coccoidea, Hemiptera). (In Latin & Czech). Acta Soc. Sci. Nat. Moravicae 13(6):1–17.

Šulc, K. 1942. Diagnosis differentialis inter gn. *Pulvinaria* et *Phyllostroma* n. gn., Folia Entomol. 5(–2):5:8.

Šulc, K. 1943. External morphology, metamorphosis and life history of *Phenacoccus aceris* Sign. (Coccoidea). (In Czech). Acta Soc. Sci. Nat. Moravicae 15(12):52 pp.

Šulc, K. 1944a. *Antoninella* n. gn. synon. *Antonina sulci* Green 1934 (Coccoidea, Homoptera). Mem. Soc. Zool. Bohemoslov. 9:148–170.

Šulc, K. 1944b. External morphology, metamorphosis and life history of *Peukinococcus* n. gn. *piceae* Loew 1883 (Coccoidea). (In Czech). Acta Soc. Sci. Nat. Moravicae 16(11):50 pp.

Šulc, K. 1953. External morphology, metamorphosis and life history of the scale insect *Cerococcus cycliger* Goux 1932. (In Czech). Rospr. Ceske Akad. Ved. Umeni Trida II 62:18 pp.

Sulzer, J. H. 1776. Abgekürzte Geschichte der Insecten nach dem Linaeischen System. Winterthur, Steiner, Pt. 1, 28, 274 (107–112):Pt. 2, 58 pp.

Szelegiewicz, H. 1971. Autapomorphous wing characters in the recent subgroups of Sternorrhyncha and their significance in the interpretation of the Paleozoic members of the group (In Polish). Ann. Zool. 29:1–67.

Szelényi, G. I. 1936. Research on the development and epidemiology of *Aspidiotus pyri* (Licht.) Reh, on the generation of the year 1935. (In Hungarian). Kísérlet. Közl. 39:159–170.

Szulczewski, J. W. 1921. Contributions to the study of the Coccidae fauna of Great Poland. Soc. Sci. Poznan, Taav. Comm. Sci. Math. & Nat. (Ser. B) 1:78–84.

Takagi, S. 1961. A contribution to the knowledge of the Diaspidini of Japan (Homoptera: Coccoidea). Pt. II. Insecta Matsumurana 24:4–42.

Takagi, S. 1969. Diaspididae of Taiwan based on material collected in connection with the Japan–U.S. Cooperative Science Programme, 1965 (Homoptera: Coccoidea). Part I. Insecta Matsumurana 32(1):1–110.

Takagi, S. 1970. Diaspididae of Taiwan based on material collected in connection with the Japan–U.S. Cooperative Science Programme, 1965 (Homoptera: Coccoidea). Part II. Insecta Matsumurana 33(1):1–146.

Takagi, S. & S. Kawai. 1967. The genera *Chionaspis* and *Pseudaulacaspis* with a criticism on *Phenacaspis* (Homoptera: Coccoidea). Insecta Matsumurana 30(1):29–43.

Takahashi, R. 1955. *Lepidosaphes* of Japan. (Diaspididae, Coccoidea, Homoptera). Bull. Univ. Osaka Prefecture, (Ser. B) 5:67–78.

Takahashi, R. 1955a. *Pulvinaria* of Japan (Coccidae, Homoptera). Kontyu 23(4):148–154.

Takahashi, R. 1955b. Key to the genera of Coccidae in Japan, with descriptions of two new genera and a little-known species. Insecta Matsumurana 19(1–2):23–28.

Takahashi, R. 1958. Key to the genera of Pseudococcidae in Japan, with descriptions of three new genera and two new species. Bull. Univ. Osaka Prefecture 7(Ser. B):1–8.

Takahashi, R., & S. Kanda. 1939. A new species of Coccidae (Homoptera) from Corea. Annot. Zool. Jap. 18:185–187.

Taketani, A. 1972. Studies on a Margarodid scale, *Matsucoccus matsumurae* (Kuwana) (Hemiptera: Coccoidea). 1. Bionomics. (In Japanese). Bull. Govern. Forest Exp. Sta. Tokyo. 246:1–9.

Talitskij, V. I., E. S. Sugonyaev, & I. K. Goantsa. 1966. Insect parasites and predators of the black locust soft scale, *Parthenolecanium corni* Bouché in the Moldavian Soviet Socialist Republic. (In Russian). Trudy Moldavian Nauch-Issled. Inst. Vinograd i Vinod. Entomol. 13:317–359.

Targioni-Tozzetti, A. 1866. Come certe Cocciniglie Sieno Cagione di Alcune Melate delle Piante, e di Alcune Ruggini; e come la Cocciniglia del Fico dia in Abbondanza una specie di cera. Atti R. Accad. Georg. (Firenze) (n.s.) 13:115–137; App. 138–146.
Targioni-Tozzetti, A. 1867. Studii sulle Cocciniglie. Me. Soc. Ital. Sci. Nat. 3(3):1–87.
Targioni-Tozzetti, A. 1868. Introduzione alla seconda Memoria per gli studj sulle Cocciniglie, e Catalogo dei generi e delle specie della famiglia dei Coccidi. Atti. Soc. Ital. Sci. Nat. 11:694–758.
Targioni-Tozzetti, A. 1869. Sopra due generi di Coccinigle (Coccidae). Bull. Soc. Entomol. Ital. 1:257–267.
Targioni-Tozzetti, A. 1879. *Diaspis blankenhornii.* Bull. Soc. Entomol. 1879, p. 17, 32.
Targioni-Tozzetti, A. 1884. Relazione intorno ai lavori della R. Stazione di entomologia agraria di Firenze per gli Anni 1879–80–81–82. Article V.–Omotteri. *In* Italy Min. Agr. Indus. Com., Ann. Agr. 1884 (Nos 86–89), p. 383–414.
Targioni-Tozzetti, A. 1885. Note sopra alcune Cocciniglie (Cocciidei). Bull. Soc. Entomol. Ital. 17:100–120.
Targioni-Tozzetti, A. 1886. Sull'insetto che Danneggia i aelsi. Riv. Bachicol. 18(1885):1–3.
Targioni-Tozzetti, A. 1893. Cocciniglie nuove, Critiche o poco Note. Bull. Soc. Entomol. Ital. 24(1892):285–312.
Teague, M. M. 1925. A review of the genus *Aclerda* (Hemiptera: Coccidoidea). Ann. Entomol. Soc. Amer. 18(4):432–441.
Tereznikova, E. M. 1959a. Distribution of scale insects (Insecta, Homoptera, Coccoidea) by life zones in the forests of the Transcarpathian Region. (In Ukrainian). Dopovidi Akad. Nauk Ukr. RSR 4:446–450.
Tereznikova, E. M. 1959b. Zoogeographical characteristics of the scale insect fauna (Insecta, Homoptera, Coccoidea) of the Transcarpathian Region. (In Ukrainian). Dopovidi Akad. Nauk Ukr. RSR 7:795–799.
Tereznikova, E. M. 1959c. On Coccoidea (Insecta, Homoptera) fauna of the Ukraine. (In Ukrainian) Dopovidi Akad. Nauk Ukr. RSR 6:683–685.
Tereznikova, E. M. 1962. A new species of mealybug of the genus *Brevennia* Goux (Coccoidea, Pseudococcidae). (In Ukrainian). Dopovidi Akad. Nauk Ukr. RSR 1:122–124.
Tereznikova, E. M. 1963. Ecologic-faunistic survey of scale insects (Homoptera, Coccoidea) of the Ukrainian woodlands. (In Ukrainian). *In* Materials for Ukrainian Entomology. Akad. Nauk Ukr. RSR Trudy Inst. Zool. 19:41–57.
Tereznikova, E. M. 1965. New species of the genus *Greenisca* Borchs. (Coccoidea, Eriococcidae). (In Ukrainian). Dopovidi Akad. Nauk Ukr. RSR 7:957–959.
Tereznikova, E. M. 1967. Geographical characteristics of coccids in the Crimea (Insecta, Homoptera, Coccoidea). (In Ukrainian). Dopovidi Akad. Nauk Ukr. RSR 1967: 473–476.
Tereznikova, E. M. 1968. A new species of *Paroudablis* Ckll. (Coccoidea, Pseudococcidae). (In Ukrainian). Dopovidi Akad. Nauk Ukr. RSR (Ser. B) 5:472–475.
Tereznikova, E. M. 1975. Scale Insects. Families: Ortheziidae, Margarodidae, Pseudococcidae. – *In* the Fauna of Ukraine. Vol 20. Part 18. (In Ukrainian). Akad. Nauk. Ukr. RSR Inst. Zool.: 295 pp.
Tereznikova, E. M. 1977. The species *Ortheziola veydovskyi* Šulc (Homoptera: Coccoidea: Ortheziidae) new for USSR Fauna (In Ukrainian). Dopovidi Akad. Nauk Ukr. RSR, (Ser. B) Geol Geogr. Khim. Biol. (9):854–856.
Tereznikova, E. M. 1981. Scale Insects. Families: Eriococcidae, Kermesidae, Asterolecaniidae, Coccidae. *In* the Fauna of Ukraine. Vol. 20. Part 19. (In Ukrainian). Akad. Nauk. Ukr. RSR Zool. Inst. 215 pp.
Ter-Grigorian, M. A. 1973. Soft and armored scales (Coccoidea), Mealybugs (Pseudococcidae). *In* "Fauna of Armenian SSR". (In Russian). Izv. Akad. Nauk Armianskoi SSR. 246 pp.
Theron, J. G. 1958. Comparative studies on the morphology of male scale insects (Hemiptera: Coccoidea). Ann. Univ. Stellenbosch 34 (Section A. No. 1):1–71.
Theron, J. G. 1962. Structure and relationships of the male of *Phenacoleachia zealandica* (Maskell) (Hemiptera: Coccoidea). Proc. R. Entomol. Soc. London (A) 37. Pts. 10–12:145–153.
Thiem, H. 1931. Eine rote Kommaschildlaus der deutschen Coccidenfauna (*Lepidosaphes rubri* n. sp.). Gartenbauwiss. 5:557–567.
Thiem, H. 1933. Über ein- und zweigeschlechtliche Kommaschildläuse (*Lepidosaphes ulmi unisexualis* und *bisexualis, L. rubri* und *L. newsteadi*) der deutschen Coccidenfauna. Zeitschr. f. Pflanzenkrank, 43:638–657.
Thiem, H., & R. Gerneck. 1934a. Verbreitung, Entwicklung und Bestimmung der bisher in Deutschland aufgefundenen Austernschihdläuse (Aspidiotini) unter Einschluss der roten Austernschildlaus (*Epidiaspis betulae*) und der San José-Schildlaus (*Aspidiotus perniciosus*). Zeitschr. f. Pflanzenkrank. 44:529–555.

Thiem, H., & R. Gerneck. 1934b. Untersuchungen an deutschen Austernschildläusen (Aspidiotini) im Vergleich mit der San José-Schildlaus (*Aspidiotus perniciosus* Comst.). Arb. Phys. angew. Entomol. 1:130–158; 208–238.

Tiensuu, L. 1951. Notes on the small reed coccid, *Eriopeltis lichtensteini* Sign. (Hem., Coccoidea, Lecanoidae) and its natural enemies in Finland. Ann. Entomol. Fennici 17:3–10.

Tranfaglia, A. 1973. Orientamenti per una lotta chimica razionale contro la Cocciniglia emisferica del Susino (*Sphaerolecanium prunastri* Fonsc.). Bol. Portici Lab. di Entomol. Agr. "Filippo Silvestri" (1972–1973) 30:145–151.

Tranfaglia, A. 1974. Studi sugli Homoptera Coccoidea II. Notizie Biologiche sull' *Asterolecanium arabidis* (Signoret) su *Hedera helix* L. Bol. Portici Lab. Entomol. Agr. "Filippo Silvestri" 31:54–60.

Tremblay, E. 1977. Advances in endosymbiont studies in Coccoidea. Virginia Polytech. Inst. State Univ. Res. Div. Bull. 127:23–33.

Trjapitzin, V. A. 1978. Superfamily Chalcidoidea. *In* Insect determination manual for the European part of USSR, by Medvedeva, G. S. Publ. Nauka, Leningrad. Hymenoptera, Vol 3, 2nd part (In Russian). p. 28–538.

Tryon, H. 1897. Scale insects–Coccidae. Queensland Agr. J. 1(2):118–129.

Tsalev, M. 1964. *Kermes* new for the fauna of Bulgaria. (In Bulgarian). Rast. Zashch. 12(2):23–26.

Tsalev, M. 1965. Parasites of scale insects of the genus *Parthenolecanium* Šulc in Bulgaria. (In Bulgarian). Rast. Zashch. 13(5):25.

Tsalev, M. 1966. Contribution to the morphology, biology and control of *Rhodococcus bulgariensis* Wünn (Homoptera, Coccoidea) in Bulgaria. (In Bulgarian). Plant Sci. 3(7):3–14.

Tsalev, M. 1967. Contributions to the biology of *Lepidosaphes populi* Săv., (Coccoidea, Diaspidinae) and its control (In Bulgarian). Forest Sci. 4(3):59–66.

Tsalev, M. 1968. Contribution to the scale insect fauna (Homoptera, Coccoidea) of the park and ornamental plants in Bulgaria (In Bulgarian). Acad. Bulg. Sci., Bull. Inst. Zool. & Mus. 28:205–218.

Tsalev, M. 1972. Entomophags of the scale insects (Homoptera, Coccoidea) in Bulgaria. (In Bulgarian). Acad. Agr. Sci. Inst. Plant Prot. 1:79–94.

Tsalev, M., & M. Valcheva. 1965. The *Quadraspidiotus marani* Zahr. (In Bulgarian). Rast. Zashch. 13(4):22–25.

Uichanco, L. B. & F. E. Villanueva. 1932. Biology of the pink mealybug of sugar cane, *Trionymus sacchari* (Cockerell), in the Philippines. Philippine Agr. 21:205–276.

Umnov, M. P. 1940. Quarantine and other pest scales (Coccidae) Krimea. Simferopol' (In Russian). 1:64 pp.

Varga, I. 1971. Present distribution of mulberry scale (*Pseudaulacaspis pentagona* Targ.) in Hungary. (In Hungarian). Növényvédelem 7(9):424–426.

Varshney; R. K. 1964. On the homonymy of *Ortheziola signoreti* (Haller) (Homoptera, Ortheziidae), and its valid name. Entomologist 97:280.

Vashchinskaya, N. V. 1969. On the biology of the soft scale *Palaeolecanium bituberculatum* (Targ.) (Homoptera, Coccoidea) in Armenia. Entomol. Obozr. 48(4):745–751, also in Entomol. Rev. 48(4):472–476.

Vayssière, P. 1914. Note sur quelques coccides nouveaux ou peu connus. Bull. Soc. Entomol. France No. 7:206–208.

Vayssière, P. 1924. Troix nouveaux Coccides de L'Afrique Française (Hem.). Bull. Soc. Entomol. France 2:28–32.

Vayssière, P. 1926. Contribution à l'étude Biologique et Systématique des Coccidae. Ann. Épiphyties 12(4–5):197–382.

Vinis, G. 1977. The spindle-berry scale, *Unaspis euonymi* (Comst.) in Hungary. (In Hungarian). Növényvédelem 13(1):5–10.

Vinis, G. 1981. *Acanthomytilus hungaricus* sp. n. and some new scale insects in the Hungarian fauna (Homoptera: Coccoidea). Folia Entomol. Hung. (n.s.) 34(1):203–207.

Vinis, G., & F. Kozár. 1981. Polymorphism in the male of *Polystomophora ostiaplurima* Kiritchenko, 1940 (Homoptera: Coccoidea), description of all stages and biology. Folia Entomol. Hung. (n.s.) 34(1):209–221.

Vu Quang Kon. 1974. Morphological and biological characteristics of the immature stages of *Encyrtus infidus* Rossi (Hymenoptera Chalcidoidea) – a parasite of the caragana scale (*Eulecanium caraganae* Borchs.). Entomol. Obozr. 53(4):732–751, also in Entomol. Rev. 53(4):8–21.

Vu Quang Kon, & E. S. Sugonjaev. 1975a. Morphological and biological adaptations of the immature stage of *Encyrtus infidus* Rossi (Hymenoptera, Chalcidoidea) to the seasonal cycle of the host, the caragana scale (*Eulecanium caraganae* Borchs.). Entomol. Obozr. 54(4):705–719, also in Entomol. Rev. 54(4):1–10.

Vu Quang Kon, & E. S. Sugonjaev. 1975b. Influence of adaptional reactions of the parasite *Encyrtus infidus* (Hymenoptera: Chalcidoidea) on its host *Eulecanium caraganae*. (In Russian). Zool. Zhur. (Moscow) 54(10):1488–1494.

Way, M. J. 1963. Mutualism between ants and honeydew-producing Homoptera. Annu. Rev. Entomol. 8:307–344.

Weidner, H. 1970. Masseninvasion der Schildlaus *Rhizoecus halophilus* (Hardy, 1860) in einem Wohnhaus auf Sylt. Anz. Schädlingskunde und Pflanzenschutz 43(8):117–119.

Westwood, J. O. 1870. The Camellia coccus – *Coccus flocciferus* Westg.. Gard. Chron. & Agr. Gaz. No. 10:308.

Wetzel, T. 1971. Tierische Schädlinge. Schildläuse, Coccina. *In* E. Muhle: Krankheiten und Schädlinge der Futtergräser. Leipzig. S. Hirzel Verlag (1971):226–368; spec. p. 297.

White, F. F. 1877–1878. Note on the zoology and botany of Glen Tilt. Scot. Nat. 4:160–163.

Wilkey, R. F. 1962. A simplified technique for clearing, staining and permanently mounting small arthropods. Ann. Entomol. Soc. Amer. 55:606.

Williams, D. J. 1958a. Mealybugs (Pseudococcidae: Homoptera) described by W. M. Maskell, R. Newstead, T. D. A. Cockerell and E. E. Green from the Ethiopian region. Bull. Brit. Must. (Nat. Hist.) Entomol. 6:205–236.

Williams, D. J. 1958b. The mealybugs (Pseudococcidae-Homoptera) described by W. J. Hall, F. Laing and A. H. Strickland from the Ethiopian region. Bull. Brit. Mus. (Nat. Hist.) Entomol. 7:1–37.

Williams, D. J. 1961. Notes on the genus *Heterococcus* Ferris (Coccoidea, Homoptera) with a description of a new species injurious to guineacorn *(Sorghum vulgare)* in Nigeria. Bull. Entomol. Res. 51(4):671–675.

Williams, D. J. 1962. The British Pseudococcidae (Homoptera: Coccoidea). Bull. Brit. Mus. (Nat. Hist.) Entomol. 12(1):1–79.

Williams, D. J. 1969. The family-group names of the scale insects (Hemiptera: Coccoidea). Bull. Brit. Mus. (Nat. Hist.) Entomol. 23(8):317–341.

Williams, D. J. 1970. The mealybugs (Homoptera, Coccoidea, Pseudococcidae) of sugar-cane, rice and sorghum. Bull. Entomol. Res. 60(1969):109–188.

Williams, D. J. 1971. Synoptic discussion of *Lepidosaphes* Shimer and its allies with a key to genera (Homoptera, Coccoidea, Diaspididae). Bull. Entomol. Res. 61(1):7–11.

Williams, D. J. 1984. Some aspects of the zoogeography of scale insects (Homoptera: Coccoidea). Verh. SIEEC X. (1983) Budapest. p. 331–333.

Williams, M. L., & M. Kosztarab. 1970. A morphological and systematic study on the first instar nymphs of the genus *Lecanodiaspis* (Homoptera: Coccoidea: Lecanodiaspididae). Va. Polytech. Inst. & State Univ. Res. Div. Bull. 52:96 pp.

Williams, M. L. & M. Kosztarab. 1972. Morphology and systematics of the Coccidae of Virginia with notes on their biology (Homoptera: Coccoidea). Va. Polytech. Inst. & State Univ. Res. Div. Bull. 74:215 pp.

Willoughby, P. & M. Kosztarab. 1974. Morphological and biological studies on two species of *Chionaspis* (Homoptera: Coccoidea: Diaspididae). Va. Polytech. Inst. & State Univ. Res. Div. Bull. 92:83 pp.

Wünn, H. 1921. *Physokermes graniformis* n. sp. Zeitschr. f. Wiss. Insektenbiol. Beilage: Neue Beitr. System. Insektenkunde 2(4):29.

Wünn, H. 1939. *Eulecanium bulgariense* n. sp., ein Schädling der bulgarischen Ölrosenkulturen aus der Familie der Cocciden. Zeitschr. f. angew. Entomol. 25:703–708.

Yang, S. P., & M. Kosztarab. 1967. A morphological and taxonomical study on the immature stages of *Antonina* and of the related genera. (Homoptera: Coccoidea). Va. Polytech. Inst. & State Univ. Res. Div. Bull. 3:73 pp.

Young, B. L., J. N. Hu, & Z. Y. Ren. 1976. Pine blast scales from China. Acta Entomol. Sinica 19(2):199–204.

Zahradnik, J. 1951. Contribution to the study of scale insects. (Homopt. Coccoidea). (In Czech). Acta Soc. Entomol. Cechosl. 48:198–200.

Zahradnik, J. 1952a. Eine neue Schildlausart-*Quadraspidiotus marani* n. sp. (Coccoidea: Diaspididae). Beitr. z. Entomol. 2(4/5):449–451.

Zahradnik, J. 1952b. Two diaspidids new to the fauna of Czechoslovakia. (Rhynch. Cocc.) (In Czech). Acta Soc. Entomol. Cechosl. 49:92–97.

Zahradnik, J. 1952c. Revision der Tschechoslowakischen Arten der Schildläuse aus der Unterfamilie der Diaspidinae. Acta Entomol. Mus. Nat. Pragae, 27(1951):89–200.

Zahradnik, J. 1955a. Über die wichtigsten artdifferenzierenden Merkmale der Schildlaus *Quadraspidiotus marani* Zahradnik (= *Q. schneideri* Bachmann). Zeitschr. angew. Entomol., 37:125–127.

Zahradnik, J. 1955b. Zur Nomenklatur von *Quadraspidiotus marani* Zahradnik (*Quadraspidiotus schneideri* Bachmann) (Homoptera: Coccoidea). Beitr. z. Entomol. 5:88–90.

Zahradnik, J. 1956a. Remarques sur la présence de quelques espèces des Cochenilles–Pseudococcidae en Tchécoslovaquie (Hom. Coccoidea, Pseudococcidae). Acta Faun. Entomol. Mus. Nat. Pragae 1:49–52.

Zahradnik, J. 1957. Über die geographische Verbreitung der Schildläuse (Homoptera, Coccoidea) in Mitteleuropa. Entomol. Nachrichtenblatt Österr. u. Schweiz. Entomol. 8(3–5):3–7.

Zahradnik, J. 1959a. An annotated bibliography of Coccoidea in Czechoslovakia (Homoptera, Coccinea). (In Czech). Acta Faun. Entomol. Mus. Nat. Pragae. Suppl. 1:69 pp.

Zahradnik, J. 1959b. Scale insects–Coccinea. (In Czech). Praha Klič Zviřeny Českoslov. Socialni Repub. 3:527–552.

Zahradnik, J. 1962. Diaspidides nouveaux ou peu connus dans la faune Tchécoslovaque. (Homoptera, Diaspididae). Acta Faun. Entomol. Mus. Nat. Pragae 8:91–94.

Zahradnik, J. 1968. Schildläuse unserer Gewächshäuser. Neue Brehm-Bücherei. Wittenberg. 42 pp.

Zahradnik, J., 1972. Überfamilie Archaeococcoidea; Neococcoidea; Pseudococcidae, Eriococcidae, Diaspididae. *In* Die Forstschädlinge Europas by W. Schwenke. Paul Parey Publ. Hamburg–Berlin 1:391–405, 422–446.

Zahradnik, J. 1974. Coccinea, *In* Treatise concerning the Tatra National Park (In Czech) Publ. by "Osveta" Nat. Enterprise, Martin, Czech. 16:141–142.

Zahradnik, J. 1977. Aleyrodinea–Coccinea. *In* Enumeratio Insectorum Bohemoslovakiae (In German). Acta Faun. Entomol. Mus. Nat. Pragae 1 (suppl. 4):117–122.

Zak-Ogaza, B. 1961. Studien über Zehrwespen (Hymenoptera, Chalcidoidea) die auf der Fauna Polens bekannten Schildläusen (Homoptera, Coccoidea) schmarotzen. Polskie Pismo Entomol. 31:349–410.

Zak-Ogaza, B. 1966. A note on some scale insects (Homoptera, Coccoidea) of the Hungarian fauna. Acta Agron. Acad. Sci. Hung. 15:79–83.

Zak-Ogaza, B. and J. Koteja. 1964. Investigations on scale insects (Homoptera, Coccoidea) of the Pieniny Mountains. Acta Zool. Cracoviensia 9(6):417–439.

Zetterstedt, J. W. 1828. Coccoids. *In* Fauna Insectorum Lapponica. Pars I, 563 p. Libraria Schulziana, Hammone (Hamburg, Germany). Spec. p. 562–563.

Zimmerman, E. C. 1948. Insects of Hawaii. Vol. 5. Homoptera: Sternorhyncha. Univ. Hawaii Press, Honolulu. 464 pp.

V. DISTRIBUTION OF SCALE INSECTS DISCUSSED IN THIS BOOK

The authors kept the locality names as were in the cited references. Therefore, records from the United Kingdom may be listed under British Isles, England, Great Britain, Scotland, UK and Wales. The numbers refer to the number given to the species.

Afghanistan 121
Algeria 68, 208, 221, 222
Argentina 215
Australia 186, 191, 212, 219, 223, 225, 227
Au (Austria) 1, 2, 3, 4, 5, 10, 11, 22, 30, 37, 53, 62, 65, 98, 99, 102, 106, 113, 123, 124, 125, 126, 127, 128, 130, 131, 133, 136, 142, 146, 148, 151, 152, 153, 155, 156, 157, 159, 167, 178, 184, 185, 186, 187, 188, 189, 190, 191, 192, 193, 195, 197, 198, 202, 203, 209, 212, 213, 214, 218, 224, 225, 226, 229, 231, 233
Azores Isls. 4
Belgium 2, 4, 227
British Isles 152
Bu (Bulgaria) 2, 3, 5, 14, 47, 53, 62, 67, 68, 80, 82, 89, 98, 99, 102, 104, 106, 107, 115, 116, 121, 122, 124, 125, 126, 127, 128, 131, 133, 134, 136, 137, 138, 142, 146, 151, 152, 153, 155, 156, 157, 166, 167, 169, 174, 176, 178, 180, 182, 183, 184, 185, 186, 188, 189, 190, 191, 195, 201, 202, 203, 205, 206, 207, 209, 210, 212, 213, 214, 215, 218, 219, 220, 221, 223, 224, 225, 226, 227, 229, 231, 232, 233
California 208
Canada 53, 125, 167, 201, 221
Canary Isls. 223, 230
Central Asia 205, 207, 208
Channel Islands 96
Chile 208
China 146
Corsica 11, 68, 105, 181, 193, 195, 198, 207, 216, 222
Cz (Czechoslovakia) 1, 2, 3, 4, 5, 6, 8, 10, 11, 23, 30, 34, 44, 50, 53, 62, 65, 66, 67, 72, 79, 88, 89, 94, 97, 98, 99, 100, 102, 103, 104, 105, 106, 108, 111, 112, 113, 115, 116, 117, 119, 120, 121, 123, 124, 125, 126, 127, 128, 130, 131, 133, 134, 136, 139, 140, 142, 143, 145, 146, 147, 152, 153, 155, 156, 157, 166, 167, 168, 170, 171, 172, 178, 180, 182, 185, 186, 188, 189, 191, 192, 194, 195, 198, 199, 202, 203, 205, 209, 212, 213, 214, 215, 217, 218, 221, 222, 224, 225, 226, 227, 228, 229, 230, 231, 232
Cyprus 134, 216
Denmark 26, 33, 79, 97, 98, 99, 100, 106, 108, 121, 123, 124, 125, 127, 136, 139, 152, 159, 178, 195, 223
Egypt 91, 127, 208
England 2, 5, 9, 16, 17, 18, 20, 26, 32, 33, 40, 59, 69, 77, 79, 82, 88, 90, 93, 95, 96, 97, 98, 99, 102, 103, 104, 106, 108, 111, 112, 113, 115, 116, 119, 121, 123, 124, 125, 127, 128, 136, 138, 139, 153, 159, 161, 162, 165, 170, 171, 178, 180, 190, 195, 205
Europe 8, 60, 133, 155, 156, 167, 186, 188, 189, 190, 191, 203, 207, 212, 213, 214, 215, 218, 219, 221, 225, 229, 230
Far East 195, 205, 225
Finland 57, 98, 99, 136, 152
Fr (France) 1, 2, 3, 4, 5, 6, 7, 8, 11, 12, 13, 21, 22, 23, 30, 31, 33, 34, 37, 38, 39, 44, 49, 53, 57, 59, 62, 65, 66, 67, 68, 69, 75, 76, 82, 89, 95, 97, 98, 99, 102, 104, 106, 108, 111, 115, 116, 119, 120, 121, 122, 123, 124, 125, 126, 127, 128, 133, 134, 136, 137, 138, 139, 140, 146, 148, 149, 152, 153, 155, 156, 157, 159, 161, 163, 165, 166, 167, 169, 171, 178, 181, 182, 184, 186, 188, 189, 190, 191, 193, 195, 196, 197, 198, 199, 203, 205, 208, 209, 212, 213, 214, 215, 216, 217, 218, 219, 220, 221, 222, 223, 224, 225, 226, 227, 229, 230, 232, 233
Ge (Germany) 1, 2, 3, 4, 5, 7, 8, 9, 10, 11, 13, 18, 22, 23, 30, 31, 32, 33, 34, 39, 40, 50, 53, 57, 62, 65, 67, 68, 69, 70, 71, 72, 74, 75, 78, 79, 82, 88, 89, 94, 97, 98, 99, 100, 102, 105, 106, 108, 111, 112, 113, 115, 116, 117, 120, 121, 122, 123, 124, 125, 127, 128, 130, 131, 133, 136, 138, 139, 146, 151, 152, 153, 154, 155, 156, 157, 159, 164, 165, 166, 167, 171, 178, 180, 182, 184, 186, 188, 189, 191, 192, 195, 197, 198, 202, 203, 205, 209, 212, 213, 214, 218, 220, 221, 222, 224, 225, 226, 227, 229, 230, 231
Great Britain 31, 184
Greece 22, 28, 48, 51, 59, 68, 69, 112, 134, 146, 183, 195, 205, 206, 207, 222, 223
Hebrides 4, 103
Holland 17, 32, 33, 57, 74, 88, 89, 93, 97, 98, 99, 102, 121, 125, 127, 136, 152, 157, 159, 195, 209, 227

VI. INDEX TO HOST PLANTS

The hosts are usually given as in the cited source; sometimes as common names and often as scientific names. Therefore, readers may find more page citations for the same host under both names, e.g. apple, also under *Malus* Often only the plant genera are listed in this index, but the original sources cited by the authors usually give species names. Numbers refer to page numbers.

VII. INDEX TO PARASITES, PREDATORS, ANTS AND OTHER ASSOCIATED ORGANISMS

VIII. INDEX TO SCALE INSECT NAMES

Valid names are in Roman type, synonyms, homonyms and misidentifications in *italics*. Beginning page numbers for descriptions of taxa (families, genera and species) are in **bold face**. Color illustrations are indicated as figures 1 to 35.

GPSR Compliance

The European Union's (EU) General Product Safety Regulation (GPSR) is a set of rules that requires consumer products to be safe and our obligations to ensure this.

If you have any concerns about our products, you can contact us on ProductSafety@springernature.com

In case Publisher is established outside the EU, the EU authorized representative is:

Springer Nature Customer Service Center GmbH
Europaplatz 3
69115 Heidelberg, Germany

Batch number: 10815974

Printed by Printforce, the Netherlands